**ESSENTIAL
PHYSICS**

Essential Physics
Volume I
2013

Andrew Duffy

Table of Contents for *Essential Physics, Volume 1*

Accompanying web site: http://physics.bu.edu/~duffy/EssentialPhysics/

The web site accompanying the book will have several features, including:

- Simulations and animations to accompany the material in the printed text. Some things, such as the connection between uniform circular motion and simple harmonic motion, are made significantly clearer when the entire motion is shown as opposed to showing still frames from the motion as can only be done in a book.

- Additional examples. Example problems are very important for students, but they also take up a significant fraction of the pages in a textbook. To keep the number of pages down in this book, some examples will appear only on the web site.

- Additional topics. The book itself will include the core topics. Additional sections will be included on the web site, an example being a section on orbits and energy to accompany the Gravitation chapter.

A Few Words of Encouragement

Many people start a physics course with the idea that physics is hard. If you view physics as a large number of loosely connected topics, and with the idea that equations are the key to unlocking the secret of a typical physics problem, then you may well find physics to be hard. It does not need to be that way, however. One of the beautiful things about physics is its simplicity, in fact. As you continue through the book, try to appreciate the fact that topics that may initially seem to be separate are actually linked by one concept (conservation of energy being one example of such a concept). Also, try to keep in mind that while the equations are useful tools, understanding the various concepts and when to apply them is more important for your success in your physics class.

Being an overview of various topics, this book does cover a lot of ground. Much is expected of you, in fact, not the least of which is to understand 2500 years worth of science, including ideas that scholars wrestled with for centuries, and for you to do that in a few months. Through this book, however, you have an advantage over great thinkers like Aristotle and Galileo, being able to draw on the work of, and experimental evidence compiled by, many scientists over the years. This book was written with the aim of taking all that material and presenting an overview of it in a clear, understandable way.

Because physics is the study of the world around you, and you have interacted with the world around you over all the years you have been alive, you come into your physics course with many ideas about how the world works. Most likely, some of these ideas match the way the world actually works, while others do not. So, try to keep an open mind, and be aware that many of your ideas actually match those of the great thinkers from ages past. You're following in the footsteps of giants, but you have the advantage of living at a time when we know much more about how everything works.

Finally, have fun learning physics! This is really a great opportunity you have, being able to spend the next few months learning how things work, from toasters to CD players, from the physics of walking to the physics governing atoms and nuclei. Enjoy the ride.

DRAWING THE BOW.

Physics is all around you. From the moment your alarm clock wakes you up in the morning to when you flip the light switch before you get into bed at night, you are immersed in a world that is governed by the laws of physics. Understanding physics can help you drive your car or bicycle more safely, improve your results in sports and in the playing of musical instruments, as well as giving you insights into how your iPod and cell phone work. Archery is just one example of an activity that involves physics. How many physics-related concepts can you come up with in looking at this picture of an archer getting ready to release an arrow from a bow?

Picture credit: public-domain image from Wikimedia Commons, originally published in 1908 in the book The Witchery of Archery, by Maurice Thompson.

Chapter 1 – Introduction to Physics
CHAPTER CONTENTS

1-1 Physics, Models, and Units
1-2 Unit Conversions, and Significant Figures
1-3 Trigonometry, Algebra, and Dimensional Analysis
1-4 Vectors
1-5 Adding Vectors
1-6 Coordinate Systems
1-7 The Quadratic Formula

In this chapter, we will discuss some of the mathematical tools we often apply in working through physics problems. These tools include the SI system of units, trigonometry, algebra, and vectors. You should be familiar with how to use these tools because we will use them throughout the book.

In Chapter 1, we will also go over how we use models in physics to help us understand the world around us. Modeling is a key feature of science in general, and the history of science is really the story of the models scientists created to understand the world around them, and of how these models evolved over time. We will discuss some of this history, too, as we go along.

1-1 Physics, Models, and Units

You will most likely be devoting several months to learn physics. Physics is the study of something, but of what? Take a minute and write one sentence describing what you think physics is.

Physics encompasses many different topics, but a nice one-sentence description is that physics is the study of how the world works. Physics could just as easily be described as the study of how the universe works, or the study of how things work. Physics can be very practical, explaining how a toaster works, for instance. It can also be mind-blowing stuff, as we will see when we talk about the quantum nature of the atom and Einstein's theory of relativity. It is also important to keep in mind that physics is a science. Physics can, in some sense, also be thought of as a logical, systematic approach to analyzing physical situations.

Another important question to ask is, what is this book, *Essential Physics*? This is your guide to specific areas of physics. In some sense it is a history book as well as a science book, covering much of the same ground that was covered by natural philosophers, scientists, and physicists over the past 2500 years or so. This book is certainly not a comprehensive look at all of physics – you can always dig deeper to find more information – but it should at least give you a reasonable basis for understanding the world around you.

Models in Physics

In this book, we will see plenty of questions about spaceships in outer space, balls, cars, people, etc. When we come to analyze situations involving such objects, however, we will often use simplifying assumptions (such as assuming that no air resistance acts on a ball in a particular case) and we will often use models in which we replace the object by something simpler.

There are several reasons for using simplifying assumptions, including:
- The assumptions can allow us to solve a particular problem in a straightforward way.
- Anything we neglect should only have a minor impact on the result if we were to account for it.
- We may not know enough about the situation to include whatever it is we're neglecting. In some cases we will address that as we go through the book, such as neglecting friction initially and then learning how to account for friction.
- The mathematical methods that would be required to solve the problem more generally are above the level of this book.

When we use a model that neglects something like air resistance, the answer we come up may not match what really happens. If the model is good, though, the answer should be a reasonable approximation of what happens. It's important to think about how an answer would change if other factors were included. For instance, if we neglect air resistance and calculate that a particular baseball hit by Josh Hamilton just clears the outfield fence for a home run, that ball would, in reality, probably be caught by an outfielder, because air resistance tends to slow an object down and reduce how far it travels.

If you start to think that we neglect too much in this book and you're interested in doing more, that's terrific. This book is merely an introduction to physics, and there is plenty of exciting physics involved with going above and beyond what we'll cover here.

Applying a model often means treating complicated objects, such as the car shown in Figure 1.1, as a simpler object, such as a particle. A modern car is rather complicated. Its shape is designed to reduce air resistance; it generally has anti-lock brakes and air bags to increase passenger safety; its engine operates under the laws of thermodynamics; and its onboard computer and accompanying electric circuits are the analog of a human's brain and central nervous system. To understand the motion of such a car along a road, however, we can generally ignore many of these complicated systems. In the early chapters of this book, for instance, we will treat cars, people, etc., as particles. We will use this **particle model** a great deal, mainly so we can focus on the big picture rather than on subtle details that depend on the precise object. As we continue through the book, our models will become more sophisticated, but when we use models we're keeping a saying of Albert Einstein's in mind: make the problem as simple as possible, but no simpler.

Figure 1.1: A photograph of a 2006 Volkswagen Jetta. A modern car is an incredibly complex object, but to understand the motion of a car like this along a straight road, we can treat the car as a particle. Photo credit: Wikimedia Commons.

Units

Physics is an experimental science. Each time we measure something, we need to be aware of the units of the measurement. Take a minute and measure the length of this page. What do you get? It would be meaningless to say 8.5, or 21.2, or 212, but if you said 8.5 inches, or 21.2 centimeters, or 212 millimeters, that would be fine. Those three measurements are equivalent, and you could measure the length of the page in other length units, too, such as feet, miles, furlongs, kilometers, or light-years. Just remember that a measurement requires both a number and a unit.

In this book we will primarily use SI (système international) units. SI has several base units, including the meter (m) as the unit of length, the kilogram (kg) as the unit of mass, and the second (s) as the unit of time. Being based on powers of ten, SI is easy to use, unlike the English system of units in which you have to remember conversion factors such as how many cups are in a gallon. In the metric system the prefix tells you which power of 10 to use. Table 1.1 gives some examples of conversion factors, with a more complete table inside the front cover of the book.

Name	Prefix	Power of ten	Example
mega	M	$\times 10^6$	92.9 MHz – frequency of an FM radio station
kilo	k	$\times 10^3$	110 km/h – speed limit on some Canadian highways
centi	c	$\times 10^{-2}$	30 cm – approximately equal to 1 foot
milli	m	$\times 10^{-3}$	500 mg – mass of Vitamin C in a Vitamin C capsule
micro	μ	$\times 10^{-6}$	150 μm – diameter of a human hair
nano	n	$\times 10^{-9}$	400 to 700 nm – wavelength range of visible light

Table 1.1: Common prefixes in the metric system.

Essential Question1.1: What is a good definition of a physical model?

Note that each section in the book ends with an Essential Question. The answer to each Essential Question is given at the top of the following page, but you should resist the temptation to immediately turn the page to look at the answer. Spending some time yourself thinking about the answer will really help you to learn the material.

Answer to Essential Question 1.1: A model is simplified version of physical situation. Using a model enables us to focus on the key elements of a particular situation, and is one way of getting a good idea of what is going on without having to consider every fine detail.

1-2 Unit Conversions, and Significant Figures

It is often necessary to convert a value from one set of units to another. To do this, we need to know the appropriate conversion factors. For instance, in Example 1.2 we will make use of these conversion factors:

- 1 hour = 3600 seconds
- 1 km = 1000 m
- 1 mile = 1.609344 km

EXAMPLE 1.2 – Unit conversions

At the 2009 World Championships in Athletics, held in Berlin, the Jamaican sprinter Usain Bolt set a world record for the 200-meter dash by running that distance in a time of 19.19 s. Assuming he ran exactly 200 m in this time, what was Usain Bolt's average speed during the race in   (a) m/s; (b) km/h; (c) miles per hour?

SOLUTION

(a) The first thing we need to do is to understand what an average speed is. Average speed is the total distance covered divided by the time in which it was covered. If we divide the given distance by the given time we'll get the answer we're looking for:

$$\text{Usain Bolt's average speed was:} \quad \frac{200 \text{ m}}{19.19 \text{ s}} = 10.422094841063 \text{ m/s}$$

This brings up the idea of **significant figures**, because you certainly do not want to quote an answer with 14 significant figures, as is shown above. Instead, round off the answer to four significant figures, because there are four in the time of 19.19 s. The rule is, ***when you multiply or divide numbers you look at the number of significant figures in the values going into the calculation and round off to the smallest number of significant figures***. Here, we're saying that the distance of 200 m is exact (see the assumption stated in the example), so that number has an infinite number of significant figures, while the time has four significant figures.

It would be more realistic to make the following argument. Lengths on a track, particularly at a major international competition such as the World Championships, are measured very accurately. For argument's sake, let's say the 200 meter distance is accurate to within 1 centimeter. Thus, the distance Usain Bolt ran was 200.00 m, seeing as 1 cm = 0.01 m. There are five significant figures in 200.00, so when dividing a number with five significant figures by one with four, we should round off our final answer to four significant figures.

Thus, Usain Bolt's average speed was 10.42 m/s.

(b) To convert from m/s to km/h, we need to know that there are 1000 m in 1 km, and that there are 3600 s in 1 hour. Then, we simply set these conversion factors up as ratios so that the units cancel properly, as follows:

$$10.4221 \frac{\text{m}}{\text{s}} \times \frac{1 \text{ km}}{1000 \text{ m}} \times \frac{3600 \text{ s}}{1 \text{ h}} = 37.52 \text{ km/h} .$$

We treat conversion factors as having an infinite number of significant figures and we remember that the minimum number of significant figures in the factors going into the average speed in m/s was four. Thus, our final answer in this case should also have four significant figures. In carrying out the calculation, however, six digits are shown for the average speed in m/s, even though we know the last two are not significant (this is why the final answer is rounded off to four significant figures in part (a)). We could even keep the 14 digits we had originally – the reason for keeping at least a couple of extra digits, and *only rounding off at the end of the calculation when you state the final answer*, is to state your answer as accurately as possible.

37.52 km/h does not differ by much from the 37.51 km/h we would get if we had started the conversion process with 10.42 m/s, but the 37.52 km/h value is more accurate.

(c) To state the average speed in miles per hour, we could start with the average speed in m/s and convert; however, it requires less work to start from km/h, so let's do that. Again, let's add an extra couple of digits for the intermediate values and round off to four significant figures at the end.

$$37.51956 \frac{km}{h} \times \frac{1 \text{ mile}}{1.609344 \text{ km}} = 23.31 \text{ miles/h}.$$

So, we have now stated Usain Bolt's average speed in three equivalent ways, all with different units. Don't forget to be amazed by how fast that is!

Significant figures

If we add or subtract numbers, the rules are a little different from what we do when we multiply or divide. Let's add the following three distances: 341.2 m, 25 cm, and 0.3367 m. First we need to convert everything to the same units. We could convert everything to meters, for instance. Then, do the addition:

341.2 m + 0.25 m + 0.3367 m = 341.7867 m.

At this point, we need to round off correctly. Here, we look at decimal places, not significant figures. The first number goes to 1 decimal place, the second number to 2 decimal places, and the third number goes to 4 decimal places. Round off the final answer to 1 decimal place, because that's the smallest number of decimal places in any of the numbers going into the sum. *When adding or subtracting, round off to the smallest number of decimal places.* In this case, our final answer would be 341.8 m.

Many people get confused by zeroes, and whether to count them as significant figures. Leading zeroes do not count, but trailing zeroes do count as significant figures. If you forget, just convert a value to scientific notation and count the significant digits.

Related End-of-Chapter Exercises: 1, 2, 3, 11, 17.

Essential Question 1.2: How many significant figures are there in the value 0.0035 m? How many are in the value 35.00 m?

Answer to Essential Question 1.2: There are only two significant figures in the value 0.0035 m, because it can be written as 3.5×10^{-3} m, which has only two significant figures. There are four significant figures in the value 35.00 m, because it can be written as 3.500×10^1 m, which has four significant figures. Trailing zeroes are very important! 3.5×10^1 m and 3.500×10^1 m represent the same length, but in the second case we know the length with greater precision than we do in the first case.

1-3 Trigonometry, Algebra, and Dimensional Analysis

Solving a physics problem often involves the geometry of right-angled triangles. Such a triangle is shown in Figure 1.2. In a right-angled triangle there are several relationships between the angle shown in the diagram and the different sides of the triangle, including:

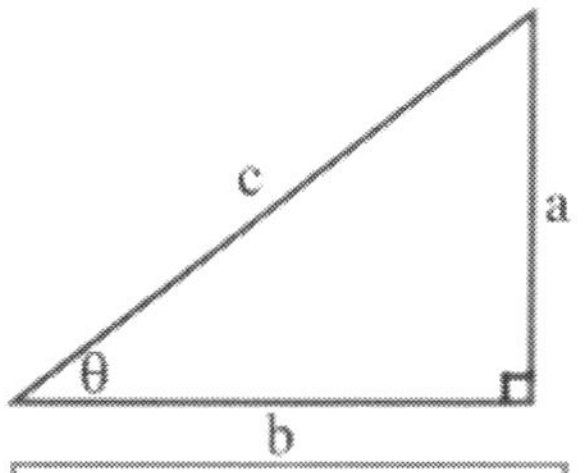

$$\sin\theta = \frac{opposite}{hypotenuse} = \frac{a}{c} \; ; \quad \cos\theta = \frac{adjacent}{hypotenuse} = \frac{b}{c} \; ; \quad \tan\theta = \frac{opposite}{adjacent} = \frac{a}{b} \; .$$

In a right-angled triangle, the Pythagorean Theorem relates the three sides:

$$c^2 = a^2 + b^2 \; .$$ (Equation 1.1: **The Pythagorean theorem**).

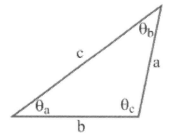

Figure 1.2: A right-angled triangle.

A few special right-angled triangles include:
- The 3-4-5 triangle in which the sides are in a 3:4:5 ratio.
- The 5-12-13 triangle in which the sides are in a 5:12:13 ratio.
- The 30°-60°-90° triangle in which the sides are in a $1:\sqrt{3}:2$ ratio.

Many triangles do not have a 90° angle. For a general triangle, such as that in Figure 1.3, if we know the length of two sides and one angle, or the length of one side and two angles, we can use the Sine Law and the Cosine Law to find the other sides and angles.

$$\frac{\sin\theta_a}{a} = \frac{\sin\theta_b}{b} = \frac{\sin\theta_c}{c} \; .$$ (Equation 1.2: **Sine Law**)

$$c^2 = a^2 + b^2 - 2ab\cos\theta_c \; .$$ (Equation 1.3: **Cosine Law**)

Figure 1.3: A general triangle.

Algebra

In addition to understanding what concepts to apply in solving a particular physics problem, you will need to know how to manipulate equations to solve for a particular unknown. In other words, you'll need to do algebra.

EXAMPLE 1.3A – Solving an equation using algebra

Solve for v in the following equation: $4v^2 - 7 = 3 - v^2$. Take a minute to solve the equation on your own before looking at the solution.

SOLUTION

To solve for a particular variable, you generally isolate that variable on one side and place everything else on the other side. Taking a step-by-step approach gives:

1. Bring all v terms to the left by adding v^2 to both sides: $5v^2 - 7 = 3$

2. Isolate the v term on the left by adding 7 to both sides: $5v^2 = 10$

3. Divide by 5: $v^2 = 2$

4. Solve for v: $v = \pm\sqrt{2}$

It is tempting to say that $v = \sqrt{2}$, but it is important to remember that the negative square root is also a possibility.

We did not concern ourselves with units above, but whenever you come up with an equation it is a good idea to do some dimensional analysis (that is, check your units). If the units check out, that does not necessarily mean your equation is correct. If your units do not check out, however, you know for sure there is something wrong with the equation.

EXAMPLE 1.3B – Using dimensional analysis

You're trying to solve for the velocity of a ball, 3 seconds after you throw it straight up in the air. You know that the velocity v has units of m/s, and you know the following parameters (defining the positive direction to be up): the initial velocity is $v_i = 20$ m/s; the acceleration is $a = -10$ m/s^2; and the time is $t = 3$ s. Your friend Sara says the equation connecting these variables is: $v = v_i + at/2$. Your friend Bob claims the equation is: $v = v_i + at^2$. Can dimensional analysis (checking the units) help you to rule out one or both of these equations as incorrect?

SOLUTION

Let's try both equations, keeping careful track of the units as we go.

Sara's method: $v = v_i + \dfrac{1}{2}at = 20\dfrac{m}{s} + \dfrac{1}{2}\left(-10\dfrac{m}{s^2}\right)(3\ s) = 20\dfrac{m}{s} - 15\dfrac{m}{s}$.

For Sara's equation, the left-hand side (v) has units of m/s, and both terms on the right-hand side also have units of m/s. This is good. ***Quantities that are added or subtracted must have the same units***, and ***the units on one side of an equation must match the units on the other***.

Bob's method: $v = v_i + at^2 = 20\dfrac{m}{s} + \left(-10\dfrac{m}{s^2}\right)(3\ s)^2 = 20\dfrac{m}{s} - 90\ m$.

Bob's equation is incorrect, because the two terms on the right do not have the same units, and the units of the last term do not match the units of the left side of the equation.

In fact, *neither Sara nor Bob has the correct equation.* As we will see in chapter 2, the correct equation relating the velocity to the initial velocity, acceleration, and time is $v = v_i + at$.

Dimensional analysis let us know that Bob's equation was incorrect, but it could not tell us that Sara's equation had an extra factor of ½ in one term, because that extra factor had no units associated with it. Dimensional analysis can be helpful, but it is just one tool in our problem-solving toolkit, and it needs to be used appropriately.

Related End-of-Chapter Exercises: 38, 45.

Essential Question 1.3: What is the connection between the Pythagorean theorem and the Cosine law?

Answer to Essential Question 1.3: The Pythagorean theorem is a special case of the Cosine Law that applies to right-angled triangles. With an angle of 90° opposite the hypotenuse, the last term in the Cosine Law disappears because cos(90°) = 0, leaving $c^2 = a^2 + b^2$.

1-4 Vectors

It is always important to distinguish between a quantity that has only a magnitude, which we call a **scalar**, and a quantity that has both a magnitude and a direction, which we call a **vector**. When we work with scalars and vectors we handle minus signs quite differently. For instance, temperature is a scalar, and a temperature of +30°C feels quite different to you than a temperature of –30°C. On the other hand, velocity is a vector quantity. Driving at +30 m/s north feels much the same as driving at –30 m/s north (or, equivalently, +30 m/s south), assuming you're going forward in both cases, at least! In the two cases, the speed at which you're traveling is the same, it's just the direction that changes. So, ***a minus sign for a vector tells us something about the direction of the vector; it does not affect the magnitude (the size) of the vector.***

When we write out a vector we draw an arrow on top to represent the fact that it is a vector, for example $\vec{A}$. A, drawn without the arrow, represents the magnitude of the vector.

EXPLORATION 1.4 – Vector components

Consider the vectors $\vec{A}$ and $\vec{B}$ represented by the arrows in Figure 1.4. The vector $\vec{A}$ lines up exactly with one of the points on the grid. The vector $\vec{B}$ has a magnitude of 4.00 m and is directed at an angle of 63.8° below the positive x-axis. It is often useful (if we're adding the vectors together, for instance) to find the **components** of the vectors. In this Exploration, we'll use a two-dimensional coordinate system with the positive x-direction to the right and the positive y-direction up. Finding the x and y components of a vector involves determining how much of the vector is directed right or left, and how much is directed up or down, respectively.

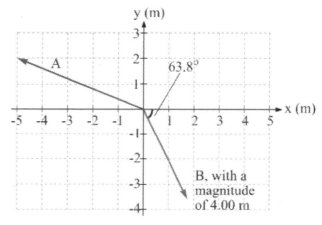

Figure 1.4: The vectors $\vec{A}$ and $\vec{B}$.

Step 1 - *Find the components of the vector $\vec{A}$.* The x and y components of $\vec{A}$ ($\vec{A}_x$ and $\vec{A}_y$, respectively) can be determined directly from Figure 1.4. Conveniently, the tip of $\vec{A}$ is located at an intersection of grid lines. In this case, we go exactly 5 m to the left and exactly 2 m up, so we can express the x and y components as:

$\vec{A}_x = +5$ m to the left, or $\vec{A}_x = -5$ m to the right.

$\vec{A}_y = +2$ m up.

This makes it look like we know the components of $\vec{A}$ to an accuracy of only one significant figure. The components are known far more precisely than that, because $\vec{A}$ lines up exactly with the grid lines. The components of $\vec{A}$ are shown in Figure 1.5.

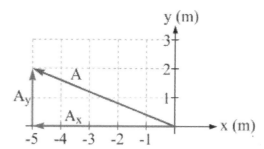

Figure 1.5: Components of the vector $\vec{A}$.

Step 2 – *Express the vector $\vec{A}$ in unit-vector notation.* Any vector is the vector sum of its components. For example, $\vec{A} = \vec{A}_x + \vec{A}_y$. This is shown graphically in Figure 1.5. It is rather long-winded to say $\vec{A} = -5$ m to the right $+ 2$ m up. We can express the vector in a more compact form by using **unit vectors**. A unit vector is a vector with a magnitude of 1 unit. We will draw a unit vector with a carat (^) on top, rather than an arrow, such as $\hat{x}$. This notation looks a bit like a hat, so we say $\hat{x}$ as "x hat". Here we make use of the following unit vectors:

$\hat{x}$ = a vector with a magnitude of 1 unit pointing in the positive x-direction

$\hat{y}$ = a vector with a magnitude of 1 unit pointing in the positive y-direction

We can now express the vector $\vec{A}$ in the compact notation: $\vec{A} = (-5$ m$)$ $\hat{x} + (2$ m$)$ $\hat{y}$.

Step 3 - *Find the components of the vector $\vec{B}$.* We will handle the components of $\vec{B}$ differently from the method we used for $\vec{A}$, because $\vec{B}$ does not conveniently line up with the grid lines like $\vec{A}$ does. Although we could measure the components of $\vec{B}$ carefully off the diagram, we will instead use the trigonometry associated with right-angled triangles to calculate these components because we know the magnitude and direction of the vector.

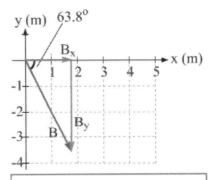

Figure 1.6: Components of the vector $\vec{B}$.

As shown in Figure 1.6, we draw a right-angled triangle with the vector as the hypotenuse, and with the other two sides parallel to the coordinate axes (horizontal and vertical, in this case). The x-component can be found from the relationship:

$\cos\theta = \dfrac{B_x}{B}$.   So $B_x = B\cos\theta = (4.00$ m$)\cos(63.8°) = 1.77$ m .

We can use trigonometry to determine the magnitude of the component and then check the diagram to get the appropriate sign. From Figure 1.6, we see that the x-component of $\vec{B}$ points to the right, so it is in the positive x-direction. We can then express the x-component of $\vec{B}$ as:

$\vec{B}_x = (+1.77$ m$)\,\hat{x}$.

The y-component can be found in a similar way:

$\sin\theta = \dfrac{B_y}{B}$.   So, $B_y = B\sin\theta = 4.00\sin(63.8°) = 3.59$ m .

The y-component of $\vec{B}$ points down, so it is in the negative y-direction. Thus:

$\vec{B}_y = -(3.59$ m$)\,\hat{y}$.

The vector $\vec{B}$ can now be expressed in unit-vector notation as:

$\vec{B} = \vec{B}_x + \vec{B}_y = (1.77$ m$)\hat{x} - (3.59$ m$)\hat{y}$.

Key ideas for vectors: It can be useful to express a vector in terms of its components. One convenient way to do this is to make use of unit vectors; a unit vector is a vector with a magnitude of 1 unit.   **Related End-of-Chapter Exercises: 6, 18.**

Essential Question 1.4: Temperature is a good example of a scalar, while velocity is a good example of a vector. List two more examples of scalars, and two more examples of vectors.

Answer to Essential Question 1.4: Other examples of scalars include mass, distance, and speed. Examples of vectors, which have directions associated with them, include displacement, force, and acceleration.

1-5 Adding Vectors

EXAMPLE 1.5 – Adding vectors

Let's define a vector $\vec{C}$ as being the sum of the two vectors $\vec{A}$ and $\vec{B}$ from Exploration 1.4. A vector that results from the addition of two or more vectors is called a **resultant vector**.

    (a) Draw the vectors $\vec{A}$ and $\vec{B}$ tip-to-tail to show geometrically the resultant vector $\vec{C}$.

    (b) Use the components of vectors $\vec{A}$ and $\vec{B}$ to find the components of $\vec{C}$.

    (c) Express $\vec{C}$ in unit-vector notation.

    (d) Express $\vec{C}$ in terms of its magnitude and direction.

SOLUTION

    (a) To add the vectors geometrically we can move the tail of $\vec{B}$ to the tip of $\vec{A}$, or the tail of $\vec{A}$ to the tip of $\vec{B}$. The order makes no difference. If we had more vectors, we could continue the process, drawing them tip-to-tail in sequence. The resultant vector always goes from the tail of the first vector to the tip of the last vector, as is shown in Figure 1.7.

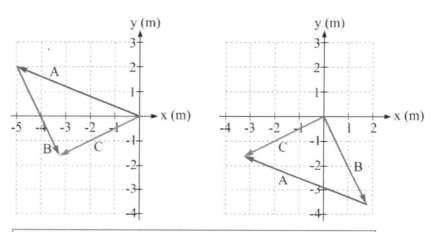

Figure 1.7: Adding vectors geometrically, tip-to-tail. In (a), the tail of vector $\vec{B}$ is placed at the tip of $\vec{A}$; in (b), the tail of vector $\vec{A}$ is placed at the tip of $\vec{B}$. The same resultant vector $\vec{C}$ is produced - the order does not matter.

    (b) Now let's add the vectors using their components. We already know the x and y components of $\vec{A}$ and $\vec{B}$ (see Exploration 1.4), so we can use those to find the components of the resultant vector $\vec{C}$. Table 1.2 demonstrates the process. Note that the components of $\vec{A}$ are shown here to two decimal places, even though we know them with more precision. Because we'll be adding the components of $\vec{A}$ to the components of $\vec{B}$, which we know to two decimal places, our final answers should also be expressed with two decimal places.

Vector	x-component	y-component
$\vec{A}$	$\vec{A}_x = -(5.00\,\text{m})\,\hat{x}$	$\vec{A}_y = +(2.00\,\text{m})\,\hat{y}$
$\vec{B}$	$\vec{B}_x = (1.77\,\text{m})\,\hat{x}$	$\vec{B}_y = -(3.59\,\text{m})\,\hat{y}$
$\vec{C} = \vec{A} + \vec{B}$	$\vec{C}_x = \vec{A}_x + \vec{B}_x$	$\vec{C}_y = \vec{A}_y + \vec{B}_y$
	$\vec{C}_x = -(5.00\,\text{m})\,\hat{x} + (1.77\,\text{m})\,\hat{x}$	$\vec{C}_y = +(2.00\,\text{m})\,\hat{y} - (3.59\,\text{m})\,\hat{y}$
	$\vec{C}_x = -(3.23\,\text{m})\,\hat{x}$	$\vec{C}_y = -(1.59\,\text{m})\,\hat{y}$

Table 1.2: Adding the vectors $\vec{A}$ and $\vec{B}$ using components. The process is shown pictorially in Figure 1.8.

Note that we are solving this two-dimensional vector-addition problem by using a technique that is very common in physics – splitting a two-dimensional problem into two separate one-dimensional problems. It is very easy to add vectors in one dimension, because the vectors can be added like scalars with signs. To find $\vec{C}_x$, for instance, we simply add the x-components of $\vec{A}$ and $\vec{B}$ together. To find $\vec{C}_y$, we carry out a similar process, adding the y-components of $\vec{A}$ and $\vec{B}$. After finding the individual components of $\vec{C}$, we then combine them, as in parts (c) and (d) below, to specify the vector $\vec{C}$.

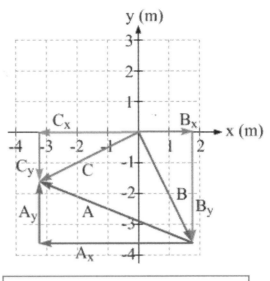

Figure 1.8: This figure illustrates the process of splitting the vectors into components when adding. Each component of the resultant vector, $\vec{C}$, is the vector sum of the corresponding components of the vectors $\vec{A}$ and $\vec{B}$.

(c) Using the bottom line in Table 1.2, the vector $\vec{C}$ can be expressed in unit-vector notation as:

$\vec{C} = \vec{C}_x + \vec{C}_y = -(3.23 \text{ m})\,\hat{x} - (1.59 \text{ m})\,\hat{y}$.

(d) If we know the components of a vector we can draw a right-angled triangle (see Figure 1.9) in which we know the lengths of two sides. Applying the Pythagorean theorem gives the length of the hypotenuse, which is the magnitude of the vector $\vec{C}$.

$$C = \sqrt{C_x^2 + C_y^2} = \sqrt{3.23^2 + 1.59^2} = \sqrt{12.961} = 3.60 \text{ m}$$

To find the angle between $\vec{C}$ and $\vec{C}_x$ we can use the relationship:

$$\tan\theta = \frac{\text{opposite}}{\text{adjacent}} = \frac{C_y}{C_x}.$$

This gives $\theta = \tan^{-1}\left(\dfrac{C_y}{C_x}\right) = \tan^{-1}\left(\dfrac{1.59}{3.23}\right) = 26.2°$.

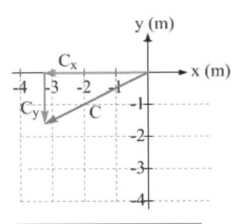

Figure 1.9: The components of the vector $\vec{C}$.

We have dropped the signs from the components, but, in stating the vector $\vec{C}$ correctly in magnitude-direction form, we can check the diagram to make sure we're accounting for which way $\vec{C}$ points: $\vec{C} =$ 3.60 m at an angle of 26.2° below the negative x-axis. The phrase "below the negative x-axis" accounts for the fact that the vector $\vec{C}$ has negative x and y components.

Related End-of-Chapter Exercises: 24 – 30.

Essential Question 1.5: Consider again the vectors $\vec{A}$ and $\vec{B}$ from Exploration 1.4 and Example 1.5. If the vector $\vec{D}$ is equal to $\vec{A} - \vec{B}$, express $\vec{D}$ in terms of its components.

Answer to Essential Question 1.5: $\vec{D} = -(6.77 \text{ m})\,\hat{x} + (5.59 \text{ m})\,\hat{y}$.

1-6 Coordinate Systems

Now that we have looked at an example of the component method of vector addition, in Example 1.5, we can summarize the steps to follow.

A General Method for Adding Vectors Using Components
1. Draw a diagram of the situation, placing the vectors tip-to-tail to show how they add geometrically.
2. Show the coordinate system on the diagram, in particular showing the positive direction(s).
3. Make a table showing the x and y components of each vector you are adding together.
4. In the last line of this table, find the components of the resultant vector by adding up the components of the individual vectors.

Coordinate systems

A coordinate system typically consists of an x-axis and a y-axis that, when combined, show an origin and the positive directions, as in Figure 1.10. A coordinate system can have just one axis, which would be appropriate for handling a situation involving motion along one line, and it can also have more than two axes if that is appropriate. An important part of dealing with vectors is to think about the coordinate system or systems that is/are appropriate for dealing with a particular situation. Let's explore this idea further.

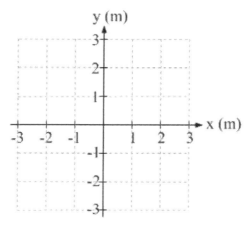

Figure 1.10: A typical x-y coordinate system.

EXPLORATION 1.6 – Buried treasure

While stranded on a desert island you find a note sealed inside a bottle that is half-buried near a big tree. Unfolding the note, you read: "Start 1 pace north of the big tree. Walk 10 paces northeast, 5 paces southeast, 6 paces southwest, 7 paces northwest, 4 paces southwest, and 2 paces southeast. Then dig." Realizing that your paces might differ in length from the paces of whoever left the note, rather than actually pacing out the distances you begin by drawing an x-y coordinate system in the sand, with positive x directed east and positive y directed north. After struggling to split the six vectors into components, however, you wonder whether there is a better way to solve the problem.

Step 1 - *Is there only one correct coordinate system, or can you choose from a number of different coordinate systems to calculate a single resultant vector that represents the vector sum of the six vectors specified in the note?* Any coordinate system will work, but there may be one coordinate system that makes the problem relatively easy, while others involve significantly more work to arrive at the answer. It's always a good idea to spend some time thinking about which coordinate system would make the problem easiest.

In fact, you should also think about whether the component method is even the easiest method to use to solve the problem. Adding vectors geometrically would also be a relatively easy way of solving this problem. Thinking about adding them geometrically (it might help to look at the six displacements, as sketched in Figure 1.11), in fact, leads us straight to the most appropriate coordinate system.

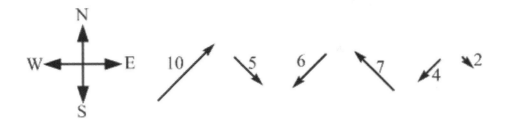

Figure 1.11: A sketch of the six displacements specified on the treasure map.

Step 2 - *What would be the simplest coordinate system to use to find the resultant vector?* One thing to notice is that the directions given are northeast, southeast, southwest, or northwest. An appropriate coordinate system is one that is aligned with these directions. For instance, we could point the positive *x*-direction northeast, and the positive *y*-direction northwest. In that case, out of the six different displacements, three are entirely in the *x*-direction and the other three are entirely in the *y*-direction. This makes the problem straightforward to solve. Figure 1.12 shows the vectors grouped by whether they are parallel to the *x*-axis or parallel to the *y*-axis.

Figure 1.12: Choosing a coordinate system that fits the problem can make the problem easier to solve. In this case we have three vectors aligned with the *x*-axis and three vectors aligned with the *y*-axis.

Step 3 - *Where should you dig?* To determine where to dig, focus first on the displacements that are either in the +*x* direction (10 paces northeast) or the –*x* direction (6 paces southwest, and 4 paces southwest). Because the total of 10 paces southwest exactly cancels the 10 paces northeast, there is no net displacement along the *x*-axis.

Now turn to the *y*-axis, where we have 7 paces northwest (the +*y* direction) and a total of 5 + 2 = 7 paces southeast (the –*y* direction). Once again these exactly cancel. Because the two components are zero, the resultant displacement vector has a magnitude of zero. You should dig at the starting point, 1 pace north of the tree (assuming you can figure out which way north is!). Digging at that spot, you find a box with a few car batteries, a 12-volt lantern, a solar cell, several wires, and a physics textbook. Reading through the book you figure out how to wire the solar cell to the batteries so the batteries are charged up while the sun shines, and you then figure out how to wire the batteries to the lantern to create a bright light you can use to signal passing planes. Using this system, you are rescued just a few days later, although you make sure to bury everything again carefully near the tree, and place the map back in the bottle, to help the next person who gets stranded there.

Key ideas for coordinate systems: Thinking carefully about the coordinate system to use can save a lot of work. Any coordinate system will work, but, in some cases, choosing the most appropriate coordinate system can make a problem considerably easier to solve.
Related End-of-Chapter Exercises: 5, 31, 41, 42.

Essential Question 1.6: In Exploration 1.6, the six displacements of 10 paces, 5 paces, 6 paces, 7 paces, 4 paces, and 2 paces happen to completely cancel one another because of their particular directions. If you could adjust the directions of each of the six vectors to whatever direction you wanted, what is the maximum distance they could take you away from the starting point?

Answer to Essential Question 1.6: If you lined up all six vectors in the same direction, you would end up 34 paces away from the starting point. When the vectors point in the same direction (and only in this case) you can add their magnitudes. $10 + 5 + 6 + 7 + 4 + 2 = 34$ paces.

1-7 The Quadratic Formula

EXAMPLE 1.7 – Solving a quadratic equation

Sometime, such as in some projectile-motion situations, we will have to solve a quadratic equation, such as $2.0x^2 = 7.0 + 5.0x$. Try solving this yourself before looking at the solution.

SOLUTION

The usual first step is to write this in the form $ax^2 + bx + c = 0$, with all the terms on the left side. In our case we get: $2.0x^2 - 5.0x - 7.0 = 0$.

We could graph this on a computer or a calculator to find the values of x (if there are any) that satisfy the equation; we could try factoring it out to find solutions; or we can use the quadratic formula to find the solution(s). Let's try the quadratic formula:

$$x = \frac{-b \pm \sqrt{b^2 - 4ac}}{2a}.$$ (Equation 1.4: **The quadratic formula**)

In our example, with $a = 2.0$, $b = -5.0$, and $c = -7.0$, the two solutions work out to:

$$x_1 = \frac{-b + \sqrt{b^2 - 4ac}}{2a} = \frac{+5.0 + \sqrt{25 + 56}}{4.0} = \frac{+5.0 + 9.0}{4.0} = +3.5 \text{, with appropriate units.}$$

$$x_2 = \frac{-b - \sqrt{b^2 - 4ac}}{2a} = \frac{+5.0 - \sqrt{25 + 56}}{4.0} = \frac{+5.0 - 9.0}{4.0} = -1.0 \text{, with appropriate units.}$$

These values agree with the graph of the function shown in Figure 1.13. The graph crosses the x-axis at two points, at $x = -1$ and also at $x = +3.5$.

Related End-of-Chapter Exercises: 36, 46.

Essential Question 1.7: Could you have a quadratic equation in the form $ax^2 + bx + c = 0$ that had no solutions for x (at least, no real solutions)? If so, what would happen when you tried to solve for x using the quadratic formula? What would the graph look like?

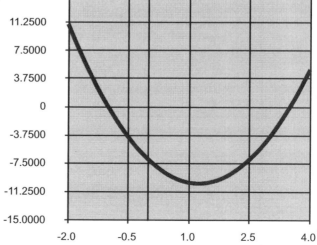

Figure 1.13: A graph of the quadratic equation $2.0x^2 - 5.0x - 7.0 = 0$, for Example 1.7.

Answer to Essential Question 1.7: Yes, you could have an equation with no real solutions. In that case when you applied the quadratic formula you would get a negative under the square root, while the graph would still be parabolic but would not cross (or touch) the *x*-axis.

Chapter Summary

Essential Idea
Physics is the study of how things work, and in analyzing physical situations we will try to apply a logical, systematic approach. Some of the basic tools we will use include:

Units
Our primary set of units is the système international (SI), based on meters, kilograms, and seconds, and four other base units. SI is widely accepted in science worldwide, and convenient because conversions are based on powers of ten. Converting between units is straightforward if you know the appropriate conversion factor(s).

Significant Figures
Three useful guidelines to follow when rounding off include:
1. Round off only at the end of a calculation when you state the final answer.

2. When you multiply or divide, round your final answer to the smallest number of significant figures in the values going into the calculation.

3. When adding or subtracting, round your final answer to the smallest number of decimal places in the values going into the calculation.

Trigonometry
In a right-angled triangle, we use the following relationships:

$$\sin\theta = \frac{opposite}{hypotenuse} = \frac{a}{c}; \qquad \cos\theta = \frac{adjacent}{hypotenuse} = \frac{b}{c}; \qquad \tan\theta = \frac{opposite}{adjacent} = \frac{a}{b}.$$

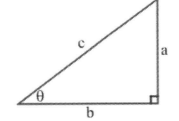

We relate the three sides using: $c^2 = a^2 + b^2$.   (Eq. 1.1: **The Pythagorean Theorem**)

Many triangles do not have a 90° angle. For a general triangle, such as that in Figure 1.3, if we know the length of two sides and one angle, or the length of one side and two angles, we can use the sine law and the cosine law to find the other sides and angles.

$$\frac{\sin\theta_a}{a} = \frac{\sin\theta_b}{b} = \frac{\sin\theta_c}{c}.$$ (Equation 1.2: **Sine Law**)

$$c^2 = a^2 + b^2 - 2ab\cos\theta_c.$$ (Equation 1.3: **Cosine Law**)

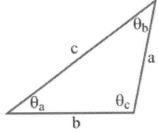

Vectors

A vector is a quantity with both a magnitude and a direction. Vectors can be added geometrically (drawn tip-to-tail), or by using components.

A unit vector is a vector with a length of one unit. A unit vector is denoted by having a carat on top, which looks like a hat, like $\hat{x}$ (pronounced "x hat").

A vector can be stated in unit-vector notation or in magnitude-direction notation.

A Method for Adding Vectors Using Components

1. Draw a diagram of the situation, placing the vectors tip-to-tail to show how they add geometrically.
2. Show the coordinate system on the diagram, in particular showing the positive direction(s).
3. Make a table showing the x and y components of each vector you are adding together.
4. In the last line of this table, find the components of the resultant vector by adding up the components of the individual vectors.

Algebra and Dimensional Analysis

Dimensional analysis can help check the validity of an equation. Units must be the same for values that are added or subtracted, as well as the same on both sides of an equation.

A quadratic equation in the form $ax^2 + bx + c = 0$ can be solved by using the quadratic formula:

$$x = \frac{-b \pm \sqrt{b^2 - 4ac}}{2a}.$$

(Equation 1.4: **The quadratic formula**)

End-of-Chapter Exercises

Exercises 1 – 10 are conceptual questions that are designed to see if you have understood the main concepts of the chapter.

1. You can convert back and forth between miles and kilometers using the approximation that 1 mile is approximately 1.6 km. (a) Which is a greater distance, 1 mile or 1 km? (b) How many miles are in 32 km? (c) How many kilometers are in 50 miles?

2. (a) How many significant figures are in the number 0.040 kg? (b) How many grams are in 0.040 kg?

3. You have two numbers, 248.0 cm and 8 cm. Rounding off correctly, according to the rules of significant figures, what is the (a) sum, and (b) product of these two numbers?

4. Figure 1.14 shows an 8-15-17 right-angled triangle. For the angle labeled θ in the triangle, express (as a ratio of integers) the angle's (a) sine, (b) cosine, and (c) tangent.

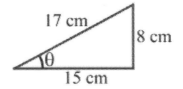

Figure 1.14: An 8-15-17 triangle, for Exercise 4.

5. You are adding two vectors by breaking them up into components. Your friend is adding the same two vectors, but is using a coordinate system that is rotated by 40° with respect to yours. Assuming you both follow the component method correctly, which of the following do the two of you agree on and which do you disagree about? (a) The magnitude of the resultant vector. (b) The direction of the resultant vector. (c) The x-component of the resultant vector. (d) The y-component of the resultant vector. (e) The angle between the x-axis and the resultant vector.

6. Three vectors are shown in Figure 1.15, along with an x-y coordinate system. Find the x and y components of (a) $\vec{A}$, (b) $\vec{B}$, and (c) $\vec{C}$.

7. The vectors $\vec{A}$ and $\vec{B}$ are specified in Figure 1.15. What is the magnitude and direction of the vector $\vec{A} + \vec{B}$?

8. You have two vectors, one with a length of length 4 m and the other with a length of 7 m. Each can be oriented in any direction you wish. If you add these two vectors together what is the (a) largest-magnitude resultant vector you can obtain? (b) smallest-magnitude resultant vector you can obtain?

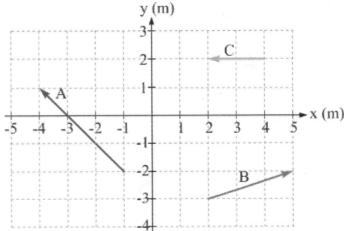

Figure 1.15: The vectors $\vec{A}$, $\vec{B}$, and $\vec{C}$, for Exercises 6 and 7.

9. You have three vectors, with lengths of 4 m, 7 m, and 9 m, respectively. If you add these three vectors together, what is the (a) largest-magnitude resultant vector you can obtain? (b) smallest-magnitude resultant vector you can obtain?

10. You have three vectors, with lengths of 4 m, 7 m, and 15 m, respectively. If you add these three vectors together what is the (a) largest-magnitude resultant vector you can obtain? (b) smallest-magnitude resultant vector you can obtain?

Exercises 11 – 17 deal with unit conversions.

11. Using the conversion factors you find in some reference (such as on the Internet), convert the following to SI units. In other words, express the following in terms of meters, kilograms, and/or seconds. (a) 1.00 years (b) 1.00 light-years (c) 8.0 furlongs (d) 12 slugs (e) 26 miles, 385 yards (the length of a marathon).

English Unit	Metric Unit
1 inch	2.54 cm
1 foot	_____ cm
1 foot	_____ m
_____ feet	1 m
1 mile	_____ km
_____ miles	1 km

Table 1.3: A table of conversion factors for length units.

English Unit	Metric Unit
1 ounce	28.35 g
1 lb.	_____ g
1 lb.	_____ kg
_____ lbs.	1 kg
1 stone	_____ kg
_____ stones	1 kg

Table 1.4: A table of conversion factors for mass units.

12. Using the fact that 1 inch is precisely 2.54 cm, fill in Table 1.3 to create your own table of conversion factors for various length units.

13. If someone were to give you a 50-carat diamond, what would be its mass in grams?

14. Fill in Table 1.4 to create your own table of conversion factors for various mass units.

15. (a) Which is larger, 1 acre or 1 hectare? (b) If you own a plot of land that has an area of exactly 1 hectare and it is square, what is the length of one of its sides, in meters? (c) If your 1-hectare lot is rectangular with a width of 20 m, how long is it?

16. What is your height in (a) inches? (b) cm? What is your mass in (c) pounds? (d) kg?

17. Firefighters using a fire hose can spray about 1.0×10^2 gallons of water per minute on a fire. What is this in liters per second? Assume the firefighters are in the USA. (Why is this assumption necessary?)

Exercises 18 – 26 deal with various aspects of vectors and vector components.

18. In Exploration 1.4, we expressed the vector $\vec{A}$ in terms of its components. Assuming the magnitude of each component is known to three significant figures, express $\vec{A}$ in terms of its magnitude and direction.

19. Using the result of Exercise 18 to help you, and aided by Figure 1.7, use the sine law and/or the cosine law to determine the magnitude and direction of the vector $\vec{C}$ shown in Figure 1.7. The vectors $\vec{A}$ and $\vec{B}$ are defined in Exploration 1.4. Hint: you may find it helpful to use geometry to first determine the angle between the vectors $\vec{A}$ and $\vec{B}$. Show your work.

20. Two vectors $\vec{Q}$ and $\vec{R}$ can be expressed in unit-vector notation as follows:
$\vec{Q} = (3.0 \text{ m})\hat{x} + (4.0 \text{ m})\hat{y}$ and $\vec{R} = (5.0 \text{ m})\hat{x} - (7.0 \text{ m})\hat{y}$. Express the following in unit-vector notation: (a) $6\vec{Q}$ (b) $6\vec{Q} - 4\vec{R}$ (c) $4\vec{R} - 6\vec{Q}$.

21. Repeat Exercise 20, but express your answers in magnitude-direction format instead.

22. See Exercise 20 for the definitions of the vectors $\vec{Q}$ and $\vec{R}$. (a) Is it possible to solve for the number a in the equation $\vec{Q} + a\vec{R} = 0$? If it is possible, then solve for a; if not, explain why not. (b) How many different values of b are there such that the sum $\vec{Q} + b\vec{R}$ has only an x-component? Find all such values of b.

23. Three vectors are shown in Figure 1.16, along with an x-y coordinate system. Use magnitude-direction format to specify vector (a) $\vec{A}$ (b) $\vec{B}$ (c) $\vec{C}$.

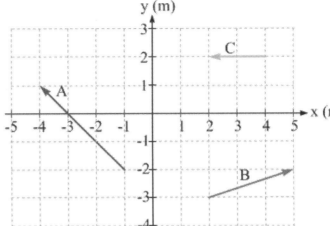

24. The vectors $\vec{A}$ and $\vec{C}$ are shown in Figure 1.16. Consider the following vectors: 1. $\vec{A}$; 2. $\vec{C}$; 3. $\vec{A} + \vec{C}$; 4. $\vec{A} - \vec{C}$. Rank those four vectors by their magnitude, from largest to smallest. Use notation such as $3 > 2 = 4 > 1$.

25. Three vectors are shown in Figure 1.16. (a) Use the geometric method of vector addition (add vectors tip-to-tail) to draw the vector representing $\vec{B} + \vec{C}$. (b) Specify that resultant vector in unit-vector notation.

Figure 1.16: The vectors $\vec{A}$, $\vec{B}$, and $\vec{C}$, for Exercises 23 – 31.

26. Three vectors are shown in Figure 1.16. (a) Use the geometric method of vector addition (add vectors tip-to-tail) to draw the vector representing $\vec{A} + \vec{B} + \vec{C}$. (b) Specify that resultant vector in unit-vector notation. (c) Specify that resultant vector in magnitude-direction notation.

Exercises 27 – 31 are designed to give you practice in applying the component method of vector addition. For each exercise, start with the following: (a) Draw a diagram showing how the vectors add geometrically (the tip-to-tail method). (b) Show the coordinate system (we'll use the standard coordinate system shown in Figure 1.16 above). (c) Make a table showing the x and y components of each vector you're adding together. (d) In the last row of the table, find the components of the resultant vector.

In addition to the vectors $\vec{A}$, $\vec{B}$, and $\vec{C}$ shown in Figure 1.16, let's make use of these two vectors: the vector $\vec{D} = (-6.00 \text{ m})\hat{x} + (-2.00 \text{ m})\hat{y}$, and the vector $\vec{E}$ with a magnitude of 5.00 m at an angle of 30° above the positive x-axis.

27. Find the resultant vector representing $\vec{A} + \vec{B} + \vec{C}$. Answer parts (a) – (d) as specified above. (e) State the resultant vector in magnitude-direction notation.

28. Find the resultant vector representing $\vec{A} + \vec{D}$. Answer parts (a) – (d) as specified above. (e) State the resultant vector in unit-vector notation.

29. Find the resultant vector representing $\vec{C} - \vec{E}$. Answer parts (a) – (d) as specified above. (e) State the resultant vector in magnitude-direction format.

30. Find the resultant vector representing $\vec{B} + \vec{D}$. Answer parts (a) – (d) as specified above. (e) State the resultant vector in unit-vector notation. (f) State the resultant vector in magnitude-direction format.

31. Repeat Exercise 30, but now use an x-y coordinate system that is rotated so the positive x-direction is the direction of the vector $\vec{B}$.

Exercises 32 – 36 involve applications of the physics concepts addressed in this chapter.

32. In 1999, NASA had a high-profile failure when it lost contact with the Mars Climate Orbiter as it was trying to put the spacecraft into orbit around Mars. Do some research and write a paragraph or two about what was responsible for this failure, and how much the project cost.

33. In 1983, an Air Canada Boeing 767 airplane was nicknamed "The Gimli Glider." Discuss the events that led to the plane getting this nickname, and how they relate to the topics in this chapter.

34. One way to travel from Salt Lake City, Utah, to Billings, Montana, is to first drive 660 km north on interstate 15 to Butte, Montana, and then drive 370 km east to Billings on interstate 90. If you did this trip by plane, instead, traveling in a straight line between Salt Lake City and Billings, how far would you travel?

35. In the sport of orienteering, participants must plan carefully to get from one checkpoint to another in the shortest possible time. In one case, starting at a particular checkpoint, Sam decides to take a path that goes west for 600 meters, and then go northeast for 400 meters on another path to reach the next checkpoint. Between the same two checkpoints, Mary decides to take the shortest distance between the two checkpoints, traveling off the paths through the woods instead. What is the distance Mary travels along her route, and in what direction does she travel between the checkpoints?

36. You throw a ball almost straight up, with an initial speed of 10 m/s, from the top of a 20-meter-high cliff. The approximate time it takes the ball to reach the base of the cliff can be found by solving the quadratic equation $-20 \text{ m} = (10 \text{ m/s})t - (5.0 \text{ m/s}^2)t^2$. Solve the equation to find the approximate time the ball takes to reach the base of the cliff.

General problems and conceptual questions

37. Prior to the 2004 Boston Marathon, the Boston Globe newspaper carried a story about a running shoe called the Nike Mayfly. According to the newspaper, the shoes were designed to last for 62.5 miles. Does anything strike you as odd about this distance? If so, what?

38. Do the following calculations make any sense? Why or why not? If any make sense, what could they represent?

(a) $a = 17\dfrac{m}{s} \times 3.2\,s$ (b) $b = 17\dfrac{m}{s} + 3.2\,s$ (c) $c = \dfrac{751\,m}{\cos\left(23°\right)}$ (d) $d = \left(751\,m\right) - \cos\left(23°\right)$.

39. The distance from Dar es Salaam, Tanzania to Nairobi, Kenya, is 677 km, while it is 1091 km from Dar es Salaam to Kampala, Uganda. (a) Using these numbers alone, can you determine the distance between Nairobi and Kampala? Briefly justify your answer. (b) Using only these numbers, what is the minimum possible distance between Nairobi and Kampala? (c) What is the maximum possible distance?

40. Use the information given in Exercise 39, combined with the fact that it is 503 km from Nairobi to Kampala, to construct a triangle with the three cities at the vertices. What is the angle between the two lines that meet at (a) Dar es Salaam? (b) Nairobi? (c) Kampala?

41. Figure 1.17 shows overhead views of similar sections of two different cities where the blocks are marked out in a square grid pattern. In City A, the streets run north-south and east-west, while in City B, the streets are at some angle with respect to those in City A. In City A, Anya intends to walk from the lower left corner (marked by a blue dot) to the upper right corner (marked by a yellow dot). In City B, Boris will walk a similar route between the colored dots, ending up due north of his starting point. For both cities, use a coordinate system where positive x is east and positive y is north. (a) Assuming Anya goes along the streets, marked in black, choose a route for Anya to follow that involves her changing direction only once. Express her route in unit-vector notation. (b) How many blocks does she travel? (c) Assuming Boris goes along the streets, choose a route for Boris to follow that involves him changing direction only once. Break Boris' trip into two parts, the first ending at the corner where he changes direction and the second starting there, and express each part as a vector in unit-vector notation. (d) What do you get when you add the two vectors from part (c)?

42. Return to the situation described in Exercise 39, where for City B we used a coordinate system where positive x is east and positive y is north. Comment on the relative advantages and disadvantages of that coordinate system over one in which the coordinate axes are aligned parallel to the streets.

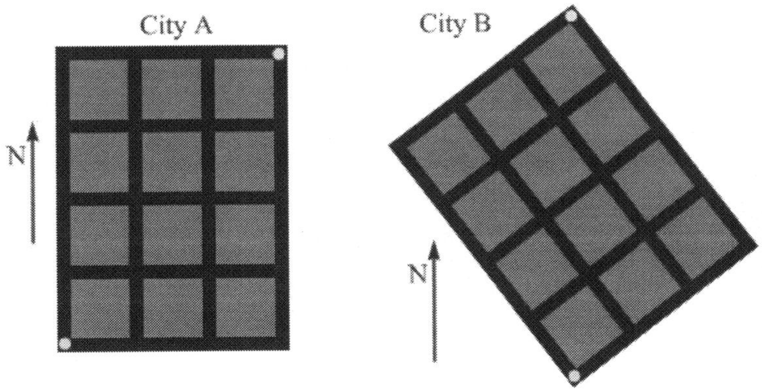

43. The top of a mountain is 2100 m north, 3300 m west, and 2300 m vertically above the initial location of a mountain climber. (a) What is the straight-line distance between the top of the mountain and the climber? (b) Later in the climb, the climber finds that she is 1200 m south, 900 m east, and 1100 m vertically below the top of the mountain. What is the minimum distance she has traveled from her starting point? (c) Checking her handheld GPS (global positioning system) receiver, she finds she has actually traveled a distance 2.5 times larger than the answer to part (b). How is this possible?

Figure 1.17: Overhead views of a 3 block by 4 block region in two different cities, for Exercises 41 and 42.

44. In Figure 1.18, four successive moves are shown near the end of a chess match. First, Black moves his Pawn (P); then, White moves her Queen (Q); then Black moves his Knight (K); and White moves her Rook (R). Using a traditional *x-y* coordinate system with positive *x* to the right and positive *y* up, we can express the movement of the Queen as +4 units in the *x*-direction and -4 units in the *y*-direction. Using similar notation, express the movement of the (a) Pawn; (b) Knight; and (c) Rook.

45. Solve for *x* in the following expressions:

    (a) $5x - 7 = 2x + 5$. (b) $3 = \dfrac{4}{x} + \dfrac{4}{2x}$.

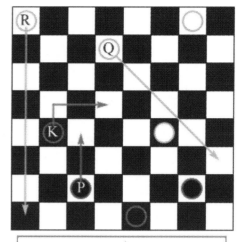

Figure 1.18: Four successive moves in a chess match.

46. Often, when we solve a one-dimensional motion problem, which we will spend some time doing in the next chapter, we need to solve a quadratic equation to find the time when some event happens. For instance, solving for the time it takes a police officer to catch a speeding motorist could involve solving an equation of the form:

$$\left(2 \text{ m/s}^2\right)t^2 - \left(10 \text{ m/s}\right)t - 100 \text{ m} = 0,$$

where the *m* stands for meters and *s* stands for seconds. (a) What are the two possible solutions for *t*, the time when the police officer catches up to the motorist? (b) Which solution is the one we want to keep as the solution to the problem?

47. In the optics section of this book, we will use an equation to relate the position of an image formed by a lens (known as the image distance, d_i) to the position of the object with respect to the lens (the object distance, d_o) and the focal length, *f*, of the lens. (a) Use dimensional analysis to determine which of the following three equations could correctly relate those three lengths.

    Equation 1: $d_i = \dfrac{d_o - f}{d_o f}$; Equation 2: $d_i = \dfrac{d_o f}{d_o - f}$; Equation 3: $d_i = \dfrac{d_o f - f}{d_o}$.

    If you think any of the above are dimensionally incorrect, explain why. (b) If you found one or more of the three equations to be dimensionally correct, does this guarantee that the equation is the correct way to relate these three lengths? Explain.

48. Three students are having a conversation. Comment on each of their statements.

Ruben: The question says, what's the magnitude of the resultant vector obtained by adding a vector of length 3 units to a vector of length 4 units. That's just 7 units, right?

Marta: I think it is 5 units, because you can make a 3-4-5 triangle.

Kaitlyn: It depends on what direction the vectors are in. If the 3 and the 4 are in the same direction, you get 7 when you add them. If you have the 3 and the 4 perpendicular to one another, you get the 3-4-5 triangle. I think you can get a resultant of anything between 5 units and 7 units, depending on the angle between the 3 and the 4.

The photo shows runners in a 100-meter race on a track, in which each runner runs in a straight line until they reach the finish line. The motion of each runner is an excellent example of motion in one dimension. Photo credit: Ana Abejon/ iStockPhoto.

Chapter 2 – Motion in One Dimension

CHAPTER CONTENTS

In Chapters 2 – 5, we will investigate how things move as well as why things move. In Chapter 3, we will examine forces in one dimension, which explain why objects move in one dimension as they do. In Chapters 4 and 5, we then extend these ideas to two-dimensional situations.

We being our investigations of motion in this chapter, in which we describe motion in one dimension (also known as one-dimensional kinematics). Applications of these ideas include the motion of a ball that is either dropped from rest, or thrown straight up into the air; the motion of a car or truck along a straight road; and your motion as you walk in a straight line along the sidewalk.

2-1 Position, Displacement, and Distance

In describing an object's motion, we should first talk about position – where is the object? A position is a vector because it has both a magnitude and a direction: it is some distance from a zero point (the point we call **the origin**) in a particular direction. With one-dimensional motion, we can define a straight line along which the object moves. Let's call this the x-axis, and represent different locations on the x-axis using variables such as $\vec{x}_0$ and $\vec{x}_1$, as in Figure 2.1.

Figure 2.1: Positions $\vec{x}_0 = +3$ m and $\vec{x}_1 = -2$ m, where the + and – signs indicate the direction.

If an object moves from one position to another we say it experiences a **displacement**.

> **Displacement:** a vector representing a change in position. A displacement is measured in length units, so the MKS unit for displacement is the meter (m).
> We generally use the Greek letter capital delta (Δ) to represent a change. If the initial position is $\vec{x}_i$ and the final position is $\vec{x}_f$ we can express the displacement as:
>
> $$\Delta \vec{x} = \vec{x}_f - \vec{x}_i .$$ (Equation 2.1: **Displacement in one dimension**)

In Figure 2.1, we defined the positions $\vec{x}_0 = +3$ m and $\vec{x}_1 = -2$ m. What is the displacement in moving from position $\vec{x}_0$ to position $\vec{x}_1$? Applying Equation 2.1 gives $\Delta \vec{x} = \vec{x}_1 - \vec{x}_0 = -2$ m $- (+3$ m$) = -5$ m. This method of adding vectors to obtain the displacement is shown in Figure 2.2. Note that the negative sign comes from the fact that the displacement is directed left, and we have defined the positive x-direction as pointing to the right.

Figure 2.2: The displacement is –5 m when moving from position $\vec{x}_0$ to position $\vec{x}_1$. Equation 2.1, the displacement equation, tells us that the displacement is $\Delta \vec{x} = \vec{x}_1 - \vec{x}_0$, as in the figure. The bold arrow on the axis is the displacement, the vector sum of the vector $\vec{x}_1$ and the vector $-\vec{x}_0$.

To determine the displacement of an object, you only have to consider the change in position between the starting point and the ending point. The path followed from one point to the other does not matter. For instance, let's say you start at $\vec{x}_0$ and you then have a displacement of 8 meters to the left followed by a second displacement of 3 meters right. You again end up at $\vec{x}_1$, as shown in Figure 2.4. The total distance traveled is the sum of the magnitudes of the individual displacements, 8 m + 3 m = 11 m. The net displacement (the vector sum of the individual displacements), however, is still 5 meters to the left: $\Delta \vec{x} = -8$ m $+ (+3$ m$) = -5$ m $= \vec{x}_1 - \vec{x}_0$.

Figure 2.3: The net displacement is still –5 m, even though the path taken from $\vec{x}_0$ to $\vec{x}_1$ is different from the direct path taken in Figure 2.2.

EXAMPLE 2.1 – Interpreting graphs

Another way to represent positions and displacements is to graph the position as a function of time, as in Figure 2.4. This graph could represent your motion along a sidewalk.

    (a) What happens at a time of $t = 40$ s?

    (b) Draw a diagram similar to that in Figure 2.3, to show your motion along the sidewalk. Add circles to your diagram to show your location at 10-second intervals, starting at $t = 0$.

    Using the graph in Figure 2.4, find (c) your net displacement and (d) the total distance you covered during the 50-second period.

SOLUTION

(a) At a time of $t = 40$ s, the graph shows that your motion changes from travel in the positive x-direction to travel in the negative x-direction. In other words, at $t = 40$ s you reverse direction.

Figure 2.4: A graph of the position of an object versus time over a 50-second period. The graph represents your motion in a straight line as you travel along a sidewalk.

(b) Figure 2.5 shows one way to turn the graph in Figure 2.4 into a vector diagram to show how a series of individual displacements adds together to a net displacement. Figure 2.5 shows five separate displacements, which break your motion down into 10-second intervals.

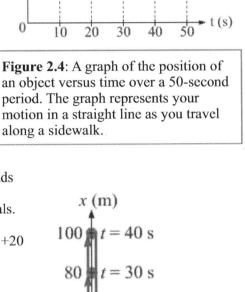

(c) The displacement can be found by subtracting the initial position, +20 m, from the final position, +60 m. This gives a net displacement of
$$\Delta \vec{x}_{net} = \vec{x}_f - \vec{x}_i = +60 \text{ m} - (+20 \text{ m}) = +40 \text{ m} .$$

    A second way to find the net displacement is to recognize that the motion consists of two displacements, one of +80 m (from +20 m to +100 m) and one of –40 m (from +100 m to +60 m). Adding these individual displacements gives $\Delta \vec{x}_{net} = \Delta \vec{x}_1 + \Delta \vec{x}_2 = +80 \text{ m} + (-40 \text{ m}) = +40 \text{ m}$.

(d) The total distance covered is the sum of the magnitudes of the individual displacements. Total distance = 80 m + 40 m = 120 m.

Related End-of-Chapter Exercises: 7 and 9

Figure 2.5: A vector diagram to show your displacement, as a sequence of five 10-second displacements over a 50-second period. The circles show your position at 10-second intervals.

Essential Question 2.1: In the previous example, the magnitude of the displacement is less than the total distance covered. Could the magnitude of the displacement ever be larger than the total distance covered? Could they be equal? Explain. *(The answer is at the top of the next page.)*

Answer to Essential Question 2.1: The magnitude of the net displacement is always less than or equal to the total distance. The two quantities are equal when the motion occurs without any change in direction. In that case, the individual displacements point in the same direction, so the magnitude of the net displacement is equal to the sum of the magnitudes of the individual displacements (the total distance). If there is a change of direction, however, the magnitude of the net displacement is less than the total distance, as in Example 2.1.

2-2 Velocity and Speed

In describing motion, we are not only interested in where an object is and where it is going, but we are also generally interested in how fast the object is moving and in what direction it is traveling. This is measured by the object's velocity.

Average velocity: a vector representing the average rate of change of position with respect to time. The SI unit for velocity is m/s (meters per second).

Because the change in position is the displacement, we can express the average velocity as:

$$\bar{v} = \frac{\Delta \vec{x}}{\Delta t} = \frac{\text{net displacement}}{\text{time interval}} \, . \qquad \text{(Equation 2.2: \textbf{Average velocity})}$$

The bar symbol (–) above a quantity means the average of that quantity. The direction of the average velocity is the direction of the displacement.

"Velocity" and "speed" are often used interchangeably in everyday speech, but in physics we distinguish between the two. Velocity is a vector, so it has both a magnitude and a direction, while speed is a scalar. Speed is the magnitude of the instantaneous velocity (see the next page). Let's define average speed.

Average Speed $= \bar{v} = \dfrac{\text{total distance covered}}{\text{time interval}}$ $\qquad$ (Equation 2.3: **Average speed**)

In Section 2-1, we discussed how the magnitude of the displacement can be different from the total distance traveled. This is why the magnitude of the average velocity can be different from the average speed.

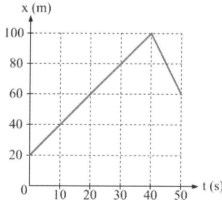

EXAMPLE 2.2A – Average velocity and average speed
Consider Figure 2.6, the graph of position-versus-time we looked at in the previous section. Over the 50-second interval, find:
(a) the average velocity, and (b) the average speed.

SOLUTION
(a) Applying Equation 2.2, we find that the average velocity is:

$$\bar{v} = \frac{\Delta \vec{x}}{\Delta t} = \frac{+40 \,\text{m}}{50 \,\text{s}} = +0.80 \,\text{m/s} \, .$$

Figure 2.6: A graph of your position versus time over a 50-second period as you move along a sidewalk.

The net displacement is shown in Figure 2.7. We can also find the net displacement by adding, as vectors, the displacement of +80 meters, in the first 40 seconds, to the displacement of –40 meters, which occurs in the last 10 seconds.

(b) Applying Equation 2.3 to find the average speed,

$$\bar{v} = \frac{\text{total distance covered}}{\text{time interval}} = \frac{80\,\text{m} + 40\,\text{m}}{50\,\text{s}} = \frac{120\,\text{m}}{50\,\text{s}} = 2.4\,\text{m/s}.$$

The average speed and average velocity differ because the motion involves a change of direction. Let's now turn to finding instantaneous values of velocity and speed.

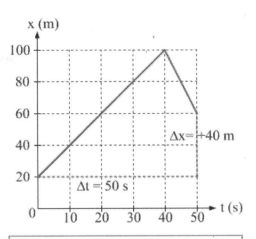

Figure 2.7: The net displacement of +40 m is shown in the graph.

Instantaneous velocity: a vector representing the rate of change of position with respect to time at a particular instant in time. A practical definition is that the instantaneous velocity is the slope of the position-versus-time graph at a particular instant. Expressing this as an equation:

$$\vec{v} = \frac{\Delta \vec{x}}{\Delta t}.$$ (Equation 2.4: **Instantaneous velocity**)

Δt is sufficiently small that the velocity can be considered to be constant over that time interval.

Instantaneous speed: the magnitude of the instantaneous velocity.

EXAMPLE 2.2B – Instantaneous velocity

Once again, consider the motion represented by the graph in Figure 2.6. What is the instantaneous velocity at (a) $t = 25$ s? (b) $t = 45$ s?

SOLUTION

(a) Focus on the slope of the graph, as in Figure 2.8, which represents the velocity. The position-versus-time graph is a straight line for the first 40 seconds, so the slope, and the velocity, is constant over that time interval. Because of this, we can use the entire 40-second interval to find the value of the constant velocity at any instant between $t = 0$ and $t = 40$ s. Thus, the velocity at t = 25 s is

$$\vec{v}_1 = \frac{\text{rise}}{\text{run}} = \frac{\text{displacement}}{\text{time}} = \frac{+80\,\text{m}}{40\,\text{s}} = +2.0\,\text{m/s}.$$

(b) We use a similar method to find the constant velocity between $t = 40$ s and $t = 50$ s:
At $t = 45$ s, the velocity is

$$\vec{v}_2 = \frac{\text{displacement}}{\text{time}} = \frac{-40\,\text{m}}{10\,\text{s}} = -4.0\,\text{m/s}.$$

Related End-of-Chapter Exercises: 2, 3, 8, 10, and 11

Essential Question 2.2: For the motion represented by the graph in Figure 2.6, is the average velocity over the entire 50-second interval equal to the average of the velocities we found in Example 2.2B for the two different parts of the motion? Explain.

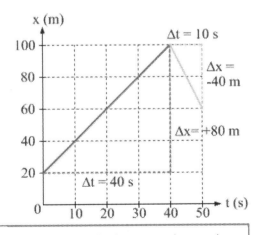

Figure 2.8: The velocity at any instant in time is determined by the slope of the position-versus-time graph at that instant.

Answer to Essential Question 2.2: If we take the average of the two velocities we found in Example 2.2B, $\bar{v}_1 = +2.0\,\text{m/s}$ and $\bar{v}_2 = -4.0\,\text{m/s}$, we get $-1.0\,\text{m/s}$. This is clearly not the average velocity, because we found the average velocity to be +0.80 m/s in Example 2.2A . The reason the average velocity differs from the average of the velocities of the two parts of the motion is that one part of the motion takes place over a longer time interval than the other (4 times longer, in this case). If we wanted to find the average velocity by averaging the velocity of the different parts, we could do a weighted average, weighting the velocity of the first part of the motion four times more heavily because it takes four times as long, as follows:

$$\bar{\bar{v}} = \frac{4 \times (+2.0\,\text{m/s}) + 1 \times (-4.0\,\text{m/s})}{4+1} = \frac{+4.0\,\text{m/s}}{5} = +0.8\,\text{m/s}.$$

2-3 Different Representations of Motion

There are several ways to describe the motion of an object, such as explaining it in words, or using equations to describe the motion mathematically. Different representations give us different perspectives on how an object moves. In this section, we'll focus on two other ways of representing motion, drawing motion diagrams and drawing graphs. We'll do this for motion with constant velocity - motion in a constant direction at a constant speed.

EXPLORATION 2.3A – Learning about motion diagrams

A motion diagram is a diagram in which the position of an object is shown at regular time intervals as the object moves. It's like taking a video and over-laying the frames of the video.

Step 1 - *Sketch a motion diagram for an object that is moving at a constant velocity.* An object with constant velocity travels the same distance in the same direction in each time interval. The motion diagram in Figure 2.9 shows equally spaced images along a straight line. The numbers correspond to times, so this object is moving to the right with a constant velocity.

Figure 2.9: Motion diagram for an object that has a constant velocity to the right.

Step 2 - *Draw a second motion diagram next to the first, this time for an object that is moving parallel to the first object but with a larger velocity.* To be consistent, we should record the positions of the two objects at the same times. Because the second object is moving at constant velocity, the various images of the second object on the motion diagram will also be equally spaced. Because the second object is moving faster than the first, however, there will be more space between the images of the second object on the motion diagram – the second object covers a greater distance in the same time interval. The two motion diagrams are shown in Figure 2.10.

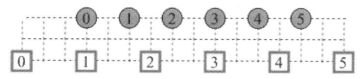

Figure 2.10: Two motion diagrams side by side. These two motion diagrams show objects with a constant velocity to the right but the lower object (marked by the square) has a higher speed, and it passes the one marked by the circles at time-step 3.

Key ideas: A motion diagram can tell us whether or not an object is moving at constant velocity. The farther apart the images, the higher the speed. Comparing two motion diagrams can tell us which object is moving fastest and when one object passes another.

Related End-of-Chapter Exercises: 23 and 24

EXPLORATION 2.3B – Connecting velocity and displacement using graphs
As we have investigated already with position-versus-time graphs, another way to represent motion is to use graphs, which can give us a great deal of information. Let's now explore a velocity-versus-time graph, for the case of a car traveling at a constant velocity of +25 m/s.

Step 1 - *How far does the car travel in 2.0 seconds*? The car is traveling at a constant speed of 25 m/s, so it travels 25 m every second. In 2.0 seconds the car goes 25 m/s × 2.0 s, which is 50 m.

Step 2 – *Sketch a velocity-versus-time graph for the motion. What on the velocity-versus-time graph tells us how far the car travels in 2.0 seconds?* Because the velocity is constant, the velocity-versus-time graph is a horizontal line, as shown in Figure 2.11.

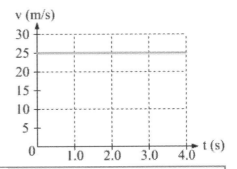

To answer the second question, let's re-arrange Equation 2.2,

$\bar{v} = \dfrac{\Delta \vec{x}}{\Delta t}$, to solve for the displacement from the average velocity.

$$\Delta \vec{x} = \bar{v}\, \Delta t .$$ (Equation 2.5: **Finding displacement from average velocity**)

Figure 2.11: The velocity-versus-time graph for a car traveling at a constant velocity of +25 m/s.

When the velocity is constant, the average velocity is the value of the constant velocity. This method of finding the displacement can be visualized from the velocity-versus-time graph. The displacement in a particular time interval is the area under the velocity-versus-time graph for that time interval. "The area under a graph" means the area of the region between the line or curve on the graph and the *x*-axis. As shown in Figure 2.12, this area is particularly easy to find in a constant-velocity situation because the region we need to find the area of is rectangular, so we can simply multiply the height of the rectangle (the velocity) by the width of the rectangle (the time interval) to find the area (the displacement).

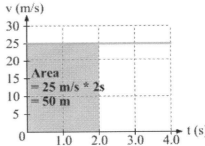

Key idea: The displacement is the area under the velocity-versus-time graph. This is true in general, not just for constant-velocity motion.

Deriving an equation for position when the velocity is constant
Substitute Equation 2.1, $\Delta \vec{x} = \vec{x}_f - \vec{x}_i$, into Equation 2.5,

$\Delta \vec{x} = \bar{v}\, \Delta t$.

This gives: $\vec{x}_f - \vec{x}_i = \bar{v}\, \Delta t = \bar{v}\left(t_f - t_i\right)$.

Generally, we define the initial time t_i to be

Figure 2.12: The area under the velocity-versus-time graph in a particular time interval equals the displacement in that time interval.

zero: $\vec{x}_f - \vec{x}_i = \bar{v}\, t_f$.

Remove the "f" subscripts to make the equation as general as possible: $\vec{x} - \vec{x}_i = \bar{v}\, t$.

$\vec{x} = \vec{x}_i + \bar{v}\, t$.   (Equation 2.6: **Position for constant-velocity motion**)

Such a position-as-a-function-of-time equation is known as an **equation of motion**.

Related End-of-Chapter Exercises: 3, 17, and 48.

Essential Question 2.3: What are some examples of real-life objects experiencing constant-velocity motion? *(The answer is at the top of the next page.)*

Answer to Essential Question 2.3: Some examples of constant velocity (or at least almost-constant velocity) motion include (among many others):
- A car traveling at constant speed without changing direction.
- A hockey puck sliding across ice.
- A space probe that is drifting through interstellar space.

2-4 Constant-Velocity Motion

Let's summarize what we know about constant-velocity motion. We will also explore a special case of constant-velocity motion - that of an object at rest.

EXPLORATION 2.4 – Positive, negative, and zero velocities

Three cars are on a straight road. A blue car is traveling west at a constant speed of 20 m/s; a green car remains at rest as its driver waits for a chance to turn; and a red car has a constant velocity of 10 m/s east. At the time $t = 0$, the blue and green cars are side-by-side at a position 20 m east of the red car. Take east to be positive.

Step 1 – *Picture the scene: sketch a diagram showing this situation.* In addition to showing the initial position of the cars, the sketch at the middle left of Figure 2.13 shows the origin and positive direction. The origin was chosen to be the initial position of the red car.

Step 2 - *Sketch a set of motion diagrams for this situation.* The motion diagrams are shown at the top of Figure 2.13, from the perspective of someone in a stationary helicopter looking down on the road from above. Because the blue car's speed is twice as large as the red car's speed, successive images of the blue car are twice as far apart as those of the red car. The cars' positions are shown at 1-second intervals for four seconds.

Step 3 - *Write an equation of motion (an equation giving position as a function of time) for each car.* Writing equations of motion means substituting appropriate values for the initial position $\vec{x}_i$ and the constant velocity $\vec{v}$ into Equation 2.5, $\vec{x} = \vec{x}_i + \vec{v}t$. The equations are shown above the graphs in Figure 2.13, using the values from Table 2.1.

	Blue car	**Green car**	**Red car**
Initial position, $\vec{x}_i$	+20 m	+20 m	0
Velocity, $\vec{v}$	-20 m/s	0	+10 m/s

Table 2.1: Organizing the data for the three cars.

Step 4 - *For each car sketch a graph of its position as a function of time and its velocity as a function of time for 4.0 seconds.* The graphs are shown at the bottom of Figure 2.13. Note that the position-versus-time graph for the green car, which is at rest, is a horizontal line because the car maintains a constant position. An object at rest is a special case of constant-velocity motion: the velocity is both constant and equal to zero.

Key ideas: The at-rest situation is a special case of constant-velocity motion. In addition, all we have learned about constant-velocity motion applies whether the constant velocity is positive, negative, or zero. This includes the fact that an object's displacement is given by $\vec{x} = \vec{x}_i + \vec{v}t$; the displacement is the area under the velocity-versus-time graph; and the velocity is the slope of the position-versus-time graph. **Related End-of-Chapter Exercise: 43**

(a) Description of the motion in words: Three cars are on a straight road. A blue car has a constant velocity of 20 m/s west; a green car remains at rest; and a red car has a constant velocity of 10 m/s east. At *t* = 0 the blue and green cars are side-by-side, 20 m east of the red car.

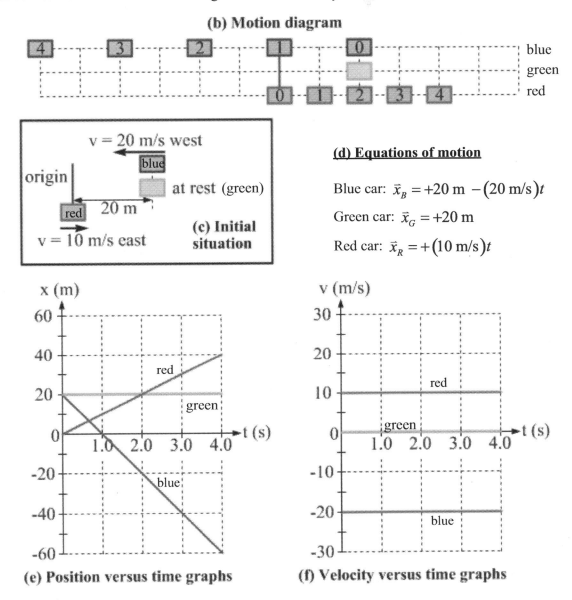

(b) Motion diagram

(d) Equations of motion

Blue car: $\vec{x}_B = +20 \text{ m} - (20 \text{ m/s})t$

Green car: $\vec{x}_G = +20 \text{ m}$

Red car: $\vec{x}_R = +(10 \text{ m/s})t$

(c) Initial situation

(e) Position versus time graphs

(f) Velocity versus time graphs

Figure 2.13: Multiple representations of the constant-velocity motions of three cars. These include (a) a description of the motion in words; (b) a motion diagram; (c) a diagram of the initial situation, at *t* = 0 (this is shown in a box); (d) equations of motion for each car; (e) graphs of the position of each car as a function of time; and (f) graphs of the velocity of each car as a function of time. Each representation gives us a different perspective on the motion.

Essential Question 2.4: Consider the graph of position-versus-time that is part of Figure 2.13. What is the significance of the points where the different lines cross? *(The answer is at the top of the next page.)*

Answer to Essential Question 2.4: The crossing points tell us the time and location at which one car passes another. For instance, the red car passes the green car at *t* = 2 s, +20 m from the origin.

2-5 Acceleration

Let's turn now to motion that is not at constant velocity. An example is the motion of an object you release from rest from some distance above the floor.

EXPLORATION 2.5A – Exploring the motion diagram of a dropped object

Step 1 - Sketch a motion diagram for a ball that you release from rest from some distance above the floor, showing its position at regular time intervals as it falls. The motion diagram in Figure 2.14 shows images of the ball that are close together near the top, where the ball moves more slowly. As the ball speeds up, these images get farther apart because the ball covers progressively larger distances in the equal time intervals. How do we know how much space to include between each image? One thing we can do is to consult experimental evidence, such as strobe photos of dropped objects. These photos show that the displacement from one time interval to the next increases linearly, as in Figure 2.14.

Step 2 - At each point on the motion diagram, add an arrow representing the ball's velocity at that point. Neglect air resistance. The arrows in Figure 2.14 represent the velocity of the ball at the various times indicated on the motion diagram. Because the displacement increases linearly from one time interval to the next, the velocity also increases linearly with time.

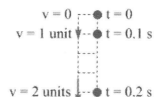

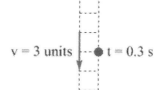

> **Key idea**: For an object dropped from rest, the velocity changes linearly with time.
> **Related End-of-Chapter Exercises: 57 - 60**

Another way to say that the ball's velocity increases linearly with time is to say that the rate of change of the ball's velocity, with respect to time, is constant:

$$\frac{\Delta \vec{v}}{\Delta t} = \text{constant} .$$

This quantity, the rate of change of velocity with respect to time, is referred to as the **acceleration**. Acceleration is related to velocity in the same way velocity is related to position.

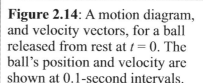

Figure 2.14: A motion diagram, and velocity vectors, for a ball released from rest at *t* = 0. The ball's position and velocity are shown at 0.1-second intervals.

> **Average acceleration:** a vector representing the average rate of change of velocity with respect to time. The SI unit for acceleration is m/s².
>
> $$\bar{a} = \frac{\Delta \vec{v}}{\Delta t} = \frac{\text{change in velocity}}{\text{time interval}} .$$
>
> (Equation 2.7: **Average acceleration**)
>
> The direction of the average acceleration is the direction of the change in velocity.

Instantaneous acceleration: a vector representing the rate of change of velocity with respect to time at a particular instant in time. The SI unit for acceleration is m/s².

A practical definition of instantaneous acceleration at a particular instant is that it is the slope of the velocity-versus-time graph at that instant. Expressing this as an equation:

$$\vec{a} = \frac{\Delta \vec{v}}{\Delta t}.$$ (Equation 2.8: **Instantaneous acceleration**)

Δt is small enough that the acceleration can be considered to be constant over that time interval.

Note that, on Earth, objects dropped from rest are observed to have accelerations of 9.8 m/s² straight down. This is known as $\vec{g}$, the acceleration due to gravity.

EXPLORATION 2.5B – Graphs in a constant-acceleration situation
The graph in Figure 2.15 shows the velocity, as a function of time, of a bus moving in the positive direction along a straight road.

Step 1 – *What is the acceleration of the bus? Sketch a graph of the acceleration as a function of time.* The graph in Figure 2.15 is a straight line with a constant slope. This tells us that the acceleration is constant because the acceleration is the slope of the velocity-versus-time graph. Using the entire 10-second interval, applying equation 2.8 gives:

$$\vec{a} = \frac{\Delta \vec{v}}{\Delta t} = \frac{+25\,\text{m/s} - (+5\,\text{m/s})}{10\,\text{s}} = \frac{+20\,\text{m/s}}{10\,\text{s}} = +2.0\,\text{m/s}^2.$$

The acceleration graph in Figure 2.16 is a horizontal line, because the acceleration is constant. Compare Figures 2.15 and 2.16 to the graphs for the red car in Figure 2.13. Note the similarity between the acceleration and velocity graphs in a constant-acceleration situation and the velocity and position graphs in a constant-velocity situation.

Step 2 - *What on the acceleration-versus-time graph is connected to the velocity?* The connection between velocity and acceleration is similar to that between position and velocity - the area under the curve of the acceleration graph is equal to the change in velocity, as shown in Figure 2.17. This follows from Equation 2.8, which in a constant acceleration situation can be written as $\Delta \vec{v} = \vec{a}\,\Delta t$.

Key ideas: The acceleration is the slope of the velocity-versus-time graph, while the area under the acceleration-versus-time graph for a particular time interval represents the change in velocity during that time interval.

Related End-of-Chapter Exercises: 16 and 32.

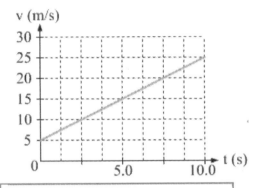

Figure 2.15: A graph of the velocity of a bus as a function of time.

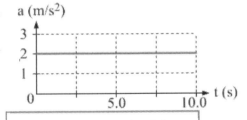

Figure 2.16: A graph of the acceleration of the bus as a function of time.

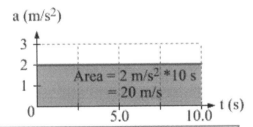

Figure 2.17: In 10 seconds, the velocity changes by +20 m/s. This is the area under the acceleration-versus-time graph over that interval.

Essential Question 2.5: Consider the bus in Exploration 2.5B. Would the graph of the bus' position as a function of time be a straight line? Why or why not?

Answer to Essential Question 2.5: The position-versus-time graph is not a straight line, because the slope of such a graph is the velocity. The fact that the bus' velocity increases linearly with time means the slope of the position-versus-time graph also increases linearly with time. This actually describes a parabola, which we will investigate further in the next section.

2-6 Equations for Motion with Constant Acceleration

In many situations, we will analyze motion using a model in which we assume the acceleration to be constant. Let's derive some equations that we can apply in such situations. In general, at some initial time $t_i = 0$, the object has an initial position $\vec{x}_i$ and an initial velocity of $\vec{v}_i$, while at some (usually later) time t, the object's position is $\vec{x}$ and its velocity is $\vec{v}$.

Acceleration is related to velocity the same way velocity is related to position, so we can follow a procedure similar to that at the end of Section 2-3 to derive an equation for velocity.

Substitute $\vec{v}_f - \vec{v}_i$ for $\Delta \vec{v}$ in the rearrangement of Equation 2.8, $\Delta \vec{v} = \vec{a}\,\Delta t$.

This gives: $\vec{v}_f - \vec{v}_i = \vec{a}\,\Delta t = \vec{a}\left(t_f - t_i\right)$.

Generally, we define the initial time t_i to be zero: $\vec{v}_f - \vec{v}_i = \vec{a}\,t_f$.

Remove the "f" subscripts to make the equation as general as possible: $\vec{v} - \vec{v}_i = \vec{a}\,t$. We also generally remove the vector symbols, although we must be careful to include signs.

$$v = v_i + at. \qquad \text{(Equation 2.9: \textbf{Velocity for constant-acceleration motion})}$$

A second equation comes from the definition of average velocity (Equation 2.2):

Average velocity = $\bar{v} = \dfrac{\Delta \vec{x}}{\Delta t} = \dfrac{\vec{x} - \vec{x}_i}{t - t_i} = \dfrac{\vec{x} - \vec{x}_i}{t}$.

If the acceleration is constant the average velocity is simply the average of the initial and final velocities. This gives, after again dropping the vector symbols:

$$\frac{v_i + v}{2} = \frac{x - x_i}{t}. \qquad \text{(Equation 2.10: \textbf{Connecting average velocity and displacement})}$$

Equation 2.10 is sometimes awkward to work with. If we substitute $v_i + at$ in for v (see Equation 2.9) in Equation 2.10, re-arranging produces an equation describing a parabola:

$$x = x_i + v_i t + \frac{1}{2}at^2. \qquad \text{(Equation 2.11: \textbf{Position for constant-acceleration motion})}$$

We can derive another useful equation by combining equations 2.9 and 2.10 in a different way. Solving equation 2.9 for time, to get $t = \dfrac{v - v_i}{a}$, and substituting the right-hand side of that expression in for t in equation 2.10, gives, after some re-arrangement:

$$v^2 = v_i^2 + 2a\,\Delta x. \qquad \text{(Eq. 2.12: \textbf{Connecting velocity, acceleration, and displacement})}$$

Motion with constant acceleration is an important concept. Let's summarize a general, systematic approach we can apply to situations involving motion with constant acceleration.

A General Method for Solving a One-Dimensional Constant-Acceleration Problem
1. **Picture the scene.** Draw a diagram of the situation. Choose an origin to measure positions from, and a positive direction, and show these on the diagram.
2. **Organize what you know, and what you're looking for.** Making a table of data can be helpful. Record values of the variables used in the equations below.
3. **Solve the problem.** Think about which of the constant-acceleration equations to apply, and then set up and solve the problem. The three main equations are:

$v = v_i + at$.   (Equation 2.9: **Velocity for constant-acceleration motion**)

$x = x_i + v_i t + \dfrac{1}{2}at^2$.   (Equation 2.11: **Position for constant-acceleration motion**)

$v^2 = v_i^2 + 2a\,\Delta x$.   (Equation 2.12: **Connecting velocity, acceleration, and displacement**)

4. **Think about the answer(s).** Check your answers to see if they make sense.

EXAMPLE 2.6 – Working with variables
In physics, being able to work with variables as well as numbers is an important skill. This can also produce insights that working with numbers does not. Let's say an object is dropped from rest from the top of a building of height H, while another object is dropped from rest from the top of a building of height $4H$. Assuming both objects fall under the influence of gravity alone (that is, they have the same acceleration), compare the times it takes them to reach the ground.

SOLUTION
$v_i = 0$, because the objects are dropped from rest. Take the initial position to be the top of the building in each case, so $x_i = 0$. This reduces Equation 2.11 to $x = at^2/2$. Because the acceleration is the same in each case, the equation tells us the position is proportional to the square of the time. To quadruple the final position, as we are doing, we need to increase the time by a factor of two. The fall from the building that is four times as high takes twice as long. This is illustrated by the motion diagram in Figure 2.18.

Related End-of-Chapter Exercises: 52 and 55

Essential Question 2.6: Return to the situation described in Example 2.6. Compare the velocities of the objects just before they hit the ground.

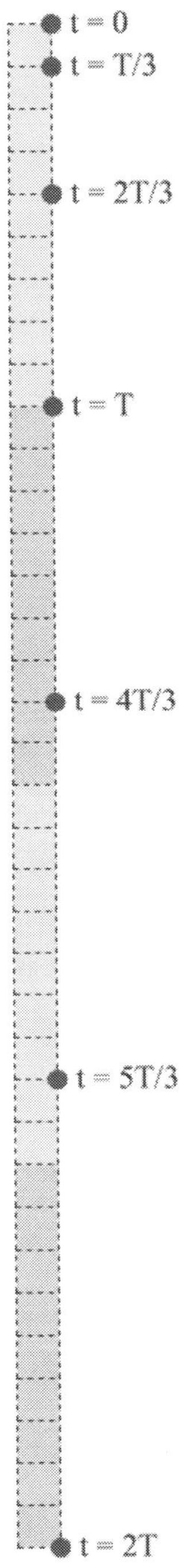

t = 0
t = T/3
t = 2T/3
t = T
t = 4T/3
t = 5T/3
t = 2T

Figure 2.18: Falling from rest for double the time quadruples the distance traveled. The height of the diagram is $4H$, four times the height of the smaller building in Example 2.6.

Answer to Essential Question 2.6: Using $v_i = 0$ reduces Equation 2.9 to $v = at$. Doubling the time doubles the final velocity, so the object dropped from the building that is four times higher has a final velocity twice as large as that of the other object. Equation 2.12 gives the same result.

2-7 Example Problem

Let's look at the various representations of motion with constant acceleration, considering the example of a ball tossed straight up in the air.

EXPLORATION 2.7 – A ball tossed straight up

You toss a ball straight up into the air. The ball takes 2.0 s to reach its maximum height, and an additional 2.0 s to return to your hand. You catch the ball at the same height from which you let it go, and the ball has a constant acceleration because it is acted on only by gravity. Consider the motion from the instant just after you release the ball until just before you catch it.

Step 1 – *Picture the scene – draw a diagram of the situation.* The diagram in Figure 2.19 shows the initial conditions, the origin, and the positive direction. We are free to choose either up or down as the positive direction, and to choose any reference point as the origin. In this case let's choose the origin to be the point from which the ball was released, and choose up to be positive.

Step 2 – *Organize the data.* Table 2.2 summarizes what we know, including values for the acceleration and the initial position. We need these values for the constant-acceleration equations. Because the ball moves under the influence of gravity alone, and we can assume the ball is on the Earth, the acceleration is the acceleration due to gravity, 9.8 m/s^2 directed down. Because down is the negative direction, we include a negative sign: $a = -9.8\,\text{m/s}^2$.

Parameter	Value
Origin	Launch point
Positive direction	up
Initial position	$x_i = 0$
Initial velocity	$v_i = +$_____ m/s
Acceleration	$a = -9.8\,\text{m/s}^2$
Position at $t = 4.0$ s	$x_{t=4s} = x_i = 0$

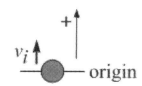

Figure 2.19: A diagram of the initial situation.

Table 2.2: Summarizing the information that was given about the ball.

Step 3 – *Solve the problem.* In this case, we want to draw graphs of the acceleration, velocity, and position of the ball, all as a function of time. To do this we should first write equations for the acceleration, velocity, and position. The acceleration is constant at $a = -9.8\,\text{m/s}^2$. Knowing the acceleration allows us to solve for the initial velocity. One way to do this is to re-arrange Equation 2.9 to give $v_i = v - at$. Because $v = 0$ at $t = 2.0$ s (the ball is at rest for an instant when it reaches its maximum height) we get $v_i = v - at = 0 - \left(-9.8\,\text{m/s}^2\right)\left(2.0\,\text{s}\right) = +19.6\,\text{m/s}$.

Knowing the initial velocity enables us to write equations for the ball's velocity (using Equation 2.9) and position (using Equation 2.11) as a function of time. The equations, and corresponding graphs, are part of the multiple representations of the motion shown in Figure 2.20. Note that the position versus time graph is parabolic, but the motion is confined to a line.

Step 4 - *Sketch a motion diagram for the ball.* The motion diagram is shown on the right in Figure 2.15. Note the symmetry of the up and down motions (this is also apparent from the graphs). The motion of the ball on the way down is a mirror image of its motion on the way up.

(a) Description of the motion: A ball you toss straight up into the air reaches its maximum height 2.0 s after being released, taking an addition 2.0 s to return to your hand. It experiences a constant acceleration from the moment you release it until just before you catch it.

(b) Equations (up is positive): Acceleration-versus-time: $a = -9.8 \text{ m/s}^2$

Velocity-versus-time: $v = +19.6 \text{ m/s} - (9.8 \text{ m/s}^2)t$

Position-versus-time: $x = +(19.6 \text{ m/s})t - (4.9 \text{ m/s}^2)t^2$

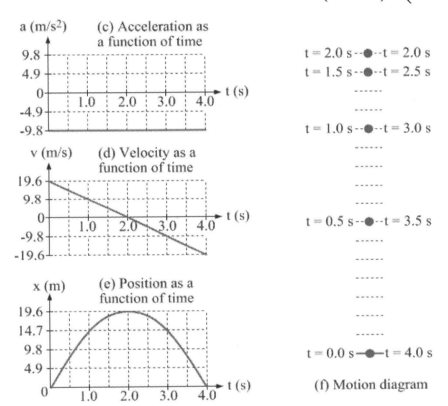

Figure 2.20: Multiple representations of a ball thrown straight up, including (a) a description in words; (b) equations for the ball's acceleration, velocity, and position; graphs giving the ball's (c) acceleration, (d) velocity, and (e) position as a function of time; and (f) a motion diagram. These different perspectives show how various aspects of the motion evolve with time.

Key ideas: At the Earth's surface the acceleration due to gravity has a constant value of $g = 9.8 \text{ m/s}^2$ directed down. We can thus apply constant-acceleration methods to situations involving objects dropped or thrown into the air. For an object that is thrown straight up the downward part of the trip is a mirror image of the upward part of the trip.
Related End-of-Chapter Exercises: 33, 61, and 62.

Essential Question 2.7: Consider again the ball in Exploration 2.7. The ball comes to rest for an instant at its maximum-height point. What is the ball's acceleration at that point?

Answer to Essential Question 2.7: A common misconception is that the ball's acceleration is zero at the maximum-height point. In fact, the acceleration is $\vec{g}$, 9.8 m/s² down, during the entire trip.

This is what is shown on the acceleration graph, for one thing – the graph confirms that nothing special happens to the acceleration at $t = 2.0$ s, even though the ball is momentarily at rest. One reason for this is that the ball is under the influence of gravity the entire time.

2-8 Solving Constant-Acceleration Problems

Consider one more example of applying the general method for solving a constant-acceleration problem.

EXAMPLE 2.8 – Combining constant-acceleration motion and constant-velocity motion

A car and a bus are traveling along the same straight road in neighboring lanes. The car has a constant velocity of +25.0 m/s, and at $t = 0$ it is located 21 meters ahead of the bus. At time $t = 0$, the bus has a velocity of +5.0 m/s and an acceleration of +2.0 m/s².

When does the bus pass the car?

SOLUTION

1. **Picture the scene – draw a diagram.** The diagram in Figure 2.21 shows the initial situation, the positive direction, and the origin. Let's choose the positive direction to be the direction of travel, and the origin to be the initial position of the bus.

2. **Organize the data.** Data for the car and the bus is organized separately in Table 2.3, using subscripts C and B to represent the car and bus, respectively.

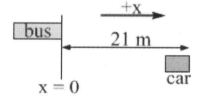

	Car	Bus
Initial position	$x_{iC} = +21$ m	$x_{iB} = 0$
Initial velocity	$v_{iC} = +25$ m/s	$v_{iB} = +5.0$ m/s
Acceleration	$a_C = 0$	$a_B = +2.0$ m/s²

Table 2.3: Summarizing the information that was given about the car and the bus.

Figure 2.21: A diagram showing the initial positions of the car and the bus, the position of the origin, and the positive direction.

3. **Solve the problem.** Let's use Equation 2.8 to write expressions for the position of each vehicle as a function of time. Because we summarized all the data in Table 2.3 we can easily find the values of the variables in the equations.

For the car: $x_C = x_{iC} + v_{iC}\,t + \dfrac{1}{2}a_C\,t^2 = +21\,\text{m} + (25\,\text{m/s})t$.

For the bus: $x_B = x_{iB} + v_{iB}\,t + \dfrac{1}{2}a_B\,t^2 = 0 + (5.0\,\text{m/s})t + (1.0\,\text{m/s}^2)t^2$.

The bus passes the car when the vehicles have the same position. At what time does $x_C = x_B$? Set the two equations equal to one another and solve for this time (let's call it t_1).

$$+21\,\text{m} + (25\,\text{m/s})t_1 = (5.0\,\text{m/s})t_1 + (1.0\,\text{m/s}^2)t_1^2$$

Bringing everything to the left side gives: $(1.0\,\text{m/s}^2)t_1^2 - (20\,\text{m/s})t_1 - 21\,\text{m} = 0$.

We can solve this with the quadratic equation, where $a = 1.0 \text{ m/s}^2$, $b = -20 \text{ m/s}$, and $c = -21 \text{ m}$.

$$t_1 = \frac{-b \pm \sqrt{b^2 - 4ac}}{2a} = \frac{+(20 \text{ m/s}) \pm \sqrt{(400 \text{ m}^2/\text{s}^2) + (84 \text{ m}^2/\text{s}^2)}}{2.0 \text{ m/s}^2} = \frac{+(20 \text{ m/s}) \pm (22 \text{ m/s})}{2.0 \text{ m/s}^2}.$$

The equation gives us two solutions for t_1: Either $t_1 = +21$ s or $t_1 = -1.0$ s. Clearly the answer we want is +21 s. The other number does have a physical significance, however, so let's try to make some sense of it. The negative answer represents the time at which the car would have passed the bus if the motion conditions after $t = 0$ also applied to the period before $t = 0$.

4. **Think about the answer.** A nice way to check the answer is to plug $t_1 = +21$ s into the two position-versus-time equations from the previous step. If the time is correct the position of the car should equal the position of the bus. Both equations give positions of 546 m from the origin, giving us confidence that $t_1 = +21$ s is correct.

Note: This example is continued on the accompanying web site, solving for the time at which the car and the bus have the same velocity.

Related End-of-Chapter Exercises: 54 and 56.

Chapter Summary

Essential Idea: Describing motion in one dimension.
The motion of many objects (such as you, cars, and objects that are dropped) can be approximated very well using a constant-velocity or a constant-acceleration model.

Parameters used to Describe Motion
Displacement is a vector representing a change in position. If the initial position is $\vec{x}_i$ and the final position is $\vec{x}_f$ we can express the displacement as:

$$\Delta \vec{x} = \vec{x}_f - \vec{x}_i .$$ (Equation 2.1: **Displacement in one dimension**)

Average Velocity: a vector representing the average rate of change of position with respect to time. The MKS unit for velocity is m/s (meters per second).

$$\bar{v} = \frac{\Delta \vec{x}}{\Delta t} = \frac{\text{net displacement}}{\text{time interval}}.$$ (Equation 2.2: **Average velocity**)

While velocity is a vector, and thus has a direction, speed is a scalar.

$$\text{Average Speed} = \bar{v} = \frac{\text{total distance covered}}{\text{time interval}}$$ (Equation 2.3: **Average speed**)

Instantaneous Velocity: a vector representing the rate of change of position with respect to time at a particular instant in time. The MKS unit for velocity is m/s (meters per second).

$$\vec{v} = \frac{\Delta \vec{x}}{\Delta t},$$ (Equation 2.4: **Instantaneous velocity**)

where Δt is small enough that the velocity can be considered to be constant over that interval.

Instantaneous Speed: the magnitude of the instantaneous velocity.

Average Acceleration: a vector representing the average rate of change of velocity with respect to time. The MKS unit for acceleration is m/s².

$$\bar{a} = \frac{\Delta \vec{v}}{\Delta t} = \frac{\text{change in velocity}}{\text{time interval}}.$$ (Equation 2.7: **Average acceleration**)

Instantaneous Acceleration: a vector representing the rate of change of velocity with respect to time at a particular instant in time.

$$\vec{a} = \frac{\Delta \vec{v}}{\Delta t},$$ (Equation 2.8: **Instantaneous acceleration**)

where Δt is small enough that the acceleration can be considered constant over that interval.

The **acceleration due to gravity,** $\vec{g}$, is 9.8 m/s², directed down, at the surface of the Earth.

A General Method for Solving a 1-Dimensional Constant-Acceleration Problem
1. **Picture the scene.** Draw a diagram of the situation. Choose an origin to measure positions from, and a positive direction, and show these on the diagram.
2. **Organize what you know, and what you're looking for.** Making a table of data can be helpful. Record values of the variables used in the equations below.
3. **Solve the problem.** Think about which of the constant-acceleration equations to apply, and then set up and solve the problem. The three main equations are:

$$v = v_i + at.$$ (Equation 2.9: **Velocity for constant-acceleration motion**)

$$x = x_i + v_i t + \frac{1}{2}at^2.$$ (Equation 2.11: **Position for constant-acceleration motion**)

$$v^2 = v_i^2 + 2a\,\Delta x.$$ (Equation 2.12: **Connecting velocity, acceleration, and displacement**)

4. **Think about the answer(s).** Check your answers, and/or see if they make sense.

Graphs
The velocity is the slope of the position-versus-time graph; the displacement is the area under the velocity-versus-time graph. The acceleration is the slope of the velocity-versus-time graph; the change in velocity is the area under the acceleration-versus-time graph.

Constant Velocity and Constant Acceleration
The motion of an object at rest is a special case of constant-velocity motion. In constant-velocity motion the position-versus-time graph is a straight line with a slope equal to the velocity.

Constant-velocity motion is a special case of constant-acceleration motion. In one dimension, if the acceleration is constant and non-zero the position-versus-time graph is quadratic while the velocity-versus-time graph is a straight line with a slope equal to the acceleration.

End-of-Chapter Exercises

Exercises 1 – 12 are conceptual questions that are designed to see if you have understood the main concepts of the chapter.

1. Refer to the position-versus-time graph in Figure 2.22 for your motion along a straight sidewalk. Consider the following four time intervals:
   - Interval 1: between $t = 0$ and $t = 10$ s
   - Interval 2: between $t = 5$ s and $t = 10$ s
   - Interval 3: between $t = 10$ s and $t = 15$ s
   - Interval 4: between $t = 15$ s and $t = 35$ s.

   Rank these intervals, from largest to smallest, based on the (a) distance traveled during the interval; (b) magnitude of the displacement during the interval. Express your rankings in a form like 2>1=3>4.

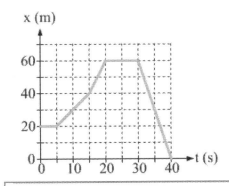

Figure 2.22: A position-versus-time graph, for Exercises 1 – 3.

2. Refer again to the position-versus-time graph in Figure 2.22 for your motion along a straight sidewalk. Consider the following four time intervals:
   - Interval 1: between $t = 0$ and $t = 10$ s
   - Interval 2: between $t = 5$ s and $t = 10$ s
   - Interval 3: between $t = 10$ s and $t = 15$ s
   - Interval 4: between $t = 15$ s and $t = 35$ s.

   Rank these intervals, from largest to smallest, based on the: (a) average speed over the interval; (b) magnitude of the average velocity over the interval. Express your rankings in a form like 2>1=3>4.

3. Compare the graph in Figure 2.22, showing your position-versus-time as you move along a straight sidewalk, to the graph in Figure 2.23, showing your velocity-versus-time for a different motion along the same straight sidewalk. (a) In a few sentences, make up a story to accompany the motion depicted in the position-versus-time graph. (b) In a few sentences, make up a story to accompany the motion depicted in the velocity-versus-time graph.

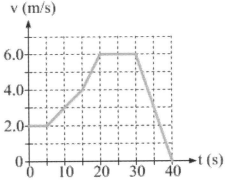

Figure 2.23: A velocity-versus-time graph, for Exercises 3 – 4.

4. Three of your friends look at the velocity-versus-time graph in Figure 2.23, depicting your motion along a straight sidewalk. They make the following comments:
   - Andy: "The graph shows that you were at rest between 0 and 5 seconds, and at rest again between 20 and 30 seconds."
   - Jennifer: "The graph shows that you were moving in one direction the entire time, until the end when you come to rest."
   - Sue: "The graph shows that you were traveling one way for a while, and then later you were traveling in the opposite direction."

   What do you agree or disagree with about each of these comments?

5. Consider the motion diagram in Figure 2.24, showing the position of an object at regular time intervals as it moves in one dimension. Is this a complete motion diagram, or is there anything missing that would help us to better determine what the object's motion is like?

Figure 2.24: Motion diagram, for Exercise 5.

6. Consider the motion diagram in Figure 2.25, showing the position of an object at 1-second time intervals as it moves in one dimension. (a) Over the time interval from $t = 0$ to $t = 10$ s, describe the object's motion. (b) When does the object experience a non-zero acceleration? (c) Describe a real-life situation that could match this motion diagram.

Figure 2.25: Motion diagram, for Exercise 6.

7. Consider the positions in Figure 2.26. What is the total distance traveled in moving from (a) $\vec{x}_0$ directly to $\vec{x}_1$? (b) $\vec{x}_1$ directly to $\vec{x}_0$? (c) $\vec{x}_0$ to $\vec{x}_2$ and then to $\vec{x}_1$? What is the displacement in moving from (d) $\vec{x}_0$ to $\vec{x}_1$? (e) $\vec{x}_1$ to $\vec{x}_0$?

(f) $\vec{x}_0$ to $\vec{x}_2$ and then to $\vec{x}_1$?

Figure 2.26: Three different positions ($\vec{x}_0$, $\vec{x}_1$, and $\vec{x}_2$) along an x-axis, for Exercise 7.

8. Describe a situation that matches each of the following, or state that it is impossible. (a) A person is traveling vertically down but with an upward acceleration. (b) A car has a velocity directed east and an acceleration directed east. (c) A baseball is at rest but has a non-zero acceleration.

9. Describe a situation that matches each of the following, or state that it is impossible. (a) An object has no acceleration and yet it is moving. (b) An object has a non-zero acceleration but it remains at rest.

10. Come up with an example that matches the following description of a motion, or state that it is impossible, assuming that the motion is in 1 dimension. (a) An object's average velocity is zero, but it did cover some distance over the time interval in question. (b) An object's velocity is positive for more than half the time, but its net displacement is negative. (c) An object's instantaneous velocity is, for some fraction of a particular time interval, equal to four times its average velocity over that time interval.

11. Consider the motion diagrams in Figure 2.27, showing the positions of two objects at 1-second intervals starting at $t = 0$ as both objects move to the right. Assume that whenever either object accelerates, its acceleration is constant and it accelerates for exactly 1 second. During the time interval depicted, which object has (a) the largest instantaneous speed? (b) the smallest instantaneous speed? (c) the average velocity with the largest magnitude over the interval $t = 0$ to $t = 9$ s?

Figure 2.27: A pair of motion diagrams for Exercises 11 – 12. The numbers shown correspond to the time in seconds (e.g., the circle with the 4 in it shows the location of the bottom object at $t = 4$ s).

12. Consider the motion diagrams in Figure 2.27, showing the positions of two different objects at 1-second intervals starting at $t = 0$ as both objects move to the right. You can assume that whenever either object accelerates its acceleration is constant and it accelerates for exactly 1 second. Note that your answers to this exercise should be of the form "At $t = 4$ s", or "At some time during the 1-s interval between $t = 3$ s and $t = 4$ s." During the time interval depicted are there any times when the objects have
    (a) the same horizontal position? If so, state when this occurs.
    (b) the same velocity? If so, state when this occurs.
    (c) the same acceleration? If so, state when this occurs.

Exercises 13 – 18 deal with calculating averages.

13. In 1973, the horse Secretariat set the record for the 1.25-mile Kentucky Derby with a time of 1:59 2/5 (one minute, 59.4 seconds). What was Secretariat's average speed in m/s?

14. The gold medalists in four events at the 2004 Athens Olympics were as follows: In the women's cycling road race, consisting of 9 laps around a 13.2 km/lap course, Sara Carrigan of Australia won in a time of 3:24:24 (3 h, 24 min, 24 s); in the 50-meter freestyle swim, Gary Hall of the USA swam one length of the pool in 21.93 s; in men's single sculls rowing, Olaf Tufte of Norway rowed the 2000 m course in 6:49.30 (6 min, 49.30 s); and in the women's 100-meter race on the track Yuliya Nesterenko of Belarus won in a time of 10.93 s. (a) Rank these athletes based on their average speed over their respective races, from largest to smallest. (b) Rank them instead based on the magnitude of their average velocity.

15. In the women's 200-meter backstroke event in swimming at the 2004 Athens Olympics, Kirsty Coventry of Zimbabwe swam the four lengths of the pool in a time of 2:09.19 (2 min., 9.19 s). At the instant Coventry touched the wall to win the race, the eighth-place swimmer in the race, Aya Terakawa of Japan, still had a few meters left to swim. (a) Which of these two swimmers had the largest average speed over the 2:09.19 time interval from the start of the race to when Kirsty Coventry touched the wall? (b) Over the same time interval, which swimmer had an average velocity with a larger magnitude? Briefly justify your answers.

16. The following times are given for the Porsche 911 Turbo Cabriolet to achieve a particular speed when accelerating from rest: 0 to 60 mph (miles per hour) in 3.8 s; 0 to 100 mph in 9.2 s; and 0 to 130 mph in 16.0 s. (a) In each of the three cases, what is the magnitude of the car's average acceleration? (b) Does the Porsche exhibit constant acceleration, or not? Briefly comment.

17. You are competing in a duathlon, an event that involves running and cycling. This duathlon involves running once around a particular loop, cycling twice around the same loop, and then finishing the race by again running once around the loop. If your average speed when running is 4.0 m/s and your average speed when cycling is 6.0 m/s, what was your average speed for the race? Assume that the time spent during the run-bike and bike-run transitions is negligible.

18. You take a trip, covering a total distance of 20 km. For the first 10 km you travel on horseback at an average speed of 20 km/h. You then switch to a different mode of transportation. What speed should you average over the second 10 km of the trip if you want your average speed for the entire trip to be (a) 10 km/h? (b) 30 km/h? (c) 40 km/h?

Exercises 19 - 26 deal with interpreting graphs.

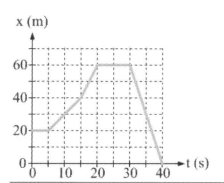

19. The graph in Figure 2.28 shows your motion as you move along a straight sidewalk. Over the 40-second interval what is (a) your net displacement? (b) the total distance traveled?

20. The graph in Figure 2.28 shows your motion as you move along a straight sidewalk. Over the 40-second interval what is (a) your average velocity? (b) your average speed?

21. Refer again to the position-versus-time graph in Figure 2.28 for your motion along a straight sidewalk. (a) Sketch the corresponding velocity-versus-time graph for the motion. (b) What is your instantaneous velocity at $t = 25$ s? (c) What is your average velocity over the interval between $t = 0$ and $t = 25$ s? (d) What is your instantaneous velocity at $t = 35$ s? (e) What is your average velocity over the interval between $t = 0$ and $t = 35$ s?

Figure 2.28: A position-versus-time graph, for Exercises 19 – 21.

22. If all you were given was the graph of velocity-versus-time in Figure 2.29, and you knew the motion depicted was for your motion along a straight sidewalk, could you answer the following questions? Simply write "Yes" if you can answer a particular question, and if you cannot explain why not. (a) What is your velocity at $t = 25$ s? (b) What is your acceleration at $t = 25$ s? (c) What is your position at $t = 25$ s? (d) What is your displacement between $t = 0$ and $t = 25$ s? (e) Are you walking east or west? (f) Are you walking forward or backward?

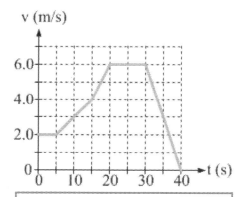

23. Refer again to the velocity-versus-time graph in Figure 2.29 for your motion along a straight sidewalk. If your position at $t = 0$ is +20 m from some convenient origin, determine your position, velocity, and acceleration at the following times: (a) $t = 10$ s, (b) $t = 25$ s, (c) $t = 35$ s.

Figure 2.29: A velocity-versus-time graph, for Exercises 22 – 26.

24. Refer again to the velocity-versus-time graph in Figure 2.29 for your motion along a straight sidewalk. If your position at $t = 0$ is +20 m from some convenient origin, sketch corresponding graphs of your position as a function of time as well as your acceleration as a function of time.

25. Refer again to the velocity-versus-time graph in Figure 2.29 for your motion along a straight sidewalk. Over the 40-second interval, what is your (a) total distance traveled? (b) net displacement? (c) average speed? (d) average velocity?

26. Refer again to the velocity-versus-time graph in Figure 2.29 for your motion along a straight sidewalk. Sketch the corresponding motion diagram showing your position, which you can represent by circles or X's, at 5-second intervals starting from $t = 0$.

Exercises 27 – 34 deal with understanding and interpreting motion diagrams.

Figure 2.30: Motion diagram for Exercises 27 – 29.

27. Consider the motion diagram in Figure 2.30, showing the position of an object at regular time intervals as it moves in one dimension. Assume that the object's acceleration is constant throughout the time interval covered by the motion diagram and that the object does not reverse direction during the motion. (a) In what direction is the object moving? (b) In what direction is the acceleration? (c) In one or two sentences, describe a real-life situation that could match this motion diagram.

28. Consider the motion diagram in Figure 2.30, showing the position of an object at regular time intervals as it moves in one dimension. The object starts from rest from the left-most point and then has a constant acceleration directed right for the entire time interval covered by the motion diagram (and continues to have this acceleration afterwards). If the vertical lines in the picture are 0.20 m apart and the time it takes the object to go from the initial position to the final position is 6.0 s, determine (a) the object's speed as it passes through the last position shown in the diagram, and (b) its acceleration.

29. Consider the motion diagram in Figure 2.30, showing the position of an object at regular time intervals as it moves in one dimension. Let's say that the object is moving from right to left, and experiencing a constant acceleration that brings it instantaneously to rest at the left-most point. The successive positions of the object are shown at 0.20 s intervals, starting from the right-most point at $t = 0$. The vertical lines in the picture are 0.20 m apart. (a) Make a copy of the diagram, labeling the positions shown on the diagram with the times. (b) Assuming the object still experiences the same constant acceleration after $t = 1.0$ s, add to the diagram the positions of the object at 0.20 s intervals between $t = 1.0$ s and $t = 2.0$ s. (c) What is the object's acceleration? (d) What is the object's velocity at $t = 0$? (e) What is the object's average velocity over the interval from $t = 0$ to $t = 1.0$ s?

30. Sketch a motion diagram for an object on the x-axis that (a) has a constant position; (b) has a constant acceleration of 1.0 m/s^2 in the negative x direction.

31. (a) Sketch a motion diagram for an object on the x-axis that is moving with a constant non-zero velocity in the positive x direction. (b) Add a motion diagram for a second object moving on a path that is parallel to the first object, but with a velocity 1.5 times larger than that of the first object.

32. Consider the motion diagram in Figure 2.31, showing the position of an object at 1-second time intervals as it moves in one dimension. Assume that the time is known precisely. If the object's velocity at $t = 4$ s is 4.0 m/s to the right, determine the object's average velocity over the time intervals (a) $t = 0$ to $t = 4$ s; (b) $t = 4$ s to $t = 7$ s; (c) $t = 7$ s to $t = 10$ s; (d) $t = 0$ to $t = 10$ s.

Figure 2.31: Motion diagram for Exercises 32 – 35.

33. Repeat Exercise 32, but now determine the object's average acceleration over the given time intervals instead of the average velocity.

34. Consider the motion diagram in Figure 2.31, showing the position of an object at 1-second time intervals as it moves in one dimension. Assume the time is known precisely. If the object's velocity at $t = 0$ is 2.0 m/s to the right, determine the object's (a) displacement, (b) average velocity, and (c) average acceleration over the time interval $t = 0$ to $t = 8$ s. (d) Would your answers to parts (a) – (c) change if you considered the time interval $t = 0$ to $t = 7$ s instead? State whether the magnitude of each of these vector quantities would increase, decrease, or stay the same.

Exercises 35 – 38 deal with transforming between various representations of motion.

35. Consider the motion diagram in Figure 2.31, showing the position of an object at 1-second time intervals as it moves along the x-axis. If, at $t = 0$, the object's velocity is 2.0 m/s to the right, and its position is $x = 0$ m, plot a graph of the object's (a) position, (b) velocity, and (c) acceleration over the interval $t = 0$ to $t = 10$ s. Assume the object has a constant non-zero acceleration lasting 1 second during this interval.

36. A plot of an object's position as a function of time is shown in Figure 2.32. (a) Briefly describe a real-life situation that would match the motion described by the graph. (b) Plot the corresponding velocity graph. (c) Draw the corresponding motion diagram.

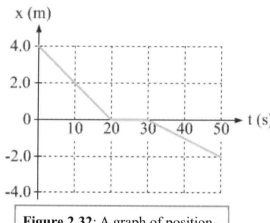

Figure 2.32: A graph of position-versus-time, for Exercise 36.

37. You are running along a straight path at constant velocity. After a few seconds, you see your friend, so you stop to chat. After a few seconds, she asks you to walk with her, so you start walking in the same direction you were originally running. (a) Sketch a graph showing your velocity as a function of time. (b) Sketch a graph showing your position as a function of time. (c) Draw a motion diagram for this motion.

38. A plot of an object's acceleration as a function of time is shown in Figure 2.33. (a) Sketch a velocity graph that would match this acceleration graph. (b) Comment on whether there is only one correct answer to part (a). (c) Briefly describe a real-life situation that would match this motion.

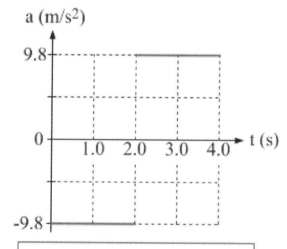

Figure 2.33: A graph of acceleration as a function of time, for Exercise 38.

Exercises 39 – 48 are designed to give you practice in solving a typical one-dimensional motion problem. Try to solve all exercises using a systematic approach. For each of these exercises start by doing the following parts. (a) **Picture the scene** - draw a diagram of the situation. Choose an origin to measure displacements from and mark that on the diagram. Choose a positive direction and indicate that with an arrow on the diagram. (b) **Organize the data** - create a table summarizing everything you know, as well as the unknowns you want to solve for.

39. You release a ball from rest, and it drops exactly 1.8 m to the floor below. Your goals in this exercise are to determine the ball's velocity just before impact and the time it takes the ball to reach the ground. Use $g = 10$ m/s^2. Carry out parts (a) – (b) as described above. (c) Which equation(s) will you use to determine the ball's velocity just before impact? (d) Solve for that velocity. (e) Which equation(s) will you use to determine the time it takes the ball to reach the ground? (f) Solve for that time.

40. When a traffic light turns green, you accelerate from rest along the road with a constant acceleration of 3.0 m/s^2. The goal here is to determine how long it takes you to reach the speed limit of 80 km/h. Carry out parts (a) and (b) as described above. (c) Which equation(s) will you use to find the time it takes to reach 80 km/h? (d) Find that time.

41. You throw a baseball straight up into the air, giving it an initial speed of 12 m/s. The baseball hits the ground 2.6 seconds later. To make the calculations easy, use $g = 10$ m/s^2. Your goals in this exercise are to solve for the initial height above the ground from which you launched the ball, and to find the maximum height above the ground reached by the ball in its motion. Carry out parts (a) and (b) as described above. (c) Which equation(s) will you use to find the initial height above the ground from which the ball was launched? (d) Solve for that initial height. (e) Which equation(s) will you use to find the maximum height above the ground reached by the ball? (f) Solve for that maximum height.

42. In this exercise you will analyze the method you used in the previous exercise, so all these questions pertain to what you did to solve the previous exercise. (a) Is there only one correct choice for the origin? Why did you make the choice you made? (b) Is there only one correct choice for the positive direction? Would your answers to (d) and (f) above change if you chose the opposite direction to be positive? (c) Find an alternative method to determine the initial height above the ground from which the ball was launched, and show that it gives the same answer as the method you used in the previous exercise. (d) Find an alternative method to determine the maximum height above the ground reached by the ball in its motion, and show that it gives the same answer as the method you used in the previous exercise.

43. A toy car is rolling down a ramp. When it passes a particular point you determine that it is traveling at a speed of 20 cm/s, and in the next 1.0 seconds it travels 40 cm. The goal of this exercise is to determine the car's acceleration, which we will assume to be constant. Carry out parts (a) and (b) as described above. (c) Which equation(s) will you use to determine the acceleration? (d) What is that acceleration?

44. Repeat parts (c) and (d) of Exercise 43, but calculate the answers another way, without using the equation(s) you used in Exercise 43.

45. You give a toy car an initial velocity of 1.00 m/s directed up a ramp. The car takes a total of 4.00 s to roll up the ramp and then roll back down again into your hand. Assuming you catch the car at the same point from which you released it, and that the acceleration is constant through the entire motion, the goal of the exercise is to determine the maximum distance the car was from your hand during the motion. Carry out parts (a) and (b) as

described above. (c) Which equation(s) will you use to determine the maximum distance the car was from your hand during the motion? (d) What is that maximum distance?

46. One of the events in the X Games is the Street Luge, in which competitors race on wheeled carts down an incline. Let's say that in one of the races a competitor starts from rest and then covers 100 m in 5.0 seconds. The goal of this exercise is to determine what the acceleration is, assuming it to be constant. Carry out parts (a) and (b) as described above. (c) What is the acceleration?

47. A professional baseball pitcher can throw a baseball with a speed of 150 km/h. Assuming the pitcher accelerates the ball from rest through a distance of 2.0 m, the goal of this exercise is to determine the ball's acceleration and the time over which this acceleration occurs. Carry out parts (a) and (b) as described above. (c) What is the ball's acceleration? (d) What is the time over which the acceleration occurs?

48. You're driving along a straight road at a speed of 48.0 km/h when you see a deer in the road 35.0 m ahead of you. After applying the brakes it takes you 2.00 s to bring your car to rest, but there is a reaction time period (the time between when you first see the deer and when you first apply the brakes) during which the car continues to travel at 48.0 km/h. The goal of this exercise is to answer the question: Assuming the acceleration is constant during the braking phase of the motion, what is the longest your reaction time can be if you are to stop the car before reaching the deer? Carry out parts (a) and (b) as described above. Note that there are two phases to the motion, a constant velocity phase and a constant-acceleration phase, so you should clearly separate the information for the two phases in your table. (c) Briefly describe the method you will use to solve the exercise. (d) Solve for your maximum possible reaction time.

General problems and conceptual questions

49. In Figure 2.13, we sketched graphs of the position and velocity of three cars as a function of time as they moved with constant velocity. Sketch a graph of the acceleration of each car as a function of time.

50. Consider the three positions shown in Figure 2.34. Starting at $\bar{x}_0$, it takes you 9.0 s to walk to $\bar{x}_2$ and then an additional 3.0 s to walk to $\bar{x}_1$. For this motion, find your (a) average speed and (b) average velocity.

Figure 2.34: Three different positions ($\bar{x}_0$, $\bar{x}_1$, and $\bar{x}_2$) along an x-axis, for Exercise 50.

51. In the men's 100-meter track race at the 2004 Athens Olympics, Justin Gatlin of the USA won the gold medal with a time of 9.85 s. Shawn Crawford (USA) ended up fourth in a time of 9.89 s. Estimate the distance separating these two runners at the finish line.

52. On August 16, 1960, Joe Kittinger of the United States Air Force jumped from a helium balloon from a height of 102,800 feet. After being in free fall for 4 minutes and 36 seconds, and falling for 85,300 feet, he opened his parachute and eventually landed safely on the ground. Assume that the acceleration was constant to answer parts (a) and (b). (a) What was the acceleration during the free fall? (b) What was Kittinger's speed just before he opened his parachute? (c) Comment on whether you think Kittinger's acceleration was constant during the fall. Depending on what source you read, Kittinger either broke, or came close to breaking, the sound barrier during this event.

53. Who was Colonel John Paul Stapp and why is he relevant to the material in this chapter? In particular, what was the magnitude of the maximum acceleration he experienced?

54. You throw a ball straight up into the air and catch it again at the same height from which you let it go. Considering the motion from the instant you release the ball until the instant you catch it again, the motion takes a time T and the ball reaches a maximum height H above the point where you release it. You now repeat the process, but this time the ball has twice the initial velocity of the previous motion. (a) How high above the release point does the ball go this time? (b) How long does this motion take?

55. You throw a ball straight up into the air, releasing it with a speed of 20 m/s. Assuming $g = 10$ m/s^2, you catch the ball 4.0 seconds later at the same point from which you let it go. Consider the motion from just after you release the ball until just before it returns to your hand. For the round trip, determine the ball's: (a) average velocity; (b) average speed; (c) average acceleration.

56. While a commuter train is stopped to pick up passengers, a freight train goes by at a constant speed of 36 km/h. Exactly 1 minute after the front of the freight train passes the front of the commuter train, the commuter train starts to move. After another 1 minute of constant acceleration, the commuter train reaches a speed of 72 km/h, and then moves at constant speed. Both trains are going the same way on parallel tracks. How much more time passes until the front of the commuter train passes the front of the freight train?

57. Two balls are launched at the same time. Ball A is released from rest from the top of a tall building of height H. Ball B is fired straight up from the ground with an initial velocity such that it just reaches the top of the same building. Neglect air resistance. (a) Which ball has the largest magnitude acceleration at the point they pass one another? (b) If ball A takes a time T to reach the ground, and ball B takes the same time T to reach the top of the building, which ball has the highest speed at time $T/2$? (c) How far from the ground are the two balls when they pass one another? Express your answer in terms of H. (d) Sketch a graph showing the velocity of ball A, and the velocity of ball B, as a function of time. Start from when the balls are launched, and end when ball A reaches the ground.

58. A cat and a dog are having a 100 m race. When the starting gun goes off, the dog lies down for a nap. The cat moves forward with a constant acceleration, reaching a speed of 2.0 m/s when she is 20 m from the starting line. After reaching that point, the cat travels at a constant velocity of 2.0 m/s until crossing the finish line. After 45 seconds, the dog wakes up from his nap and covers the 100 m with a constant acceleration of 2.0 m/s^2. (a) Who wins the race? Clearly justify your answer. (b) What is the distance between the animals when the winner crosses the finish line? (c) What is the distance between the animals at the only time (other than at the start of the race) they have the same velocity?

59. Consider the motion diagrams in Figure 2.35, showing the positions of two different objects at 1-second intervals starting at $t = 0$. Assume the top object, which has its position at regular intervals denoted by squares, is at $x = 0$ at $t = 0$, and that the vertical lines in the diagram are 1.0 m apart. Plot a graph of this object's (a) position, (b) velocity, and (c) acceleration over the interval between $t = 0$ and $t = 9$ s. You can assume that whenever the object accelerates its acceleration is constant and it accelerates for exactly 1 second.

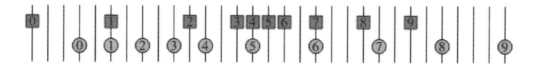

Figure 2.35: A pair of motion diagrams for Exercises 59 – 60. The numbers shown correspond to the time in seconds (e.g., the circle with the 4 in it shows the location of the bottom object at $t = 4$ s).

60. Consider the motion diagrams in Figure 2.35, showing the positions of two different objects at 1-second intervals starting at $t = 0$ as both objects move to the right. Which of the objects has the average velocity with the largest magnitude over the time interval (a) $t = 0$ to $t = 4$ s, (b) $t = 0$ to $t = 6$ s, (c) $t = 2$ s to $t = 7$ s?

61. You overhear two of your classmates discussing the issue of the acceleration of a ball that is tossed straight up into the air. Comment on each of their statements.

Jim: As the ball goes up, it slows down, so the acceleration is negative, while as the ball comes down again it speeds up, so the acceleration is positive. In other words, the acceleration changes sign when the ball changes direction.

Karen: What about the magnitude of the acceleration? I think it decreases to zero as the ball travels up, and then increases again as the ball comes down.

Jim: That sounds like the velocity. Remember that the acceleration is just the acceleration due to gravity, so that has a constant magnitude the whole time.

62. Referring back to the previous question, the conversation continues with two more classmates joining the discussion. Again, comment on each statement as they discuss the ball's acceleration at the instant it reaches the highest point.

Maria: Karen, isn't the ball's acceleration constant the whole time, even at the highest point? Are you saying that the acceleration is zero at that point?

Jose: I think Karen's right about that. Acceleration is velocity over time, so when the velocity goes to zero the acceleration must be zero, too.

Maria: Actually, isn't acceleration the change in velocity over some time interval? So, don't we have to consider how the velocity changes rather than worrying about what the value of the velocity is?

A 100-meter sprint is a good example of one-dimensional motion, but what is it that gets a sprinter moving in the first place? That's what we'll investigate in this chapter.

Photo credit: a photo of the American runner Jesse Owens at the Berlin Olympic Games in 1936, from Wikimedia Commons.

Chapter 3 – Forces and Newton's Laws

CHAPTER CONTENTS

In Chapter 2, we talked about how things move, at least in one dimension. In this chapter, we'll begin talking about why things move. The explanation for why an object moves (or why it doesn't move, if it remains at rest) revolves around the force or forces that the object experiences, so we will spend some time discussing what forces are.

Sir Isaac Newton made several key contributions to our understanding of forces. For instance, some of the basic rules about forces are summarized by Newton's laws of motion, so we will devote some time to understanding what these laws are all about and how they are applied in various situations. A key contribution of Newton's, however, is the understanding that the same laws that govern how soccer balls and cars move here on Earth also determine how planets move around the Sun and how stars interact with one another within galaxies. Prior to Newton, most people thought of the heavens and the Earth as being completely separate – we now know that the laws of physics apply to objects in the heavens just as well as they apply here on Earth. Thus, although our discussions in this chapter concern everyday objects, remember that the conclusions can be applied much more generally.

3-1 Making Things Move

Let's say a pen is lying on the desk in front of you, at rest. How could you make the pen move? There are many things you could do, such as:

- Pushing the pen with your hand.
- Picking the pen up and then dropping it.
- Tilting the desk, or the desktop, so the pen slides.

In all these cases, you are either directly applying a force to the pen or you are setting up a situation where something else applies an unbalanced force to the pen. What is a force?

> ***A force is simply a push or a pull.*** A force is a vector, so it has a direction. The SI unit of force is the newton (N). 1 N = 1 kg m/s².
>
> In addition, ***a force represents an interaction between objects***. For instance, the Earth exerts a force on the pen and the pen exerts a force on the Earth. Each object involved in an interaction experiences a force.

Question: Can an inanimate object like a pen or a desk exert a force? If so, describe how it is possible for inanimate objects to exert forces.

Answer: Yes, inanimate objects can exert forces. For instance, if you stand on a trampoline, the trampoline deforms under your weight (see Figure 3.1), and the trampoline exerts an upward force on you to prevent you from falling through it. The forces between atoms and molecules are much like the springs and stretchy material that make up a trampoline, so when a pen is on a desk both the pen and the desk are deformed a little (see Figure 3.2). The deformation is too small for you to observe easily, but the forces associated with it prevent the pen from falling through the desk.

Figure 3.1: Note how the surface of the trampoline deforms when the man exerts a force down on the trampoline, in the top picture, but it is not deformed when the man is in mid-air, in the bottom picture. Photo credit: public-domain image from Wikimedia Commons.

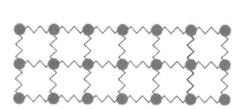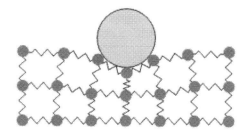

Figure 3.2: A diagram of an array of balls and springs, a model of a solid. A solid deforms when it is supporting an object. In this case, we'll use a model in which the object on top is considerably harder than the solid, so the object on top does not deform. In reality, both the object and the supporting solid would deform.

Some of the forces we'll make use of in the first part of the book include:

The Force of Tension (F_T)

Tension is the force exerted on an object by a string or rope. Remember that you can't push with a string or rope! The force of tension, F_T, on an object always points away from the object along the string or rope.

The Contact Force (F_C)

A contact force, F_C, arises when objects are in contact with one another. We often divide the contact force into its components, which we call the normal force and the force of friction.

The Normal Force (F_N)

The normal force is associated with objects in contact with one another, the normal force being the component of the contact force that is perpendicular to the surface of contact. When a book lies on a horizontal desktop, for instance, the desktop exerts an upward normal force on the book while the book exerts a downward normal force on the desktop. The symbol we will use to represent the normal force is F_N. In physics, "normal" generally means "perpendicular."

The Force of Friction (F_K or F_S)

The force of friction is also associated with objects being in contact with one another, being the component of the contact force that is parallel to the surface of contact. The force of kinetic friction, F_K, applies to situations where one object is sliding over another, while the force of static friction, F_S, applies to situations where the objects do not move relative to one another. We'll get into much more detail about the force of friction in Chapter 5.

The Force of Gravity (F_G)

Unlike the other forces above, which require contact, the force of gravity acts at a distance. The Sun and the Earth, or a book and the Earth, exert a force of gravity on one another without the objects having to be in direct contact. We will explore gravity in more detail later, but let's begin by saying that the force of gravity that one object exerts on another is F_G, and points toward the object exerting the force.

Four Fundamental Forces

As we do above, we often list several forces. When we classify them, it turns out that there are only four fundamental forces (although there is excitement these days about a possible fifth force, associated with an increase in the expansion rate of the universe). The four forces are:

1. The force of gravity – an attractive force between objects that have mass.
2. The electromagnetic force – a force between charged objects, which we'll discuss in the second half of the book). The contact force, and its components the normal force and the force of friction, arise from interactions between tiny objects (e.g., electrons) that have charge – these forces are all manifestations of the electromagnetic force.
3. The nuclear force – the force that holds nuclei together (a nucleus is the collection of protons and neutrons at the center of every atom).
4. The weak nuclear force – associated with radioactive decay (such as when an atom of americium-241 in a smoke detector emits an alpha particle to ionize air molecules, a process that is explained in more detail toward the end of the book).

The electromagnetic force, the nuclear force, and the weak nuclear force, are actually all associated with a single force called the electroweak force. Physicists are currently working on a grand unified theory, attempting to unify gravity and the electroweak force into a single force.

Essential Question 3.1: Jump up into the air. While you are in mid-air, not in contact with the ground, do any of the forces listed above act on you? If so, which?

Answer to Essential Question 3.1: The primary force acting on you once you have left the ground is the force of gravity. You are attracted toward the Earth by the force of gravity even though you are not in contact with the Earth.

3-2 Free-Body Diagrams

When analyzing a particular physical situation, it can be helpful to draw what is called a **free-body diagram**. This is a diagram in which arrows are attached to an object to represent the various forces applied to that object by external influences. The direction of an arrow is the same as the direction of the force the arrow represents, and the length of the arrow is proportional to the magnitude of that force. Each arrow is labeled with an appropriate symbol denoting the force the arrow represents.

When drawing a free-body diagram, it is helpful to keep two questions in mind:
1. For each force shown on the free-body diagram, what exerts the force?
2. Is the motion that would result from the set of forces acting on the object consistent with the actual motion of the object?

The following Explorations should help us learn how to answer those questions.

EXPLORATION 3.2A – Drawing a free-body diagram for an object at rest

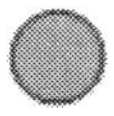

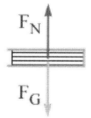

Step 1 – *Sketch a free-body diagram for an object that is at rest in outer space, billions of kilometers away from anything.* The free-body diagram in Figure 3.3 shows no forces, because the object does not interact with anything.

Figure 3.3: Free-body diagram for an object that is not interacting with anything.

Step 2 - *Sketch a free-body diagram for a book is at rest on a horizontal tabletop.* The free-body diagram for this very common situation is shown in Figure 3.4. The Earth applies a downward force of gravity to the book, but the book remains at rest because there is an upward normal force applied to the book by the table. How do you think the magnitudes of these two forces compare? For the book to remain at rest, these two forces must cancel one another exactly, so the two forces have the same magnitude.

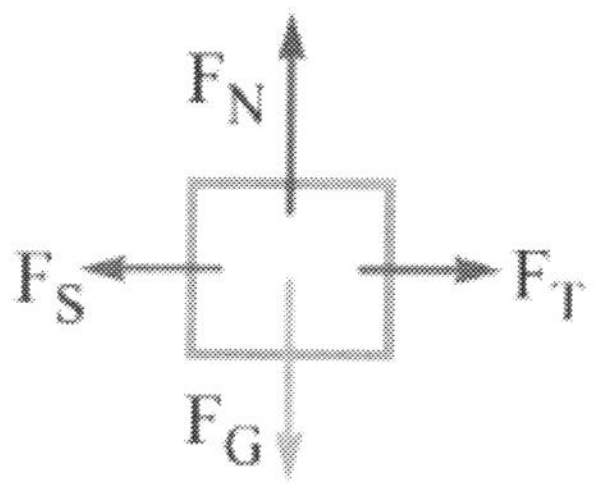

F_N

F_G

Figure 3.4: Free-body diagram for a book sitting at rest on a table.

Step 3 - *Sketch a free-body diagram for a box that remains at rest on a horizontal tabletop even though you exert a horizontal force by pulling on a string tied to the box.* The force you exert is transferred to the box by the string, so it is shown as a force of tension in Figure 3.5. When drawing the free-body diagram consider this question: Why doesn't the box move? If you pull hard enough, the box will move. In this case, though, the force you exert on the box is small enough that it can be balanced by another force in the opposite direction. This balancing force is the force of static friction, which we discuss in detail in chapter 5. Because the box does not move horizontally these two forces must cancel one another, so their magnitudes are equal.

F_N

F_S — F_T

F_G

As in step 2, the Earth applies a downward force of gravity to the box that is exactly balanced by the upward normal force applied to the box by the table.

Figure 3.5: Free-body diagram for a box at rest on a table while you pull on a string to the right.

Step 4 – *What, if anything, is common to the three situations discussed above?* First, in each case, the object remains at rest – its motion does not change. Second, although the three free-body diagrams clearly are different, in each case there is no net force acting on the object. ***The net force is the sum of all the forces acting on an object.*** Because forces are vectors, we must account for both the directions and magnitudes of forces when we add them. There is no net force in any of the cases above because either there is no force acting at all or all the forces cancel out. Based on this, let's theorize that when no net force acts on an object that is at rest, the object remains at rest.

Key ideas for an object at rest: The net force is the vector sum of all the forces acting on an object. When no net force acts on an object that is at rest, the object remains at rest.
Related End-of-Chapter Exercises: 1, 15.

EXPLORATION 3.2B – The motion diagram and the free-body diagram
Let's now start connecting forces to the motion ideas from chapter 2.

Step 1 - *Sketch a motion diagram for a ball you release from rest from some distance above the floor, showing its position at regular time intervals as it falls.* The motion diagram in Figure 3.6 shows images of the ball that are close together near the top, where the ball moves slowly. As the ball speeds up, these images gradually get farther apart as the ball covers progressively larger distances in equal time intervals.

Step 2 - *Sketch the ball's free-body diagram, showing the forces acting on the ball as it falls. Neglect air resistance.* If we can neglect air resistance, the Earth is the only object applying a force to the ball, and that force is a downward force of gravity. The free-body diagram is shown near the bottom right of in Figure 3.6.

Step 3 - *Does the free-body diagram show a net force acting on the ball as it falls? Does this net force increase substantially, decrease substantially, or stay reasonably constant as the ball falls?* A downward net force acts on the ball because there is nothing to balance the force of gravity. This force is associated with the interaction between the ball and the Earth, and the strength of that interaction depends on the distance between the ball and the center of the Earth. If we drop a ball from a typical height of 1-2 meters, the distance between the ball and the center of the Earth changes by a very small fraction, so we can assume that the net force acting on the ball is constant.

Step 4 - *What is the connection between the motion diagram and the free-body diagram?* The motion diagram shows that the ball's motion changes, because the successive images of the ball are not drawn equally spaced. When a net force acts on an object, the motion of the object changes.

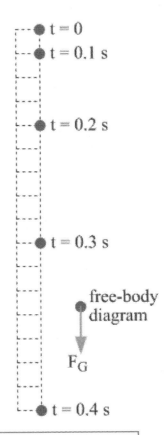

Figure 3.6: Motion diagram and free-body diagram for a ball dropped from rest.

Key idea for motion and force: When a net force acts on an object, the object's motion changes.   **Related End-of-Chapter Exercises: 29, 30, 49.**

Essential Question 3.2: What if the ball in Exploration 3.2B had been thrown straight up, so it came to rest for an instant 0.4 s after leaving your hand? What would its motion diagram, and free-body diagram, look like?

Answer to Essential Question 3.2: After the ball leaves your hand, the only force acting on the ball is the force of gravity, so the free-body diagram would be the same as in Figure 3.6. The successive locations of the ball on the motion diagram would also be the same, with the times increasing from bottom to top instead of from top to bottom.

3-3 Constant Velocity, Acceleration, and Force

If an object is moving at constant velocity, is there a net force acting in the direction of motion? Let's explore that idea.

EXPLORATION 3.3A – Forces in a constant-velocity situation

Step 1 - *Sketch a motion diagram for an object that is drifting through space with a constant velocity to the right, billions of kilometers from anything.* The motion diagram in Figure 3.7 shows images of the object that are equally spaced along a straight line.

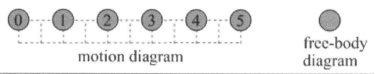

motion diagram   free-body diagram

Figure 3.7: Motion diagram and free-body diagram for an object drifting to the right through space.

Step 2 - *Sketch the object's free-body diagram, showing all the forces acting on the object as it drifts through space.* Many people think that the free-body diagram should show a force acting to the right. Here is where you ask yourself, though, "What would apply that force?" The object is not interacting with anything, so there are no forces to show on the free-body diagram in Figure 3.7.

Step 3 - *Repeat steps 1 and 2 for a box you drag with a constant velocity to the right across a horizontal table. You drag the box by pulling on a string tied to the box. The string is horizontal because of the force you exert on it. There is some friction acting on the box.* The motion diagram for the box, which is shown in Figure 3.8, is similar to that for the object drifting through space – the images of the object that are equally spaced. The free-body diagram is similar to that for the box at rest on the table, when you pulled on the string to the right, except that in this case the friction force is the kinetic force of friction. If the box was initially at rest, the tension force you exert via the string would have to be larger than the friction force to start the box moving. Once the box is moving, the tension force simply has to balance the kinetic friction force to maintain the motion.

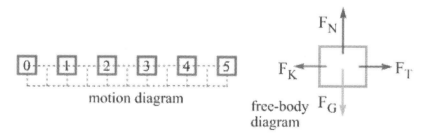

motion diagram   free-body diagram

Figure 3.8: Motion diagram and free-body diagram for a box being dragged to the right, by means of a string, across a flat surface.

Step 4 - *Compare and contrast the free-body diagrams you drew in steps 2 and 3.* The free-body diagrams are quite different, with one having no forces and the other having four. However, the net force in both cases is zero, because for the box the vertical forces cancel one another and the horizontal forces cancel one another.

Step 5 - *What is the connection between the motion diagrams in these cases and the free-body diagrams?* Even though we considered objects in motion, the free-body diagrams here are equivalent to those we drew in Exploration 3.2A for objects at rest. In all cases, here and in Exploration 3.2A, the net force is zero. The lesson for us is that when no net force acts on an object, the object's velocity is unchanged.

> **Key ideas for constant-velocity motion**: Again, we see the connection between an object at rest and an object in motion with constant velocity – as far as forces are concerned the situations are the same, with no net force in either case. When no net force acts on an object, its velocity is constant. **Related End-of-Chapter Exercises: 2, 5.**

In Explorations 3.2A and 3.3A, we learned that when an object experiences no net force its velocity is constant, which means its acceleration is zero. An object with a net force has an acceleration. To see how force is connected to acceleration, let's start by doing some dimensional analysis. The SI unit of force is the newton (N). 1 N = 1 kg m/s^2. The SI unit of acceleration is m/s^2, so the units differ by a factor of kilograms. The mass of an object has units of kg, so we can speculate that force and acceleration are connected by a factor a mass.

This connection is easy to verify experimentally. When we apply a particular net force to an object of mass m and measure the acceleration $\vec{a}$, the acceleration turns out to be given by:

$$\vec{a} = \frac{\sum \vec{F}}{m}$$ (Equation 3.1: **Connecting acceleration and force**)

The symbol Σ represents a sum, so $\sum \vec{F}$ is the net force (the sum of all the forces) acting on the object. Note that the right-hand side of Equation 3.1 has units of N/kg, so the units of acceleration can be stated as N/kg or as m/s^2.

EXPLORATION 3.3B – A race

Take two objects of different mass and hold one in one hand and one in the other. If you simultaneously release them from rest from the same distance above the floor, which object hits the ground first? Be sure to use objects that will not be affected much by air resistance, such as a pen and a ball. A piece of paper is a poor choice, for instance, unless you crumple it into a tight ball.

When you do the experiment, you should observe that the objects reach the ground at the same time. This contradicts a common belief (and a belief that was held by Greek thinkers such as Aristotle) that objects with more mass fall faster.

> **Key idea for an object moving under the influence of gravity alone**: If air resistance can be neglected, the acceleration of an object moving under the influence of gravity is independent of the mass of the object. **Related End-of-Chapter Exercises: 34, 35.**

Essential Question 3.3: If a free-body diagram shows a net force acting on an object at all times, can the object be at rest?

Answer to Essential Question 3.3: Yes, but only for an instant. If an object experiences a net force, its motion changes. If the object is initially at rest, it will start to move. If the object is initially moving, the net force might bring it to rest for an instant and then reverse its direction.

3-4 Connecting Force and Motion

Why would someone think that an object with more mass falls faster than an object with less mass? One reason is that an object with more mass has a larger force of gravity acting on it. Consider the following statement, made by a student before taking a physics course: "The force of gravity is what makes an object fall to the floor when we let go of it, so an object with more mass should fall faster." That sounds logical, but it is incorrect. What directly determines how much time something takes to fall to the ground is the acceleration, not the force. Force and acceleration are directly related, but they are not the same.

The force of gravity acting on an object is proportional to its mass. If we are dropping objects relatively small distances when we are at, or near, the surface of the Earth, the force of gravity is constant and is given by:

$$\vec{F}_G = m\vec{g},$$ (Equation 3.2: **Force of gravity at the surface of the Earth**)

where $\vec{g}$ is commonly referred to as the acceleration due to gravity. As we discussed in chapter 2, at the surface of the Earth, the value of $\vec{g}$ is about 9.8 N/kg directed down. A better name for $\vec{g}$ is "the value of the local gravitational field," but we will address that in chapter 8 when we talk about gravity in detail, and when we show where the value of 9.8 N/kg comes from.

Figure 3.9 shows the free-body diagrams of two objects, one with a mass of m and the other with a mass of $3m$, when they are in free fall. Because the objects move under the influence of gravity alone, only one force, the force of gravity, appears on each free-body diagram. The force of gravity is proportional to mass, so the force of gravity acting on the second object is three times larger than the force acting on the first object. Figure 3.9 is correct, but these free-body diagrams reinforce the incorrect idea that an object with more mass falls faster. Let's go beyond the free-body diagrams, and apply Equation 3.1.

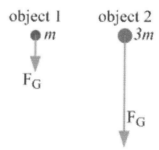

For object 1: $\vec{a}_1 = \dfrac{\sum \vec{F}}{m} = \dfrac{m\vec{g}}{m} = \vec{g}$;

For object 2: $\vec{a}_2 = \dfrac{\sum \vec{F}}{3m} = \dfrac{3m\vec{g}}{3m} = \vec{g}$.

Even though the forces have different magnitudes, the accelerations are the same. This is why both objects reach the ground at the same time.

It is easy to confuse force and acceleration, and we will consider more situations later where we must carefully distinguish the two. As a final thought, consider the following.

Figure 3.9: Free-body diagrams for two objects in free fall. Object 2 has three times the mass of object 1, so the force of gravity acting on it is three times as large as that on object 1. Despite this, the objects have equal accelerations.

Question: Take two identical objects and apply a net force to the second one that is three times larger than the net force applied to the first one, as shown in Figure 3.10. If both objects start from rest at the same time, which object has a larger acceleration?

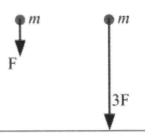

Answer: Now the obvious answer is the correct one – the second object has an acceleration three times larger than the first because its net force is three times larger. This is true only because the objects have the same mass! Compare this situation to that shown in Figure 3.9, in which we dropped objects of different mass.

Figure 3.10: Free-body diagrams for two identical objects, one experiencing a force three times larger than that experienced by the other.

Force and motion are connected by acceleration. Knowing about forces can give us an acceleration we can use in the constant-acceleration equations. Conversely, we can use the equations to find acceleration and then find the net force. Here is an example of that process.

EXAMPLE 3.4 – Combining forces with the constant-acceleration equations

Cindy kicks a soccer ball, with a mass of 0.40 kg. If the ball starts from rest on the ground and ends up with a velocity of 15 m/s directed horizontally, what is the average force exerted on the ball by Cindy's foot if her foot is in contact with the ball for 0.10 s?

SOLUTION

A diagram of the situation and a free-body diagram of the ball are shown in Figure 3.11. Table 3.1 summarizes what we know about the motion.

Initial position	$\bar{x}_i = 0$
Initial velocity	$\bar{v}_i = 0$
Final velocity	$\bar{v} = +15 \, m/s$
Time	$t = 0.10 \, s$

Table 3.1: A summary of the data for the ball.

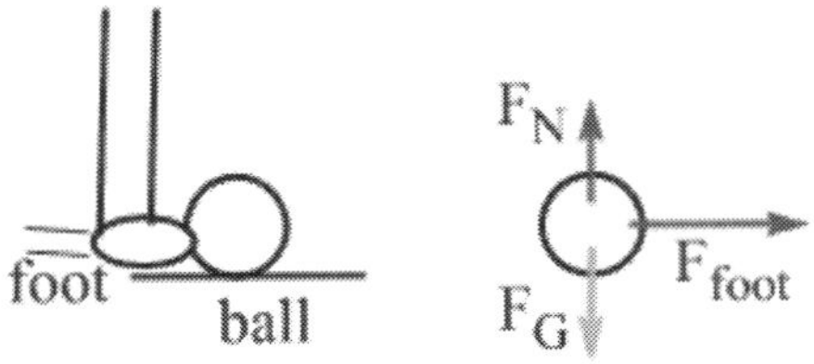

Figure 3.11: A diagram, and free-body diagram, for the ball.

There is enough information here to solve for the ball's acceleration using Equation 2.7, $\bar{v} = \bar{v}_i + \bar{a}t$. Re-arranging this equation to solve for the acceleration gives:

$$\bar{a} = \frac{\bar{v} - \bar{v}_i}{t} = \frac{+15 \text{ m/s} - 0}{0.10 \text{ s}} = +150 \text{ m/s}^2.$$

We can apply a re-arranged version of Equation 3.1, $\sum \vec{F} = m\bar{a}$, to determine the force Cindy exerts on the ball. Assume that $\vec{F}_{foot}$, the force Cindy's foot exerts on the ball, is horizontal, so the normal force the ground exerts on the ball balances the force of gravity acting on the ball.

$$\vec{F}_{foot} = (0.40 \text{ kg})(+150 \text{ m/s}^2) = +60 \text{ N}.$$

Related End-of-Chapter Exercises: 14, 26, 27, 52, 55.

Essential Question 3.4: Let's say that, in Example 3.4, the ball is rolling toward Cindy at a speed of 8.0 m/s at the instant she kicks it, and she exerts the same force on the ball for the same time period as in Example 3.4 above. What is the ball's final velocity in this new situation?

Answer to Essential Question 3.4: In this case, we can use the definition of acceleration to write Equation 3.1 as $\vec{F}_{net} = m\vec{a} = m\Delta\vec{v}/\Delta t$. The net force acting on the ball, the ball's mass, and the time interval over which the force is applied are the same in this situation as in Example 3.4, so the ball's change in velocity, +15 m/s, must also be the same. Because the ball's initial velocity is –8.0 m/s, adding the change of +15 m/s results in a final velocity of +7.0 m/s.

3-5 Newton's Laws of Motion

Sir Isaac Newton (1642 – 1727) made many contributions to mathematics and science, including three laws of motion. Previously, we looked at several situations involving objects at rest and objects moving with constant velocity. We found that, in all such cases, the net force on the object was zero. In contrast, whenever an object's velocity was changing we found that there was a non-zero net force acting on the object. These observations are summarized by:

> **Newton's First Law** - If no net force acts on an object, the object's velocity is unchanged: the object either remains at rest or it keeps moving with constant velocity. If there is a non-zero net force acting, then the object's velocity changes.
>
> Recall that the net force is the sum of all the forces acting on an object. Always remember to add forces as vectors. The net force can be symbolized by $\sum\vec{F}$.

When there is a non-zero net force acting on an object, Newton's first law is a rather qualitative statement. It tells us that the velocity of the object changes, but it does not tell us how the velocity changes. This is where Newton's second law comes in.

> $$\vec{a} = \frac{\sum\vec{F}}{m}.$$
> (Equation 3.1: **Newton's Second Law**)
>
> Equation 3.1 is often re-arranged to the following form, but it's the same equation!
>
> $$\sum\vec{F} = m\vec{a}$$
> (Equation 3.1: **Newton's Second Law**)

Thus, if we know all the forces that act on an object, and we also know the object's mass, we can determine the object's acceleration. Once we know the acceleration we can go on to analyze the object's motion using the methods, and constant-acceleration equations, of Chapter 2.

How quickly an object's velocity changes when a net force is applied depends on the magnitude of the net force as well as the object's inertia. The more mass an object has, the harder it is to change the object's velocity. In other words, an object's inertia is its mass.

Newton's third law is simple to state but it can be rather counter-intuitive. This law follows from the fact that forces are associated with interactions, and both objects involved in an interaction experience a force of the same magnitude because of that interaction.

> **Newton's Third Law** - When one object exerts a force on a second object, the second object exerts a force equal in magnitude, and opposite in direction, on the first object.

Note that Newton's laws apply for observers who are not accelerating while observing a system. For instance, you can use Newton's laws to explain the motion of an apple being tossed up and down by a person on a bus if you are at rest on the sidewalk watching the bus go by, or even if you are on the bus while the bus is moving at constant velocity. Newton's laws give an incomplete picture if you are on the bus, analyzing the motion of the apple while the bus accelerates. A non-accelerating reference frame is known as an inertial frame of reference.

Question: A fast-moving train collides with a small car that stalled as it crossed the railroad tracks. (Fortunately, the driver was able to run to safety before the collision.) Which object exerts more force on the other during the collision? Justify your answer.

Answer: Despite the fact that the train has much more mass than the car, the force the train exerts on the car is always equal in magnitude, and opposite in direction, to the force the car exerts on the train. Newton's third law addresses the fact that a force comes from an interaction, and the interacting objects are always equal partners in that interaction in the sense that they experience forces of equal magnitude.

So, why do many people think that the train exerts more force on the car than the car exerts on the train? The issue is that while the forces may be equal-and-opposite, the accelerations are different. Because the train's mass is much larger than the car's mass, the train's acceleration (the net force divided by the mass) is much less than the car's. Because the car experiences a large acceleration, so does a person in the car, and the forces exerted on a person in the car can be large enough to cause serious injury or death. Conversely, because the train experiences a small acceleration, someone on the train experiences a modest force and a person on the train may hardly even notice a collision occurred (until the engineer applies the brakes, at least). Thus, although the forces are equal-and-opposite, the effects of the forces differ greatly.

Although Newton's third law can be counter-intuitive, it is easy to verify. One way to verify it is to mount force sensors on carts and set up collisions between the carts, or have one cart push or pull the other. No matter what the situation, even if the carts have different masses and/or one of the carts is initially stationary, the result is that the force that the first cart exerts on the second is always equal in magnitude, and opposite in direction, to the force the second cart exerts on the first. A result from an actual experiment is shown in Figure 3.12, showing graphs of the force, as a function of time, experienced by two carts being pushed together.

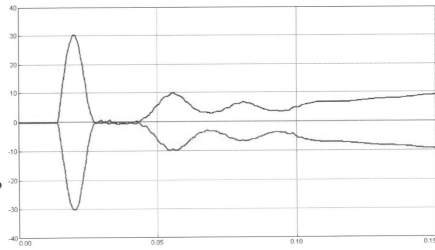

Figure 3.12: The top graph shows the force, in newtons, as a function of time, in seconds, that cart A exerts on cart B. The bottom graph shows the force cart B exerts on cart A. Such graphs are always mirror images, providing experimental evidence of Newton's third law.

Related End-of-Chapter Exercises: 56 and 57.

Essential Question 3.5: Does the Sun exert more force on the Earth, or does the Earth exert more force on the Sun?

Answer to Essential Question 3.5: Even though the Sun is enormous compared to the Earth, Newton's third law tells us that the force the Sun exerts on the Earth is equal in magnitude, and opposite in direction, to the force the Earth exerts on the Sun.

3-6 Exploring Forces and Free-Body Diagrams

A force is simply a push or a pull. A force is a vector, so it has a direction. Also, a force represents an interaction between objects. Let's build on these facts and learn more about forces.

EXPLORATION 3.6A – The normal force
Step 1 - *Sketch free-body diagrams for the following situations: (a) a book rests on a table; (b) you exert a downward force on the book as it sits on the table; (c) you tie a helium-filled balloon to the book, which remains on the table.*
The diagrams are shown in Figure 3.13. In (a), the normal force applied by the table on the book has to balance the force of gravity acting on the book. If you push down on the book, the normal force increases – it has to balance the force of gravity as well as the force you exert. Tying a helium balloon to the book reduces the normal force because the normal force and the tension in the string combine to balance the force of gravity. The normal force depends on the situation – the magnitude of the normal force is whatever is necessary to prevent the book from falling through the table.

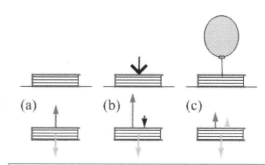

Figure 3.13: The upward normal force the table exerts on the book is larger when you exert a downward force (b) and is smaller when a string tied to a helium balloon exerts an upward force (c).

Let's say we have a scale calibrated in force units. The magnitude of the normal force is the force the scale reads if the scale is placed between the objects in contact.

Step 2 - *When does one object lose contact with another? For instance, how many helium balloons would we have to tie to the book to make it lift off the table?* Objects lose contact when the normal force between them goes to zero. The minimum number of balloons is the number needed to reduce the normal force the table exerts on the book (and the normal force the book exerts on the table) to zero.

> **Key ideas about the normal force**: The magnitude of the normal force in a particular situation is whatever is required to prevent one object from passing through another, and is equal to the scale reading if a scale were placed between the objects in contact. Objects lose contact when the normal force between them goes to zero. **Related End-of-Chapter Exercises: 4, 17.**

EXAMPLE 3.6 – Calculating the normal force
As shown in Figure 3.14, a large box (box 1), with a weight of $m_1 g = 20$ N, is at rest on the floor. A smaller box (box 2), with a weight of $m_2 g = 10$ N, sits on top of the large box. (a) Draw free-body diagrams for each box. Calculate the normal force (b) exerted on box 2 by box 1, (c) exerted on box 1 by box 2, and (d) exerted on box 1 by the floor.

Figure 3.14: Two boxes, one on top of the other, at rest on the floor.

SOLUTION
(a) The free-body diagrams are shown in Figure 3.15. For box 2, the upward normal force applied by box 1 balances the downward force of gravity acting on box 2. For box 1, the table balances both the force of gravity on box 1 and the downward normal force applied on box 1 by box 2.

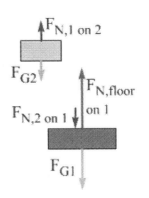

(b) For the forces to balance, the normal force applied on box 2 by box 1 is 10 N up.
(c) By Newton's third law, box 2 applies a normal force on box 1 of 10 N down.
(d) For the forces on box 1 to balance, the table applies a normal force of 30 N up.

The Contact Force (F_C)

In general, when two objects are in contact with one another, they exert contact forces that are equal in magnitude but opposite in direction. We usually split the contact force into two components, a normal force perpendicular to the surfaces in contact, and a force of friction that is parallel to the surfaces in contact. It can be useful to look at the whole vector, however.

Figure 3.15: Free-body diagrams for the two boxes of Example 3.6.

EXPLORATION 3.6B – A whole-vector approach

While unloading a truck, you place a box on a ramp leading from the truck.

Step 1 - *The box remains at rest on the ramp. What is the net force acting on the box?* The velocity of the box remains constant (at $v = 0$, in this case), so there is no net force on the box.

Step 2 - *Sketch a diagram of the situation and a free-body diagram showing the forces acting on the box. What is the magnitude and direction of the force the ramp exerts on the box?* In this case, it is simpler to use the whole contact force, rather than using the two components (a normal force perpendicular to the ramp and a static force of friction directed up the ramp). The net force on the box is zero, so the contact force is directed straight up with a magnitude equal to the force of gravity, as shown in Figure 3.16.

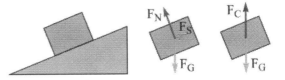

Figure 3.16: A diagram and two equivalent free-body diagrams for the box at rest on the ramp.

Key idea: It can be helpful to view the contact force as one vector, instead of breaking it into its components, a normal force and a force of friction. **Related End-of-Chapter Exercises: 38, 39.**

The Force of Gravity, Weight, and Apparent Weight

The force of gravity, F_G, does not require objects to be in contact but acts at a distance. The force of gravity is always attractive, directed toward the object exerting the force. In Chapter 8, we will look at situations in which the distance between objects changes, changing the force of gravity. For now, we will deal with situations in which the force of gravity is constant, as it is at the Earth's surface, where we use $\vec{F}_G = m\vec{g}$, with $\vec{g}$ the acceleration due to gravity. In this book, we generally use the term "force of gravity," but $m\vec{g}$ is often called the **weight**.

Your mass is the same no matter where you are. Near the surface of the Earth, the force of gravity acting on you is also constant. However, you have probably experienced feeling that you weigh more or less than usual, such as when you're on a roller coaster, or in a car going over a hill. Your weight (the force of gravity acting on you) is constant, but your apparent weight is different. Your apparent weight is, in many cases, equal in magnitude to the normal force acting on you, so you often feel a change in your apparent weight when the normal force changes.

Essential Question 3.6: Jump straight up into the air. While you are still in contact with the ground, but accelerating upward, how does the normal force applied on you by the ground compare to the force of gravity applied on you by the Earth? In what direction is the net force on you after you lose contact with the ground?

Answer to Essential Question 3.6: While you are still in contact with the ground, but accelerating upward, you must have a net force directed up. This comes from an upward normal force on you, applied by the ground, which is larger in magnitude than the downward force of gravity applied on you. After you leave the ground, there is no longer a normal force acting on you, so the downward force of gravity acting on you is the net force in that situation.

3-7 Practice with Free-Body Diagrams

Let's start by looking at the forces involved when you are in an elevator.

EXAMPLE 3.7 – An elevator at rest
(a) You are standing in an elevator that is at rest. Draw three free-body diagrams. The first should show all the forces acting on you as you stand in the elevator; the second should show all the forces acting on the elevator as you stand in it, and the third should show all the forces acting on the system consisting of you and the elevator combined. Draw them to scale, assuming that the mass of the elevator (M) is twice as large as your mass (m).

(b) Use your free-body diagrams to help determine an expression for the tension in the cable.

SOLUTION
We should start by drawing a diagram, shown below as part a of Figure 3.17. This shows the cable attached to the top of the elevator.

(a) Your free-body diagram is the same as the free-body diagram we would draw if you were simply standing on the floor. We show a downward force of gravity and an upward normal force that is exerted on you by the floor of the elevator. These two forces balance one another, to give a net force of zero, consistent with your constant velocity (of zero).

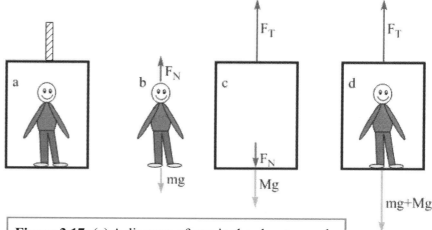

Figure 3.17: (a) A diagram of you in the elevator, and free-body diagrams for (b) you, (c) the elevator, and (d) the system consisting of you plus the elevator.

The free-body diagram of the elevator is a bit more challenging. Let's start by drawing a downward force of gravity and an upward tension force, which are the only forces that act on the elevator when the elevator is empty. Even though we draw what looks like an empty elevator, to construct the free-body diagram we have to remember that, in this case, you are in the elevator exerting a downward force on it. How do we represent this on the free-body diagram? Newton's third law is helpful. If the elevator applies an upward normal force on you, then you apply a downward normal force on the elevator of exactly the same magnitude.

A common mistake is to label the downward force the person applies to the elevator, in the elevator's free-body diagram, as *mg* instead of F_N. There are two reasons why doing so is not a good idea. First, the ***forces shown on an object's free-body diagram are forces exerted on that object by other things***. For the elevator, F_N is the force exerted on the elevator by you. *mg* is exerted on you by the Earth, so that force belongs on your free-body diagram but not the elevator's. Second, while *mg* and F_N are numerically equal in this situation, we will soon deal with a situation in which they are not, so using mg in place of F_N can actually lead to calculation errors.

The third free-body diagram, showing the forces acting on the system consisting of you and the elevator combined, is the combination of the first two free-body diagrams. We have the tension the cable exerts on the elevator, directed up, and the combined force of gravity acting on the system. When you combine the first two free-body diagrams, you also get the upward normal force the elevator exerts on you and the downward normal force you exert on the elevator. By Newton's third law, these forces are equal-and-opposite, so they cancel one another when they are combined. For this reason, as well as the fact that when we draw a free-body diagram, the forces we draw are exerted by objects external to the system we are considering, we don't include them on the free-body diagram of the combined system.

(b) *Use your free-body diagrams to determine an expression for the tension in the cable.* At this point, we apply Newton's second law. In general $\sum \vec{F} = m\vec{a}$, but, in this situation, the acceleration is zero, so we use the simplified equation $\sum \vec{F} = 0$. Choosing a coordinate system, let's define up to be positive. We will account for the fact that forces are vectors by using a plus sign if the force is directed up, and a minus sign if the force is directed down.

We could solve the problem by applying Newton's second law to the last free-body diagram, but let's consider all three diagrams to make sure everything is consistent.

Apply Newton's second law to the free-body diagram of you:
$$\sum \vec{F} = 0 \, ;$$

$+F_N - mg = 0$ which tells us that $F_N = mg$.

Apply Newton's second law to the elevator's free-body diagram:
$$\sum \vec{F} = 0 \, ;$$

$+F_T - Mg - F_N = 0$ which tells us that $F_T = Mg + F_N$.

Apply Newton's second law to the combined system's free-body diagram:
$$\sum \vec{F} = 0 \, ;$$

$+F_T - (M+m)g = 0$ which tells us that $F_T = Mg + mg$.

Everything is consistent. The second and third free-body diagrams give the same result for the tension because we know $F_N = mg$ from the first free-body diagram.

Related End-of-Chapter Exercises: 18, 43.

Essential Question 3.7: If the elevator is traveling up at constant velocity what, if anything, would change on the free-body diagrams? For example, would we need to add one or more forces to any of the free-body diagrams? Would any of the existing forces change in magnitude and/or direction?

Answer to Essential Question 3.7: If the velocity is constant, there is no acceleration. The situation is the same as in part (a). As far as forces are concerned "at rest" and "constant velocity" are equivalent. Nothing changes on any of the free-body diagrams.

3-8 A Method for Solving Problems Involving Newton's Laws

Now that we have explored some of the basic steps involved in solving a typical problem that relies on Newton's laws, let's summarize a general method that we can apply to all such problems. We have not yet addressed all the issues covered by the method (such as the various coordinate systems mentioned in step 3), but we will continue to address these issues in the remaining examples of this chapter as well as in chapter 5.

A General Method for Solving a Problem Involving Newton's Laws, in One Dimension
1. Draw a diagram of the situation.
2. Draw one or more free-body diagrams, with each free-body diagram showing all the forces acting on an object.
3. For each free-body diagram, choose an appropriate coordinate system. The different coordinate systems should be consistent with one another. A good rule of thumb is to align each coordinate system with the direction of the acceleration.
4. Apply Newton's second law to each free-body diagram.
5. Put the resulting force equations together and solve.

Let's apply the general method in the following Example.

EXAMPLE 3.8 – An elevator accelerating up
Let's say the elevator from Example 3.7, with you in it, has a constant acceleration directed up.

(a) What, if anything, would change on the free-body diagrams we drew in Example 3.7? Determine an expression for the tension in the cable now.
(b) If we know the elevator's acceleration is directed up, with a magnitude of $g/2$, what can we say about the direction the elevator is moving?

SOLUTION
(a) By Newton's second law, $\sum \vec{F} = m\vec{a}$, an upward acceleration requires a net upward force. Something must change on each free-body diagram to give us a net upward force. We have already accounted for all the interactions, so we don't add or subtract forces. Instead, one or more of the forces on each free-body diagram (see Figure 3.18) must change in magnitude.

The force of gravity cannot change, so to get a net upward force on the first free-body diagram, the normal force must increase. The normal force now has two jobs. It prevents you from falling through the floor of the elevator, and it also provides the extra upward force needed for the upward acceleration. Applying Newton's second law gives:

$$\sum \vec{F} = m\vec{a} = +\frac{1}{2}mg \ .$$

$$F_N - mg = +\frac{1}{2}mg \qquad \text{which tells us that } F_N = \frac{3}{2}mg \ .$$

On the free-body diagram of just the elevator, our previous results mean that the downward normal force increases. Applying Newton's second law gives:

$$\sum \vec{F} = M\vec{a} = +\frac{1}{2}Mg \, .$$

$$F_T - Mg - F_N = +\frac{1}{2}Mg$$

which tells us that

$$F_T = +\frac{3}{2}Mg + F_N \, .$$

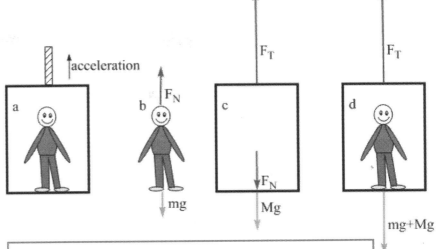

Figure 3.18: A diagram of you in the elevator (a), and free-body diagrams for you (b), the elevator (c), and the system consisting of you plus the elevator (d), when the elevator has an upward acceleration with a magnitude of g/2.

Here, it is critical to show the force that you exert on the elevator as a normal force, not a force of gravity, because the normal force is larger in magnitude than the force of gravity acting on you. Note that the increase in the tension offsets the increased normal force and also provides the net force required to accelerate the system upwards.

Applying Newton's second law to the third free-body diagram gives:

$$\sum \vec{F} = (M+m)\vec{a} = +\frac{1}{2}(M+m)g$$

$$F_T - (M+m)g = +\frac{1}{2}(M+m)g \, , \text{ which tells us that } F_T = \frac{3}{2}(Mg+mg) \cdot$$

The only change here is an increase in the tension. Check the two expressions for tension to make sure that they are consistent with one another.

(b) The elevator could actually be moving in any direction and have an upward acceleration. For instance, if the elevator is gaining speed while moving up, the acceleration is directed up. However, if the elevator is slowing while moving down, the acceleration is also directed up. A similar situation is a ball thrown straight up into the air. Whether the ball is moving up or down, once you release the ball, the acceleration is directed down because the force of gravity is the only force acting and is directed down. For one instant at the very top, the acceleration is down, yet the ball is at rest. Knowing the direction of the acceleration is not enough to determine the direction of motion.

Related End-of-Chapter Exercises: 19, 44.

We'll do another example to become more familiar with applying the method of solving a typical problem involving Newton's laws. First, we'll start with a conceptual question.

Essential Question 3.8: Three boxes are placed side-by-side on the floor (see Figure 3.19). The red box has a weight of 40 N, the green box a weight of 60 N, and the blue box a weight of 50 N. When you exert a constant force to the right on the red box, the entire system accelerates to the right. Which box experiences the largest-magnitude net force?

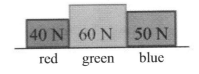

Figure 3.19: Three boxes placed side-by-side.

Answer to Essential Question 3.8: The boxes will move together as one unit, so the boxes have the same acceleration. The net force acting on a box is the mass of the box multiplied by the acceleration. Thus, the green box, with the largest mass, experiences the largest net force.

3-9 Practicing the Method

EXAMPLE 3.9 – Three boxes

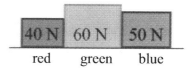

Let's look in more detail at the system of three boxes that are side-by-side on a frictionless floor. The weights of the boxes are given in Figure 3.19. By exerting a constant force of $\vec{F} = 30$ N to the right on the red box, you cause the three boxes to accelerate. Use $g = 10$ m/s².

Figure 3.19: Three boxes placed side-by-side.

(a) What is the acceleration of the boxes?

(b) Using a notation in the form F_{RG}, which denotes the magnitude of the force that the red box applies to the green box, sketch free-body diagrams for (i) the red box; (ii) the green box; (iii) the system consisting of the red and green boxes together; and (iv) the system consisting of the green and blue boxes together.

(c) Which of the four free-body diagrams above would you use to determine F_{RG}, the magnitude of the force that the red box applies to the green box? What is F_{RG} ?

SOLUTION

(a) Let's begin with a free-body diagram, but which system should we draw a free-body diagram for? Choosing the system carefully can make a problem easier to solve. The free-body diagrams for each box involve horizontal forces we don't yet know the magnitude of. Is there a free-body diagram we could draw so the only horizontal force acting is the 30 N force you apply?

As shown in Figure 3.20, the free-body diagram of the whole system involves only a downward force of gravity, an upward normal force, and the 30 N horizontal force. Let's choose a coordinate system with the positive x direction pointing to the right. Applying Newton's second law in that direction gives: $\sum F_x = (m_R + m_G + m_B)\vec{a}_x$.

The only horizontal force acting on the combined system is your 30 N force in the $+x$ direction, so: $\quad 30 \text{ N} = (m_R + m_G + m_B)\vec{a}_x$.

To find the masses, divide the weights by g, which we are taking here to be 10 m/s². The masses are $m_R = 4.0$ kg , $m_G = 6.0$ kg , and $m_B = 5.0$ kg . Solving for the acceleration gives

$$\vec{a}_x = \frac{+30 \text{ N}}{(4.0 \text{ kg} + 6.0 \text{ kg} + 5.0 \text{ kg})} = \frac{+30 \text{ N}}{15.0 \text{ kg}} = +2.0 \text{ m/s}^2.$$

(b) The free-body diagram of the red box shows the 30 N horizontal force you exert, as well as the upward normal force exerted on the red box by the floor, $\vec{F}_{FR}$, and the downward force of gravity exerted on the red box by the Earth, $\vec{F}_{ER}$. The diagram must also account for the interaction between the green and red boxes – the green box exerts a force to the left on the red box, $\vec{F}_{GR}$. This force is smaller in magnitude than your 30 N force because the red box has a net force to the right.

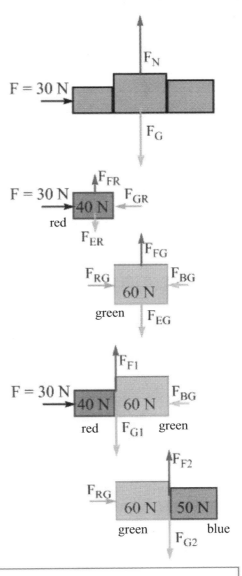

The free-body diagrams of the green and red boxes are similar. In both cases, the upward normal force balances the downward force of gravity, while the fact that there is a net force to the right means that the force to the right is somewhat larger than the force to the left. Your 30 N force is applied only to the red box, so it appears only on the free-body diagram of the red box (or on the free-body diagram of a system involving the red box). The green box experiences a force to the right, from the red box, but it is less than 30 N.

If we combine the red and green boxes into one system (system 1), the system's free-body diagram is the sum of the individual free-body diagrams. The net upward normal force $\vec{F}_{F1} = \vec{F}_{FR} + \vec{F}_{FG}$ balances the net downward force of gravity $\vec{F}_{E1} = \vec{F}_{ER} + \vec{F}_{EG}$. Horizontally, your 30 N force acts to the right, while the force $\vec{F}_{BG}$ that the blue box exerts on the green box is directed left.

We do not have to include $\vec{F}_{RG}$ and $\vec{F}_{GR}$ because, by Newton's third law, these are equal-and-opposite and thus cancel one another. Another reason to exclude this pair of forces is that they are internal forces in the system. Forces that belong on the free-body diagram come from external interactions, forces exerted on the system by things outside the system, such as by you, the floor, the Earth, and the blue box.

Combining the green and blue boxes into one system, system 2, the upward normal force from the floor $\vec{F}_{F2}$ balances the downward force of gravity $\vec{F}_{E2}$. Horizontally, there is only one horizontal force acting, the force the red box exerts on the green box $\vec{F}_{RG}$.

Figure 3.20: Various free-body diagrams for the three-box situation. In each case, the vertical forces cancel, so we can focus on the horizontal forces.

(c) To find $\vec{F}_{RG}$, we cannot use the system consisting of the red and green boxes together. $\vec{F}_{RG}$ is an internal force in that system, so it does not appear on the system's free-body diagram. $\vec{F}_{RG}$ appears on the free-body diagram of the green box, as well as on the free-body diagram of the system consisting of the green and blue boxes. In addition, $\vec{F}_{GR}$ appears on the free-body diagram of the red box so we could solve for that, because $\vec{F}_{RG}$ is equal in magnitude to $\vec{F}_{GR}$.

Let's use the last free-body diagram, because $\vec{F}_{RG}$ is the only horizontal force that appears on that diagram. Applying Newton's second law in the horizontal direction gives:

$$\sum \vec{F}_x = \left(m_G + m_B \right) \vec{a}_x \text{, so} \qquad \vec{F}_{RG} = (6.0 \text{ kg} + 5.0 \text{ kg}) * (+2.0 \text{ m/s}^2) = +22 \text{ N}.$$

Related End-of-Chapter Exercises: 36, 45, 58, 59.

Essential Question 3.9: Calculate the net force acting on each of the boxes in Example 3.9.

Answer to Essential Question 3.9: By Newton's second law, the net force acting on each box is equal to $m\vec{a}$ for each box. This gives 8.0 N to the right for the red box; 12 N to the right for the green box and 10 N to the right for the blue box. As they should, these net forces add together to 30 N to the right, matching the net force on the system (the 30 N force you apply, directed right).

Chapter Summary

Essential Idea: Forces and Newton's Laws
In this chapter, we covered one of the main methods of analyzing a physical situation, which is to think about all the forces being exerted on an object by external influences, and then apply Newton's second law to determine the acceleration.

Forces and Newton's Three Laws of Motion
A force, which is a push or a pull, acts on an object when there is an interaction between that object and another object. Newton's laws give us some guidelines to use when applying forces to solve problems. They are:

Newton's First Law: if no net force acts on an object, the object's velocity is unchanged: the object either remains at rest or it keeps moving with constant velocity. If there is a non-zero net force acting, then the object's velocity changes.

Newton's Second Law tells us that the connection between an object's net force and its acceleration is given by:

$$\vec{a} = \frac{\sum \vec{F}}{m}$$, which we can re-write as $$\sum \vec{F} = m\vec{a} .$$ (Equation 3.1)

Newton's Third Law: Whenever one object exerts a force on a second object, the second object exerts a force equal in magnitude, and opposite in direction, on the first object.

A General Method for Solving Problems Involving Newton's Laws, in One Dimension
The most important lesson to take away from this chapter is that most one-dimensional problems involving forces can be solved by applying the following method:

1. Draw a diagram of the situation.

2. Draw one or more free-body diagrams, with each free-body diagram showing all the forces acting on an object.

3. For each free-body diagram, choose an appropriate coordinate system. The coordinate systems for the different free-body diagrams should be consistent with one another. A good rule of thumb is to align each coordinate system with the direction of the acceleration.

4. Apply Newton's second law to each free-body diagram.

5. Put the resulting force equations together, and solve.

End-of-Chapter Exercises

Exercises 1 – 12 are conceptual questions that are designed to see if you have understood the main concepts of the chapter.

1. Five possible free-body diagrams of one of your friends are shown in Figure 3.21. (a) Which free-body diagram applies if your friend remains at rest? Which free-body diagram applies if your friend is moving with a constant velocity directed (b) to the right? (c) to the left? (d) straight up? (e) straight down? You can use a free-body diagram more than once if you wish.

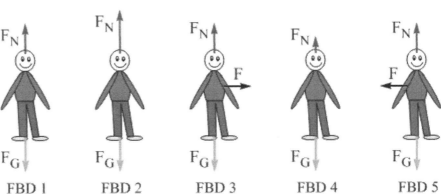

FBD 1   FBD 2   FBD 3   FBD 4   FBD 5

Figure 3.21. Five possible free-body diagrams of your friend, for Exercises 1 and 2. Note that the magnitude of $\vec{F}_N$ is equal to that of $\vec{F}_G$ in diagrams 1, 3, and 5.

2. Five possible free-body diagrams of one of your friends are shown in Figure 3.21. Describe a situation in which the applicable free-body diagram is (a) FBD 2 (b) FBD 3 (c) FBD 4.

3. Three possible free-body diagrams are shown in Figure 3.22 for a car moving to the right. $\vec{F}_{air}$ represents a resistive force due to air resistance, while $\vec{F}_{road}$ represents the force the road exerts on the car. Which free-body diagram is consistent with the car (a) moving at constant velocity? (b) speeding up? (c) slowing down? You can use a free-body diagram more than once if you wish. A resistive force is a force that opposes motion.

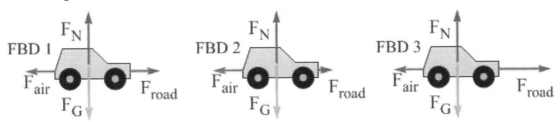

Figure 3.22: Three possible free-body diagrams for a car moving to the right. The magnitude of the force of air resistance is equal to that exerted by the road in FBD 1. For Exercises 3 and 4.

4. Consider again the three free-body diagrams shown in Figure 3.22. $\vec{F}_{air}$ represents a force that the air exerts on the car, while $\vec{F}_{road}$ represents the force the road exerts on the car. (a) Are any of the free-body diagrams consistent with the car remaining at rest? If so, which? (b) If you chose a free-body diagram in (a), describe a situation in which that free-body diagram would apply, with the car remaining at rest.

5. In class, you see a demonstration involving a penny and a feather that are dropped simultaneously from rest inside a glass tube. The tube is held so the two objects fall vertically from one end of the tube to the other. At first, the penny easily beats the feather to the lower end of the tube, but then your professor uses a vacuum pump to remove most of the air from inside the tube. When the objects are again released from rest, which object reaches the lower end of the tube first? Why?

6. Two solid steel ball bearings are dropped from rest from the same height above the floor. Ball A is somewhat larger and heavier than ball B. Assuming air resistance can be neglected, rank the balls based on (a) the magnitude of their accelerations as they fall; (b) the magnitude of the net force acting on each ball as they fall; (c) the time it takes them to reach the ground.

7. Three identical blocks are placed in a vertical stack, one on top of the other, as shown in Figure 3.23. The stack of blocks remains at rest on the floor. (a) Which block experiences the largest net force? Explain. (b) Compare the free-body diagram of block 2, in the middle of the stack, to that of block 3, at the bottom of the stack. Comment on any differences, if there are any, between the two free-body diagrams.

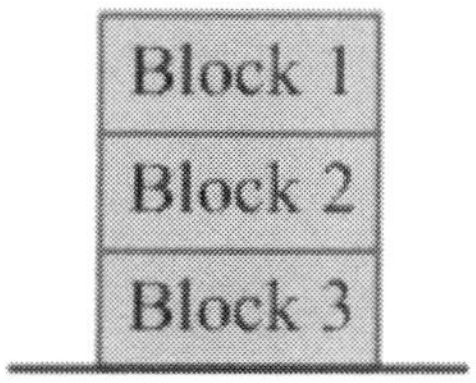

Figure 3.23: A stack of three identical blocks, for Exercise 7.

8. What situations can you think of in which one object exerts a larger-magnitude force on a second object than the second object exerts on the first object?

9. Describe a situation that matches each of the following, or state that it is impossible. (a) An object has no net force acting on it, and yet it is moving. (b) An object has a net force acting on it, but it remains at rest. (c) An object has at least one force acting on it, but it remains at rest.

10. A team of construction workers knows that the cable on a crane will break if its tension exceeds 8000 N. They then connect the cable to a load of bricks with a total weight of 7500 N. When the crane operator slowly raises the bricks off the ground, everything looks fine, and the team gives her the signal to go faster. As she increases the speed at which the bricks are being raised, however, the cable breaks, showering bricks on the ground below. Fortunately, everyone is wearing proper safety equipment, so there are no serious injuries. Using your knowledge of physics, can you come up with an explanation of the cause of the accident? Come up with two ways the accident could have been prevented.

11. Yuri, a cosmonaut on the space station, is taking a spacewalk outside of the station to fix a malfunctioning array of solar cells that provide electricity for the station. Unfortunately, he forgets to tether himself to the station, and his rocket pack also is not working, so when he finds himself drifting slowly away from the station Yuri realizes he's in a bit of trouble. Fortunately he is holding a large wrench. Based on the principles of physics we have discussed in this chapter, explain what Yuri can do to get himself back to the space station.

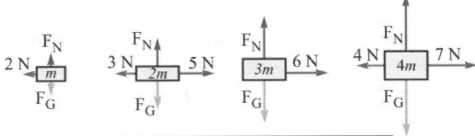

12. Four free-body
    diagrams are shown
    in Figure 3.24, for objects
    that have masses of m,
    $2m$, $3m$, and $4m$,
    respectively. Rank these
    situations, from largest to smallest, based on (a) the net force being applied to the object,
    and (b) the acceleration of the object. Your answers should have the form 3>2=4>1.

Figure 3.24: Four free-body diagrams, for Exercise 12. On each free-body diagram, the magnitude of the normal force, $\vec{F}_N$, is equal to the magnitude of the force of gravity, $\vec{F}_G$.

Exercises 13 - 20 are designed to give you practice with free-body diagrams.

13. You are at rest, sitting down with your weight completely supported by a chair. Sketch a free-body diagram for (a) you, and (b) the chair.

14. Repeat Exercise 13, except that now you and the chair are inside an elevator that has a constant velocity directed down. Sketch a free-body diagram for (a) you, and (b) the chair. (c) Comment on what, if anything, changes on the free-body diagrams here compared to those you drew in Exercise 13.

15. Repeat Exercise 13, except that now you and the chair are inside an elevator that has a constant acceleration directed down.

16. You step off a chair and allow yourself to drop, feet first, straight down to the floor below. Sketch your free-body diagram while you are (a) dropping to the floor, not in contact with anything, (b) slowing down, after your feet initially make contact with the floor, and (c) at rest, standing on the floor. (d) What would a bathroom scale, on the floor, read if you landed on it instead of the floor while you were in the three positions in (a), (b), and (c)? Would the scale reading equal your mass, or not? (e) Are your answers in (d) consistent with the free-body diagrams you drew in (a) – (c)? Explain.

17. A ball is initially at rest in your hand. You then accelerate the ball upwards, releasing it so that it goes straight up into the air. When it comes down, you catch it and bring it to rest again. Neglect air resistance. Sketch a free-body diagram for the ball when it is (a) accelerating upward in your hand; (b) moving up after you release it; (c) at rest, just for an instant, at the top of its flight; (d) moving down before you catch it; (e) slowing down after it makes contact with your hand again. (f) What is the minimum number of unique free-body diagrams that you can draw to represent the five situations described in (a) – (e)? Explain.

(a)   (b)

18. As shown in Figure 3.25 (a), two boxes are initially at rest on a frictionless horizontal surface. The mass of the large box is five times larger than that of the small box. You then exert a horizontal force F directed right on the large box. Sketch a free-body diagram for (a) the two-box system (b) the large box (c) the small box. (d) Does the large box exert more force on the small box than the small box exerts on the large box? Explain.

Figure 3.25: Two situations involving two boxes placed side-by-side on a frictionless surface, for Exercises 18 – 20.

19. Repeat Exercise 18, except that now the position of the boxes is reversed, as shown in Figure 3.25 (b).

20. In which case above, in Exercise 18 or Exercise 19, is the force that the large box exerts on the small box larger in magnitude? Explain.

Exercises 21 – 30 are designed to give you practice with applying the general method for solving a problem that involves Newton's laws. For each exercise, start by doing the following: (a) Draw a diagram of the situation. (b) Draw one or more free-body diagram(s) showing all the forces that act on various objects or systems. (c) Choose an appropriate coordinate system for each free-body diagram. (d) Apply Newton's second law to each free-body diagram.

21. You pull on a horizontal string attached to a small block that is initially at rest on a horizontal frictionless surface. The block has an acceleration of 3.0 m/s² when you exert a force of 6.0 N. Parts (a) – (d) as described above. (e) What is the mass of the block?

22. Jennifer, Katie, and Leah are attempting to push Katie's car, which has run out of gas. The car has a mass of 1500 kg. Each woman exerts a force of 200 N while pushing the car forward, and, once they get the car moving, there is a net resistive force of 570 N opposing the motion. Parts (a) – (d) as described above. (e) What is the magnitude of the car's acceleration?

23. A small box with a weight of 10 N is placed on top of a larger box with a weight of 40 N. The boxes are at rest on the top of a horizontal table. Parts (a) – (d) as described above, in part (b) drawing three free-body diagrams, one for each of the boxes and one for the two-box system. What is the magnitude and direction of the force exerted on the large box by (e) the small box? (f) the table?

24. Repeat Exercise 23, except that now the system of boxes and the table are inside an elevator that has a constant acceleration down of $g/2$.

25. A small box with a weight of 10 N is placed on top of a larger box with a weight of 40 N. The boxes are at rest on the top of a horizontal table. You apply an additional downward force of 10 N to the top of the small box by resting your hand on it. Parts (a) – (d) as described above, in part (c) drawing three free-body diagrams, one for each of the boxes and one for the two-box system. What is the magnitude and direction of the force exerted on the large box by (e) the small box? (f) the table?

26. A small block with a weight of 4 N is hung from a string that is tied to the ceiling of an elevator that is at rest. A large block, with a weight of 8 N, is hung from a second string that hangs down from the small block. Parts (a) – (d) as described above, in part (c) drawing three free-body diagrams, one for the each block and one for the two-block system. If the elevator is at rest, find the tension in (e) the string tied to the ceiling of the elevator; (f) the string between the blocks.

27. Return to the situation described in Exercise 26. Repeat the exercise with the elevator now having an acceleration of $g/4$ directed up.

28. Erin is playing on the floor with a wooden toy train consisting of an engine with a mass of 800 g and two passenger cars, each with a mass of 600 g. The three parts of the train are arranged in a line and are connected by horizontal strings of negligible mass. Erin accelerates the entire train forward at 4.0 m/s² by pulling horizontally on another string attached to the front of the engine. Neglect friction. Parts (a) – (d) as described above, in part (c) drawing four free-body diagrams, one for the engine, one for each of the passenger cars, and one for the entire train. (e) What is the tension in each of the three strings?

29. Two boxes, one with a mass two times larger than the other, are placed on a frictionless horizontal surface and tied together by a horizontal string, as shown in Figure 3.26(a). You then apply a horizontal force of 30 N to the left by pulling on another string attached to the larger box. Part (a), the diagram, is already done. Parts (b) – (d) as described above, in part (c) drawing three free-body diagrams, one for each box and one for the two-box system. (e) Find the magnitude of the tension in the string between the boxes.

Figure 3.26: Two situations involving two boxes on a horizontal surface. The boxes are connected by a string of negligible mass. For Exercises 29 and 30.

30. Repeat Exercise 29, but now you apply a horizontal force of 30 N to the right by pulling on a string attached to the smaller box, as shown in Figure 3.26(b).

Exercises 31 – 40 are designed to give you some practice connecting the force ideas from this chapter to the motion with constant acceleration ideas from Chapter 2.

31. A car with a mass of 2000 kg experiences a horizontal net force, directed east, of 4000 N for 10 seconds. What is the car's final velocity if the initial velocity of the car is (a) zero (b) 10 m/s east (c) 20 m/s west.

32. A flea, with a mass of 500 nanograms, reaches a maximum height of 50 cm after pushing on the ground for 1.3 milliseconds. What is the average force the ground exerts on the flea while the flea is in contact with the ground as it accelerates up?

33. Yolanda, having a mass of 50 kg, steps off a 2.0-meter high wall and drops down to the ground below. What is the average force exerted on Yolanda by the ground if, after first making contact with the ground, she comes to rest by bending at the knees so her upper body drops an additional distance of (a) 3 cm (b) 30 cm?

34. In a demonstration known as the vacuum bazooka, a ping-pong ball is placed inside a PVC tube, the ends of the tube are sealed with tape or foil, and most of the air is removed from the tube. The demonstrator then pierces the seal at the end of the tube where the ball is, and the in-rushing air accelerates the ball along the tube until the ball bursts through the seal at the far end and emerges from the tube at high speed. (a) If the mass of the ball is 2.5 g, the tube is approximately horizontal and has a length of 1.5 m, and the average force the air exerts on the ball is 100 N, find an upper limit for the ball's speed when it emerges from the tube. (b) In practice, the ball's speed is impressive but somewhat less than the theoretical maximum speed determined in (a). What are some factors that could reduce the ball's speed when it emerges from the tube?

35. In a tennis match, Serena Williams hits a ball that has a velocity of 20 m/s directed horizontally. If the force of her racket is applied for 0.10 s, causing the ball to completely reverse direction and acquire a velocity of 30 m/s directed horizontally, what is the average horizontal force the racket applies to the ball? A tennis ball has a mass of 57 grams.

36. Consider the motion diagram shown in Figure 3.27. If the vertical marks in the diagram are 1.0 meters apart, the object has a mass of 2.0 kg, and the images of the object are shown at 1.0-second intervals, determine the net force applied to the object if the object is moving with a constant acceleration from (a) left to right (b) right to left.

Figure 3.27: Motion diagram for Exercise 36.

37. Consider the motion diagram shown in Figure 3.28. Describe the general behavior, over the 12-second interval shown in the diagram, of a net force that could be applied to the object to produce this motion diagram.

Figure 3.28: Motion diagram for Exercise 37.

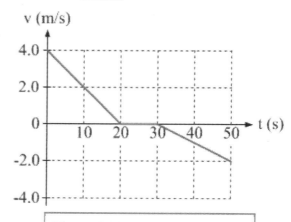

Figure 3.29: A graph of velocity versus time, for Exercise 38.

38. A plot of a cat's velocity as a function of time is shown in Figure 3.29. If the cat has a mass of 5.0 kg, plot the corresponding net force vs. time graph for the cat.

39. Starting from rest, a person on a bicycle travels 200 m in 20 s, moving in a straight line on a horizontal road. Assuming that the acceleration is constant over this time interval, determine the magnitude of the horizontal force applied to the person-bicycle system in the direction of motion if there is a constant resistive force of 20 N acting horizontally opposite to the direction of motion and the person-bicycle system has a combined mass of 80 kg.

Exercises 40 – 44 involve applications of forces in one dimension.

40. A baseball pitcher can accelerate a 150-g baseball from rest to a horizontal velocity of 150 km/h over a distance of 2.0 m. What is the average horizontal force the pitcher exerts on the ball during the throwing motion?

41. You read in the paper about a planet that has been discovered orbiting a distant star. The astrophysicist quoted in the newspaper article states that the acceleration due to gravity on this planet is about 20% larger than that here on Earth. In an attempt to simulate what it would feel like to live on this newly discovered planet, you get into an elevator on the third floor of a five-story building. (a) To have an apparent weight larger than your actual weight immediately when the elevator starts to move, should you press the button for the

first floor or the fifth floor? (b) What does the acceleration of the elevator have to be for you to feel (at least briefly!) like you are living on the newly discovered planet?

42. Modern cars are designed with a number of important safety features to protect you in a crash. These include crumple zones, air bags, and seat belts. Consider how a crumple zone (a section of the car that is designed to compress, like an accordion, as does the front of the car in the photograph of a crash test shown in Figure 3.30) and a seat belt work together in a head-on collision in which you go from a speed of 120 km/h to rest. (a) If you are not wearing a seat belt then, in the crash, you generally keep moving forward until you hit something like the windshield. If you come to rest after decelerating through a distance of 4.0 cm after hitting the windshield, what is the magnitude of your average acceleration? (b) If, instead, you are wearing your seat belt, it keeps you in your seat, and you keep moving forward as the front of the car crumples like an accordion. If the compression of the crumple zone is 80 cm, what is the magnitude of your average acceleration? (c) In which case do you think you have a better chance of surviving the crash?

Figure 3.30: In this crash-test photo, the crumple zone at the front of the vehicle has compressed like an accordion, leaving the car's passenger cabin intact. For Exercise 42. Photo credit: Douglas Waite, via Wikimedia Commons.

43. A "solar sailboat" is a space probe that is propelled by sunlight reflecting off a shiny sail with a large area. Let's say the probe and sail have a combined mass of 1000 kg and that the net force exerted on the system is 4.0 N directed away from the Sun. (a) What is the acceleration of the probe/sail system? If the system is released from rest, how fast will it be traveling after (b) 1 day? (c) 1 week? The net force will actually decrease in magnitude as the probe gets farther from the Sun, but let's not worry about that.

44. NASA's Goddard Space Center is named after Robert Goddard of Worcester, Massachusetts, who was a pioneer in the field of rocketry. After Goddard published a paper in 1919 about rockets, the New York Times, in 1920, published an editorial lambasting Goddard, and stating that everybody knows that rockets won't travel in the vacuum of space, where there is nothing to push against. (The paper retracted the statement in 1969, after the launch of Apollo 11.) How does a rocket work? How would you respond to the issue raised by the Times?

Figure 3.31: In 1964, Robert Goddard was honored by the United States Postal Service, via this image on an 8-cent stamp, for his contributions to rocketry. For Exercise 44. Photo credit: Wikimedia Commons.

General Problems and Conceptual Questions.

45. Three children are pushing a very large ball, which has a mass of 10 kg, around a field. If each child exerts a force of 12 N, determine the maximum and minimum possible values of the ball's acceleration. Assume the ball is in contact with the ground, and the normal force from the ground acting on the ball exactly balances the force of gravity acting on the ball.

46. A box with a weight of 25 N remains at rest when it is placed on a ramp that is inclined at 30° with respect to the horizontal. What is the magnitude and direction of the contact force exerted on the box by the ramp?

47. You exert a horizontal force of 10 N on a box with a weight of 25 N, but it remains at rest on a horizontal tabletop. What is the magnitude of the contact force exerted on the box by the table?

48. In the following situations, which object exerts a larger-magnitude force on the other? (a) The head of a golf club strikes a golf ball. In other words, does the club exert more force on the ball than the ball exerts on the club, is the opposite true, or is there another answer? (b) While stretching, you push on a wall. (c) A large truck and a small car have a head-on collision on the freeway. (d) The Earth orbits the Sun.

49. Three blocks are tied in a vertical line by three strings, and the top string is tied to the ceiling of an elevator that is initially at rest (see Figure 3.32). If the tension in string 3 is T what is the tension in (a) string 2? (b) string 1 (in terms of T)?

50. Return to the situation described in Exercise 49. Use $g = 10$ m/s^2. What is the magnitude of the tension in the second string if the elevator is (a) at rest? (b) moving at a constant velocity of 2.0 m/s up? (c) accelerating up at 2.0 m/s^2? (d) accelerating down at 2.0 m/s^2?

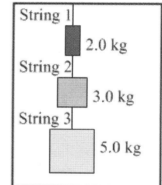

Figure 3.32: Three blocks connected by strings and tied to the ceiling of an elevator, for Exercises 49 and 50.

51. Three blocks are placed side-by-side on a horizontal frictionless surface and subjected to a horizontal force F, as shown in case 1 of Figure 3.33, which causes the blocks to accelerate to the right. The blocks are then re-arranged, as in case 2, and subjected again to the same horizontal force F. (a) In which case does the 2.0 kg block experience a larger net force? (b) In terms of F, calculate the magnitude of the net force on the 2.0 kg block in (i) Case 1 (ii) Case 2. (c) In which case does the 5.0 kg block exert more force on the 2.0 kg block? (d) In terms of F, calculate the magnitude of the force exerted by the 5.0 kg block on the 2.0 kg block in (i) Case 1 (ii) Case 2.

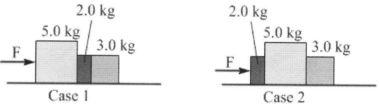

Figure 3.33: Two situations involving three blocks being pushed from the left by a horizontal force F, for Exercise 51.

52. As shown in Figure 3.34, a mobile made from ten 1.0 N balls is tied to the ceiling. Assume the other parts of the mobile (the strings and rods) have negligible mass. What is the tension in the string (a) above the green ball? (b) above the red rod? (c) tying the mobile to the ceiling?

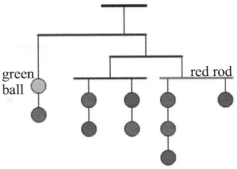

53. Repeat Exercise 52, but this time let's say that the mobile is tied to the ceiling of an elevator, and that the elevator is accelerating down at $g/2$.

Figure 3.34: A mobile made from 10 balls and some light strings and rods, for Exercises 52 and 53.

54. You give a book a small push with your hand so that, after you remove your hand, the book slides for some distance across a table before coming to rest. (a) Sketch a motion diagram for the book as it is sliding, showing the book's position at regular time intervals after you remove your hand. (b) You should see a trend in your motion diagram. Does the distance between successive images of the book on the motion diagram change as time goes by? If so, how? (c) Sketch a free-body diagram for the book, showing all forces acting on the book as it is sliding, and after the period in which your hand was pushing the book. (d) What applies each of the forces on your free-body diagram? (e) Is there a net force acting on the book as it is sliding? If so, in what direction is it? (f) Is the book's motion consistent with this net force?

55. In Exploration 3.2B, we drew a free-body diagram for a ball falling straight down toward the floor. (a) Now, consider the free-body diagram for a ball you toss straight up in the air. While it is in flight (after it leaves your hand) does the free-body diagram differ from that in Exploration 3.2B? If so, in what way(s)? (b) Consider the free-body diagram for a ball you toss across the room to your friend. While the ball is in flight, does the free-body diagram differ from that in Exploration 3.2B? If so, in what way(s)?

56. You overhear two of your classmates discussing the issue of the force experienced by a ball that is tossed straight up into the air. Comment on each of their statements.

Sarah: Once the ball leaves my hand, the only force acting on it is gravity, so the ball's acceleration changes at a steady rate.

Tasha: But, after it leaves your hand, the ball is moving up, and gravity acts down. The ball must have the force of your throw acting on it as it moves up.

Sarah: So what makes the ball slow down?

Tasha: As time goes by, the force of your throw decreases, and gravity takes over.

57. A hockey puck of mass m is sliding across some ice at a constant speed of 8 m/s. It then experiences a head-on collision with a hockey stick of mass $5m$ that is lying on the ice. The puck is in contact with the stick for 0.10 s. After the collision, the puck is traveling at a constant speed of 2 m/s in the direction opposite what it was going originally. (a) Find the magnitude of the average force exerted on the puck by the stick during the collision. (b) Find the velocity of the stick after the collision. You can assume no friction acts on either object.

58. Consider the following situations involving a car of mass m and a truck of mass $5m$. In each situation, state which force has a larger magnitude, the force the truck exerts on the car or the force the car exerts on the truck. (a) The truck collides with the car, which is parked by the side of the road. (b) The car collides with the truck, which is parked by the side of the road. (c) The vehicles have identical speeds and are going in opposite directions when they have a head-on collision. (d) The vehicles are going in opposite directions when they collide, with the car's speed being five times larger than the truck's speed.

59. Two boxes are side-by-side on a frictionless horizontal surface as shown in Figure 3.35. In Case 1, a horizontal force F directed right is applied to the box of mass M. In Case 2, the horizontal force F is instead directed left and applied to the box of mass m. Find an expression for the magnitude of F_{Mm}, the force the box of mass M exerts on the box of

mass m, in (a) Case 1 (b) Case 2. Express your answers in terms of variables given in the problem. (c) If F_{Mm} is four times larger in case 2 than it is in case 1, find the ratio of the

masses of the boxes.

60. Consider again the situation shown in Figure 3.35 and described in Exercise 59, but now let's say that $M = 2m$. In which case is the magnitude of the (a) acceleration of the two-box system larger? (b) acceleration of the box of mass m larger? (c) force that the box of mass M exerts on the box of mass m larger? (d) force that the box of mass m exerts on the box of mass M larger?

Figure 3.35: Two cases involving two boxes on a frictionless surface, with an applied force, for Exercises 59 and 60.

61. You get on an elevator on the fifth floor of a building, stand on a regular bathroom scale, and then push a button in the elevator. The elevator doors close, and the elevator moves from the fifth floor to a different floor, where it stops and the doors open again. At this point, you get off the scale and exit the elevator. A graph of the scale reading as a function of time is shown in Figure 3.36. Use $g = 10$ m/s². Based on this graph, (a) qualitatively describe the motion of the elevator; (b) determine the magnitude of the peak acceleration of the elevator; (c) determine how far, and in what direction, the elevator moved.

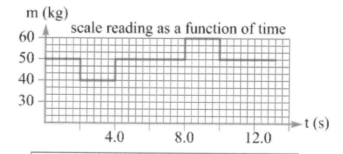

Figure 3.36: A graph of a scale reading as a function of time while you are standing on it in an elevator, for Exercise 61.

In a soccer game, the player shooting the ball uses his years of experience to gauge how fast, and in what direction, to kick the ball so that it will end up in the net. Similarly, the goalkeeper uses his experience to know when and where to dive to try to stop the ball. Understanding the parabolic trajectory followed by the soccer ball is one of the main themes of this chapter. (Photo credit: Via Wikimedia Commons, a public-domain image taken by USAF Master Sergeant Lance Cheung of 1st Lt. Richard Cullen.)

Chapter 4 – Motion in Two Dimensions

CHAPTER CONTENTS

There are a variety of applications of the kind of parabolic motion exhibited by a soccer ball during a game. Exactly how fast, and at what angle, should you launch a basketball so that it goes through the hoop? How does the center fielder on the baseball team know when to time his diving catch, and how precisely to throw the ball to home plate to throw out the runner trying to score from third base? Which club should Tiger Woods choose, and how should he swing it, so that his golf ball lands on the green? How fast, and at what angle, should Maria Sharapova hit a tennis ball so that it passes over the net and lands in play on the other side?

In each of the examples above, it is not necessary for the person involved to get out their calculator and their physics equations to calculate the ball's trajectory. Instead, years of practice have given them a feel for what to do. Applying physics to such situations, however, gives us insight into what we, or world-class athletes, have to do to achieve the intended results.

In this chapter, we make the transition from one-dimensional motion to two-dimensional motion. Our basic method is to break two-dimensional motion problems into two separate one-dimensional motion problems. What we learned in Chapter 2 will thus be critical in understanding two-dimensional motion. We will also learn some new concepts that apply in one dimension just as they apply in two dimensions. Our starting point will be constant-velocity situations, but we will then move to equations involving two-dimensional motion with constant acceleration, which describes many projectile-motion situations we encounter on a daily basis.

4-1 Relative Velocity in One Dimension

Before we generalize to two dimensions, let's consider a familiar situation involving relative velocity in one dimension. You are driving east along the highway at 100 km/h. The car in the next lane looks like it is barely moving relative to you, while a car traveling in the opposite direction looks like it is traveling at 200 km/h. This is your perception, even though the speedometers in all three vehicles say that each car is traveling at about 100 km/h.

How can we explain your observations? First, consider the velocity of your car relative to you. Even though your car is zooming along the highway at 100 km/h (with respect to the road), your car is at rest relative to you. To get a result of zero for the velocity of your car relative to you, we subtract 100 km/h east (your velocity with respect to the ground) from the velocity of the car with respect to the ground. This method of subtracting your velocity with respect to the ground also works to find the velocity of something else (such as an oncoming car) with respect to you. Subtracting your velocity from the velocity of other objects is equivalent to adding the opposite of your velocity to these velocities. Figure 4.1 illustrates the process.

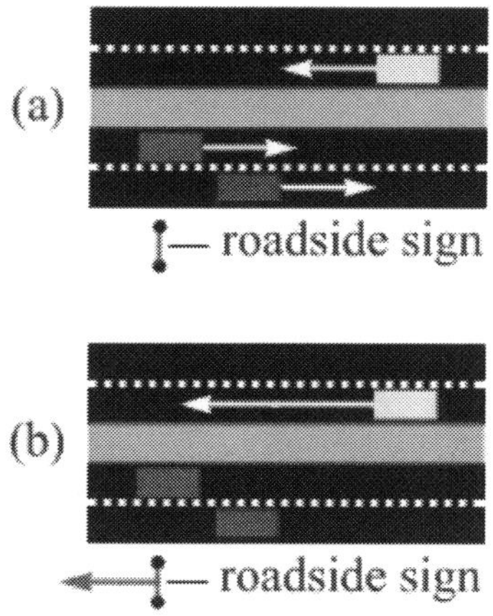

Most relative-velocity problems can be thought of as vector-addition problems. The trick is to keep track of which vectors we're adding (or subtracting). Let's explore this idea further.

(a)

— roadside sign

(b)

— roadside sign

Figure 4.1: (a) An overhead view, with arrows showing the velocities of three cars and a roadside sign, with respect to the ground. (b) The same situation, but with the velocities shown with respect to you, the driver of the bottom car.

EXPLORATION 4.1 – Crossing a river

You are crossing a river on a ferry. Assume there is no current in the river (that is, the water is at rest with respect to the shore). The ferry is traveling at a constant velocity of 7.0 m/s north with respect to the shore, while you are walking at a constant velocity of 3.0 m/s south relative to the ferry.

Step 1 - *What is the ferry's velocity relative to you?* We can use the notation $\vec{v}_{YF}$ to denote your velocity relative to the ferry, so we have $\vec{v}_{YF} = 3.0$ m/s south. The velocity of the ferry with respect to you is exactly the opposite of your velocity with respect to the ferry, so $\vec{v}_{FY} = 3.0$ m/s north. This relation is always true: the velocity of some object A with respect to another object B is the opposite of the velocity of B with respect to A ($\vec{v}_{AB} = -\vec{v}_{BA}$).

Step 2 - *What is your velocity relative to the shore?* If you stood still on the ferry, your velocity relative to the shore would match the ferry's velocity with respect to the shore. In this case, however, you are moving with respect to the ferry, so we have to add the relative velocities as vectors (see Figure 4.2). Your velocity with respect to the shore is your velocity relative to the ferry plus the ferry's velocity relative to the shore:

$$\vec{v}_{YS} = \vec{v}_{YF} + \vec{v}_{FS} = 3.0 \text{ m/s south} + 7.0 \text{ m/s north} = -3.0 \text{ m/s north} + 7.0 \text{ m/s north}.$$

$$\vec{v}_{YS} = 4.0 \text{ m/s north}.$$

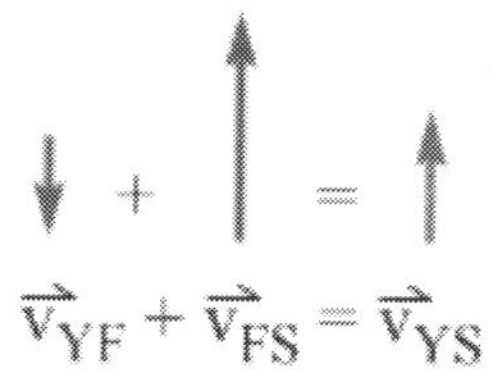

$\vec{v}_{YF} + \vec{v}_{FS} = \vec{v}_{YS}$

This vector-addition method is valid in general and works in more than one dimension.

Figure 4.2: Finding your velocity relative to the shore by vector addition.

A relative-velocity problem is a vector-addition problem. The velocity of an object A relative to an object C is the vector sum of the velocity of A relative to B plus the velocity of B relative to C:

$$\vec{v}_{AC} = \vec{v}_{AB} + \vec{v}_{BC} \qquad \text{(Equation 4.1: The vector addition underlying relative velocity)}$$

Step 3 - *Assume now that there is a current of 2.0 m/s directed south in the river, and that the ferry's velocity of 7.0 m/s north is relative to the water. What is the ferry's velocity relative to the shore? What is your velocity relative to the shore?*

To find the ferry's velocity with respect to the shore, we can use Equation 4.1 to combine the known values of the ferry's velocity with respect to the water and the water's velocity with respect to the shore.

In this case, using the subscript W to refer to the water, we get:

$$\vec{v}_{FS} = \vec{v}_{FW} + \vec{v}_{WS} = +7.0 \text{ m/s north} + 2.0 \text{ m/s south}.$$

$$\vec{v}_{FS} = +7.0 \text{ m/s north} - 2.0 \text{ m/s north} = +5.0 \text{ m/s north}.$$

Going against the current, the ferry takes longer to get to its destination, because it is moving slower with respect to the shore than when there was no current.

Now that we know the ferry's velocity with respect to the shore, we can apply Equation 4.1 to find your velocity with respect to the shore.

$$\vec{v}_{YS} = \vec{v}_{YF} + \vec{v}_{FS} = +3.0 \text{ m/s south} + 5.0 \text{ m/s north}.$$

$$\vec{v}_{YS} = -3.0 \text{ m/s north} + 5.0 \text{ m/s north} = +2.0 \text{ m/s north}.$$

An alternate approach to finding your velocity with respect to the shore is to extend the procedure of Step 2 to three vectors, as represented in Figure 4.3. In this case, we get:

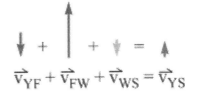

$$\vec{v}_{YF} + \vec{v}_{FW} + \vec{v}_{WS} = \vec{v}_{YS}$$

Figure 4.3: Finding your velocity relative to the shore by vector addition after adding a current in the river.

$$\vec{v}_{YS} = \vec{v}_{YF} + \vec{v}_{FW} + \vec{v}_{WS} = 3.0 \text{ m/s south} + 7.0 \text{ m/s north} + 2.0 \text{ m/s south}.$$

$$\vec{v}_{YS} = -3.0 \text{ m/s north} + 7.0 \text{ m/s north} - 2.0 \text{ m/s north} = +2.0 \text{ m/s north}.$$

Key idea for relative velocity: Solving a relative velocity problem amounts to solving a vector-addition problem. In general, the velocity of an object A relative to an object C is the vector sum of the velocity of A relative to B plus the velocity of B relative to C: $\vec{v}_{AC} = \vec{v}_{AB} + \vec{v}_{BC}$.

Related End-of-Chapter Exercises: 1, 5.

Essential Question 4.1 In Step 3 of Exploration 4.1, could you adjust your speed with respect to the ferry so that you are at rest with respect to the shore? If so, how would you adjust it?

Answer to Essential Question 4.1 Yes. Adding an additional velocity of 2.0 m/s directed south will cancel the velocity of 2.0 m/s north that you are moving with respect to the shore. You were already walking at 3.0 m/s south with respect to the ferry, so you now need to be moving at 5.0 m/s south (so you need to run now) with respect to the ferry.

4-2 Combining Relative Velocity and Motion

Let's now connect relative velocity with the one-dimensional motion situations that we considered in Chapter 2.

EXPLORATION 4.2 – Who's faster?
On their way to play soccer in the World Cup, Mia and Brandi get stranded at O'Hare Airport in Chicago because of bad weather. Late at night, with nobody else around, they decide to have a race to see who is faster. They run down to a particular point, then turn around and return to their starting point. That race turns out to be a tie.

They race again, this time with Mia running on a moving sidewalk. They start at the same time, run east at top speed a distance L through the airport terminal, and then turn around and run west back to the starting point. Brandi runs at a constant speed v, while Mia runs on a moving sidewalk that travels at a speed of $v/2$. Mia runs at a constant speed v relative to the moving sidewalk. Neglect the time it takes the women to turn around at the halfway point. A diagram is shown in Figure 4.4 to help with the analysis.

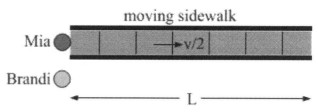

Figure 4.4: A diagram showing the positions of Mia and Brandi at the start of the race.

Step 1 - *Make a prediction – who wins this race?* Most people predict that this race is also a tie, because Mia is helped by the moving sidewalk for half the distance, and she has to work against the moving sidewalk for the remainder of the race.

Step 2 - *If Brandi takes a time T to reach the turn-around point, how long does Mia take to reach the same point?* First, let's do a relative-velocity analysis for Mia as she runs east, treating east as the positive direction. Figure 4.5 shows the vectors being added together. Using subscripts of M for Mia, S for sidewalk, and G for ground, we get:

$$\vec{v}_{MS} + \vec{v}_{SG} = \vec{v}_{MG}$$

Figure 4.5: A vector diagram showing Mia's velocity relative to the ground as she runs east.

$$\vec{v}_{MG} = \vec{v}_{MS} + \vec{v}_{SG} = +v + \frac{v}{2} = +\frac{3v}{2}$$

This relative-velocity situation is really a one-dimensional motion problem in disguise. Let's analyze the motion as the two women are moving away from the start line, choosing the origin as the start line and the positive direction to the right, in the direction the women are running. Let's summarize what we know in Table 4.1.

	Mia	Brandi
Initial position	$x_{iM} = 0$	$x_{iB} = 0$
Final position	$x_{fM} = +L$	$x_{fB} = +L$
Initial velocity	$v_{iM} = +3v/2$	$v_{iB} = +v$
Acceleration	$a_M = 0$	$a_B = 0$
Time	$t_M = ?$	$t_B = T$

Table 4.1: Organizing the data for the outbound trips.

Let's analyze Brandi's motion to see how the time T relates to the distance L and the speed v. We can use the following constant-acceleration equation, which we used previously in Section 2-5, to relate these parameters:

$$x_{fB} = x_{iB} + v_{iB}t_B + \frac{1}{2}a_B t_B^2 .$$

Substituting appropriate values from Table 4.1 gives: $L = 0 + vT + 0$, and thus $T = L/v$.

Using a similar analysis for Mia, we start with: $x_{fM} = x_{iM} + v_{iM}t_M + \frac{1}{2}a_M t_M^2 .$

Substituting appropriate values from Table 4.1, we get:

$$L = 0 + \frac{3}{2}vt_M + 0, \text{ which gives } t_M = \frac{2L}{3v} = \frac{2}{3}T.$$

So Mia has a sizable lead by the time she reaches the turn-around point.

Step 3 - *Brandi takes an equal time T for the return trip from the turn-around point to the start/finish line. How long does Mia's return trip take?* The vector addition in this case is shown in Figure 4.6. For Mia's return trip, her velocity relative to the ground is:

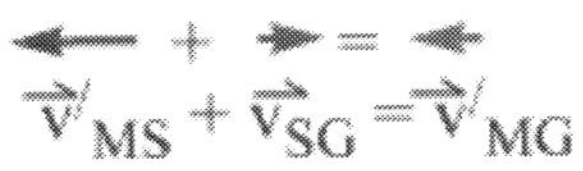

$$\vec{v}_{MG}' = \vec{v}_{MS}' + \vec{v}_{SG} = -v + \frac{v}{2} = -\frac{v}{2} .$$

$$\overleftarrow{\vec{v}_{MS}} + \overrightarrow{\vec{v}_{SG}} = \overleftarrow{\vec{v}_{MG}}$$

Figure 4.6: A vector diagram showing Mia's velocity relative to the ground as she runs west.

The primed ($'$) values represent the values of these variables on the return trip.

The data table for the return trips is shown in Table 4.2. For Brandi, the return trip is the same as the outbound trip, so there is no need to repeat that analysis. Let's solve for Mia's time for the return trip:

$$x_{fM}' = x_{iM}' + v_{iM}' t_M + \frac{1}{2}a_M' (t_M')^2 .$$

Substituting the values from Table 4.2 gives:

$$0 = +L - \frac{1}{2}vt_M' + 0, \text{ which gives } t_M' = \frac{2L}{v} = 2T.$$

	Mia	Brandi
Initial position	$x_{iM}' = +L$	$x_{iB}' = +L$
Final position	$x_{fM}' = 0$	$x_{fB}' = 0$
Initial velocity	$v_{iM}' = -v/2$	$v_{iB}' = -v$
Acceleration	$a_M' = 0$	$a_B' = 0$
Time	$t_M' = ?$	$t_B' = T$

Table 4.2: Organizing the data for the return trips.

Step 4 - *Based on your answers to steps 2 and 3, who wins the race?* Brandi's time for the entire trip is $2T$, the same time Mia takes just to come back along the moving sidewalk. Mia's total time is $2T + 2T/3$, so Brandi wins this race quite easily.

Key ideas: The methods we used to analyze one-dimensional motion situations in Chapter 2 can be combined with relative-velocity problems. **Related End-of-Chapter Exercises: 2, 31, 32.**

Essential Question 4.2 In Exploration 4.2, we concluded that Mia is at a disadvantage in the race when she runs on the moving sidewalk. What is a good conceptual explanation for this disadvantage?

Answer to Essential Question 4.2 Running on the moving sidewalk is a disadvantage for Mia because she spends considerably more time running against the moving sidewalk than she does running with it.

4-3 Relative Velocity in Two Dimensions

Let's modify Exploration 4.1, about your motion on a ferry, to see how things change when we switch to two dimensions. The major difference between one and two dimensions is that it is more challenging to add vectors in two dimensions than in one dimension; however, we simply need to follow what we learned about vector addition in Chapter 1.

EXPLORATION 4.3 – Crossing a river
You are crossing a river on a ferry. The ferry is pointing north, and it is traveling at a constant speed of 7.0 m/s relative to the water. The current in the river is 2.0 m/s directed southeast.

Step 1 - *What is the ferry's velocity relative to the shore?* The first step is to draw a vector diagram, as in Figure 4.7, showing how the relevant vectors combine to produce the velocity we're interested in.

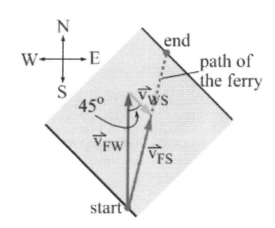

We can apply the standard relative-velocity equation, using subscripts of F for the ferry, W for the water, and S for the shore, to get:

$$\vec{v}_{FS} = \vec{v}_{FW} + \vec{v}_{WS} \, .$$

This equation is consistent with the vector diagram in Figure 4.7. Because the direction of the current is specified as southeast, we can use a 45° angle for that vector.

Here we have a typical vector-addition problem in two dimensions. One method of solving the problem is to break the vectors into components and add the components. This method involves separating the two-dimensional problem into two different one-dimensional problems, which we do quite often in physics.

Figure 4.7: Vector diagram to find the ferry's velocity relative to the shore. We add the velocity of the ferry relative to the water to the velocity of the water relative to the shore to find the velocity of the ferry relative to the shore.

As we learned in Chapter 1, we can put together a table to help us keep the x and y components separate. Doing so makes it easy to add the vectors using the component method. Define the positive x direction as east, and the positive y direction as north.

Vector	x components	y components
$\vec{v}_{FW}$	$v_{FWx} = 0$	$v_{FWy} = +7.00$ m/s
$\vec{v}_{WS}$	$v_{WSx} = +(2.00 \text{ m/s})\cos\left(45°\right) = +1.41$ m/s	$v_{WSy} = -(2.00 \text{ m/s})\sin\left(45°\right) = -1.41$ m/s
$\vec{v}_{FS} = \vec{v}_{FW} + \vec{v}_{WS}$	$v_{FSx} = v_{FWx} + v_{WSx}$ $v_{FSx} = 0 + 1.41 \text{ m/s} = +1.41$ m/s	$v_{FSy} = v_{FWy} + v_{WSy}$ $v_{FSy} = +7.00 \text{ m/s} - 1.41 \text{ m/s} = +5.59$ m/s

Table 4.3: Adding the vectors using components.

We can leave the answer in terms of components, as is done at the bottom of Table 4.3, or we can specify the velocity of the ferry with respect to the shore in terms of its magnitude and direction. To specify the velocity vector in the magnitude-direction format, start by drawing the right-angled triangle corresponding to the vector and its components, as in Figure 4.8.

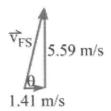

The magnitude of the vector can be found using the Pythagorean theorem:

$$v_{FS} = \sqrt{1.41^2 + 5.59^2} = 5.8 \text{ m/s}.$$

Figure 4.8: A vector diagram showing how the components add to give the net velocity vector.

The angle can be found using $\tan\theta = \dfrac{5.59}{1.41}$, so $\theta = 76°$.

This angle is measured with respect to east. The ferry is not traveling due east, but rather somewhat north of east. Thus, we can say that the ferry's velocity relative to the shore is 5.8 m/s at an angle of 76° north of east (or, equivalently, 14° east of north).

Step 2 - *You are standing on an observation deck at the top of the ferry, leaning against the railing on the starboard (right) side of the ferry. Suddenly someone yells from the port (left) side that there are some porpoises frolicking in the water. To get from one side of the ferry to the other in the shortest time to see the porpoises, in which direction should you run? Choose the best answer below.*

1. Run directly across the ferry, taking the shortest route to the other side (this means you will be carried along with the ferry, and the current).
2. Run partly toward the bow (front) of the ferry, so the ferry's velocity adds to your velocity and you travel at a higher speed.
3. Run partly toward the stern (rear) of the ferry, so the distance you travel relative to the shore is minimized.

If the ferry is docked and you want to cross from one side of the ferry to the other in the least time, you simply head straight across the ferry. If the ferry is moving, you do the same thing! The fact that the ferry is moving does not even cross your mind, which is good, because that is irrelevant. The reason the ferry's motion does not matter is that the ferry's motion is perpendicular to the direction you want to go, and a velocity in one direction does not affect motion in a perpendicular direction. The time for you to cross from one side of the ferry to the other is determined by how much of your velocity is directed across the ferry. If you aim yourself entirely in that direction, you minimize the time it takes to cover a particular distance in that direction.

Key ideas for relative velocity in two dimensions: The relative velocity equation, $\vec{v}_{AC} = \vec{v}_{AB} + \vec{v}_{BC}$, applies just as well in two or three dimensions as it does in one dimension. To solve the equation, we can apply the vector-addition methods, such as the component method, we covered in Chapter 1. Another important concept is that motion in two dimensions can be split into two independent one-dimensional motions. **Related End-of-Chapter Exercises: 7, 8.**

Essential Question 4.3 A pilot has aimed her plane north, and her airspeed indicator reads 150 km/h. The control tower reports that the wind is directed west at 50 km/h. Explain why the plane's speed relative to the ground is greater than 150 km/h, but less than 200 km/h.

Answer to Essential Question 4.3 The plane's velocity relative to the ground is the vector sum of the plane's velocity relative to the air (150 km/h north) and the air's velocity relative to the ground (50 km/h west). This sum is less than 200 km/h in a particular direction because the two vectors are not in the same direction. The plane's speed relative to the ground is 158 km/h, the length of the hypotenuse of the right-angled triangle we get by adding the vectors.

4-4 Projectile Motion

Projectile motion is, in general, two-dimensional motion that results from an object with an initial velocity in one direction experiencing a constant force in a different direction. A good example is a ball you throw to a friend. You give the ball an initial velocity when you throw it, and then the force of gravity acts on the ball as it travels to your friend. In this section, we will learn how to analyze this kind of situation.

EXPLORATION 4.4 – A race
You release one ball (ball *A*) from rest at the same time you throw another ball (ball *B*), which you release with an initial velocity that is directed entirely horizontally. You release both balls simultaneously from the same height *h* above level ground. Neglect air resistance.

Step 1 - *Which ball travels a greater distance before hitting the ground?* Ball *A* takes the shortest path to the ground, so ball *B* travels farther.

Step 2 - *Which ball reaches the ground first? Why?* We can construct a motion diagram (see Figure 4.9) by, for instance, analyzing a video of the balls as they fall. Many people think that because ball *B* travels farther it takes longer to reach the ground; however, ball *B* also has a higher speed. The reality is that both balls reach the ground at the same time. The reason is that the motion of ball *B* can be viewed as a combination of its horizontal motion and its vertical motion. The horizontal motion has no effect whatsoever on the vertical motion, so what happens vertically for ball *B* is exactly the same as what happens vertically for ball *A*.

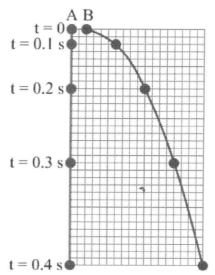

Figure 4.9: A motion diagram can be constructed from experimental evidence, such as by analyzing a video of the balls as they fall.

Key idea for projectile motion: The key idea of this chapter is the independence of *x* and *y*. The basic idea is that the motion that happens in one direction (*x*) is independent of the motion that happens in a perpendicular direction (*y*), and vice versa, as long as the force is constant.
Related End-of-Chapter Exercises: 9, 10.

The *x*-direction and *y*-direction motions are independent in the sense that each of the one-dimensional motions occurs as if the other motion is not happening. These motions are connected, though. The object's motion generally stops after a particular time, so the time is the same for the *x*-direction motion and the *y*-direction motion.

This powerful concept allows us to treat a two-dimensional projectile motion problem as two separate one-dimensional problems. We already have a good deal of experience with one-dimensional motion, so we can build on what we learned in Chapter 2. For the most part, we will deal with situations where the acceleration is constant, so all our experience with constant-acceleration situations in one dimension will be directly relevant here.

Solving a Two-Dimensional Constant-Acceleration Problem

Our general method for analyzing a typical projectile-motion problem builds on the method we used for analyzing one-dimensional constant-acceleration motion in Chapter 2. The basic idea is to split the two-dimensional problem into two one-dimensional subproblems, which we can call the x subproblem and the y subproblem.

1. Draw a diagram of the situation.
2. Draw a free-body diagram of the object showing all the forces acting on the object while it is in motion. A free-body diagram helps in determining the acceleration of the object.
3. Choose an origin.
4. Choose an x-y coordinate system, showing which way is positive for each coordinate axis.
5. Organize your data, keeping the information for the x subproblem separate from the information for the y subproblem.
6. Only then should you turn to the constant-acceleration equations. Make sure the acceleration is constant so the equations apply! We use the same three equations that we used in Chapter 2, but we customize them for the x and y subproblems, as follows:

Equation from Chapter 2	x-direction equations	y-direction equations
$v = v_i + at$ (2.7)	$v_x = v_{ix} + a_x t$ (4.2x)	$v_y = v_{iy} + a_y t$ (4.2y)
$x = x_i + v_i t + \frac{1}{2}at^2$ (2.9)	$x = x_i + v_{ix}t + \frac{1}{2}a_x t^2$ (4.3x)	$y = y_i + v_{iy}t + \frac{1}{2}a_y t^2$ (4.3y)
$v^2 = v_i^2 + 2a\Delta x$ (2.10)	$v_x^2 = v_{ix}^2 + 2a_x \Delta x$ (4.4x)	$v_y^2 = v_{iy}^2 + 2a_y \Delta y$ (4.4y)

Table 4.4: Constant acceleration equations for two-dimensional projectile motion. The equation numbers are shown in parentheses after each equation.

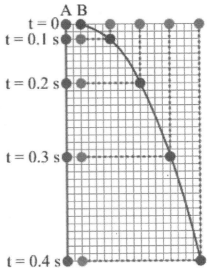

Let's apply the method above to analyze the race from Exploration 4.4. We begin by sketching a motion diagram and a free-body diagram for each ball, and continue the analysis in the next section. On the motion diagram for ball B, show the separate x (horizontal) and y (vertical) motions.

The motion diagrams are shown in Figure 4.10, while the free-body diagram of each ball is shown in Figure 4.11. Let's consider the motion from just after the balls are released until just before the balls make contact with the ground. Because the only force acting on either ball is the force of gravity, the same free-body diagram applies to both objects.

> **Figure 4.10**: Motion diagram for balls A and B. For ball B, the vertical and horizontal motions are shown separately. These two independent motions combine to give the parabolic path followed by ball B.

Essential Question 4.4 The free-body diagrams in Figure 4.11 imply that the balls have the same mass. What would happen if the balls had different masses?

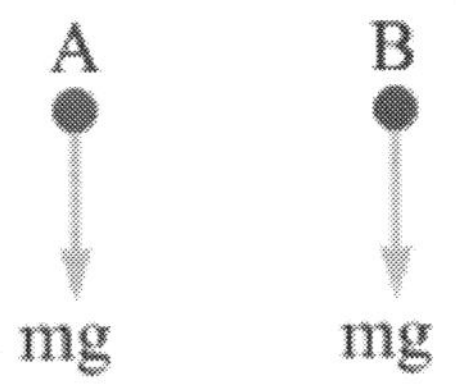

> **Figure 4.11**: Free-body diagrams for balls A and B. From the instant just after you release the balls until the instant just before the balls hit the ground, the only force acting on either ball is the force of gravity, so the balls have identical free-body diagrams. The balls travel along different paths only because their initial velocities are different.

Answer to Essential Question 4.4 Assuming that we can neglect air resistance, the relative mass of the balls is completely irrelevant. If *B*'s mass was double *A*'s mass, for instance, the force of gravity on *B* would be twice that on *A*, but both balls would still have an acceleration of $\vec{g}$, and the two balls would still hit the ground simultaneously.

4-5 The Independence of x and y

A key to understanding projectile motion is the independence of *x* and *y*, the fact that the horizontal (*x*-direction) motion is completely independent of the vertical (*y*-direction) motion. Let's exploit this concept to continue our analysis of the race from Exploration 4.4.

EXPLORATION 4.5 – Analyzing the race

Step 1 – *Find the acceleration of each ball.* The free-body diagram in Figure 4.11, combined with Newton's second law, tells us that the acceleration of each object is simply the acceleration due to gravity, $\vec{g}$. This comes from:

$$\vec{a} = \frac{\Sigma \vec{F}}{m} = \frac{m\vec{g}}{m} = \vec{g}.$$

Step 2 – *Use the general method to find the time it takes the balls to reach the ground.* Because the motion is directed right and down, let's choose positive directions as +*x* to the right and +*y* down, and set the origin for each ball to be the point from which it is released. Choosing up as positive, with an origin at ground level, would also work well.

Let's say both balls fall through a vertical distance of *h*, and that the initial velocity of ball B is directed horizontally with a velocity of v_i. Table 4.5 shows how we organize the data. Note how the data for the *x*-direction (horizontal) motion for ball *B* are kept separate from the data for the *y*-direction (vertical) motion.

Component	Ball *B*, x direction	Ball *B*, y direction	Ball *A*, y direction
Initial position	$x_{iB} = 0$	$y_{iB} = 0$	$y_{iA} = 0$
Final position	$x_B = ?$	$y_B = +h$	$y_A = +h$
Initial velocity	$v_{ixB} = +v_i$	$v_{iyB} = 0$	$v_{iyA} = 0$
Final velocity	$v_{xB} = +v_i$	$v_{yB} = ?$	$v_{yA} = ?$
Acceleration	$a_{xB} = 0$	$a_{yB} = +g$	$a_{yA} = +g$

Table 4.5: Organizing the data for ball *A* (dropped from rest) and ball *B* (with an initial velocity that is horizontal). Note that everything is the same for the two balls in the *y*-direction, which is vertical.

One of the most common errors in analyzing a projectile-motion situation is to mix up information from the *x* and *y* directions, such as by using the acceleration due to gravity as the acceleration in the horizontal direction. Organizing the data into a table like the one above makes such errors far less likely. Including the appropriate sign on all vectors is another way to reduce errors, because it reminds us to think about which sign is correct and whether we really want a + or a –. A statement like $y_B = +h$ tells us that the final vertical position of ball *B* is a distance *h* from the origin in the positive *y*-direction.

Can we use the data from Table 4.5 to justify the conclusion from Exploration 4.4 that the two balls reach the ground at the same time? Absolutely. The appropriate motion diagrams are shown in Figure 4.12. Analyzing the y-direction subproblem for ball B, we can use equation 4.3y to find an expression for the time to reach the ground.

$$\vec{y} = \vec{y}_i + \vec{v}_{iy}t + \frac{1}{2}\vec{a}_y t^2,$$

$$+h = 0 + 0 + \frac{1}{2}gt^2.$$

This gives $t^2 = +\dfrac{2h}{g}$.

Therefore, the time for ball B to reach the ground is $t = \sqrt{\dfrac{2h}{g}}$.

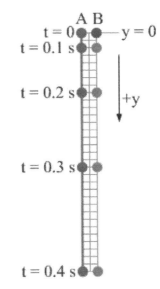

Figure 4.12: A motion diagram for the vertical components of the motion for the balls.

Does the answer make any sense? First, it does have the right units. Second, it says that if we increase h the ball takes longer to reach the ground, which makes sense. Third, it says that the larger the acceleration due to gravity the smaller the time the ball takes to reach the ground, which also sounds right. Note that if we solve for the time ball A takes to reach the ground we get exactly the same result, because A has the same initial position, final position, initial vertical velocity, and vertical acceleration as B.

Step 3 – *Find an expression for the horizontal distance traveled by ball B before it reaches the ground.* Even though we're dealing with the x subproblem, we can use the time from the y subproblem – that is often key to solving projectile motion problems. The motion diagram for the x-direction motion is shown in Figure 4.13. One way to find the horizontal distance that ball B travels is to use Equation 4.3x (see Table 4.4 in Section 4.4 for the equations).

$$x = x_i + v_{ix}t + \frac{1}{2}a_x t^2,$$

$$x = 0 + v_i t + 0 = +v_i\sqrt{\frac{2h}{g}}.$$

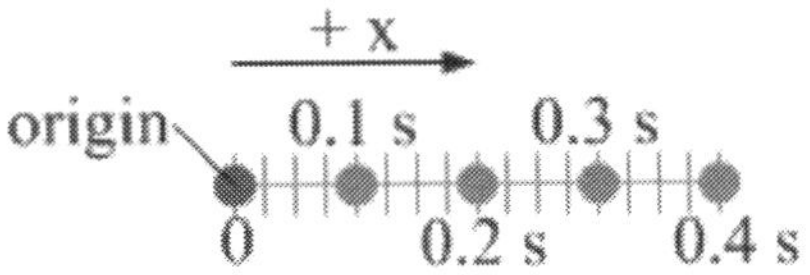

Figure 4.13: A motion diagram for the horizontal component of ball B's motion.

Again, be careful not to mix the x information with the y information. Here, for instance, we can use Table 4.5 to remind us that the acceleration in the x direction is zero. The motion diagram in Figure 4.13, showing constant-velocity motion, confirms that the acceleration in the x direction is zero.

Key idea for projectile motion: One way to solve a projectile-motion problem is to break the two-dimensional problem into two independent one-dimensional problems, ***linked by the time***, and apply the one-dimensional constant-acceleration methods from Chapter 2.
Related End-of-Chapter Exercises: 44, 45.

Essential Question 4.5 When a sailboat is at rest, a beanbag you release from the top of the mast lands in a bucket that is on the deck at the base of the mast. Will the beanbag still land in the bucket if you release the beanbag from rest when the sailboat is moving with a constant velocity?

Answer to Essential Question 4.5 Yes. From your perspective, moving with the sailboat at a constant velocity, the beanbag still drops straight down from rest. From the perspective of someone at rest on shore, the beanbag follows a parabolic motion. Its horizontal motion keeps it over the bucket at all times, though, so it still falls into the bucket because the bucket, beanbag, and boat all have the same horizontal velocity. We're neglecting air resistance here, by the way.

4-6 An Example of Projectile Motion

When a soccer goalkeeper comes off the goal line to challenge a shooter, the shooter can score by kicking the ball so that it goes directly over the goalkeeper, and then comes down in time to either bounce into, or fly into, the net. This is known as chipping the goaltender, because the shooter chips the ball to produce the desired effect. Let's analyze this situation in detail.

EXAMPLE 4.6 – Chipping the goaltender

Precisely 1.00 s after you kick it from ground level, a soccer ball passes just above the outstretched hands of a goaltender who is 5.00 m away from you, and lands on the goal line before bouncing into the net for the winning goal in a soccer match. The goaltender's fingertips are 2.50 m above the ground. Neglect air resistance, and use $g = 9.81$ m/s^2.

   (a) At what angle did you launch the ball?
   (b) What was the ball's initial speed?
   (c) Assuming the ground is completely level, how long does the ball spend in the air?

SOLUTION

As usual, we should be methodical. A diagram of the situation is shown in Figure 4.14. The diagram includes an appropriate coordinate system, with the origin at the launch point. Table 4.6 summarizes what we know, separated into x and y components.

Take the direction of the horizontal component of the ball's motion to be the positive x direction, and take up to be the positive y direction.

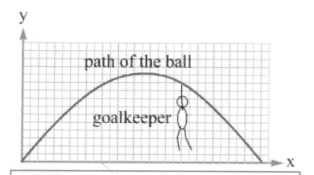

Figure 4.14: The flight of the soccer ball. The squares on the grid measure 0.25 m × 0.25 m.

Component	x direction	y direction
Initial position	$x_i = 0$	$y_i = 0$
Final position	$x = ?$	$y = 0$
Initial velocity	$v_{ix} = ?$	$v_{iy} = ?$
Final velocity	$v_x = v_{ix}$	$v_y = ?$
Acceleration	$a_x = 0$	$a_y = -g$
At $t = 1.00$ s	$x_{t=1} = +5.00$ m	$y_{t=1} = +2.50$ m

Table 4.6: Organizing the data for the problem.

(a) *At what angle did you launch the ball?* It is tempting to draw a right-angled triangle with a base of 5.00 m and a height of 2.50 m and take the angle between the base and the hypotenuse to be the launch angle, but doing so is incorrect. The ball follows a parabolic path that curves down, not a straight path to just above the goaltender's fingertips. Thus, the launch angle is larger than the angle of that particular right-angled triangle.

The launch angle is the angle between the ball's initial velocity and the horizontal, so let's work on determining the initial velocity. We can use what we know about the ball at $t = 1.00$ s to help us. Once again, we will make use of the equations in Table 4.4.

To find the x component of the initial velocity, use equation 4.3x: $x_{t=1} = x_i + v_{ix}t + \frac{1}{2}a_x t^2$.

$$x_{t=1} = 0 + v_{ix}t + 0.$$

This gives $v_{ix} = \dfrac{x_{t=1}}{t} = \dfrac{+5.00 \text{ m}}{1.00 \text{ s}} = +5.00 \text{ m/s}$.

To find the y component of the initial velocity, use equation 4.3y: $y_{t=1} = y_i + v_{iy}t + \frac{1}{2}a_y t^2$.

$$y_{t=1} = 0 + v_{iy}t - \frac{1}{2}gt^2.$$

This gives $\vec{v}_{iy} = \dfrac{y_{t=1} + gt^2/2}{t} = \dfrac{+2.50 \text{ m} + 4.90 \text{ m}}{1.00 \text{ s}} = +7.40 \text{ m/s}$.

From the two components of the initial velocity, we can determine the launch angle θ and the initial speed. The geometry of the situation is shown in Figure 4.15.

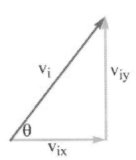

$$\tan\theta = \frac{v_{iy}}{v_{ix}} = \frac{7.40 \text{ m/s}}{5.00 \text{ m/s}}, \text{ so } \theta = 56.0°.$$

Figure 4.15: The right-angled triangle that we use to find the launch angle and launch speed of the ball.

(b) **What was the initial speed of the ball?** The launch speed is the magnitude of the initial velocity. Applying the Pythagorean theorem to the triangle in Figure 4.15 gives:

$$v_i = \sqrt{v_{ix}^2 + v_{iy}^2} = \sqrt{(5.00 \text{ m/s})^2 + (7.40 \text{ m/s})^2} = 8.93 \text{ m/s}.$$

(c) **Assuming the ground is completely level, how long does the ball spend in the air?** We should not assume that the ball passes over the goaltender when the ball is at its maximum height. Here is a good rule of thumb: If you do not have to assume something, don't assume it! Instead, let's make use of Equation 4.3y to determine the time for the entire flight:

$$y = y_i + v_{iy}t + \frac{1}{2}a_y t^2.$$

$$0 = 0 + v_{iy}t - \frac{1}{2}gt^2.$$

At first glance it looks as though we have to use the quadratic formula to solve this equation, but we can simply divide through by a factor of t. Doing so gives:

$$0 = +v_{iy} - \frac{1}{2}gt.$$

Solving for t gives $t = \dfrac{2v_{iy}}{g} = \dfrac{2 \times 7.405 \text{ m/s}}{9.81 \text{ m/s}^2} = \dfrac{14.81 \text{ m/s}}{9.81 \text{ m/s}^2} = 1.51 \text{ s}.$

Related End-of-Chapter Exercises: 13, 17, 18.

Essential Question 4.6 Consider again the ball from Example 4.6. How does the time it takes the ball to reach maximum height compare to the time for the entire flight? Explain.

Answer to Essential Question 4.6 The result we found in Example 4.6, that the total time for the flight is $t_{impact} = 2v_{iy}/g$, is a special case that applies only when the projectile lands at the same height from which it was launched. In this special case, the time for the entire flight is twice the time to reach maximum height (thus, $t_{max\ height} = v_{iy}/g$). This result comes about because of the symmetry of the motion. The second half of the motion, when the projectile is coming down, is a mirror image of the first half of the motion, when the projectile is going up. These two halves take exactly the same time (neglecting air resistance). The projectile's speed on the way up at a particular height is equal to its speed on the way down as it passes through that same height.

4-7 Graphs for Projectile Motion

Graphs can give us a lot of information about a particular motion. Let's consider how to draw a set of graphs for the motion of the soccer ball in Example 4.6.

EXAMPLE 4.7 – Graphs for the ball chipped over the goalkeeper
Draw graphs showing, as a function of time, the *x*-component of position, *x* component of velocity, and *x* component of acceleration for the soccer ball in Example 4.6. Take *x* to be horizontal, with the positive *x* direction pointing from the launch point toward the goalkeeper and the net. Draw another set of graphs for the *y* component, taking the positive *y* direction to be up.

SOLUTION
A graph can be thought of as a picture of an equation. To graph these various parameters as a function of time, it can be helpful to write down the corresponding equation. Graphs for the *x* components are shown in Figure 4.16.

To graph the *x* component of the position as a function of time, take what we know about the ball and substitute these known quantities into Equation 4.3x (see Table 4.4 in Section 4.4):

$$x = x_i + v_{ix}t + \frac{1}{2}a_x t^2$$

$$x = 0 + (+5.00\ m/s)t + 0 .$$

$$x = +(5.00\ m/s)t$$

This is a linear function starting at the origin and having a slope of $+5.00\ m/s$.

To graph the *x* component of the velocity as a function of time, take what we know about the ball and fill in equation 4.2x:

$$v_x = v_{ix} + a_x t$$

$$v_x = +5.00\ m/s + 0 .$$

$$v_x = +5.00\ m/s$$

The *x* component of the velocity is constant, so the graph of this velocity component is a horizontal line, as shown in the middle graph in Figure 4.16.

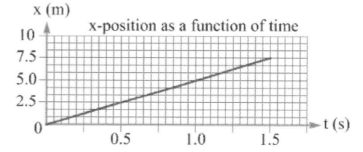

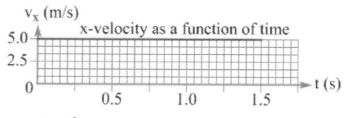

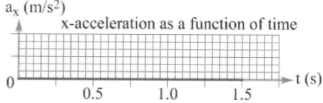

Figure 4.16: Graphs of the horizontal components of the ball's position, velocity, and acceleration.

The graph of the acceleration in the x direction is also a horizontal line, being drawn along the time axis, because there is no acceleration in the x direction.

In the y direction, we can do something similar. The three y-component graphs are drawn in Figure 4.17. To graph the y component of the position as a function of time, take what we know about the ball and fill in equation 4.3y:

$$y = y_i + v_{iy}t + \frac{1}{2}a_y t^2$$

$$y = 0 + (7.40 \text{ m/s})t - (4.90 \text{ m/s}^2)t^2.$$

The graph representing this equation is a parabola curving downward. Note the symmetry of the graph, with the second half of the motion being a mirror-image of the first half. Equation 4.4y can be used to find the maximum height the ball rises above the ground, knowing that at maximum height the y component of the velocity is zero. From Equation 4.4y we get:

$$v_y^2 = v_{iy}^2 + 2a_y \Delta y,$$

which gives $\Delta y = \dfrac{v_y^2 - v_{iy}^2}{2a_y} = \dfrac{0 - (7.40 \text{ m/s})^2}{-2(9.81 \text{ m/s}^2)} = \dfrac{54.76 \text{ m}^2/\text{s}^2}{19.62 \text{ m/s}^2} = 2.79 \text{ m}.$

It is helpful to know the value of the maximum height when we graph the y component of the position.

To graph the y component of the velocity as a function of time, we take what we know about the ball and fill in Equation 4.2y:

$$v_y = v_{iy} + a_y t = +(7.40 \text{ m/s}) - (9.81 \text{ m/s}^2)t.$$

A graph of the y component of the velocity is a straight line with a y-intercept of +7.40 m/s and a slope of –9.81 m/s^2. Note that we can tell when the ball reaches maximum height from both the y-position graph and the y-velocity graph.

The y component of the acceleration is constant at –9.81 m/s^2, so that graph is a horizontal line.

Related End-of-Chapter Exercises: 47 – 49.

Essential Question 4.7 Compare the two sets of graphs in Example 4.7 to the graphs for motion with constant velocity, and motion with constant acceleration, that we drew in Chapter 2. What similarities are there?

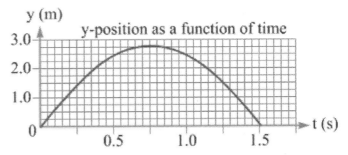

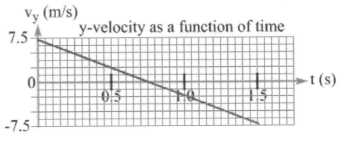

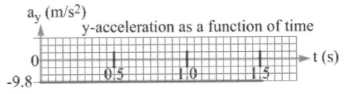

Figure 4.17: Graphs of the vertical components of the ball's position, velocity, and acceleration.

Answer to Essential Question 4.7 Because the ball has constant velocity in the *x* direction, the graphs for the *x* components of the ball's motion match the constant-velocity graphs from Chapter 2. Similarly, because the ball has constant acceleration in the *y* direction, the graphs for the *y* components of the ball's motion match the constant-acceleration graphs from Chapter 2.

4-8 Range, and One Final Example

The **range** of a projectile is the horizontal distance it travels before striking the ground. If a projectile lands at the same height from which it was launched, what launch angle gives the maximum range if the initial speed of the projectile is constant? Let's answer this question with the aid of the answer to Essential Question 4.6, that the time-of-flight when a projectile lands at the height from which it was launched is $t_{impact} = 2v_{iy}/g$. Substituting this time into equation 4.3x, taking the origin to be the launch point and recalling that the acceleration is zero, gives:

$$\vec{x} = \vec{x}_i + \vec{v}_{ix}t + \frac{1}{2}\vec{a}_x t^2 = 0 + \vec{v}_{ix}t + 0 = \vec{v}_{ix}t .$$

Therefore, $\text{range} = v_{ix}\, t_{impact} = \dfrac{2v_{ix}v_{iy}}{g}$.

If the launch angle θ is measured from the horizontal, we have $v_{ix} = v_i \cos\theta$ and $v_{iy} = v_i \sin\theta$. Substituting these expressions into the range equation above gives:

$$\text{range} = \frac{2v_{ix}v_{iy}}{g} = \frac{2(v_i\cos\theta)(v_i\sin\theta)}{g} = \frac{2v_i^2 \sin\theta\, \cos\theta}{g} .$$

This can be simplified with the trigonometric identity $2\sin\theta\, \cos\theta = \sin(2\theta)$:

$$\text{range} = \frac{v_i^2 \sin(2\theta)}{g} .$$

This equation applies only when the projectile lands at the height from which it was launched.

So, the range is proportional to the square of the initial speed, is inversely proportional to *g*; and depends on the launch angle in an interesting way. A graph of $\sin(2\theta)$ is shown in Figure 4.18. If the launch angle θ is between 0 and 90°, then 2θ is between 0 and 180°. Keeping v_i and *g* constant, the maximum range is achieved when $\sin(2\theta)$ is maximized. This occurs when $2\theta = 90°$, so $\theta = 45°$ is the launch angle that gives the maximum range when the projectile lands at the same height from which it was launched.

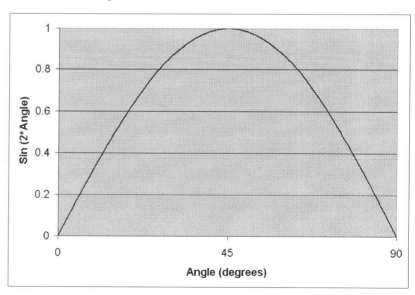

Figure 4.18: A graph of $\sin(2\theta)$ as a function of θ.

Related End-of-Chapter Exercises: 57 – 62.

EXAMPLE 4.8 – A corner kick
A corner kick, in which a player directs the ball into a crowd of players in front of the net, is one of the most exciting plays in soccer. You are taking the kick, and you want the ball, while it is on the way down, to be precisely 2.00 m above the ground when it reaches your teammate, who will attempt to head the ball into the net. Your teammate is 30.0 m away, and you kick the ball from ground level at an angle of 25.0° with respect to the horizontal. What initial speed should you give the ball? Use $g = 9.81$ m/s². (a) Sketch a diagram of the situation, and organize what you know in a data table, keeping the x (horizontal) information separate from the y (vertical) information. (b) Use the x information to find an expression for the time it takes the ball to reach your teammate. (c) Use the y information, and the time, to find the ball's initial speed.

SOLUTION (a) Define the origin as the launch point, and the positive directions so that toward your teammate is the positive x direction, and up is the positive y direction. Figure 4.19 shows a diagram of the situation, and the data is given in Table 4.7.

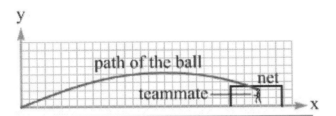

Figure 4.19: The flight of the soccer ball. The squares on the grid measure 1 m × 1 m.

Component	x-direction	y-direction
Initial position	$x_i = 0$	$y_i = 0$
Final position	$x_f = +30.0$ m	$y_f = +2.00$ m
Initial velocity	$v_{ix} = +v_i \cos(25°)$	$v_{iy} = +v_i \sin(25°)$
Acceleration	$a_x = 0$	$a_y = -9.81$ m/s²

Table 4.7: Organizing the data for the problem.

(b) Here, we can use Equation 4.3x, $x = x_i + v_{ix}t + (0.5)a_x t^2$, which reduces to $x = v_{ix}t$. Solving

for the time the ball takes to reach your teammate, we get: $t = \dfrac{x}{v_{ix}} = \dfrac{+30.0 \text{ m}}{v_i \cos(25°)}$.

(c) We can now substitute our expression for time into equation 4.3y:

$$y = y_i + v_{iy}t + \frac{1}{2}a_y t^2 : \quad +2.00 \text{ m} = v_i \sin(25°) \times \left(\frac{+30.0 \text{ m}}{v_i \cos(25°)} \right) + \frac{1}{2}(-9.81 \text{ m/s}^2) \times \left(\frac{+30.0 \text{ m}}{v_i \cos(25°)} \right)^2 .$$

Simplifying, we get: $+2.00 \text{ m} = +13.989 \text{ m} - \dfrac{5374.40 \text{ m}^3/\text{s}^2}{v_i^2}$.

This gives $+\dfrac{5374.40 \text{ m}^3/\text{s}^2}{v_i^2} = +11.989 \text{ m}$, so $v_i = \sqrt{\dfrac{5374.40 \text{ m}^3/\text{s}^2}{11.989 \text{ m}}} = 21.2$ m/s.

This is fast, but not unusual for a typical soccer game.

Related End-of-Chapter Exercises: 19, 28, 63.

Essential Question 4.8 Return again to the situation described in Example 4.8. Could we have found the answer more easily by applying the range equation from the previous page?

Answer to Essential Question 4.8 No. The range equation applies only in the special case when a projectile's initial and final heights are the same. In this case, the final position is 2.00 m higher than the initial position.

Chapter Summary

Essential Idea - The Independence of x and y

This powerful concept enables us to split a two-dimensional problem into two one-dimensional problems, in a situation like projectile motion when the force applied to the projectile is constant. The motion in one direction is completely unaffected by the motion in the perpendicular direction, except that the two motions share the same time.

Relative Velocity

A relative velocity problem is a vector addition problem. The velocity of A to C is the vector sum of the velocity of A relative to B plus the velocity of B relative to C:

$$\vec{v}_{AC} = \vec{v}_{AB} + \vec{v}_{BC} \qquad \text{(Equation 4.1)}$$

Rules of Thumb for Projectile Motion (motion under the influence of gravity alone)

- The motion is symmetric, with the downward part of the motion being a mirror image of the upward part of the motion.
- The larger the initial vertical velocity, the higher the projectile goes.
- The higher a projectile goes, the longer the time of flight.
- Range is maximized when the launch angle is 45°, as long as the projectile lands at the same height from which it was launched.

General Method for Solving a Two-Dimensional Motion Problem

The basic method is to split a two-dimensional motion problem into two one-dimensional subproblems, which we can call the x subproblem and the y subproblem.
1. Draw a diagram of the situation.
2. Draw a free-body diagram of the object showing all the forces acting on the object while it is in motion. A free-body diagram helps in determining the acceleration of the object.
3. Choose an origin.
4. Choose an x-y coordinate system, showing which way is positive for each direction.
5. Organize your data, keeping the information for the x subproblem separate from the information for the y subproblem.
6. As long as the acceleration is constant, apply the constant-acceleration equations. These equations are based on the equations from Chapter 2:

x equations		y equations	
$v_x = v_{ix} + a_x t$	(Equation 4.2x)	$v_y = v_{iy} + a_y t$	(Equation 4.2y)
$x = x_i + v_{ix}t + \dfrac{1}{2}a_x t^2$	(Equation 4.3x)	$y = y_i + v_{iy}t + \dfrac{1}{2}a_y t^2$	(Equation 4.3y)
$v_x^2 = v_{ix}^2 + 2a_x \Delta x$	(Equation 4.4x)	$v_y^2 = v_{iy}^2 + 2a_y \Delta y$	(Equation 4.4y)

End-of-Chapter Exercises

Exercises 1 – 12 are conceptual questions designed to see whether you understand the main concepts of the chapter.

1. You are going to paddle your kayak from one side of a river to the other side. The current in the river is directed downstream, as is usual for a river. How should you point your kayak (upstream, perpendicular to the riverbank, or downstream) so that you cross from one side of the river to (a) anywhere on the other side in the shortest possible time? (b) the point directly across the river from your starting point? Briefly justify your answers.

2. A ballistic cart is a wheeled cart that can launch a ball in a direction perpendicular to the way the cart moves and can then catch the ball again if it falls back down on the cart. Holding the cart stationary on a horizontal track, you confirm that the ball does indeed land in the cart after it is launched. You now give the cart a quick push so that, after you release it, the cart rolls along the track with a constant velocity. If the ball is launched while the cart is rolling, will it land in front of the cart, behind the cart, or in the cart? Briefly justify your answer.

3. On an assignment, you are asked to find the time it takes for a ball launched with a particular initial velocity to reach the surface of the water some distance below. When setting up the exercise, you choose the origin to be at the base of the cliff. Your friend chooses the origin to be at the top of the cliff, from where the ball was launched. Who is right, or is there no one right place to choose as the origin? What do you and your friend agree on? What do you disagree on? Explain.

4. On an assignment, you are asked to find the time it takes for a ball launched with a particular initial velocity to reach the surface of the water some distance below. When setting up the exercise, you choose an *x-y* coordinate system in which the positive *y* direction is up, while your friend chooses an *x-y* coordinate system in which the positive *y* direction is down. Who is right, or is there no one correct direction for the positive *y*-axis? What do you and your friend agree on? What do you disagree on? Explain.

5. Consider the trajectories of three objects, labeled *A*, *B*, and *C*, shown in Figure 4.20. Rank these objects from largest to smallest, based on (a) their times of flight, (b) the *y* component of their initial velocities, (c) the *x* component of their initial velocities, and (d) their launch speeds.

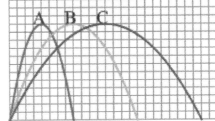

Figure 4.20: The trajectories of three projectiles, for Exercise 5.

6. The trajectories of two objects, *D* and *E*, are shown in Figure 4.21. The grid shown on the diagram is square. (a) Which object has the longer time of flight? Briefly explain. (b) If object *E* has a time of flight of *T*, what is object *D*'s time of flight? (c) If object *E* has a constant horizontal velocity of v_{ix} to the right, what is the constant horizontal velocity of object *D*?

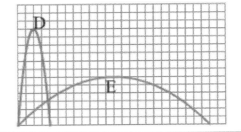

Figure 4.21: The trajectories of two projectiles, for Exercise 6.

7. The trajectories of three objects, *E*, *F*, and *G*, are shown in Figure 4.22. Assume that the grid shown on the diagram is square, and note that object *E* is launched at an angle of 45° with respect to the horizontal. Rank these three projectiles from largest to smallest, based on (a) their times of flight, (b) the *y* component of their initial velocities, (c) the *x* component of their initial velocities, and (d) their launch speeds.

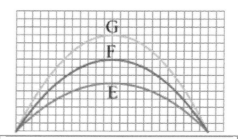

Figure 4.22: The trajectories of three projectiles, for Exercise 7.

8. The motion diagram in Figure 4.23 shows a ball's position at regular intervals as the ball flies from left to right through the air. Copy the motion diagram onto a sheet of graph paper. (a) On the same graph, sketch the motion diagram for a ball that also starts at the lower left corner, has the same initial vertical velocity as the ball in Figure 4.23, but has a horizontal velocity 50% larger than that of the ball in Figure 4.23. (b) On the same graph, sketch the motion diagram for a ball that has the same starting point, half the initial vertical velocity, and twice the horizontal velocity, of the ball in Figure 4.23.

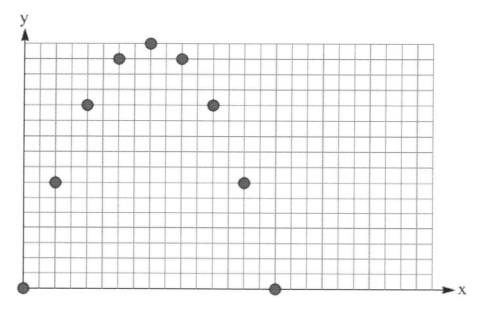

Figure 4.23: A motion diagram showing the position of a ball at regular time intervals, as the ball moves from left to right, for Exercises 8 and 9.

9. Consider the motion diagram shown in Figure 4.23. Assuming the ball's position is shown at 1-second intervals and that $g = 10$ m/s^2, sketch graphs of the ball's: (a) *y* velocity as a function of time, (b) *y* position as a function of time, (c) *x* velocity as a function of time, and (d) *x* position as a function of time.

10. To answer a typical projectile-motion problem, it is not necessary to know the mass of the projectile. Why is this?

11. You find that when you throw a ball with a particular initial velocity, the ball reaches a maximum height *H*, has a time of flight *T*, and covers a range *R* when landing at the same height from which it was launched. If you throw the ball again but now you double the launch speed, keeping everything else the same, what are the ball's (a) maximum height, (b) time of flight, and (c) range?

12. You find that when you throw a ball on Earth with a particular initial velocity, the ball reaches a maximum height H, has a time-of-flight T, and covers a range R when landing at the same height from which it was launched. You now travel to the Moon, where the acceleration due to gravity is one-sixth of what it is on Earth, and throw the ball with the same initial velocity. What are (a) the maximum height, (b) the time of flight, and (c) the range of the ball on the Moon?

Exercises 13 – 20 deal with relative-velocity situations.

13. You are driving your car on a highway, traveling at a constant velocity of 110 km/h north. Ahead of you on the road is a truck traveling at a constant velocity of 90 km/h north. On the other side of the road, coming toward you, is a motorcycle traveling at a constant velocity of 100 km/h south. (a) What is your velocity relative to the truck? (b) What is the truck's velocity relative to you? (c) What is your velocity relative to the motorcycle? (d) What is the motorcycle's velocity relative to the truck? (e) After you pass the truck and the motorcycle roars past on the other side of the road, do any of the answers above change? Explain.

14. You paddle your canoe at a constant speed of 4.0 km/h relative to the water. You are canoeing along a river that is flowing at a constant speed of 1.0 km/h. If you paddle for 60 minutes downstream (with the current) and then turn around, how long does it take you to get back to your starting point?

15. Return to the situation described in Exercise 14, and let's say that the river is 300 m wide. (a) If you paddle your canoe so you cross from one side of the river to the other in the shortest possible time, how long does it take to cross the river? (b) If instead you paddle your canoe so that you land at the point directly across the river from where you started, following a straight line, how long does it take you?

16. Return again to the situation described in Exercises 14 and 15. Compare what happens if, in case 1, you aim your canoe at an angle of 15° upstream to what happens if, in case 2, you aim your canoe 15° downstream. (a) In which case does it take you longer to cross the river? How much time does it take you to cross in that case? (b) How far upstream or downstream are you when you reach the far side in (i) case 1? (ii) case 2?

17. Four people are moving along a sidewalk. Ron's velocity relative to Susan is 2.0 m/s east; Susan's velocity relative to Tamika is 6.0 m/s west; and Tamika's velocity relative to Ulrich is 3.0 m/s east. What is Ulrich's velocity relative to (a) Susan? (b) Ron? (c) a lamppost at rest with respect to the sidewalk?

18. Return to the situation described Exercise 17, except now Ron is crossing the street and his velocity relative to Susan has components of 3.0 m/s north and 2.0 m/s east. Find Ulrich's velocity relative to Ron now, expressing it in (a) component form and (b) magnitude-and-direction form.

19. You are flying your airplane at a constant speed of 200 km/h relative to the air. As you pass over Zurich, Switzerland, you check your map and find that Frankfurt, Germany, is 300 km due north, so you point your plane due north. (a) If there is no wind blowing, how long should it take you to get to Frankfurt? (b) After that amount of time has passed you are shocked to find that, instead of being over Frankfurt, you are near the town of Tenneville, Belgium, precisely 225 km due west of Frankfurt. What is the velocity of the wind? (c) You now immediately change direction so you can fly to Frankfurt. In which direction should you point your plane, and how long does it take you to reach Frankfurt? (Let's hope you have plenty of fuel!)

20. Having learned from your experience of Exercise 19, the next time you fly from Zurich to Frankfurt you account for the wind, which is blowing west at a speed of 120 km/h. (a) If your plane travels at a constant speed of 200 km/h relative to the air, in which direction should you point your plane so that you travel due north, toward Frankfurt? (b) How long does it take you to fly from Zurich to Frankfurt this time?

Exercises 21 – 24 deal with the concept of the independence of x and y.

21. You climb 20 meters up the mast of a tall ship and release a ball from rest so it hits the deck of the ship at the base of the mast. Use $g = 10$ m/s^2 to simplify the calculations. (a) How long does it take the ball to reach the deck of the ship? (b) From the same point, 20 meters up the mast, you launch a second ball so that its initial velocity is 1.0 m/s directed horizontally and east. How long does this ball take to reach the deck of the ship? (c) How far is the second ball from the base of the mast when it hits the deck?

22. Returning to the situation described in Exercise 21, assume now that the ship is traveling with a constant velocity of 4.0 m/s east, and that the water is calm so the mast remains vertical. Once again you climb 20 meters up the mast and release one ball from rest, relative to you, and give a second ball an initial velocity of 1.0 m/s, directed east, relative to you. (a) Does the fact that the ship is moving affect the time it takes either ball to fall to the deck of the ship? Explain. (b) How far from the base of the mast does the first ball land now? (c) How far from the base of the mast does the second ball land now? (d) From the point of view of a seagull sitting motionless on a buoy watching the tall ship sail past, what is the horizontal displacement of the second ball as it falls to the deck of the ship?

23. You are standing inside a bus that is traveling at a constant velocity of 8 m/s along a straight horizontal road. You toss an apple straight up into the air and then catch it again 1.2 seconds later at the same height from which you let it go. Use $g = 10$ m/s^2 to simplify the calculations. Answer these questions based on what you see. In other words, consider the motion of the apple relative to you. (a) What is the maximum height reached by the apple? (b) What is the initial velocity of the apple? (c) How far does the apple move horizontally (relative to you)?

24. Repeat Exercise 23, but now answer the questions from the perspective of a person at rest outside the bus, looking in through the bus windows. For part (c), in particular, how far does the apple move horizontally relative to the person at rest outside the bus?

Exercises 25 – 36 are designed to help you practice applying the general method for solving projectile-motion problems. For each exercise, (a) draw a diagram. (b) Sketch a free-body diagram, showing the forces acting on the object. (c) Choose an origin, and mark it on your diagram. (d) Choose an *x-y* coordinate system and draw it on your diagram, showing which way is positive for each coordinate axis. (e) Organize your data in a table, keeping the information for the *x* subproblem separate from the information for the *y* subproblem.

25. Before game 1 of the 2004 World Series, Johnny Damon and Trot Nixon are playing catch in the outfield at Fenway Park. The two players are 23.0 m apart. Damon throws the ball to Nixon so that it reaches a maximum height of 15.5 m above the point he released it. Assume that Nixon catches the ball at the same level from which Damon releases it, that air resistance is negligible, and that $g = 9.81$ m/s^2. Answer parts (a) through (e), as described above. (f) At what initial velocity did Damon throw the ball?

26. Working as an accident reconstruction expert, you find that a car that was driven off a horizontal road over the edge of a 15-m-high cliff traveled 42 m horizontally before impact. Use $g = 9.8$ m/s^2. Answer parts (a) through (e) as described above. (f) At what speed was the car moving when it left the road? Express the speed in m/s, km/h, and miles/h. (g) At what speed was the car moving just before impact?

27. You toss a set of keys up to your friend, who is leaning out a window above you. You are at ground level in front of the window, and you release the keys 2.0 m horizontally from, and 6.0 m vertically below, the point where your friend catches them. The keys happen to be at their maximum height point when they are caught by your friend. Answer parts (a) through (e) as described above. (f) With what initial velocity did you launch the keys?

28. You throw a ball with an initial speed v_i at an angle of θ above the horizontal. The ball lands at a point some distance away that is a height *h* below the point from which it was launched. Answer parts (a) through (e) as described above. (f) In terms of g, v_i, θ, and h, determine an expression for the time the ball is in flight. Think about what would happen if *h* was zero, or if the landing point was some height *h* above the launch point instead.

29. You throw a ball so that it is in the air for 4.56 s and travels a horizontal distance of 50.0 m. Use $g = 9.81$ m/s^2, and assume the ball lands at the same height from which it was launched. Answer parts (a) through (e) as described above. (f) What was the ball's initial velocity?

30. Repeat Exercise 29, but now assume the ball lands at a height 20.0 m below the point from which it was launched.

31. Taking a corner kick in a soccer game, you kick the ball from ground level so that, after reaching a maximum height of 3.5 m above the ground, the ball reaches your teammate 22 meters away. Your teammate makes contact with the ball when the ball is 2.0 m off the ground and heads the ball into the net. Set the origin to be the point from which you kicked the ball, take up to be the positive *y* direction, and point the positive *x* direction from you toward your teammate. Answer parts (a) through (e) as described above. (f) Determine the *x* and *y* components of the ball's initial velocity. (g) Plot graphs of the ball's *x* position, *y* position, *x* velocity, and *y* velocity, all as a function of time. (h) Plot a graph of the ball's *y* position as a function of its *x* position.

32. You throw a ball from the edge of a vertical cliff overlooking the ocean. The ball is launched with an initial velocity of 12.0 m/s at an angle of 34.0° above the horizontal, from a height of 55.0 m above the water. Neglect air resistance, and use $g = 9.81$ m/s^2. Answer parts (a) through (e), as described above. (f) How long does it take the ball to reach the water? (g) Assuming you throw the ball so the horizontal component of its velocity is perpendicular to the cliff face, how far is it from the base of the cliff when it hits the water?

33. Repeat parts (a) through (e) of Exercise 32, but, this time, choose a different origin. Show that you still get the same answers for parts (f) and (g).

34. Repeat parts (a) through (e) of Exercise 32, but, this time, reverse the direction you take to be positive for the vertical coordinate axis. Show that you still get the same answers for parts (f) and (g).

35. A spacecraft with a mass of 8000 kg is drifting through deep space with a constant velocity of 5.0 m/s in the positive y direction. The thrusters are then turned on so the craft experiences a constant acceleration of 4.0 m/s^2 in the positive x direction for a period of 5.0 s (assume the mass lost in this process is negligible). The thrusters are then turned off. Take the origin to be the position of the spacecraft when the thrusters are first turned on. Answer parts (a) through (e) as described above. (f) How far is the spacecraft from the origin when the thrusters are turned off? (g) How far is the spacecraft from the origin after another 5.0 s has passed?

36. Repeat Exercise 35, but this time assume that, when the thrusters are turned on, the spacecraft experiences a constant acceleration of 4.0 m/s^2 in the positive x direction as well as a constant acceleration of 3.0 m/s^2 in the positive y direction. Once again, the thrusters are turned off after 5.0 s.

Exercises 37 – 40 deal with common applications of projectile motion.

37. For a typical kickoff in a football game, the "hang time" (the time the ball spends in the air) is 4.0 s, and the ball travels 50 m. Neglecting air resistance, determine (a) the vertical component of the initial velocity, (b) the horizontal component of the initial velocity, (c) the magnitude and direction, relative to the horizontal, of the initial velocity.

38. A basketball hoop is 3.05 m above the floor, and the horizontal distance from the free-throw line to a point directly below the center of the hoop is 4.60 m. Assuming that a basketball player releases the ball at an angle of 60.0° with respect to the horizontal, from a point that is located exactly 1.70 m above the free-throw line, what should the launch speed be so the ball passes through the center of the hoop?

39. In 1983, Chris Bromham set a world record by jumping a motorcycle over 18 double-decker buses. If Mr. Bromham launched the motorcycle off a ramp with a launch speed of 42 m/s and traveled a horizontal distance of 63 m, landing at the same height from which he left the ramp, at what angle was the launch ramp inclined? Note that such ramps are inclined at angles considerably less than 45°.

40. The Russian shot-putter Natalya Lisovskaya set a world record, in 1987, for the shot put that was still unbroken at the time this book went to press. Neglect air resistance. Assume the shot was launched from a height of 1.800 m above the ground, at an angle of 43.00° above the horizontal, and with a speed of 14.32 m/s. Use $g = 9.810$ m/s^2, and determine the distance of the record throw.

General Problems and Conceptual Questions

41. In Exploration 4.2, let's say $L = 36.0$ m and $v = 8.0$ m/s. By what distance does Brandi win the race?

42. In Exploration 4.2, let's say $L = 36.0$ m and $v = 8.0$ m/s. (a) Aside from the very start of the race, at how many different instants are the two women the same distance from the start line at the same time, between the time the race starts and the time Brandi arrives at the finish line? (b) When are these instants, and how far from the start line are the two women when these instants occur?

43. You are driving your car at a constant velocity of 110 km/h north. As you pass under a bridge, a train is passing over the bridge, traveling at a constant speed of 60 km/h. What is your velocity relative to the train if (a) the train is traveling due east? (b) the train is traveling at an angle of 15° south of east?

44. A ballistic cart is a wheeled cart that can launch a ball in a direction perpendicular to the way the cart moves and can then catch the ball again if it falls back down on the cart. Holding the cart stationary on a horizontal track, you confirm that the ball does indeed land in the cart after it is launched. Let's say that the cart launches the ball with an initial velocity of 4.00 m/s up relative to the cart while the cart is rolling with a constant velocity of 3.00 m/s to the right. Using $g = 9.81$ m/s², determine (a) the time it takes the ball to return to the height from which it was launched, (b) the displacement of the cart during this time, and (c) the displacement of the ball during this time.

45. In 1991, Mike Powell of the United States set a world record of 8.95 m in the long jump. With this jump, Powell broke Bob Beamon's record of 8.90 m, which was set at the Mexico City Olympic Games in 1968. Estimate how fast Powell was going when he left the ground, knowing that he was trying to jump as far as he could. How does your value compare to the top speed of a world-class sprinter?

46. While playing catch with your friend, who is located due north of you, you throw a ball such that its initial velocity components are 20 m/s up and 7.0 m/s north. Some time later, your friend catches the ball at the same height from which you released it. Considering the motion from the instant just after you released it until just before your friend caught it, what is (a) the ball's average velocity over this interval? (b) the ball's average acceleration over this interval?

47. Consider the trajectory of object B in Figure 4.24. At what angle from the horizontal was it launched? Assume the grid shown on the diagram is square.

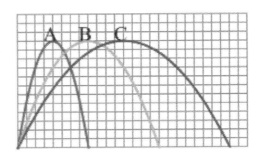

Figure 4.24: The trajectories of three projectiles. For Exercise 47, just focus on object B.

48. The motion diagram in Figure 4.25 shows a ball's position at regular intervals as the ball flies from left to right through the air. (a) On the x-axis, draw the motion diagram corresponding to the x subproblem. What does this tell you about the ball's velocity and acceleration in the x direction? (b) On the y-axis, draw the motion diagram corresponding to the y subproblem. What does this tell you about the ball's velocity and acceleration in the y direction? If the ball's position is shown at 1-second intervals and $g = 10$ m/s^2, (c) what is the maximum height reached by the ball? (d) How far does the ball travel horizontally?

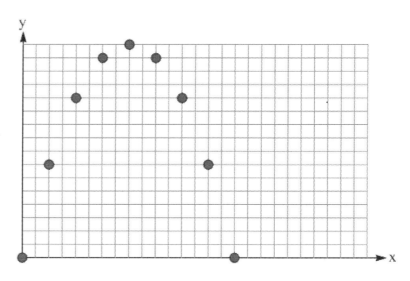

Figure 4.25: A motion diagram showing the position of a ball at regular time intervals, as the ball moves from left to right, for Exercise 48.

49. A ball is launched and travels for exactly 3 seconds before being caught. Graphs of the ball's x velocity and y velocity are shown in Figure 4.26, where the acceleration due to gravity is assumed to be $g = 10$ m/s^2. (a) Is the ball caught at a level that is higher, lower, or the same as the level from which it was launched? How do you know? (b) How far does the ball travel horizontally during the 3-second period? (c) At what instant does the ball reach its maximum height? Justify your answer.

50. Consider again the situation described in Exercise 49, and the graphs of the ball's x and y velocity shown in Figure 4.26. Sketch corresponding graphs of the ball's (a) x acceleration, (b) y acceleration, (c) x position, and (d) y position. Draw each graph as a function of time.

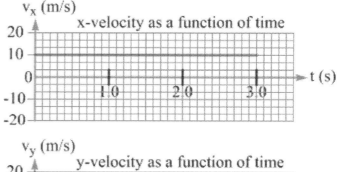

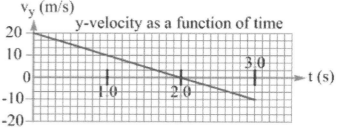

Figure 4.26: Graphs of the x-velocity and y-velocity for a ball thrown into the air. For Exercises 49 – 51.

51. On a piece of graph paper, sketch the motion diagram corresponding to the velocity graphs in Figure 4.26, showing the position of the ball at 0.50-second intervals between $t = 0$ and $t = 3$ seconds.

52. A ball is launched with an initial velocity of 12.0 m/s at an angle of 30.0° above the horizontal. It lands some time later at the same height from which it was launched. Using the standard component method of analysis, and using $g = 9.81$ m/s^2, determine the time the ball is in flight.

53. You kick a soccer ball from ground level with an initial velocity of 7.50 m/s at an angle of 30.0° above the horizontal. The ball hits a wall 8.00 m away. How far up the wall is the point of impact?

54. You kick a soccer ball from ground level with an initial velocity of 7.50 m/s at an angle of 30.0° above the horizontal. When the ball strikes a wall that is some distance from you, the ball is 50.0 cm off the ground. How far away is the wall? Find all possible answers, assuming that the ball does not bounce before hitting the wall. Use $g = 9.80$ m/s^2.

55. Standing in the middle of a perfectly flat field, you kick a soccer ball from ground level, launching it with an initial speed of 20 m/s. Using $g = 10$ m/s^2 to simplify the calculations, and assuming you can use any launch angle between 0 and 90°, measured from the horizontal, determine the following: (a) the shortest and longest times the ball can spend in the air before hitting the ground, (b) the maximum height the ball can reach in its flight, and (c) the maximum range of the ball.

56. Returning to the situation of Exercise 55, what launch angle(s) give a range that is 75% of the maximum range?

57. You are playing catch with your friend, who is standing 20.0 m from you. When you first throw the ball, you launch it at an angle of 65.0° and the ball falls 5.00 m short of reaching your friend. (a) At what speed did you launch the ball? (b) If you kept the launch angle the same, at what initial speed should you launch the ball on the next throw so the ball reaches your friend? (c) If instead you kept the launch speed the same as in part (a), at what launch angle should you launch the ball so it reaches your friend? Assume the ball lands at the same height from which it was launched for all parts of this exercise.

58. You are having a snowball fight with your friend, who is 7.0 m away from you. Knowing some physics, you throw one snowball at an angle of 70° above the horizontal, launching it so that the snowball will hit your friend at the same height from which you let it go. You wait for a short time interval, and then launch a second snowball from the same point, but at an angle of 20° above the horizontal. If you want both snowballs to hit your friend in the same place simultaneously, how long should the time interval be between throwing the two snowballs?

59. You repeat the two-snowball situation described in Exercise 58, but this time with different launch angles. You observe that the maximum height (measured from the launch point) of one snowball is exactly 5 times higher than the maximum height reached by the other snowball. At what angles were the two snowballs launched, assuming they were launched with the same speed?

60. You repeat the two-snowball situation described in Exercise 58, but this time with different launch angles. You observe that the time of flight for one snowball is exactly 5 times longer than the time of flight of the other snowball. At what angles were the two snowballs launched, assuming they were launched with the same speed?

61. You flick a coin from a tabletop that is 1.30 m above the floor, giving the coin an initial speed of 2.40 m/s when it leaves the table. Calculate the horizontal distance between the launch point and the point of impact if the launch angle, relative to the horizontal, is (a) 0°, (b) 30°, and (c) 45°.

Figure 4.27: This decorative fountain shows the water following parabolic trajectories. After being sprayed from a pipe in the fountain, each stream of water is acted upon by gravity and follows a two-dimensional path as the stream deflects toward the Earth. (Photo credit: Edward Hor / iStockphoto.)

62. The photograph in Figure 4.27 shows the parabolic paths followed by water in a decorative fountain. If the water emerging from one of the pipes has an initial speed of 4.0 m/s, and is projected at an angle of 60° above the horizontal, determine the maximum height reached by the water above the level of the top end of the pipe.

63. Three students are trying to solve a problem that involves a ball being launched, at a 30° angle above the horizontal, from the top of a cliff, and landing on the flat ground some distance below. The students know the launch speed, the acceleration due to gravity, and the height of the cliff. They are looking for the time the ball spends in the air. Comment on the part of their conversation that is recorded below.

Avi : I think we have to do the problem in two steps. First, we find the point where the ball reaches its maximum height, and then we go from that point down to the ground.

T.J. : I think we can do it all in one step. The equations can handle it, just going all the way from the initial point to the ground.

Kristin: I think we need two steps, too, but I would do it differently than Avi. What if we first find the point where the ball comes down to the same height from where it was launched, and then we go from that point down to the ground?

What keeps these people from falling down the vertical wall of the old carnival ride called the Rotor as it whirls them around in a circle? Friction. In particular, it is the type of friction known as static friction.

There are a number of common beliefs about friction that are simply plain wrong. In this chapter, we will spend some time investigating the correct ways to analyze problems involving friction. We will also be looking at circular motion situations, such as the one shown here.

Photo courtesy of David Burton.

Chapter 5 – Applications of Newton's Laws

CHAPTER CONTENTS

We begin this chapter by learning about friction. By accounting for friction, we are refining our model of how things work. When two objects are in contact, the friction force, if there is one, is the component of the contact force between the objects that is parallel to the surfaces in contact. ***Friction tends to oppose <u>relative</u> motion between objects.***

After understanding friction, we will go on to investigate how to apply forces and Newton's laws to situations involving straight-line motion and circular motion. At first glance, it might appear that friction and circular motion are very different things. The link between them, in fact, is that, with both friction and circular motion, we get another opportunity to learn about how to apply Newton's second law to a physical system.

Circular motion is particularly interesting, as it represents the motion of a car moving around the banked turn of a freeway exit or a racetrack, the motion of a water bucket on a string that you whirl in a circle, as well as, essentially, the motion of the Earth as it orbits the Sun.

5-1 Kinetic Friction

When two objects are in contact, the friction force, if there is one, is the component of the contact force between the objects that is parallel to the surfaces in contact. (The component of the contact force that is perpendicular to the surfaces is the normal force.) Friction tends to oppose relative motion between objects. When there is relative motion, the friction force is the kinetic force of friction. For instance, when a book slides across a table, kinetic friction slows the book.

> If there is relative motion between objects in contact, the force of friction is the kinetic force of friction (F_K). We will use a simple model of friction that assumes the force of kinetic friction is proportional to the normal force. A dimensionless parameter, called the coefficient of kinetic friction, μ_K, represents the strength of that frictional interaction.
>
> $$\mu_K = \frac{F_K}{F_N} \qquad \text{so} \qquad F_K = \mu_K F_N. \qquad \text{(Equation 5.1: \textbf{Kinetic friction})}$$
>
> Note that some typical values for the coefficient of kinetic friction, as well as for the coefficient of static friction, which we will define in Section 5-2, are given in Table 5.1. The coefficients of friction depend on the materials that the two surfaces are made of, as well as on the details of their interaction. For instance, adding a lubricant between the surfaces tends to reduce the coefficient of friction. There is also some dependence of the coefficients of friction on the temperature.

Interacting materials	Coefficient of kinetic friction (μ_K)	Coefficient of static friction (μ_S)
Rubber on dry pavement	0.7	0.9
Steel on steel (unlubricated)	0.6	0.7
Rubber on wet pavement	0.5	0.7
Wood on wood	0.3	0.4
Waxed ski on snow	0.05	0.1
Friction in human joints	0.01	0.01

Table 5.1: Approximate coefficients of kinetic friction, and static friction (see Section 5-2), for various interacting materials.

EXPLORATION 5.1 – A sliding book

You slide a book, with an initial speed v_i, across a flat table. The book travels a distance L before coming to rest. What determines the value of L in this situation?

Step 1 – *Sketch a diagram of the situation and a free-body diagram of the book.* These diagrams are shown in Figure 5.1.

The Earth applies a downward force of gravity on the book, while the table applies a contact force. Generally, we split the contact force into components and show an upward normal force and a horizontal force of kinetic friction, $\vec{F}_K$, acting on the book. $\vec{F}_K$ points to the right on the book, opposing the relative motion between the book (moving left) and the table (at rest).

Figure 5.1: A diagram of the sliding book and a free-body diagram showing the forces acting on the book as it slides.

Step 2 – *Find an expression for the book's acceleration.* We are working in two dimensions, and a coordinate system with axes horizontal and vertical is convenient. Let's choose up to be positive for the y-axis and, because the motion of the book is to the left, let's choose left to be the positive x-direction. We will apply Newton's Second Law twice, once for each direction. There is no acceleration in the y-direction, so we have: $\sum \vec{F}_y = m\vec{a}_y = 0$.

Because we're dealing with the y sub-problem, we need only the vertical forces on the free-body diagram. Adding the vertical forces as vectors (with appropriate signs) tells us that:

$$+F_N - F_G = 0, \quad \text{so} \quad F_N = F_G = mg.$$

Repeat the process for the x sub-problem, applying Newton's Second Law, $\sum \vec{F}_x = m\vec{a}_x$. Because the acceleration is entirely in the x-direction, we can replace $\vec{a}_x$ by $\vec{a}$. Now, we focus only on the forces in the x-direction. All we have is the force of kinetic friction, which is to the right while the positive direction is to the left. Thus: $-F_K = m\vec{a}$.

We can now solve for the acceleration of the book, which is entirely in the x-direction:

$$\vec{a} = \frac{-F_K}{m} = \frac{-\mu_K F_N}{m} = \frac{-\mu_K mg}{m} = -\mu_K g.$$

Step 3 – *Determine an expression for length, L, in terms of the other parameters.* Let's use our expression for the book's acceleration and apply the method for solving a constant-acceleration problem. Table 5.2 summarizes what we know. Define the origin to be the book's starting point and the positive direction to be the direction of motion.

Initial position	$x_i = 0$
Final position	$x = +L$
Initial velocity	$+v_i$
Final velocity	$v = 0$
Acceleration	$a = -\mu_K g$

We can use equation 2.10 to relate the distance traveled to the coefficient of friction:

$$v^2 = v_i^2 + 2a\Delta x, \qquad \text{so} \qquad \Delta x = \frac{v^2 - v_i^2}{2a}.$$

Table 5.2: A summary of what we know about the book's motion to help solve the constant-acceleration problem.

In this case, we get $\qquad L = \dfrac{0 - v_i^2}{-2\mu_K g} = \dfrac{v_i^2}{2\mu_K g}.$

Let's think about whether the equation we just derived for L makes sense. The equation tells us that the book travels farther with a larger initial speed or with smaller values of the coefficient of friction or the acceleration due to gravity, which makes sense. It is interesting to see that there is no dependence on mass – all other parameters being equal, a heavy object travels the same distance as a light object.

> **Key idea**: A useful method for solving a problem with forces in two dimensions is to split the problem into two one-dimensional problems. Then, we solve the two one-dimensional problems individually. **Related End-of-Chapter Exercises 13 – 15, 31.**

Essential Question 5.1: You are standing still and you then start to walk forward. Is there a friction force involved here? If so, is it the kinetic force of friction or the static force of friction?

Answer to Essential Question 5.1: When asked this question, most people are split over whether the friction force is kinetic or static. Think about what happens when you walk. When your shoe (or foot) is in contact with the ground, the shoe does not slip on the ground. Because there is no relative motion between the shoe and the ground, the friction force is static friction.

5-2 Static Friction

If there is no relative motion between objects in contact, then the friction force (if there is one) is the static force of friction (F_S). An example is a box that remains at rest on a ramp. The force of gravity acting down the ramp is opposed by a static force of friction acting up the ramp. A more challenging example is when the box is placed on the floor of a truck. When the truck accelerates and the box moves with the truck (remaining at rest relative to the truck), it is the force of static friction that acts on the box to keep it from sliding around in the truck.

Consider again the question about the friction force between the sole of your shoe and the floor, when you start to walk. In which direction is the force of static friction? Many people think this friction force is directed opposite to the way you are walking, but the force of static friction is actually directed the way you are going. To determine the direction of the force of static friction, think about the motion that would result if there were no friction. To start walking, you push back with your foot on the floor. Without friction, your foot would slide back, moving back relative to the floor, as shown in Figure 5.2. Static friction opposes this motion, the motion that would occur if there was no friction, and thus static friction is directed forward.

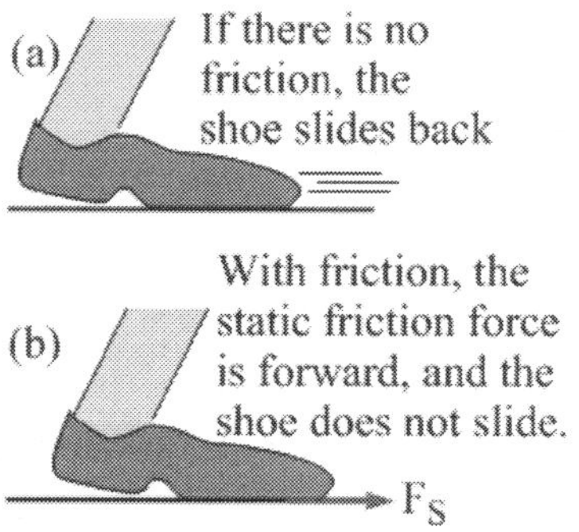

Figure 5.2: On a frictionless floor, your shoe slides backward over the floor when you try to walk forward (a). Static friction opposes this motion, so the static force of friction, applied by the ground on you, is directed forward (b).

The static force of friction opposes the relative motion that would occur if there were no friction. Another interesting feature is that ***the static force of friction adjusts itself to whatever it needs to be to prevent relative motion between the surfaces in contact.*** Within limits, that is. The static force of friction has a maximum value, $F_{S,max}$, and the coefficient of static friction is defined in terms of this maximum value:

$$\mu_S = \frac{F_{S,max}}{F_N} \quad \text{so} \quad F_S \leq \mu_S F_N. \qquad \text{(Equation 5.2: \textbf{Static friction})}$$

Let's now explore a situation that involves the adjustable nature of the force of static friction.

EXPLORATION 5.2 – A box on the floor
A box with a weight of $mg = 40$ N is at rest on a floor. The coefficient of static friction between the box and the floor is $\mu_S = 0.50$, while the coefficient of kinetic friction between the box and the floor is $\mu_K = 0.40$.

Step 1 - *What is the force of friction acting on the box if you exert no force on the box?* Let's draw a free-body diagram of the box (see Figure 5.3b) as it sits at rest. Because the box remains at rest, its acceleration is zero and the forces must balance. Applying Newton's Second Law tells us that $F_N = mg = 40$ N. There is no tendency for the box to move, so there is no force of friction.

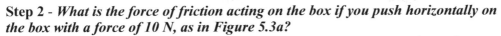

Step 2 - *What is the force of friction acting on the box if you push horizontally on the box with a force of 10 N, as in Figure 5.3a?*
Nothing has changed vertically, so we still have $F_N = mg = 40$ N . To determine whether or not the box moves, let's use equation 5.2 to determine the maximum possible force of static friction in this case. We get:

$$F_S \leq \mu_S F_N = 0.50 \times 40 \text{ N} = 20 \text{ N} .$$

The role of static friction is to keep the box at rest. If we exert a horizontal force of 10 N on the box, the force of static friction acting on the box must be 10 N in the opposite direction, to keep the box from moving. The free-body diagram of this situation is shown in Figure 5.3c. 10 N is below the 20 N maximum value, so the box will not move.

Step 3 - *What is the force of friction acting on the box if you increase your force to 15 N?* This situation is similar to step 2. Now, the force of static friction adjusts itself to 15 N in the opposite direction of your 15 N force. 15 N is still less than the maximum possible force of static friction (20 N), so the box does not move.

Step 4 - *What is the force of friction acting on the box if you increase your force to 20 N?* If your force is 20 N, the force of static friction matches you with 20 N in the opposite direction. We are now at the maximum possible value of the force of static friction. Pushing even a tiny bit harder would make the box move.

Step 5 - *What is the force of friction acting on the box if you increase your force to 25 N?* Increasing your force to 25 N, which is larger in magnitude than the maximum possible force of static friction, makes the box move. Because the box moves, the friction is the kinetic force of friction, which is in the direction opposite to your force with a magnitude of

$$F_K = \mu_K F_N = 0.40 \times 40 \text{ N} = 16 \text{ N} .$$

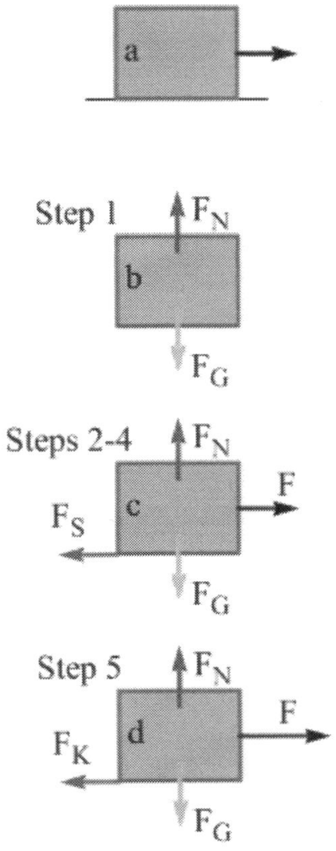

Figure 5.3: (a) The top diagram shows the box, and the force you exert on it. (b) The free-body diagram for step 1, in which you exert no force. (c) The free-body diagram that applies to steps 2 – 4, in which the force you exert is less than or equal to the maximum possible force of static friction. (d) The free-body diagram that applies to step 5, in which your force is large enough to cause the box to move.

Key ideas for static friction: The static force of friction is whatever is required to prevent relative motion between surfaces in contact. The static force of friction is adjustable only up to a point. If the required force exceeds the maximum value $F_{S,\text{max}} = \mu_S F_N$, then relative motion will occur.
Related End-of-Chapter Exercises: 32, 34.

A microscopic model of friction
Figure 5.4 shows a magnified view of two surfaces in contact. Surface irregularities interfere with the motion of one surface left or right with respect to the other surface, giving rise to friction.

Figure 5.4: A magnified view of two surfaces in contact. The irregularities in the objects prevent smooth motion of one surface over the other, giving rise to friction.

Essential Question 5.2: What is the magnitude of the net contact force exerted by the floor on the box in step 5 of Exploration 5.2?

Answer to Essential Question 5.2: The two components of the contact force are the upward normal force and the horizontal force of kinetic friction. These two components are at right angles to one another, so we can use the Pythagorean theorem to find the magnitude of the contact force:

$$F_C = \sqrt{F_K^2 + F_N^2} = \sqrt{(16 \text{ N})^2 + (40 \text{ N})^2} = 43 \text{ N} .$$

5-3 Measuring the Coefficient of Friction

Let's connect the force ideas to the one-dimensional motion situations from Chapter 2.

EXPLORATION 5.3 – Measuring the coefficient of static friction

Coefficients of static friction for various pairs of materials are given in Table 5.1. Here's one method for experimentally determining these coefficients for a particular pair of materials. Take an aluminum block of mass m and a board made from a particular type of wood (we could also use a block of the wood and a piece of inflexible aluminum). Place the block on the board and slowly raise one end of the board. The angle of the board when the block starts to slide gives the coefficient of static friction. How?

To answer this question, let's extend the general method for solving a problem that involves Newton's laws.

Step 1 – *Draw a diagram of the situation.* The diagram is shown in Figure 5.5a.

Step 2 – *Draw a free-body diagram of the block when it is at rest on the inclined board.* Two forces act on the block, the downward force of gravity and the upward contact force from the board. We generally split the contact force into components, the normal force perpendicular to the incline, and the force of static friction acting up the slope. This free-body diagram is shown in Figure 5.5b.

Step 3 – *Choose an appropriate coordinate system.* In this case, if we choose a coordinate system aligned with the board (one axis parallel to the board and the other perpendicular to it), as in Figure 5.5c, we will only have to split the force of gravity into components.

Step 4 – *Split the force of gravity into components.* If the angle between the board and the horizontal is θ, the angle between the force of gravity and the y-axis is also θ. The component of the force of gravity acting parallel to the slope has a magnitude of $F_{Gx} = F_G \sin\theta = mg \sin\theta$. The perpendicular component is $F_{Gy} = F_G \cos\theta = mg \cos\theta$. These components are shown on the lower free-body diagram, in Figure 5.5d.

Step 5 – *Apply Newton's second law twice, once for each direction.* Again, we break a two-dimensional problem into two one-dimensional problems. With no acceleration in the y-direction we get: $\sum \vec{F}_y = m\vec{a}_y = 0$. Looking at the lower diagram in Figure 5.5 for the y-direction forces:

$+F_N - mg \cos\theta = 0$, which tells us that

$F_N = mg \cos\theta$.

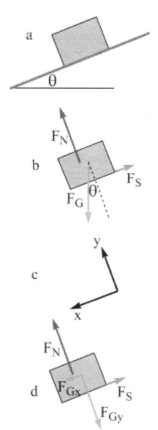

Figure 5.5: (a) A diagram showing the block on the board. (b) The initial free-body diagram of the block. (c) An appropriate coordinate system, aligned with the incline. (d) A second free-body diagram, with the forces aligned parallel to the coordinate axes.

While the box is at rest, there is no acceleration in the x-direction so: $\sum \vec{F}_x = m\vec{a}_x = 0$.
Looking at the lower free-body diagram, Figure 5.5d, for the forces in the x-direction:

$$mg\sin\theta - F_S = 0, \qquad \text{which tells us that} \quad F_S = mg\sin\theta.$$

These two equations, $F_N = mg\cos\theta$ and $F_S = mg\sin\theta$, tell us a great deal about what happens as the angle of the incline increases. When the board is horizontal, the normal force is equal to mg and the static force of friction is zero. As the angle increases, $\cos\theta$ decreases from 1 and $\sin\theta$ increases from zero. Thus, as the angle increases, the normal force (and the maximum possible force of static friction) decreases, while the force of static friction required to keep the block at rest increases. At some critical angle θ_C, the force of static friction needed to keep the block at rest equals the maximum possible force of static friction. If the angle exceeds this critical angle, the block will slide. Using the definition of the coefficient of static friction:

$$\mu_S = \frac{F_{S,max}}{F_N} = \frac{mg\sin\theta_C}{mg\cos\theta_C} = \tan\theta_C.$$

Thus, it is easy to determine coefficients of static friction experimentally. Take two objects and place one on top of the other. Gradually tilt the objects until the top one slides off. The tangent of the angle at which sliding occurs is the coefficient of static friction.

> **Key idea regarding the coefficient of static friction**: The coefficient of static friction between two objects is the tangent of the angle beyond which one object slides down the other.
> **Related End-of-Chapter Exercises: 7, 36.**

The steps we used to solve the problem in Exploration 5.3 can be applied generally to most problems involving Newton's laws. Let's summarize the steps here. Then, we will get some more practice applying the method in the next section.

A General Method for Solving a Problem Involving Newton's Laws in Two Dimensions
1. Draw a diagram of the situation.

2. Draw one or more free-body diagrams, with each free-body diagram showing all the forces acting on an object.

3. For each free-body diagram, choose an appropriate coordinate system. Coordinate systems for different free-body diagrams should be consistent with one another. A good rule of thumb is to align each coordinate system with the direction of the acceleration.

4. Break forces into components that are parallel to the coordinate axes.

5. Apply Newton's second law twice to each free-body diagram, once for each coordinate axis. Put the resulting force equations together and solve.

Related End-of-Chapter Exercises 17, 19.

Essential Question 5.3: Could we modify the procedure described in Exploration 5.3 to measure the coefficient of kinetic friction? If so, how could we modify it?

Answer to Essential Question 5.3: Yes. With the top object at rest on the inclined second object, give the top object a little push to get it moving. If it slides down the second object with constant velocity, the tangent of the angle of the incline equals the coefficient of kinetic friction.

5-4 A System of Two Objects and a Pulley

EXAMPLE 5.4 – Working with more than one object, and a pulley
    A red box of mass $M = 10$ kg is placed on a ramp that is a 3-4-5 triangle, with a height of 3.0 m and a width of 4.0 m. The red box is tied to a green block of mass $m = 1.0$ kg by a string passing over a pulley, as shown in Figure 5.6. The coefficients of friction for the red box and the incline are $\mu_s = 0.50$ and $\mu_k = 0.25$. Use $g = 10$ m/s². When the system is released from rest, what is the acceleration?

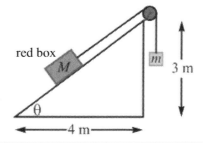

Figure 5.6: A diagram for the system of two objects and a pulley.

SOLUTION
    To determine if there is an acceleration, and to find the direction of any acceleration, think about what happens if there is no friction. With no friction, we have the free-body diagrams in Figure 5.7. Choose a coordinate system aligned with the slope for the red box, with the positive x-direction down the slope and a positive y-direction perpendicular to the incline. If the red box moves down the slope, the green block moves up, so a consistent coordinate system for the green block has the positive direction up.

    To align all the forces with the coordinate axes, only the force of gravity acting on the red box needs to be split into components. As shown in Figure 5.7c, we get $Mg\sin\theta$ directed down the slope and $Mg\cos\theta$ perpendicular to the slope.

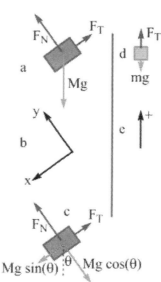

Applying Newton's second law, $\sum \vec{F} = m\vec{a}$, to the green block gives:

$+F_T - mg = +ma$,   which tells us that $F_T = mg + ma$.

Applying Newton's second law in the x-direction for the red box gives

$+Mg\sin\theta - F_T = Ma$,   which gives $Mg\sin\theta = F_T + Ma$.

    Combining the equation from the green block with the equation from the red box tells us that: $Mg\sin\theta = mg + ma + Ma$. This can be re-arranged into

$(M\sin\theta - m)g = (m + M)a$.

    The acceleration of the system, with no friction, is positive if $M\sin\theta$ exceeds m. In this case, $M\sin\theta = 10$ kg×(3/5) = 6.0 kg is larger than the mass $m = 1.0$ kg of the green block. Thus, if the system accelerates, the red box accelerates down the slope. We don't know for sure that the system accelerates, because the force of static friction could prevent any motion. The force of static friction would be directed up the ramp, as in Figure 5.8, to stop the box from sliding down the ramp. How large must the force of static friction be to prevent the system from moving?

Figure 5.7: Free-body diagrams if there is no friction. (a) The free-body diagram of the red box. (b) An appropriate coordinate system for the red box. (c) The free-body diagram of the red box, with force components aligned with the coordinate system. (d) and (e), a free-body diagram and coordinate system for the green block.

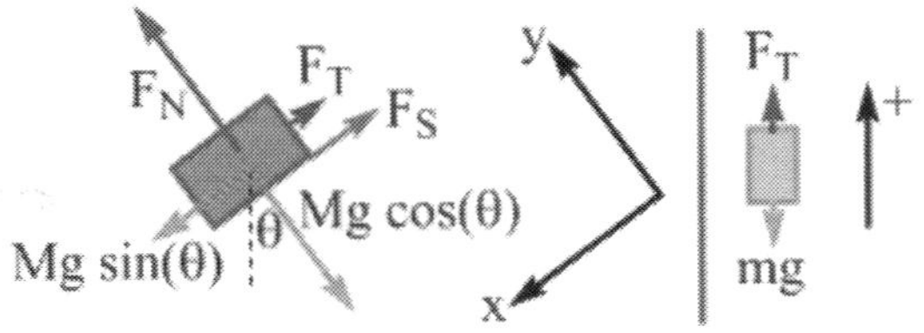

If the system remains at rest, the acceleration is zero. Applying Newton's second law, $\sum \vec{F}_x = M\vec{a} = 0$, to the red box in the x-direction gives $+Mg\sin\theta - F_S - F_T = 0$.

With no acceleration, the force equation for the green block is $F_T = mg$. Using $F_T = mg$ in the previous equation gives $+Mg\sin\theta - F_S - mg = 0$. Thus, the force of static friction needed to prevent motion is:

$$F_S = Mg\sin\theta - mg = (M\sin\theta - m)g = [10\text{ kg}\times(3/5) - 1.0\text{ kg}]g = (5.0\text{ kg})\,g.$$

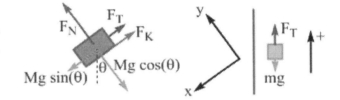

Figure 5.8: The free-body diagrams if static friction prevents motion are similar to those in Figure 5.7, except that, in this case, the forces balance and there is a force of static friction, on the red box, directed up the slope.

Let's compare this value to the maximum possible force of static friction. Applying Newton's second law to the red box in the y-direction gives $F_N = Mg\cos\theta$, because there is no acceleration in that direction. Thus, the maximum possible force of static friction in this case is:

$$F_{S,\max} = \mu_S F_N = \mu_S Mg\cos\theta = 0.5\times(10\text{ kg})\times(4/5)\times g = (4.0\text{ kg})\,g.$$

The force of static friction required to prevent the system from accelerating is larger than its maximum possible value, which cannot happen. Thus, the system does accelerate. To find the acceleration, we again draw free-body diagrams, as in Figure 5.9. Now kinetic friction acts up the slope on the red box, and each object has a net force acting on it.

Applying Newton's second law, to the green block gives $+F_T - mg = +ma$, which tells us that $F_T = mg + ma$. A common error is to assume that $F_T = mg$, which is true only when the acceleration is zero.

Applying Newton's second law to the red box in the x-direction gives $+Mg\sin\theta - F_k - F_T = Ma$.

Figure 5.9: Free-body diagrams for the situation of the red box accelerating down the slope and the green block accelerating up.

Substituting for the force of tension (using our result from analyzing the green block), we get $+Mg\sin\theta - F_k - mg - ma = Ma$, which we re-write as $+Mg\sin\theta - F_k - mg = (M + m)a$.

We can now substitute for the force of kinetic friction and find the acceleration:

$$a = \frac{+Mg\sin\theta - \mu_K F_N - mg}{M + m} = \frac{+Mg\sin\theta - \mu_K Mg\cos\theta - mg}{M + m}.$$

$$a = \frac{+10\text{ kg}\times(10\text{ m/s}^2)(3/5) - 0.25\times(10\text{ kg})(10\text{ m/s}^2)(4/5) - (1.0\text{ kg})\times(10\text{ m/s}^2)}{10.0\text{ kg} + 1.0\text{ kg}}.$$

$$a = \frac{+60\text{ N} - 20\text{ N} - 10\text{ N}}{11.0\text{ kg}} = \frac{+30\text{ N}}{11.0\text{ kg}} = +2.7\text{ m/s}^2.$$

Note that the role of the pulley in this situation is simply to redirect the force of tension.

Related End-of-Chapter Exercises: 18, 41, 47.

Essential Question 5.4: If an object moves in a circle at constant speed, is there an acceleration?

Answer to Essential Question 5.4: Yes. An object has an acceleration whenever its velocity changes. Although the magnitude of the velocity is constant, the direction of the velocity changes, so there must be an acceleration. We'll investigate this further in the next section.

5-5 Uniform Circular Motion

Uniform circular motion is motion in a circle with constant speed. Let's define T, the period of the uniform circular motion, to be the time it takes an object to travel around one complete circle. Because the speed is constant, we can relate the speed to the distance traveled very simply: $v = \text{distance/time} = 2\pi r / T$.

As with straight-line motion, the magnitude of the acceleration is related to the speed the same way that the speed is related to the distance: $a = 2\pi v / T$.

If we combine the two equations above, we get what's called the *centripetal acceleration*.

When an object is traveling in a circular path, the object has an acceleration directed toward the center of the circle. This acceleration is known as the centripetal acceleration :

$$a_c = \frac{v^2}{r}.$$ (Equation 5.3: **Centripetal acceleration**)

The direction of the centripetal acceleration is toward the center of the circle, because the change in velocity is toward the center, as illustrated in Figure 5.10.

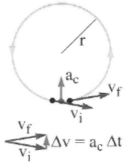

Figure 5.10: Subtracting the velocity just before the object is at the bottom of the circle ($\vec{v}_i$) from the velocity just after that point ($\vec{v}_f$) gives the change in velocity $\Delta\vec{v}$, which is directed toward the center of the circle. The centripetal acceleration is proportional to this change in velocity and thus is also directed toward the center.

Many people have heard the term "centripetal force." Is this a new force that arises because something goes in a circle? No, it is not. Let's investigate this idea. Which free-body diagram in Figure 5.11 correctly shows the force(s) acting on the Earth (E) as it orbits the Sun, when the Earth is at the position shown, to the right of the Sun? F_G is the gravitational force exerted on the Earth by the Sun, while F_C stands for centripetal force.

The correct free-body diagram is diagram 3, which shows only the force of gravity applied by the Sun on the Earth. The word "centripetal" means "directed toward the center." When an object experiences uniform circular motion, the object has a centripetal acceleration directed toward the center of the circle. The centripetal acceleration requires a net force directed toward the center, but this net force comes from one or more real forces (such as gravity, tension, or friction) or their components. There is no magical new centripetal force responsible for this motion. Thus, we will avoid the term "centripetal force" altogether, and talk about the force, or forces, responsible for the centripetal acceleration, instead.

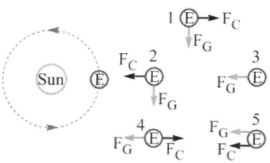

Figure 5.11: Five possible free-body diagrams for the Earth (E), as it orbits the Sun.

Figure 5.12: To avoid confusion, we will avoid the term "centripetal force" and we will not to draw a centripetal force on a free-body diagram.

EXPLORATION 5.5 – Identifying the force(s) responsible for the centripetal acceleration
Which force gives rise to the centripetal acceleration in the following situations?

Step 1 - *While you are riding your bike, you turn a corner, following a circular arc. What is the force acting on your bike that is associated with the centripetal acceleration, keeping you going in a circle?*

Let's sketch a free-body diagram (see Figure 5.14). As usual, there is a downward force of gravity acting on the system consisting of you and the bike, balanced by an upward normal force applied by the road. If there is no friction acting on the bike tires, the bike would keep going in a straight line, moving away from the center of the circle. The force of friction acting on the bike tires is the force pointing toward the center of the circular arc, opposing the tendency for the bike to move out from the center of the circle. As long as the tires do not skid on the road surface, the friction force is a static force of friction.

Figure 5.13: A photograph of motorcycle racer Steve Martin rounding a bend. Which force is directed in toward the center of the circular arc as the motorcycle rounds the bend? Photo credit: John Edwards, via http://www.publicdomainpictures.net.

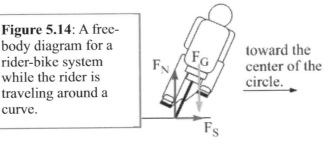

Figure 5.14: A free-body diagram for a rider-bike system while the rider is traveling around a curve.

toward the center of the circle.

Step 2 - *On a carnival ride called the Rotor, shown in the opening photo of this chapter, once the ride is spinning quickly, the riders are pinned to the wall and the floor is removed from under them. Which force is directed toward the center of the circle?*
Let's sketch the free-body diagram of a person on the moving ride. As usual, the force of gravity is acting down. The person's velocity tries to carry them farther from the center of the circle, but the wall gets in the way. There is a contact force associated with the person-wall interaction. The wall exerts a normal force, directed toward the center of the circle, on the person. This is the force we're looking for. The complete free-body diagram, in Figure 5.15, also shows an upward force of friction opposing the force of gravity. This force of friction is static friction because there is no relative motion between the person and the wall.

Figure 5.15: A free-body diagram for a person on the Rotor.

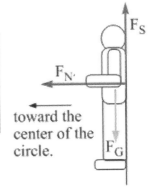

toward the center of the circle.

Key ideas for circular motion: In uniform circular motion, there is a net force directed toward the center of the circle. We do not need to invent a magical new force to act as the net force. Instead, this net force toward the center of the circle is associated with one or more of the standard forces we already know about. **Related End-of-Chapter Exercises 11, 59.**

Essential Question 5.5: Why does the Rotor have to be spinning fast before the floor is removed?

Answer to Essential Question 5.5: The faster the ride goes, the larger the normal force that the wall exerts on the rider. The larger the normal force is, the larger the maximum possible force of static friction that the wall can exert on the rider. The upward static friction force must balance the downward force of gravity. To provide a margin of safety, the maximum possible force of static friction should exceed the rider's weight by a significant amount. If the ride is too slow, the normal force is reduced, reducing the maximum possible force of static friction. If the maximum possible force of static friction drops below the rider's weight, the rider will slide down the wall. A similar situation could occur if the coefficient of static friction, associated with the interaction between the wall and the rider's clothes, is too small.

5-6 Solving Problems Involving Uniform Circular Motion

Let's investigate a typical circular-motion situation in some detail, although first we should slightly modify our general approach to solving problems using forces. Usually, the method that we follow in a uniform circular motion situation is identical to the approach that we use for other problems involving Newton's second law, where we apply the equation $\sum \vec{F} = m\vec{a}$. However, for uniform circular motion, the acceleration has the special form of Equation 5.3, $a_c = v^2/r$. Thus, when we apply Newton's second law, it has a special form.

The special form of Newton's second law for uniform circular motion is:

$$\sum \vec{F} = \frac{mv^2}{r}$$ (Eq. 5.4: **Newton's second law for uniform circular motion**)

where the net force, and the acceleration, is directed toward the center of the circle.

EXAMPLE 5.6 – A ball on a string

You are whirling a ball of mass m in a horizontal circle at the end of a string of length L. The ball has a constant speed v, and the string makes an angle θ with the vertical.
  (a) What is the tension in the string? Express your answer in terms of m, g, and θ.
  (b) What is v? Express your answer in terms of m, L, g, and/or θ.

SOLUTION

Let's apply the general method for solving problems using Newton's laws. The first step is to draw a diagram (see Figure 5.16a) showing the ball, the string, and the circular path followed by the ball. The next step is to draw a free-body diagram showing the forces acting on the ball. Although the ball is going in a horizontal circle, the string is at an angle. As shown in Figure 5.16b, only two forces act on the ball, the downward force of gravity and the force of tension that is directed away from the ball along the string.

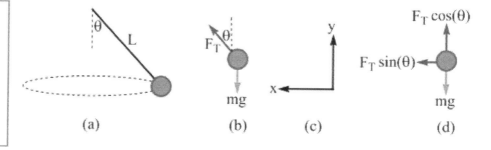

Figure 5.16: (a) A diagram and (b) free-body diagram for a ball being whirled in a horizontal circle at the end of a string. (c) An appropriate coordinate system. (d) A free-body diagram, with force components aligned with the coordinate system.

(a) (b) (c) (d)

Now, choose an appropriate coordinate system. The key is to align the coordinate system with the acceleration. Because the ball is experiencing uniform circular motion, the acceleration is directed horizontally toward the center of the circle. We can choose a coordinate system with axes that are horizontal and vertical, as in Figure 5.16c. Finally, split the tension into components, with $F_T\cos\theta$ vertically up and $F_T\sin\theta$ toward the center of the circle, as in Figure 5.16d.

(a) To find an appropriate expression for the tension, we can apply Newton's second law in the y-direction. Because there is no acceleration vertically, we have:

$$\sum \vec{F}_y = m\vec{a}_y = 0 .$$

Looking at the free-body diagram to evaluate the left-hand side of this equation gives:

$$+F_T\cos\theta - mg = 0 .$$

Solving for the tension gives: $F_T = \dfrac{mg}{\cos\theta}$.

(b) To find an expression for the speed of the ball, let's apply Newton's second law in the x-direction. The positive x-direction is toward the center of the circle, in the direction of the centripetal acceleration, so we apply the special form of Newton's second law that is appropriate for use in circular motion situations. The general equation is:

$$\sum \vec{F}_x = \dfrac{mv^2}{r} ,$$ where the acceleration is directed toward the center of the circle.

Looking at the free-body diagram in Figure 5.16d, we see that there is only one force in the x-direction, so:

$$+F_T\sin\theta = \dfrac{mv^2}{r} .$$

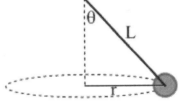

A common error in this situation is to assume that r, the radius of the circular path, is equal to L, the length of the string. Referring to Figure 5.17, however, it can be seen that $r = L\sin\theta$.

Substituting that into our equation gives:

$$F_T\sin\theta = \dfrac{mv^2}{L\sin\theta} , \quad \text{so} \quad v^2 = \dfrac{F_T L\sin^2\theta}{m} .$$

Figure 5.17: Note that r, the radius of the circular path, is not the same as L, the length of the string.

Using our result from part (a) to eliminate F_T gives:

$$v^2 = \dfrac{mgL\sin^2\theta}{m\cos\theta} = \dfrac{gL\sin^2\theta}{\cos\theta} .$$

Taking the square root of both sides gives: $v = \sin\theta\sqrt{\dfrac{gL}{\cos\theta}}$.

Related End-of-Chapter Exercises 21, 57.

Essential Question 5.6: If the speed of the ball in Example 5.6 is increased, what happens to θ, the angle between the string and the vertical?

Answer to Essential Question 5.6: As the speed increases the angle θ increases. One way to see this is to consider what happens to the tension. The vertical component of the tension remains the same, balancing the force of gravity, while the horizontal component of the tension increases (increasing the angle), because it provides the force toward the center needed for the ball to go in a circle, and that force increases as v increases. We can also convince ourselves that the angle increases, as the speed increases, by looking at the final equation in part (b) of Example 5.6.

5-7 Using Whole Vectors

The standard method of solving a problem involving Newton's laws is to break the forces into components. However, using whole vectors is an alternate approach. Let's see how whole vectors can be applied in a particular situation.

EXAMPLE 5.7 – Using whole vectors
(a) A box is placed on a frictionless ramp inclined at an angle θ with the horizontal. The box is then released from rest. Find an expression for the normal force acting on the box in this situation. What is the role of the normal force? What is the acceleration of the box?
(b) A box truck is traveling in a horizontal circle around a banked curve that is inclined at an angle θ with the horizontal. The curve is covered with ice and is effectively frictionless, so the truck can make it safely around the curve only if it travels at a particular constant speed (known as the *design speed* of the curve). Find an expression for the normal force acting on the truck in this situation. What is the role of the normal force? What is the design speed of the curve?
(c) Compare and contrast these two situations.

SOLUTION
(a) As usual, our first step is to draw a diagram, and then a free-body diagram showing the forces acting on the box, as in Figure 5.18. The two forces are the downward force of gravity, and the normal force applied by the ramp to the box. Because we are using whole vectors, we don't need to worry about splitting vectors into components. It is crucial, however, to think about the direction of the acceleration, which in this case is directed down the slope.

Let's apply Newton's second law, $\sum \vec{F} = m\vec{a}$, adding the forces as vectors. In this case, we get the right-angled triangle in Figure 5.18c. Each side of the triangle represents one vector in the equation $m\vec{g} + \vec{F}_N = m\vec{a}$. The vector $m\vec{a}$ (the net force) is parallel to the ramp, while the normal force is perpendicular to the ramp and the force of gravity is directed straight down. θ, the angle of the ramp, is the angle at the bottom of the triangle.

Because the force of gravity is on the hypotenuse of the triangle, we get:

$$\cos\theta = \frac{F_N}{mg}, \quad \text{so} \quad F_N = mg\cos\theta .$$

The role of the normal force is simply to prevent the box from falling through the ramp.

We can find the acceleration from the geometry of the right-angled triangle:

$$\sin\theta = \frac{ma}{mg} = \frac{a}{g} .$$

This relationship gives $\vec{a} = g\sin\theta$ directed down the slope.

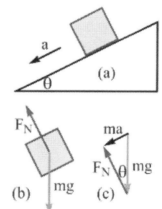

Figure 5.18: (a) A diagram, (b) free-body diagram, and (c) a right-angled triangle to show how the force of gravity and normal force combine to give the net force on the box.

(b) The situation of the box truck traveling around the banked curve resembles the box on the incline. A diagram of the situation, showing the back of the truck, is illustrated in Figure 5.19. The same forces, the force of gravity and the normal force, appear on this free-body diagram as in the free-body diagram for the box. The difference lies in the direction of the acceleration, which for the circular motion situation is directed horizontally to the left, toward the center of the circle.

Again we apply Newton's second law, $\sum \vec{F} = m\vec{a}$, adding the forces as vectors, where the magnitude of the acceleration has the special form $a_c = v^2/r$. This gives:

$$m\vec{g} + \vec{F}_N = \frac{mv^2}{r}$$, directed toward the center of the circle.

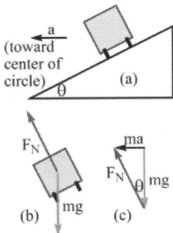

Again, each force represents one side of a right-angled triangle. Now the normal force is on the hypotenuse, and clearly must be larger than it is in part (a).

Using what we know about the geometry of right-angled triangles, we get:

$$\cos\theta = \frac{mg}{F_N}, \quad \text{so now } F_N = \frac{mg}{\cos\theta}.$$

If the angle of the ramp is larger than zero, then $\cos\theta$ is a number less than 1 and $F_N > mg$. In part (a) we had $F_N < mg$. The normal force for the box truck is larger because it has two roles. Not only does the normal force prevent the truck from falling through the incline, it must also provide the force directed toward the center to keep the truck moving around the circle.

Figure 5.19: (a) A diagram, (b) free-body diagram, and (c) right-angled triangle to show how the force of gravity and normal force combine to give the net force on the truck.

To find the design speed of the curve, we can use the other side of the triangle:

$$\sin\theta = \frac{ma}{F_N} = \frac{mv^2}{rF_N} = \frac{mv^2 \cos\theta}{rmg} = \frac{v^2 \cos\theta}{rg}.$$

Re-arranging the preceding equation gives a design speed of $v = \sqrt{\dfrac{rg\sin\theta}{\cos\theta}} = \sqrt{rg\tan\theta}$.

This is an interesting result. First, there is a design speed, a safest speed to negotiate the curve. At the design speed, the vehicle needs no assistance from friction to travel around the circle. Going significantly faster is dangerous because the vehicle is only prevented from sliding toward the outside of the curve by the presence of friction - the faster you go, the larger the force of friction required. Second, the design speed does not depend on the vehicle mass, which is fortunate for the road designers. The same physics applies to a Mini Cooper as to a large truck.

(c) A key similarity is the free-body diagram: in both cases there is a downward force of gravity and a normal force perpendicular to the slope. The key difference is that the accelerations are in different directions, requiring a larger normal force in the circular motion situation.

Related End-of-Chapter Exercises: 25 – 27.

Essential Question 5.7: Consider again the situation of the truck on the banked curve. In icy conditions, is it safest to drive very slowly around the curve or to drive at the design speed?

Answer to Essential Question 5.7: Even in very low-friction conditions, it is safer to travel at the design speed than at a slower speed! If you go too slowly around a banked curve, there is a tendency for your vehicle to slide down the slope and run off the road on the inside of the curve.

5-8 Vertical Circular Motion

A common application of circular motion is an object moving in a vertical circle. Examples include roller coasters, cars on hilly roads, and a bucket of water on a string. The bucket and roller coaster turn completely upside down as they travel, so they differ a little from the situation of the car on the road, which (we hope) remains upright.

EXAMPLE 5.8A – Whirling a bucket of water
A bucket of water is being whirled in a vertical circle of constant radius r at a constant speed v. What is the minimum speed required for the water to remain in the bucket at the top of the circle?

SOLUTION
Let's apply the general method, starting with the diagram in Figure 5.20. We then draw a free-body diagram, although we have to decide whether to analyze the bucket or the water. If we consider the bucket, two forces act on it, the force of gravity and the tension in the string, both of which are directed down when the bucket is at the top of the circle. If we consider the water, there is a downward force of gravity, and a downward normal force from the bucket takes the place of the tension. The analysis is the same in both cases, so let's consider the water.

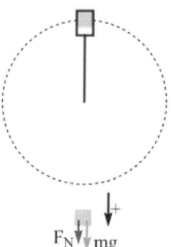

Figure 5.20: A bucket of water whirled in a vertical circle and a free-body diagram showing the forces acting on the water at the top of the loop.

Next, we choose an appropriate coordinate system. It is generally best to choose the positive direction as the direction of the acceleration, which points toward the center of the circle. When the bucket is at the top of the path, the acceleration, and the positive direction, points down. We don't need to split any forces into components. Let's apply Newton's second law, $\sum \vec{F} = m\vec{a}$.

Have a look at the free-body diagram to evaluate the left-hand side, and write the right-hand side in the usual circular-motion form. This gives:

$$+mg + F_N = +\frac{mv^2}{r}.$$

Solving for the normal force gives: $F_N = \frac{mv^2}{r} - mg$.

As long as the first term on the right exceeds the second term (in other words, as long as the normal force is positive), we're in no danger of having the water fall on us. Objects lose contact with one another (the water starts to fall out) when the normal force goes to zero. Setting the normal force to zero gives us the minimum safe speed of the bucket at the top of the circle:

$$0 = \frac{mv_{min}^2}{r} - mg, \qquad \text{which leads to } v_{min} = \sqrt{gr}.$$

Related End-of-Chapter Exercises: 12, 61.

EXAMPLE 5.8B – Apparent weight on a roller coaster
You are riding on a roller coaster that is going around a vertical circular loop. What is the expression for the normal force on you at the bottom of the circle?

SOLUTION
Once again, we apply the general method, starting with a diagram and a free-body diagram in Figure 5.21. We then draw a free-body diagram, which shows an upward normal force and a downward force of gravity. The system can be you or the car – the analysis is the same. Here we choose a coordinate system with a positive direction up, in the direction of the acceleration (toward the center of the circle). There is no need to split forces into components, so we can go straight to step 5 of the general method and apply Newton's second law:

$$\sum \vec{F} = m\vec{a} .$$

Have a look at the free-body diagram to evaluate the left-hand side, and remember that the right-hand side can be written in the usual circular-motion form. This gives:

$$+F_N - mg = +\frac{mv^2}{r} .$$

Solving for the normal force at the bottom of the circle gives:

$$F_{N,bottom} = \frac{mv^2}{r} + mg .$$

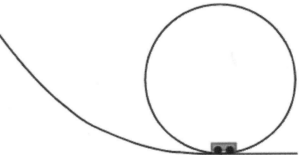

Let's compare our expression for the normal force on the car (or you) at the bottom of the loop to the expression for the normal force when the car (or you) is at the top. We can use the expression that we derived for the bucket at the top, in Example 5.8A, because the free-body diagram is the same in the two situations at the top of the loop.

Figure 5.21: A car on a roller-coaster track (left), as well as (right) the free-body diagram when the car is at the bottom of the loop.

$$F_{N,top} = \frac{mv^2}{r} - mg .$$

Note that the normal force at the bottom is larger than it is at the top. This difference is enhanced by the fact that the speed of the roller coaster at the bottom of the loop is larger than the speed at the top. Does this change in the normal force match the experience of a rider, who feels that she is lighter than usual at the top of the loop and heavier than usual at the bottom? Yes, because the normal force is the rider's apparent weight. Roller coasters are generally designed to have non-zero but fairly small normal forces at the top, so a rider feels almost weightless. At the bottom of the loop, the apparent weight can be considerably larger than *mg*, so a rider feels much heavier than usual.

Related End-of-Chapter Exercises: 20, 62.

Essential Question 5.8: You are on a roller coaster that reaches a top speed of 120 km/h at the bottom of a circular loop of radius 30 m. If you have a mass of 50 kg (and therefore a weight of 490 N), what is your apparent weight at the bottom of the loop? If the roller coaster's speed at the top of the loop has dropped to 80 km/h, what is your apparent weight at the top of the loop?

Answer to Essential Question 5.8: Let's first convert the speed from km/h to m/s, so all the units are SI units. The 120 km/h is equal to 33.3 m/s. Your apparent weight at the bottom of the loop is the normal force acting on you:

$$F_{N,bottom} = \frac{mv^2}{r} + mg = \frac{(50 \text{ kg})(33.3 \text{ m/s})^2}{30 \text{ m}} + 490 \text{ N} = +1850 \text{ N} + 490 \text{ N} = 2340 \text{ N} .$$

For the top of the loop, the speed of 60 km/h is equal to 22.2 m/s. Once again, your apparent weight is the normal force acting on you. Using the equation we found for the normal force at the top of the loop:

$$F_{N,top} = \frac{mv^2}{r} - mg = \frac{(50 \text{ kg})(22.2 \text{ m/s})^2}{30 \text{ m}} - 490 \text{ N} = +820 \text{ N} - 490 \text{ N} = 330 \text{ N} .$$

Chapter Summary

Essential Idea Regarding Applications of Newton's Laws
In this chapter we extended our understanding of what we can do with Newton's second law by applying it to situations involving friction and circular-motion situations.

Friction
The force of friction comes into play when objects are in contact. The force of friction is the component of the contact force that is parallel to the surfaces in contact (the normal force is the perpendicular component). The force of friction ***opposes relative motion*** between objects in contact. If there is relative motion between surfaces in contact (when one object is sliding over another), the friction force is the force of kinetic friction. The strength of the interaction is measured by the coefficient of kinetic friction, defined as:

$$\mu_K = \frac{F_K}{F_N} \qquad \text{so} \qquad F_K = \mu_K F_N . \qquad \text{(Equation 5.1: \textbf{The kinetic force of friction})}$$

If the force of friction prevents relative motion between surfaces in contact, the friction force is the force of static friction. The maximum strength of this interaction is measured by the coefficient of static friction, defined as:

$$\mu_S = \frac{F_{S,max}}{F_N} \qquad \text{so} \qquad F_S \leq \mu_S F_N . \qquad \text{(Equation 5.2 \textbf{The static force of friction})}$$

The coefficients of friction depend on the pair of interacting materials.

Be careful when working with static friction because, if the friction force that is required to ensure no relative motion is less than the maximum possible force of static friction, the static force of friction adjusts itself to whatever it needs to be to prevent relative motion. Two points to keep in mind about static friction are:

- we usually use Equation 5.2 with the equals sign only when we are certain we are dealing with a limiting case and the force of static friction is the maximum value;

- even the direction of the static force of friction can be counter-intuitive. A good rule of thumb is that the static force of friction opposes the relative motion that would occur if there were no friction.

A General Method for Solving a Problem Involving Newton's Laws, in Two Dimensions

The standard method of solving force problems involves splitting the forces into components, applying Newton's second law once for each direction, and combining the results. However, it should be kept in mind that some problems lend themselves to being solved with whole vectors. The steps in the standard method include:

1. Draw a diagram of the situation.

2. Draw one or more free-body diagrams, with each free-body diagram showing all the forces acting on an object.

3. For each free-body diagram, choose an appropriate coordinate system. Coordinate systems for different free-body diagrams should be consistent with one another. A good rule of thumb is to align each coordinate system with the direction of the acceleration.

4. Break forces into components that are parallel to the coordinate axes.

5. Apply Newton's second law twice to each free-body diagram, once for each coordinate axis. Put the resulting force equations together and solve.

Uniform Circular Motion – motion in a circle at constant speed

When an object travels with a speed v in a circular path of radius r, there is a centripetal acceleration directed toward the center of the circle, given by:

$$a_c = \frac{v^2}{r}.$$ (Equation 5.3: **Centripetal acceleration**)

To solve a problem involving uniform circular motion, we apply the method we use for other problems involving Newton's laws. The only change is in the form of Newton's second law:

$$\sum \vec{F} = \frac{mv^2}{r} \quad , \quad$$ (Eq. 5.4: **Newton's Second Law for uniform circular motion**)

where the acceleration is directed toward the center of the circle.

We avoid the term "centripetal force" because many people think of it as a new force that appears when an object travels in a circle - there is no such force! In uniform circular motion, there is a force directed toward the center of the circle but it is not a magical new force that arises just because something goes in a circle. Instead, the force is a familiar force, such as the force of gravity, the normal force, tension, or friction; or, the force is a force that we will investigate later in the book, such as the electric force between charged particles or the magnetic force.

End-of-Chapter Exercises

For all these exercises, assume that all strings are massless and all pulleys are both massless and frictionless. We will improve our model and learn how to account for the mass of a pulley in Chapter 10.

Exercises 1 – 12 are conceptual questions designed to see whether you understand the main concepts of the chapter. The first six exercises, in particular, are intended to give you more experience with free-body diagrams.

1. A box is placed in the middle of the floor of a truck. From the five free-body diagrams shown in Figure 5.22, which free-body diagram corresponds to the following situations? Assume the floor of the truck is horizontal at all times, and that for parts (a) - (d) the box does not slip on the floor of the truck. (a) The truck is moving at constant velocity to the right. (b) The truck is moving at constant velocity to the left. (c) The truck, while moving right, is speeding up. (d) The truck, while moving right, is slowing down. (e) The truck, while moving right, is stopping so quickly that the box slides over the floor of the truck. (f) The truck, while moving right, is speeding up so rapidly that the box slides over the floor of the truck.

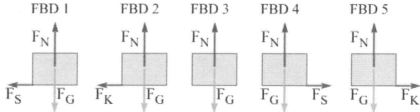

2. Various possible free-body diagrams are shown in Figure 5.22 for a box on the floor of a moving truck. Imagine that you are in the back of the truck, looking forward at the box. As the truck makes a turn to the right, gently enough that the box does not slide over the floor of the truck, which free-body applies? Explain your choice.

Figure 5.22: Possible free-body diagrams for a box in a truck, for Exercises 1 and 2.

3. Figure 5.23 shows five possible free-body diagrams (labeled FBD 1 through FBD 5) for a box on a ramp. (a) On the free-body diagrams, four different forces are shown. Briefly describe each of these forces and state what applies the force to the box. (b) Determine which free-body diagram, if any, matches each of the following situations: (i) the box remains at rest on the ramp; (ii) the box is sliding down the ramp, with some friction acting; (iii) the box is sliding up the ramp, which is frictionless.

4. Figure 5.23 shows five possible free-body diagrams (labeled FBD 1 through FBD 5) for a box on a ramp. Describe a situation in which the correct free-body diagram is FBD 3.

5. Let's say that the free-body diagrams shown in Figure 5.23 are possible free-body diagrams for a box placed on the floor of a truck that is traveling on a hill. Identify the free-body diagram that could apply in each of the following situations. For parts (a) – (d),

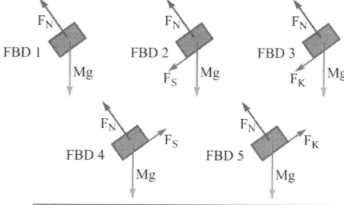

Figure 5.23: Five possible free-body diagrams for a box on a ramp, for Exercises 3 – 6.

assume the box does not slide over the floor of the truck. (a) The truck is traveling at constant velocity up the hill. (b) The truck is traveling at constant velocity down the hill. (c) The truck is at rest at a red light halfway up the hill. (d) The truck is speeding up as it climbs the hill. (e) The truck's acceleration up the hill is so large that the box slides over the floor of the truck.

6. Consider again the situation described in Exercise 5, regarding a box on the floor of a truck traveling up a hill that has a constant angle θ with respect to the horizontal. Now the truck is traveling down the hill and its speed is increasing. Which of the free-body diagrams shown in Figure 5.23 could apply in that situation, assuming the box does not slide over the floor of the truck? Select all possible answers, and explain in what situation(s) each free-body diagram applies.

7. A box with a weight of mg is initially at rest on a board that is initially horizontal. The angle between the board and the horizontal, θ, is gradually increased until the box starts to slide on the board. The coefficient of static friction between the box and the board is $\mu_s = 1.00$. (a) Plot a graph of the magnitude of the normal force applied to the box by the board, as a function of θ from $\theta = 0$ to $\theta = 90°$, when θ is gradually increased from zero. (b) What else could that graph represent? (c) On the same graph, plot the magnitude of the component of the force of gravity acting on the box, as a function of θ, as θ is gradually increased. (d) What else could the graph from part (c) represent? (e) On the same graph, plot the magnitude of the maximum possible force of static friction that could act on the box if the coefficient of static friction is reduced to 0.50. (f) Briefly describe how the graphs can be used to find the angle at which the block first starts to slide on the board.

8. Three situations involving a block with a weight of 10 N and a string are shown in Figure 5.24. In cases 2 and 3, the string passes through a hook on top of the block. (a) Rank these situations based on the tension in the string, from largest to smallest (e.g., 3>1=2). (b) What is the tension in the string in case 3?

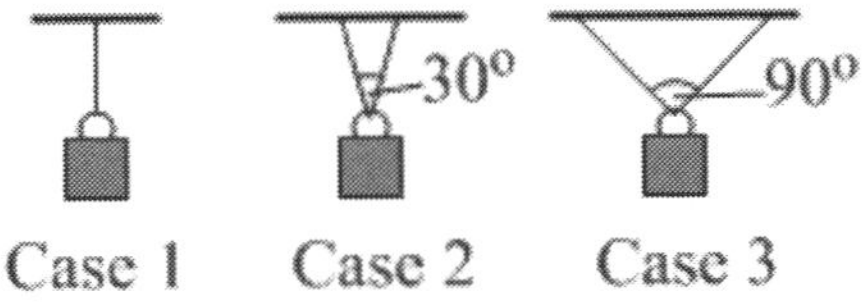

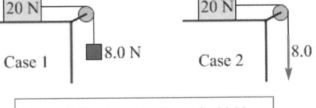

Case 1   Case 2   Case 3

Figure 5.24: Three different situations of a block being supported by a string, for Exercise 8.

9. As shown in Figure 5.25, identical 20-N boxes are placed on identical horizontal surfaces. Each box has a string attached that passes over a pulley. In Case 1, the other end of the string is attached to a block with a weight of 8.0 N. In case 2, you simply exert an 8.0 N force straight down on the string. In both cases, enough friction acts on the box to keep the box at rest. (a) In which case is the force of friction acting on the box larger? (b) In which case is the tension in the string larger? (c) Calculate the force of friction acting on the box, and the tension in the string, in case 1. (d) What, if anything, can you say about the coefficient of static friction between the box and the surface in case 1?

Figure 5.25: Two situations of a 20-N box on a level surface, for Exercise 9.

10. A box of mass m is placed on a frictionless ramp inclined at an angle of θ with respect to the horizontal. The ramp has a mass M and is on a frictionless horizontal surface. If the ramp remained at rest, the box would slide down the incline, but if the ramp is given a

horizontal acceleration of just the right magnitude, the box will remain at rest with respect to the ramp. (a) Using whole vectors, construct a right-angled triangle showing the normal force applied to the box, the force of gravity on the box, and the net force on the box. (b) Use the triangle to determine the acceleration of the system.

11. A disk is placed on a turntable at some distance from the center of the turntable. When the turntable rotates, the disk moves along with it without slipping. Which force, forces, or force components give rise to the disk's centripetal acceleration?

12. Take a metal coat hanger and stretch it out, as shown in Figure 5.26. It is best if a straight line through the hook of the hanger passes through the top vertex of the hanger, as shown by the dashed line in the figure. Holding the hanger vertically, balance a penny on the hook. After getting the hanger to rock back and forth on your finger, make the penny and coat hanger rotate so that at some times the penny is completely upside down. The penny should remain on the hanger at all times! (Bonus points if you can bring the system to a stop without the penny falling off.) Explain how this is possible.

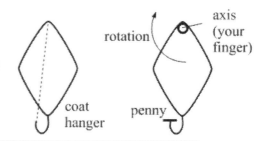

Figure 5.26: Spin a penny on the end of a stretched-out metal coat hanger, for Exercise 12.

Exercises 13 – 19 are designed to give you practice with the general method for solving a problem involving Newton's laws. For each of these exercises, start by doing the following: (a) Draw a diagram of the situation. (b) Draw one or more free-body diagram(s) showing all the forces that act on various objects or systems. (c) Choose an appropriate coordinate system for each free-body diagram. (d) If necessary, split forces into components that are parallel to the coordinate axes. (e) Apply Newton's second law enough times to obtain the necessary equations to solve the problem.

13. You pull a box, which has a weight of 20 N, across a horizontal floor at a constant velocity of 5.0 m/s. You do this by exerting a 10 N force horizontally on a string attached to the box. The goal here is to find the coefficient of kinetic friction between the box and the floor. Parts (a) – (e) as described above. (f) Find the coefficient of kinetic friction between the box and the floor.

14. Repeat Exercise 13, except that now the string between the box and your hand is at an angle of 30° with the horizontal, so your 10 N force has a component directed vertically up.

15. A box with a mass of 2.5 kg is initially at rest on a horizontal frictionless surface. You then pull on a string attached to the box, exerting a constant force F at an angle of 45° above the horizontal. This pull causes the box to travel 4.0 m horizontally in 2.0 s. Parts (a) – (e) as described above. (f) Find the box's acceleration by treating this as a one-dimensional constant-acceleration problem. (g) Find the magnitude of the force F. (h) Find the magnitude of the normal force acting on the box. Use $g = 9.8$ m/s^2.

16. In Atwood's machine (see Figure 5.27), the two blocks connected by a string passing over the pulley have masses of 1.0 kg (on the left) and 3.0 kg (on the right). The system is initially held at rest by means of a second string tied from the lighter block to the ground. Use $g = 9.8$ m/s^2. Parts (a) – (e) as described above. (f) What is the tension in the string connecting the two blocks? (g) What is the tension in the second string? (h) Without doing any calculations, describe whether the tension in the first string will increase,

decrease, or remain the same when the second string is cut. (i) Calculate the tension in the first string when the second string is cut.

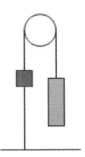

17. While moving out of your apartment, you push a box of textbooks, which has a mass of 20 kg, up the ramp of your moving truck. The coefficient of kinetic friction between the ramp and the box is 0.20, and the ramp is inclined at 30° with respect to the horizontal. Our goal here is to determine what force you push the box with, if the box travels at constant speed up the ramp and the force you exert is parallel to the ramp. Parts (a) – (e) as described above. (f) What is the magnitude of the force you exert on the box? Use $g = 9.8$ m/s².

Figure 5.27: Atwood's machine, held at rest by a string connecting the smaller mass to the ground, for Exercise 16.

18. Three blocks, A, B, and C, are connected by massless strings, as shown in Figure 5.28. The two strings pass over massless frictionless pulleys. The mass of block A is $m_A = 4.00$ kg. The mass of block B is $m_B = 8.00$ kg. Use $g = 10.0$ m/s². Our first goal will be to determine the acceleration of the system if the mass of block C is $m_C = 8.00$ kg. Let's start by assuming that the table supporting block B is frictionless. Parts (a)-(e) as described above. (f) What is the acceleration of the system if $m_C = 8.00$ kg? (g) What would the mass of the block C have to be for the system to have no acceleration? (h) If the coefficient of static friction between block B and the table is $\mu_S = 0.300$, the system will remain at rest if the mass of block C is between a maximum value m_{Cmax} and a minimum value m_{Cmin}. What are m_{Cmax} and m_{Cmin}? (i) If the mass of block C is 5.00 kg, what is the force of friction exerted on block B by the table? Assume the system remains at rest.

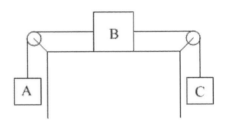

Figure 5.28: A diagram of three blocks connected by strings, for Exercise 18.

19. A box of mass $m = 3.0$ kg is placed in the center of a flatbed truck. The coefficients of friction between the box and the truck are $\mu_S = 0.40$ and $\mu_K = 0.30$. Our first goal will be to determine the force of friction acting on the box when the truck accelerates forward on a level road, starting from rest. Assume the box remains at rest relative to the truck. Parts (a) – (e) as described above. (f) If the truck's acceleration is 2.0 m/s², what is the magnitude and direction of the force of friction acting on the box? Is it static friction or kinetic friction? (g) What is the largest magnitude that the truck's acceleration can be for the box to remain at rest relative to the truck? (h) Which of the values stated in the problem did you not need to solve the problem?

Exercises 20 – 22 are designed to give you practice with the general method for solving a uniform circular motion problem. For each of these exercises, start by doing the following: (a) Draw a diagram of the situation. (b) Draw one or more free-body diagram(s) showing all the forces that act on various objects or systems. (c) Choose an appropriate coordinate system for each free-body diagram. (d) If necessary, split forces into components that are parallel to the coordinate axes. (e) Apply Newton's second law enough times to get the equations necessary to solve the problem, and solve the resulting equations.

20. While driving your car on a hilly road, you pass over a hill that, at the top, is a circular arc with a radius of 20 m. Parts (a) – (e) as described above. (f) How fast are you traveling if, at the top of the hill, the normal force exerted by the road on your car is 30% less than it was the previous day, when you stopped your car there to admire the view? Assume the mass of the car is the same on the two days.

21. A *whirligig conical pendulum* consists of a ball of mass M being whirled in midair in a horizontal circle at a constant speed v (see Figure 5.29). The ball is tied to a string that passes over a pulley, which rotates on frictionless bearings as the ball travels around the circle, and down through a hollow tube. A block of mass m is tied to the other end of the string. As the ball travels around the circle, the string makes an angle of θ with the horizontal, and the part of the string that goes from the ball to the pulley has a fixed length L. Parts (a) – (e) as described above. (f) In terms of the variables M, m, and/or g only, what is the magnitude of the tension in the string? (g) In terms of M, m, g, L, and/or θ only, what is v, the constant speed of the ball?

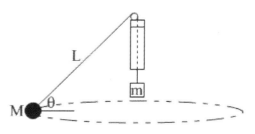

Figure 5.29: A whirligig conical pendulum, for Exercise 21.

22. Mary is whirling around on a merry-go-round at the playground that her friend Caitlin is gradually spinning faster and faster. The goal of the exercise is to determine the maximum speed Mary can travel at, if the coefficient of static friction between her shoes and the floor is μ_S. Assume that Mary is not hanging on, but is simply balancing on her feet at a distance r from the center. Parts (a) – (e) as described above. (f) Find Mary's maximum speed. (g) Which, if any, of the following would help Mary stay on the merry-go-round? (i) Moving closer to the center of the turntable. (ii) Moving closer to the edge of the turntable. (iii) Wearing a heavy backpack that increases her mass.

Exercises 23 – 27 are designed to give you practice with applying whole vectors, so take a whole-vector approach to the situations described.

23. A box with a weight of 20 N is initially at rest on a horizontal surface. The coefficients of friction between the box and the surface are $\mu_S = 0.30$ and $\mu_K = 0.20$. A force is then applied to the box at an angle of 30° with the horizontal so that the box has a non-zero acceleration directed entirely horizontally and no friction acts on the box. What is the magnitude of this applied force?

24. A box with a weight of 40 N remains at rest on a horizontal surface, even though a 25 N horizontal force is being applied to it. What are the magnitude and direction (the angle with respect to the horizontal) of the contact force exerted on the box by the surface?

25. A box with a weight of 40 N remains at rest on a flat surface inclined at 30° with respect to the horizontal, even though a 25 N horizontal force is being applied to it. (a) What are the magnitude and direction (the angle with respect to the horizontal) of the contact force exerted on the box by the surface? What are the magnitude of the (b) normal force and (c) friction force exerted on the box by the surface?

26. As shown in Figure 5.30, a disk with a weight of 2.0 N lies on a rotating turntable such that the disk does not slip on the turntable. The turntable spins at a steady rate of 1 revolution every second and the disk is 50 cm from the turntable's center. (a) Draw a right-angled triangle showing the contact force applied to the disk by the turntable, the force of gravity acting on the disk, and the disk's net force. (b) Using $g = 9.8$ m/s^2, find the angle between the horizontal and the contact force.

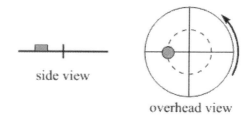

Figure 5.30: A disk on a rotating turntable, for Exercise 26.

27. A 5.0 kg box is at rest on a ramp that is at an angle of 15° with the horizontal. The coefficient of static friction between the box and the ramp is $\mu_S = 0.40$. Begin by sketching a free-body diagram for the box, showing the force of gravity and the contact force. (a) What are the magnitude and direction of the contact force applied by the ramp on the box? (b) What is the magnitude of the normal force applied by the ramp on the box? (c) What is the magnitude of the force of friction applied by the ramp on the box?

General problems and conceptual questions

28. Return to the situation described in Exercise 27, but this time use an approach in which you choose an appropriate coordinate system and split forces into components. A 5.0 kg box is at rest on a ramp that is at an angle of 15° with the horizontal. The coefficient of static friction between the box and the ramp is $\mu_S = 0.40$. Determine (a) the normal force applied by the ramp on the box, (b) the force of friction applied by the ramp on the box, and (c) the magnitude and direction of the contact force applied by the ramp on the box.

29. Return to the situation described in Exercises 27 and 28. If you would like the box to start moving, what is the minimum force you can apply if your force is directed (a) parallel to the slope? (b) perpendicular to the slope? (c) horizontally?

30. A shuffleboard disk with a mass m is sliding on a horizontal surface. The coefficient of kinetic friction between the disk and the surface is μ_K, and the acceleration due to gravity is g. In terms of the variables stated here, determine the magnitude of (a) the normal force acting on the disk, (b) the force of friction acting on the disk, and (c) the net contact force acting on the disk from the surface. (d) What is the angle between the contact force and the surface?

31. Return to the situation described in Exercise 30. If the disk has an initial speed of 4.0 m/s, $\mu_K = 0.20$, and $g = 9.8$ m/s², how far does the disk slide?

32. A 4.0 kg box is initially at rest on a flat surface. The coefficient of static friction between the box and the surface is $\mu_S = 0.50$. Determine the minimum force you have to apply to start the box moving if your force is directed (a) horizontally; (b) at an angle of 20° from the horizontal with a vertical component down; (c) at an angle of 20° from the horizontal with a vertical component up.

33. Return to the situation described in Exercise 32. If the angle between the force you apply and the horizontal is larger than a particular critical angle, the box will not move, no matter what magnitude your force is. (a) Is this true when your force has a downward vertical component, an upward vertical component, or in both of these cases? (b) Find an expression for the critical angle θ_C in terms of the coefficient of static friction μ_S. (c) Solve for θ_C when $\mu_S = 0.50$.

34. A box with a weight of 20 N is initially at rest on a flat surface. The coefficients of friction for the box and the surface are $\mu_S = 0.50$ and $\mu_K = 0.40$. A horizontal force with a constant direction is then applied to the box. The magnitude of the force increases with time according to the equation $F = (2.0 \text{ N/s})t$ until the box starts to move. Then, the force remains constant at the value it has when the box starts to move. (a) Sketch a graph showing the magnitude of this horizontal force, and the magnitude of the force of friction

acting on the box, as a function of time for the first 8.0 s after the horizontal force is first applied. (b) Sketch a graph of the speed of the box as a function of time over the same 8.0-second interval.

35. Figure 5.31 shows a box with a mass of 2.5 kg remaining at rest on a horizontal table. A string is tied to the box and held in a horizontal position by means of a system of two other strings, one that hangs vertically and supports a small block with a mass of 500 g, and the other that makes a 30° angle with the horizontal and is tied to a wall. Use $g = 9.8$ m/s². (a) What is the magnitude of the force of friction acting on the box? (b) What, if anything, can you say about the coefficient of static friction between the box and the table in this situation?

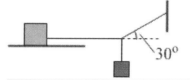

Figure 5.31: An equilibrium situation involving two blocks, for Exercise 35.

36. A box is placed at the bottom of a ramp that measures 3.00 m vertically by 4.00 m horizontally. The coefficients of friction for the box and the ramp are: $\mu_S = 0.60$ and $\mu_K = 0.50$. The box is then given an initial velocity up the ramp of 4.0 m/s. The box slows down as it slides up the ramp and eventually comes to a stop. (a) Will the box remain at rest, or will the stop just be for an instant before it slides down the ramp again? Justify your answer. (b) If you determine that the box slides down again, which takes more time, sliding up the ramp or sliding down the ramp? Justify your answer without calculating either time.

37. Return to the situation described in the previous exercise, and use $g = 10$ m/s². (a) How far does the box travel up the ramp? (b) How long does the box spend sliding up the ramp? (c) If you determine that the box slides down the ramp again, how long does the box spend sliding down the ramp?

38. As shown in Figure 5.32, identical 20-N boxes are placed on identical horizontal surfaces. Each box has a string attached that passes over a pulley. In Case 1, the other end of the string is attached to a block with a weight of 8.0 N. In case 2, you simply exert an 8.0 N force straight down on the string. The coefficients of friction for the box-surface interaction are the same in both cases. Although there is some friction, in both cases the box accelerates from rest to the right. Try to answer parts (a) – (c) without doing any calculations. (a) In which case is the force of friction acting on the box larger? (b) In which case is the tension in the string larger? (c) In which case is the acceleration larger? (d) What, if anything, can you say about the coefficient of static friction between the box and the surface?

39. Return to the situations shown in Figure 5.32. In each case, the coefficients of friction are $\mu_S = 0.20$ and $\mu_K = 0.10$. Use $g = 10$ m/s². Determine the acceleration of the 20-N box in (a) case 1 (b) case 2.

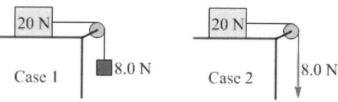

40. Consider the situation shown in case 1 in Figure 5.32. Assume now that the coefficients of friction between the box and the surface are $\mu_S = 0.25$

Figure 5.32: Two situations of a 20-N box on a level surface, for Exercises 38 – 40.

and $\mu_K = 0.20$. The 20-N box is given an initial velocity of 2.0 m/s away from the pulley.

(a) How far will the box travel before coming instantaneously to rest? (b) How much time passes until the box comes instantaneously to rest? (c) Will the box remain at rest or will it reverse direction and travel back toward the pulley? How can you tell? (d) If you decide that the box does travel back toward the pulley, determine (i) how fast it is going when it reaches its starting point; (ii) how long the return trip to the starting point takes.

41. Two identical boxes are linked by a string passing over a pulley, as in Figure 5.33. One box is on a horizontal surface, while the other is on an incline in the shape of a 3-4-5 triangle. If the surfaces are frictionless, what is the acceleration of the system?

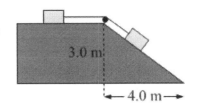

42. Return to the situation described in Exercise 41 and shown in Figure 5.33. Each of the following changes to the system is made separately – the changes are not sequential. State whether the change affects the acceleration of the system and, if so, state what happens to the acceleration. (a) Double the mass of both boxes. (b) Double the mass of the box on the incline only. (c) Double the mass of the box on the horizontal surface only. (d) Transfer the whole system to the Moon, where the acceleration due to gravity is about 1/6 of the value here on Earth.

Figure 5.33: Two boxes connected by a string, for Exercises 41 – 45.

43. Two identical boxes are linked by a string passing over a pulley, as in Figure 5.33. One box is on a horizontal surface, while the other is on an incline in the shape of a 3-4-5 triangle. If the coefficients of friction for both box-surface interactions are $\mu_S = 0.15$ and $\mu_K = 0.10$ and the system is released from rest, what is the magnitude of the acceleration of the system?

44. Two identical boxes are linked by a string passing over a pulley, as in Figure 5.33. One box is on a horizontal surface, and the other is on an incline in the shape of a 3-4-5 triangle. Both boxes are moving such that the string between them remains taut. If the acceleration of each box has a magnitude of $g/10$, what is the coefficient of kinetic friction between the boxes and the surface? Assume the coefficient of kinetic friction is the same for both boxes. How many different solutions does this problem have? Find all the solutions.

45. Two identical boxes are linked by a string passing over a pulley, as in Figure 5.33. One box is on a horizontal surface, while the other is on an incline in the shape of a 3-4-5 triangle. Both boxes are moving with a constant non-zero velocity such that the string between them remains taut. What is the coefficient of kinetic friction between the boxes and the surface? Assume the coefficient of kinetic friction is the same for both boxes. How many different solutions does this problem have? Find all the solutions.

46. Fred pulls down on a massless rope (rope 1) with a constant force F (see Figure 5.34). This rope passes over a massless, frictionless pulley and is tied to a box with a mass $m_1 = 2.00$ kg. Rope 2 connects the first box to a second box with a mass $m_2 = 4.00$ kg. The two boxes accelerate upwards with an acceleration $a = 4.00$ m/s². Use $g = 10.0$ m/s². (a) Find the tension in rope 2. (b) Find the tension in rope 1. (c) If the boxes are initially at rest, what distance have they traveled 0.500 seconds after they start moving? (d) If the ground exerts an upward normal force on Fred of 500 N, what is Fred's mass?

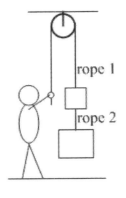

Figure 5.34: Fred accelerates a system of two boxes upward, for Exercise 46.

47. A 5.00 kg box is placed on a ramp that measures 3.00 m vertically by 4.00 m horizontally. As shown in Figure 5.33, the box is attached to a 1.00 kg ball by a massless rope that passes over a massless and frictionless pulley. A horizontal force of $F = 10.0$ N is also applied to the box. Use $g = 10.0$ m/s². (a) If there is no friction between the box and the ramp, what is the acceleration of the box? State clearly if the acceleration is directed up or down the ramp. (b) What value of the horizontal force F would result in zero acceleration for the box? (c) Let's introduce some friction between the box and the ramp, with a coefficient of static friction of $\mu_S = 1/3$. In this case, the box, if it starts from rest, will have no acceleration for all values of the horizontal force F between some minimum value F_{min} and some maximum value F_{max}. What are these minimum and maximum values?

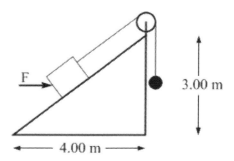

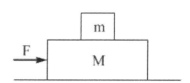

Figure 5.35: The situation described in Exercise 47.

48. A block of mass m is placed on top of a larger block of mass M and the two-block system rests on a horizontal frictionless surface. There is friction between the two blocks, with static and kinetic friction coefficients μ_S and μ_K, respectively. As shown in Figure 5.36, a horizontal force F is applied to the larger block. Express all your answers below in terms of m, M, g, F, and/or μ_S or μ_K. (a) For small values of F, the two blocks slide together on the surface. For this situation, what are: (i) the acceleration of the large block and (ii) the friction force on the small block? (b) For large values of F the small block slides with respect to the large block. What is the acceleration of the large block in this case? (c) What is the maximum value of F such that the small block does not move with respect to the large block?

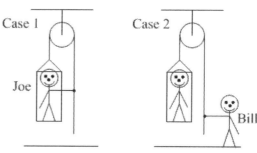

Figure 5.36: A stack of two blocks, with a force being applied to the lower block, for Exercises 48 and 49.

49. Return to the situation described in Exercise 48 and shown in Figure 5.36, but now we will use numerical values for the variables. We will also introduce some friction between the large box and the horizontal surface. Let's use $m = 5.0$ kg, $M = 10$ kg, $g = 10$ m/s², and say that $\mu_S = 0.40$ and $\mu_K = 0.30$ for the interaction between the blocks as well as the interaction between the large block and the horizontal surface. (a) If the blocks are initially at rest, what is the minimum value of the horizontal force F necessary to start the system moving? (b) Once the blocks are moving, what is the minimum value of the horizontal force F necessary to keep the system moving at constant velocity? (c) If the blocks are initially at rest, what is the maximum value the horizontal force F can be such that the small block does not slide on the large block?

50. As shown in Figure 5.37, Joe is placed in a large bucket that is suspended in mid-air by a rope that passes over a pulley. In Case 1, Joe holds the other end of the rope. In Case 2, Bill, on the ground, holds the other end of the rope. If Joe and the bucket are stationary, in which case is the tension in the rope larger?

51. Return to the situation described in Exercise 50. If Bill's mass is 70 kg, the combined mass of Joe and the bucket is 50 kg, and the bucket is accelerating up at the rate of 2.0 m/s², what is the tension in the rope in (a) Case 1? (b) Case 2?

Figure 5.37: Two cases of Joe in a bucket, for Exercises 50 and 51.

52. Two boxes are side-by-side on a horizontal floor as shown in Figure 5.38. The coefficient of kinetic friction between each box and the floor is μ_K. In Case 1, a horizontal force F directed right is applied to the box of mass M and the boxes are moving to the right. In Case 2, the horizontal force F is instead directed left and applied to the box of mass m, and the boxes are moving to the left. Find an expression for the magnitude of F_{Mm}, the force the box of mass M exerts on the box of mass m, in (a) Case 1 (b) Case 2. Express your answers in terms of variables given in the problem.

Figure 5.38: Two cases involving two boxes on a horizontal floor, with an applied force, for Exercises 52 and 53.

53. Consider again the situation described in Exercise 52 and shown in Figure 5.38. Let's use $m = 5.0$ kg, $M = 10$ kg, $g = 10$ m/s², $F = 20$ N, and say that $\mu_K = 0.40$ for the interaction between the boxes and the floor. What are the magnitude and direction of the acceleration of the system if the boxes are initially moving to the right in (a) Case 1; (b) Case 2? What are the magnitude and direction of the acceleration of the system if the boxes are initially moving to the left in (c) Case 1; (d) Case 2?

54. Three blocks are placed side-by-side on a horizontal surface and subjected to a horizontal force F, as shown in case 1 of Figure 5.39. Doing so causes the blocks to accelerate from rest to the right. The coefficients of friction between each box and the surface are $\mu_S = 0.30$ and $\mu_K = 0.20$. Use $g = 10$ m/s², and consider case 1 only. If the horizontal force F is 40 N, what are (a) the acceleration of the system? (b) the net force acting on the 2.0 kg block? (c) the force exerted by the 5.0 kg block on the 2.0 kg block?

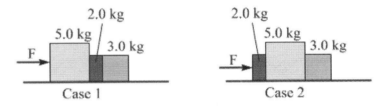

Figure 5.39: Two situations involving three blocks being pushed from the left by a horizontal force F, for Exercises 54 – 55.

55. Repeat Exercise 54, but now answer the questions for case 2, instead.

56. You get off a bus, take a plastic cube out of your pocket, and hold the cube against the front of the bus, which is vertical. When the bus starts to move, you carefully let the cube go, and step away from the bus. As the bus accelerates at 4.0 m/s², you notice that the cube remains in place on the front of the bus. What does this tell you about the coefficient of static friction between the cube and the bus?

57. While on a train, you are holding a string with a ball attached to it. At first, the train's velocity is constant and the string is vertical. When the train changes direction while going around a curve with a radius of 200 m, you measure the angle between the string and the vertical to be 20°. What is the speed of the train?

58. A hockey puck, with a mass of 160 g, is placed 75 cm from the center of a horizontal turntable. The turntable begins to rotate, gradually spinning faster as its rate of rotation increases at a steady rate. The speed of the puck, as a function of time, is given by $v = (0.10 \text{ m/s}^2)t$. The coefficients of friction between the puck and the turntable are $\mu_S = 0.60$ and $\mu_K = 0.50$. (a) Assuming the force of friction is directed toward the center of the turntable, determine at what time the puck starts to slide on the turntable. Plot graphs, as a function of time, of (b) the maximum possible force of static friction, and (c) the magnitude of the actual force of friction acting on the puck.

59. A cube of mass m is placed in a funnel that is rotating around the vertical axis shown in Figure 5.40. There is no friction between the cube and the funnel, but the funnel is rotating at just the right speed needed to keep the cube rotating with the funnel. The cube travels in a circular path of radius r, and the angle between the vertical and the wall of the funnel is θ. Express your answers in terms of m, r, g, and/or θ. (a) Which force, forces, or force components give rise to the centripetal acceleration in this situation? (b) What is the normal force acting on the cube? (c) What is the speed v of the cube?

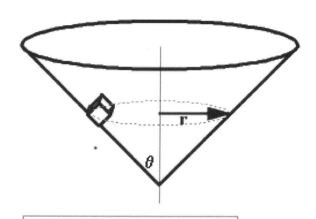

Figure 5.40: A cube in a rotating funnel, for Exercises 59 and 60.

60. Return to the situation described in Exercise 59, except now there is some friction between the cube and the funnel. Let's use the following values: $\theta = 45°$, $m = 500$ grams, $r = 20$ cm, and $g = 9.8 \text{ m/s}^2$, with coefficients of friction between the cube and the funnel of $\mu_S = 0.20$ and $\mu_K = 0.10$. (a) What is the cube's speed, if there is no force of friction acting on the cube? (b) What are the maximum and minimum values that the cube's speed can be, so that the cube will not slide up or down in the funnel?

61. Matt, with a mass of 22.0 kg, is swinging on a tire swing (a tire attached to a rope hanging from a tree). The tire itself has a mass of 5.0 kg, and the rope has a length of 4.0 m. (a) If Matt's speed is 3.0 m/s when the tire reaches the bottom of its circular arc, what is the tension in the rope at that point? (b) Matt's father, Bob, thinks that the tire swing looks like fun. If the rope will break when the tension exceeds 750 N, is it safe for Bob to jump on the swing with Matt, should Bob swing by himself, or should he not swing at all? Justify your answer. Bob's mass is 70.0 kg, and he would reach the same 3.0 m/s speed that Matt does at the bottom of the swing.

62. You are traveling on a hilly road. At a particular spot, when your car is perfectly horizontal, the road follows a circular arc of some unknown radius. Your speedometer reads 80 km/h, and your apparent weight is 30% larger than usual. (a) Are you at the bottom of a hill or the top of a hill at that instant? (b) What is the radius of the circular arc?

63. As shown in Figure 5.41, a small disk of mass *m* is placed on a rotating turntable so that the disk spins with the turntable at a constant radius *R* from the center, and with a constant speed *v*. The disk is tied to a cylinder, which has a mass of 3*m*, by a string that passes through a hole in the center of the turntable. Find an expression for the speed of the disk, in terms of the variables given here and *g*, the magnitude of the acceleration due to gravity. Neglect friction between the disk and the turntable.

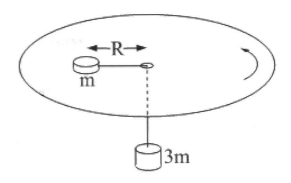

Figure 5.41: A disk of mass *m* on a rotating turntable, for Exercise 63. The disk is tied to a cylinder, of mass 3*m*, by a string that passes through a hole in the center of the turntable.

64. Why does a passenger tilt to the right when the car the passenger is riding in turns left, following a circular path? Comment on the statements made by these physics students as they wrestle with this question.

Lucy: The question mentions a circular path, so it must be a circular motion situation. In that case, there must be a net force on the passenger directed toward the center of the circle. But that direction is to the left, isn't it? Why does the passenger go right?

Brad: We've all experienced this, though. There must be a force on the passenger going to the right. How about we try to identify the forces that are directed horizontally, to see what we come up with?

Jillian: I think this is like the situation of the object on the turntable. When the turntable goes fast enough, the object starts sliding away from the center, but it is not because of a force in that direction. There is just not enough force directed inward to keep the object going in a circular path – is this situation like that one?

The photograph shows the result of a car crash staged as part of a Canadian study to investigate child restraint systems. The car has momentum and kinetic energy before the collision. The car's momentum and kinetic energy are then modified, as this photo shows so clearly, by the force applied to the car by the barrier during the collision.
Photo courtesy NHTSA/DOT.

Chapter 6 – Linking Forces to Momentum and Energy

CHAPTER CONTENTS

In this chapter, we will build on what we have learned previously. Looking at Newton's laws and one of the constant acceleration equations from a different perspective will bring in some new concepts, giving us more tools in our toolbox for analyzing physical situations such as the collision shown in the photograph above.

One of the new concepts we will examine is momentum, which is a vector quantity. Momentum is directly related to velocity, and changes in momentum are produced by applying a net force. Another new concept is energy. On the one hand, both momentum and energy differ from concepts we have examined earlier merely by a factor of mass – thus, maybe these new concepts are not so new after all. On the other hand, the way we work with momentum and energy will give us important insights into physical systems, insights that we have not had before. In that sense, these concepts are of fundamental importance. Energy, in particular, will continue to be important as we work our way through much of the material in later chapters.

6-1 Rewriting Newton's Second Law

In this chapter, we will begin by taking a look at two ideas that we are familiar with from previous chapters. Let's see what happens when we combine Newton's second law, $\vec{F}_{net} = m\vec{a}$, with the definition of acceleration, $\vec{a} = \Delta\vec{v}/\Delta t$. This gives $\vec{F}_{net} = m\Delta\vec{v}/\Delta t$. Let's be a bit creative and write this relationship in a form more as Newton himself originally did:

$$\vec{F}_{net} = \frac{\Delta(m\vec{v})}{\Delta t}$$ (Equation 6.1: **General form of Newton's Second Law**)

Equation 6.1 is more general than Newton's second law stated in the form $\vec{F}_{net} = m\vec{a}$, because equation 6.1 allows us to work with systems (such as rockets) where the mass changes. $\vec{F}_{net} = m\vec{a}$ applies only to systems where the mass is constant, although so many systems have constant mass that we find this form of the equation to be very useful.

The general form of Newton's second law connects the net force on an object with the rate of change of the quantity $m\vec{v}$. This quantity has a name, which you may already be familiar with.

> An object's **momentum** is the product of its mass and its velocity. Momentum is a vector, pointing in the direction of the velocity. The symbol we use to represent momentum is $\vec{p}$.
>
> $$\vec{p} = m\vec{v}.$$ (Equation 6.2: **Momentum**)

Equation 6.1 can be re-arranged to read: $\vec{F}_{net}\Delta t = \Delta(m\vec{v}) = \Delta\vec{p}$.

Thus, to change an object's momentum, all we have to do is to apply a net force for a particular time interval. To produce a larger change in momentum, we can apply a larger net force or apply the same net force over a longer time interval. The product of the net force and the time interval over which the force is applied is such an important quantity that we should name it, too.

> The product of the net force and the time interval over which the force is applied is called **impulse**. An impulse produces a change in momentum. An impulse is a vector.
>
> $$\vec{F}_{net}\Delta t = \Delta(m\vec{v}) = \Delta\vec{p}.$$ (Equation 6.3: **Impulse**)

EXPLORATION 6.1 – Hitting the boards

Just before hitting the boards of an ice rink, a hockey puck is sliding along the ice at a constant velocity. The components of the velocity are 3.0 m/s in the direction perpendicular to the boards and 4.0 m/s parallel to the boards. After bouncing off the boards, the puck's velocity component perpendicular to the boards is 2.0 m/s and the component parallel to the boards is unchanged. The puck's mass is 160 g.

Step 1 - *What is the impulse applied to the puck by the boards?* Let's sketch a diagram (see Figure 6.1) to help visualize what is going on. Impulse is the product of the net force and the time--but we don't know the net force so we can't get impulse that way. Impulse is also equal to the change in momentum, as we can see from equation 6.3, so let's figure out that change.

Remember that momentum, impulse, and velocity are all vectors. Let's choose a coordinate system with the positive x-direction to the right and the positive y-direction up. It is important to notice that there has been no change in the puck's velocity in the y-direction, so there is no change in momentum, no impulse, and no net force in that direction. We can focus on the x-direction to answer the question.

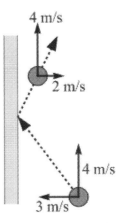

The puck's mass is constant, so the puck's momentum changes because there is a change in velocity. What is the magnitude of the change in the puck's velocity in the x-direction, perpendicular to the boards? It's tempting to say 1.0 m/s, but it is actually 5.0 m/s. That result comes from:

$$\Delta \vec{v}_x = \vec{v}_x - \vec{v}_{ix} = +2.0 \text{ m/s} - (-3.0 \text{ m/s}) = +5.0 \text{ m/s} \, .$$

Knowing the change in velocity, we can find the impulse, which is in the +x-direction:

Figure 6.1: The puck's velocity components before and after it collides with the boards.

$$\Delta \vec{p}_x = m \Delta \vec{v}_x = 0.16 \text{ kg} \times (+5.0 \text{ m/s}) = +0.80 \text{ kg m/s} \, .$$

Step 2 - *If the puck is in contact with the boards for 0.050 s, what is the average force applied to the puck by the boards?* The force varies over the 0.050 s the puck is in contact with the boards, but we can get the average force from:

$$\vec{F}_x = \frac{\Delta \vec{p}_x}{\Delta t} = \frac{+0.80 \text{ kg m/s}}{0.050 \text{ s}} = +16 \text{ N}$$

Key ideas for impulse and momentum: Analyzing situations from an impulse-momentum perspective can be very useful, as it allows us to directly connect force, velocity, and time. It is absolutely critical to account for the fact that momentum, impulse, force, and velocity are all vectors when carrying out such an analysis.   **Related End-of-Chapter Exercises: 14, 15**

Let's now summarize a general method we can use to solve a problem involving impulse and momentum. We will apply this method in Exploration 6.2.

Solving a Problem Involving Impulse and Momentum
    A typical impulse-and-momentum problem relates the net force acting on an object over a time interval to the object's change in momentum. A method for solving such a problem is:

1. Draw a diagram of the situation.
2. Add a coordinate system to the diagram, showing the positive direction(s). Keeping track of direction is important because force and momentum are vector quantities.
3. Organize what you know, and what you're looking for, such as by drawing one or more free-body diagrams, or drawing a graph of the net force as a function of time.
4. Apply equation 6.3 $\left(\vec{F}_{net} \Delta t = \Delta(m\vec{v}) \right)$ to solve the problem.

Essential Question 6.1: At some time T after a ball is released from rest, the force of gravity has accelerated that ball to a velocity v directed straight down. Taking into account impulse and momentum, what is the ball's velocity at a time $2T$ after being released?

Answer to Essential Question 6.1: The force is constant, and Equation 6.3 tells us that the velocity increases linearly with time. Thus, at a time 2*T*, the velocity will be 2*v* directed down.

6-2 Relating Momentum and Impulse

In this section, we will apply the general method from the end of Section 6-1 to solve a problem using the concepts of impulse and momentum.

EXPLORATION 6.2 – An impulsive bike ride

Suki is riding her bicycle, in a straight line, along a flat road. Suki and her bike have a combined mass of 50 kg. At *t* = 0, Suki is traveling at 8.0 m/s. Suki coasts for 10 seconds, but when she realizes she is slowing down, she pedals for the next 20 seconds. Suki pedals so that the static friction force exerted on the bike by the road increases linearly with time from 0 to 40 N, in the direction Suki is traveling, over that 20-second period. Assume there is constant 10 N resistive force, from air resistance and other factors, acting on her and the bicycle the entire time.

Step 1 - *Sketch a diagram of the situation.* The diagram is shown in Figure 6.2, along with the free-body diagram that applies for the first 10 s and the free-body diagram that applies for the 20-second period while Suki is pedaling.

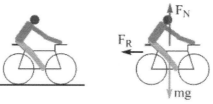 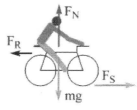

(a) Suki on her bicycle. (b) Free-body diagram while coasting. (c) Free-body diagram while pedaling.

Step 2 - *Sketch a graph of the net force acting on Suki and her bicycle as a function of time.* Take the positive direction to be the direction Suki is traveling. In the vertical direction, the normal force exactly balances the force of gravity, so we can focus on the horizontal forces. For the first 10 seconds, we have only the 10 N resistive force, which acts to oppose the motion and is thus in the negative direction. For the next 20 seconds, we have to account for the friction force that acts in the direction of motion and the resistive force. We can account for their combined effect by drawing a straight line that goes from –10 N at *t* = 10 s, to +30 N (40 N – 10N) at *t* = 30 s. The result is shown in Figure 6.3.

Figure 6.2: A diagram of (a) Suki on her bike, as well as free-body diagrams while she is (b) coasting and while she is (c) pedaling. Note that in free-body diagram (c), the static friction force $\vec{F}_S$ gradually increases because of the way Suki pedals.

Step 3 - *What is Suki's speed at t = 10 s?* Let's apply Equation 6.3, which we can write as:

$$\vec{F}_{net} \Delta t = \Delta(m\vec{v}) = m\,\Delta\vec{v} = m\left(\vec{v}_{10s} - \vec{v}_i\right).$$

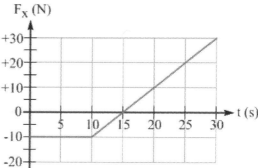

Solving for the velocity at *t* = 10 s gives:

Figure 6.3: A graph of the net force acting on Suki and her bicycle as a function of time.

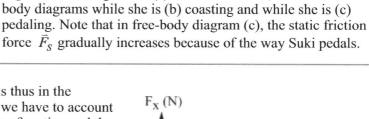

$$\vec{v}_{10s} = \vec{v}_i + \frac{\vec{F}_{net} \Delta t}{m} = +8.0 \text{ m/s} + \frac{(-10 \text{ N})(10 \text{ s})}{50 \text{ kg}} = +8.0 \text{ m/s} - 2.0 \text{ m/s} = +6.0 \text{ m/s}.$$

Thus, Suki's speed at $t = 10$ s is 6.0 m/s. We can also obtain this result from the force-versus-time graph, by recognizing that the impulse, $\vec{F}_{net}\,\Delta t$, represents the area under this graph over some time interval Δt. Let's find the area under the graph, over the first 10 seconds, shown in Figure 6.4. The area is negative, because the net force is negative over that time interval. The area under the graph is the impulse:

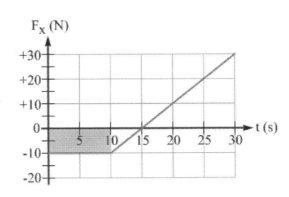

$$\vec{F}_{net}\,\Delta t = -10\text{ N}\times 10\text{ s} = -100\text{ N s} = -100\text{ kg m/s}.$$

Figure 6.4: The rectangle represents the area under the graph for the first 10 s. The area is negative, because the force is negative.

From Equation 6.3, we know the impulse is equal to the change in momentum. Suki's initial momentum is $m\vec{v}_i = 50\text{ kg}\times(+8.0\text{ m/s}) = +400\text{ kg m/s}$. Her momentum at $t = 10$ s is therefore $+400\text{ kg m/s} - 100\text{ kg m/s} = +300\text{ kg m/s}$. Dividing this by the mass to find the velocity at $t = 10$ s confirms what we found above:

$$\vec{v}_{10s} = \frac{\vec{p}_{10s}}{m} = \frac{\vec{p}_i + \Delta\vec{p}}{m} = \frac{+400\text{ kg m/s} - 100\text{ kg m/s}}{50\text{ kg}} = \frac{+300\text{ kg m/s}}{50\text{ kg}} = +6.0\text{ m/s}.$$

Step 4 - *What is Suki's speed at t = 30 s?* Let's use the area under the force-versus-time graph, between $t = 10$ s and $t = 30$ s, to find Suki's change in momentum over that 20-second period. This area is highlighted in Figure 6.5, split into a negative area for the time between $t = 10$ s and $t = 15$ s, and a positive area between $t = 15$ s and $t = 30$ s. These regions are triangles, so we can use the equation for the area of a triangle, $0.5\times base\times height$. The area under the curve, between 10 s and 15 s, is $0.5\times(5.0\text{ s})\times(-10\text{ N}) = -25\text{ kg m/s}$. The area between 15 s and 30 s is $0.5\times(15\text{ s})\times(30\text{ N}) = +225\text{ kg m/s}$. The total area (total change in momentum) is +200 kg m/s.

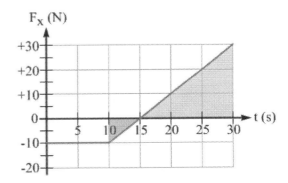

Figure 6.5: The shaded regions correspond to the area under the curve for the time interval from $t = 10$ s to $t = 30$ s.

Note that another approach is to multiply the average net force acting on Suki and the bicycle (+10 N) over this interval, by the time interval (20 s), for a +200 kg m/s change in momentum.

In step 3, we determined that Suki's momentum at $t = 10$ s is +300 kg m/s. With the additional 200 kg m/s, the net momentum at $t = 10$ s is +500 kg m/s. Dividing by the 50 kg mass gives a velocity at $t = 30$ s of +10 m/s.

> **Key idea for the graphical interpretation of impulse**: The area under the net force versus time graph for a particular time interval is equal to the change in momentum during that time interval.
> **Related End-of-Chapter Exercises: 24, 27 – 30.**

Essential Question 6.2: Return to the 30-second interval covered in Exploration 6.2. At what time during this period does Suki reach her minimum speed?

Answer to Essential Question 6.2: At $t = 15$ s. The graph in Figure 6.4 is helpful in determining when Suki reaches her minimum speed. As long as the net force is negative, Suki slows down (unless her velocity becomes negative, which never happens in this case). Suki continues to slow down until $t = 15$ s. After that time, the net force is positive, so Suki speeds up after $t = 15$ s.

6-3 Implication of Newton's Third Law: Momentum is Conserved

EXPLORATION 6.3A – Two carts collide

Let's do an experiment in which two carts, cart 1 and cart 2, collide with one another on a horizontal track, as shown in Figure 6.6. How does the momentum of each cart change? What happens to the momentum of the two-cart system? The upward normal force applied by the track on

Figure 6.6: Two carts colliding.

each cart is balanced by the downward force of gravity, so the net force experienced by each cart during the collision is that applied by the other cart.

Let's use the subscripts *i* for the initial situation (before the collision), and *f* for the final situation (after the collision).

The collision changes the momentum of cart 1 from $\vec{p}_{1i}$ to $\vec{p}_{1f} = \vec{p}_{1i} + \Delta\vec{p}_1$.

Similarly, the collision changes the momentum of cart 2 from $\vec{p}_{2i}$ to $\vec{p}_{2f} = \vec{p}_{2i} + \Delta\vec{p}_2$.

The total momentum of the system beforehand is $\vec{p}_{1i} + \vec{p}_{2i}$.

The total momentum of the system afterwards is $\vec{p}_{1f} + \vec{p}_{2f} = \vec{p}_{1i} + \Delta\vec{p}_1 + \vec{p}_{2i} + \Delta\vec{p}_2$.

Consider $\Delta\vec{p}_1$, the change in momentum experienced by cart 1 in the collision. This change in momentum comes from the force applied to cart 1 by cart 2 during the collision. Similarly, $\Delta\vec{p}_2$, cart 2's change in momentum, comes from the force applied to cart 2 by cart 1 during the collision. Newton's third law tells us that, no matter what, the force applied to cart 1 by cart 2 is equal and opposite to that applied to cart 2 by cart 1. Keeping in mind that the change in momentum is directly proportional to the net force, and that we're talking about vectors, this means:

$$\Delta\vec{p}_2 = -\Delta\vec{p}_1.$$

Substituting this result into our expression for the total momentum of the system after the collision shows that momentum is conserved (momentum remains constant):

$$\vec{p}_{1f} + \vec{p}_{2f} = \vec{p}_{1i} + \vec{p}_{2i}.$$

Key idea: The total momentum of the system after the collision equals the total momentum of the system before the collision. This **Law of Conservation of Momentum** applies to any system where there is no net external force.   **Related End-of-Chapter Exercise: 4.**

We'll spend more time on the law of conservation of momentum in Chapter 7 but, for now, consider the following exploration.

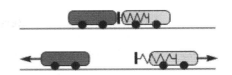

EXPLORATION 6.3B – An explosive situation

Two carts are placed back-to-back on a horizontal track. One cart contains a spring-loaded piston. When the spring is released, the piston pushes against the other cart and the two carts move in opposite directions along the track, as shown in Figure 6.7. Assume the carts are initially at rest in the center of the track and that friction is negligible.

Figure 6.7: A diagram showing the initial situation (top), and the situation after the carts have moved apart (bottom).

Step 1 - *If the two carts have equal masses, is momentum conserved in this process?* A good answer to this question is "it depends." The momentum of each cart individually is not conserved, because each cart starts with no momentum and ends up with a non-zero momentum. This is because each cart experiences a net force (applied by the other cart), so its momentum changes according to the impulse equation (Equation 6.3).

On the other hand, the law of conservation of momentum tells us that *the momentum of the two-cart system is conserved because no net external force acts on this system*. The upward normal force, exerted by the track on this system, balances the downward force of gravity. Cart 1 acquires some momentum because of the force applied by cart 2, but cart 2 acquires an equal-and-opposite momentum because of the equal-and-opposite force applied to it by cart 1. The net momentum of the two-cart system is zero, even when the carts are in motion. Momentum is a vector, so the momentum of one cart is cancelled by the momentum of the other cart.

Step 2 - *If we double the mass of one of the carts and repeat the experiment, is momentum conserved?* Yes, the momentum of this system is conserved because no net external force acts on the system. Changing the mass of one cart will change the magnitude of the momentum acquired by each cart, but the momentum of the two-cart system is always zero, both before and after the spring is released. To conserve momentum, the force applied on cart 1 by cart 2 must be equal and opposite to the force applied on cart 2 by cart 1. Newton's third law tells us that these forces are equal and opposite, no matter how the masses compare.

Step 3 - *If we make the experiment more interesting, by balancing the track on a brick before releasing the spring, will the track tip over after the spring is released?* If we tried this experiment when the masses are equal, what would happen? The track would remain balanced, even when the carts are in motion, because of the symmetry. The tendency of cart 1 to tip the track one way is balanced by the tendency of cart 2 to tip it the opposite way. We don't have the same symmetry in step 2, but the track still remains balanced. The cart with half the mass of the other cart is always twice as far from the balance point. That maintains the balance, as shown in Figure 6.8.

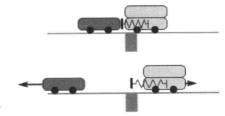

Figure 6.8: As the carts move apart, the track remains balanced on the brick even if the carts have different masses.

Key idea for momentum conservation: Even if the momenta of individual parts of a system are not conserved, the momentum of the entire system is conserved (constant), as long as no net external force acts on the system. Conservation of momentum is a consequence of Newton's third law.   **Related End-of-Chapter Exercises: 44, 45.**

Essential Question 6.3: In Exploration 6.3B, the momentum of the system is always zero. Is there anything about the two-cart system that remains at rest and that shows clearly why the track doesn't tip over when balanced on the brick?

6-4 Center of Mass

In the previous chapters, we treated everything as a particle, or, equivalently, as a ball. A ball, or particle, thrown through the air follows a parabolic path. What if you take a non-spherical object (a pen, for instance) and throw it so it spins? Most points on the object follow complicated paths, but the center of mass still follows a parabolic curve, as shown in Figure 6.9.

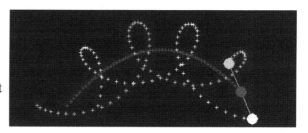

For a uniform object, the center of mass is located at the geometric center of the object. In general, the center of mass of an object, or a collection of objects, is given by Equation 6.4.

Figure 6.9: The motion of three balls on a stick. Only the red ball, located at the center of mass of the system, follows the parabolic path characteristic of free fall.

The **center of mass** is the point on an object that moves as though all the mass of the object is concentrated there. The x-coordinate of the center of mass is given by:

$$X_{CM} = \frac{x_1 m_1 + x_2 m_2 + x_3 m_3 + \ldots}{m_1 + m_2 + m_3 + \ldots}$$
(Equation 6.4: **Position of the center of mass**)

where the m's represent the masses of different objects in the system (or of various pieces of a single object) and the x's represent the x-coordinates of those objects or pieces. Similar equations give us the y and z-coordinates of the center of mass.

EXAMPLE 6.4A – Three balls on a stick

Three balls are placed on a meter stick. Ball 1, at the 0-cm mark, has a mass of 1.0 kg. Ball 2, at the 80-cm mark, has a mass of 3.0 kg. Ball 3, at the 90-cm mark, has a mass of 2.0 kg.
  (a) If the meter stick has negligible mass, where is the system's center of mass?
  (b) If the meter stick has a mass of 2.0 kg, where is the system's center of mass?

SOLUTION
(a) As usual, let's begin with a diagram of the situation (see Figure 6.10).

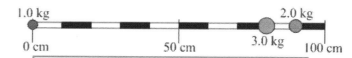

To find the center of mass, we can substitute the given values into Equation 6.4:

Figure 6.10: A diagram showing the position of the three balls on the meter stick.

$$X_{CM} = \frac{x_1 m_1 + x_2 m_2 + x_3 m_3}{m_1 + m_2 + m_3} = \frac{0 \times 1.0 \text{ kg} + (80 \text{ cm})(3.0 \text{ kg}) + (90 \text{ cm})(2.0 \text{ kg})}{1.0 \text{ kg} + 2.0 \text{ kg} + 3.0 \text{ kg}} = \frac{420 \text{ kg cm}}{6.0 \text{ kg}} = 70 \text{ cm}.$$

(b) If the stick's mass is uniformly distributed, we can treat the stick as a fourth ball, with a mass of 2.0 kg, located at the 50-cm mark. Making use of the result from part (a), which says that the first three balls are equivalent to a single 6.0-kg ball located at the 70-cm mark, we get:

$$X'_{CM} = \frac{(70 \text{ cm})(6.0 \text{ kg}) + (50 \text{ cm})(2.0 \text{ kg})}{6.0 \text{ kg} + 2.0 \text{ kg}} = \frac{520 \text{ kg cm}}{8.0 \text{ kg}} = 65 \text{ cm}.$$

Related End-of-Chapter Exercises: 31, 32.

The center of mass is particularly useful in systems experiencing no net external force. In such systems, the motion of the system's center of mass is unchanged, even if the motion of different parts of the system changes. This is a consequence of Newton's Second Law. Without a net external force acting, the acceleration of the center of mass of the system is zero. Different parts of the system can accelerate, but the forces associated with these accelerations cancel because the net force on the system is zero. Let's now consider an example of such a system.

EXAMPLE 6.4B – Canoe move the center of mass?
A man, with a mass of 90 kg, stands 2.0 from the center of a 30 kg canoe that is floating on the calm water of a lake. Both the man and the canoe are initially at rest.
(a) If the man then moves to the point 2.0 m on the opposite side of the center of the canoe from where he starts, how far does the canoe move?
(b) How far does the man actually move relative to a fixed point on the shore?

SOLUTION

(a) We could solve this problem formally, but let's solve it conceptually by looking at Before and After pictures in Figure 6.11. First, let's determine the position of the center of mass of the system in the Before picture, before the man changes position.

Define the man's initial position as the origin (you can pick a different origin if you want), and assume the canoe's center of mass to be the middle of the canoe. We get:

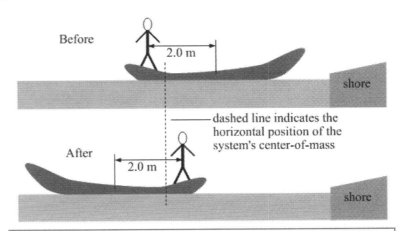

Figure 6.11: The position of the man and the canoe before and after the man moves from one end of the canoe to the other.

$$X_{CM} = \frac{x_1 m_1 + x_2 m_2}{m_1 + m_2} = \frac{0 \times 90 \text{ kg} + (2.0 \text{ m})(30 \text{ kg})}{90 \text{ kg} + 30 \text{ kg}} = \frac{60 \text{ kg} \cdot \text{m}}{120 \text{ kg}} = 0.50 \text{ m}$$

In the Before picture, the man is 50 cm to the left, and the canoe's center of mass is 1.5 m to the right, of the system's center of mass. The canoe's center of mass is three times farther from the system's center of mass than the man is because the canoe's mass is 1/3 of the man's mass. Because no net external force acts on the canoe-man system, when the man moves to the right, the canoe moves left in such a way that the system's center of mass remains at rest. The man moves to a position that is a mirror image of his initial position, so the After picture is a mirror image of the Before picture (placing the mirror at the system's center of mass). The canoe's center of mass moves from 1.5 m to the right of the system's center of mass to 1.5 m to the left of the system's center of mass, for a net displacement of 3.0 m to the left.

(b) Applying a similar analysis to the man, the man moves from 0.5 m to the left of the system's center of mass to 0.5 m to the right, a net displacement of 1.0 m right. Equivalently, the man moves 4.0 m to the right relative to the canoe while the canoe moves 3.0 m to the left with respect to the shore, so the man ends up moving just 1.0 m right relative to the shore.

Related End-of-Chapter Exercise: 33.

Essential Question 6.4: In Example 6.4B, what force makes the canoe move when the man starts to move? What force stops the canoe when the man stops?

Answer to Essential Question 6.4: In each case, the force of friction (static friction if there is no slipping) between the man's shoes and the canoe causes the changes in the canoe's motion.

6-5 Playing with a Constant Acceleration Equation

Once again, let's start with a familiar relationship and look at in a new way to come up with a powerful idea. Return to one of our constant acceleration equations: $v_x^2 = v_{ix}^2 + 2\vec{a}_x \Delta \vec{x}$. If we re-arrange this equation to solve for the acceleration, we get: $\vec{a}_x = \dfrac{v_x^2 - v_{ix}^2}{2\Delta \vec{x}}$.

Substituting this into Newton's second law, $\vec{F}_{net,x} = m\vec{a}_x$, gives, after some re-arranging:

$$\frac{1}{2}mv_x^2 - \frac{1}{2}mv_{ix}^2 = \vec{F}_{net,x}\Delta \vec{x} .$$

We can do the same thing in the *y*-direction. Adding the *x* and *y* equations gives:

$$\left(\frac{1}{2}mv_x^2 + \frac{1}{2}mv_y^2\right) - \left(\frac{1}{2}mv_{ix}^2 + \frac{1}{2}mv_{iy}^2\right) = \vec{F}_{net,x}\Delta \vec{x} + \vec{F}_{net,y}\Delta \vec{y} .$$

Recognizing that $v_x^2 + v_y^2 = v^2$, the left side of the equation can be simplified:

$$\frac{1}{2}mv^2 - \frac{1}{2}mv_i^2 = \vec{F}_{net,x}\Delta \vec{x} + \vec{F}_{net,y}\Delta \vec{y} .$$

The right side can also be simplified, because its form matches a dot product:

$$\frac{1}{2}mv^2 - \frac{1}{2}mv_i^2 = \vec{F}_{net} \bullet \Delta \vec{r} = F_{net}\Delta r \cos\theta , \qquad \text{(Equation 6.5)}$$

where θ is the angle between the net force $\vec{F}_{net}$ and the displacement $\Delta \vec{r}$.

So, we've now come up with two more useful concepts, which we name and define here.

Kinetic energy is energy associated with motion:

$$K = \frac{1}{2}mv^2 \qquad ; \qquad \text{(Equation 6.6: \textbf{Kinetic energy})}$$

Work relates force and displacement $W = \vec{F}_{net} \bullet \Delta \vec{r} = F_{net}\Delta r \cos\theta$. (Eq. 6.7: **Work**)

Both work and kinetic energy have units of joules (J), and they are both scalars.

Equation 6.5, when written in the form below, is known as the **work-kinetic energy theorem**. In this case, the work is the work done by the net force.

$$\Delta K = W_{net} = F_{net}\,\Delta r \cos\theta , \qquad \text{(Eq. 6.8: \textbf{The work-kinetic energy theorem})}$$

where θ is the angle between the net force $\vec{F}_{net}$ and the displacement $\Delta \vec{r}$.

In general, when a force is perpendicular to the displacement, the force does no work. If the force has a component parallel to the displacement, the force does positive work. If the force has a component in the direction opposite to the displacement, the force does negative work.

Compare Exploration 6.5 to Exploration 6.2, in which Suki was riding her bike.

Chapter 6 – Linking Forces to Momentum and Energy

EXPLORATION 6.5 – A hard-working cyclist

Peter is riding his bicycle in a straight line on a flat road. Peter and his bike have a total mass of 60 kg and, at $t = 0$, he is traveling at 8.0 m/s. For the first 70 meters, he coasts. When Peter realizes he is slowing down, he pedals so that the static friction force exerted on the bike by the road increases linearly with distance from 0 to 40 N, in the direction Peter is traveling, over the next 140 meters. A constant 10 N resistive force acts on Peter and the bicycle the entire time.

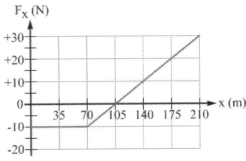

Step 1 - *Sketch a graph of the net force acting on Peter and his bicycle as a function of position.* Take the positive direction to be the direction Peter is traveling. In the vertical direction, the normal force balances the force of gravity, so we can focus on the horizontal forces. For the first 70 m, we have only the 10 N resistive force, which opposes the motion and is thus in the negative direction. For the next 140 m, we have to account for the friction force, which acts in the direction of motion, and the resistive force. We can account for their combined effect by drawing a straight line, as in Figure 6.12, that goes from –10 N at $x = 70$ m to +30 N (40 N – 10N) at $x = 210$ m.

Figure 6.12: A graph of the net force acting on Peter and his bicycle, as a function of position.

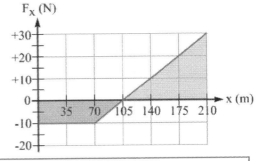

Step 2 - *What is Peter's speed at x = 210 m?* Let's use the area under the F_{net} versus position graph, between $x = 0$ and $x = 210$ m, to find the net work over that distance. This area is shown in Figure 6.13, split into a negative area for the region $x = 0$ to $x = 105$ m, and a positive area between $x = 105$ m and $x = 210$ m. Each box on the graph has an area of $10 \text{ N} \times 35 \text{ m} = 350 \text{ N m}$. The negative area covers two-and-a-half boxes on the graph, while the positive area covers four-and-a-half boxes, for a net positive area of 2 boxes, or 700 N m.

The net area under the curve in Figure 6.13 is the net work done on Peter and the bicycle, which is the change in kinetic energy ($W_{net} = \Delta K = K_f - K_i$). Thus, the final kinetic energy is:

Figure 6.13: The area within the shaded regions represents the area under the curve for the region from $x = 0$ to $x = 210$ m.

$$K_f = K_i + W_{net} = \frac{1}{2}mv_i^2 + W_{net} = \frac{1}{2}(60 \text{ kg})(8.0 \text{ m/s})^2 + 700 \text{ N m} = 1920 \text{ J} + 700 \text{ J} = 2620 \text{ J}.$$

Solving for the final speed from $K_f = (1/2)mv_f^2$ gives: $v_f = \sqrt{\dfrac{2K_f}{m}} = \sqrt{\dfrac{2 \times 2620 \text{ J}}{60 \text{ kg}}} = 9.3 \text{ m/s}.$

Key idea: The area under the net force-versus-position graph for a particular region is the work, and the change in kinetic energy, over that region. **Related End-of-Chapter Exercises: 48, 49.**

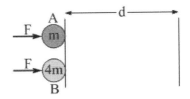

Essential Question 6.5: Initially, objects A and B are at rest. B's mass is four times larger than A's mass. Identical net forces are applied to the objects, as shown in Figure 6.14. Each force is removed once the object it is applied to has accelerated through a distance d. After the forces are removed, which object has more (a) kinetic energy? (b) momentum?

Figure 6.14: An overhead view of two objects, A and B, experiencing the same net force F as they move from rest through a distance d.

Answer to Essential Question 6.5: The two objects experience equal forces over equal displacements, so the work done is the same. Thus, the change in kinetic energy is the same for each and, because they start with no kinetic energy, **their final kinetic energies are equal.**

The change in momentum is the force multiplied by the time over which the force acts. Both objects experience the same force, but, because *B* has more mass, *B* takes more time to move through the distance *d* than *A* does. The force acts on *B* for a longer time, so **B's final momentum is larger than A's.**   **Related End-of-Chapter Exercises: 9, 10, 52 – 54.**

6-6 Conservative Forces and Potential Energy

Let's first write down a method for solving a problem involving work and kinetic energy, similar to the method we use for solving an impulse-and-momentum problem.

A General Method for Solving a Problem Involving Work and Kinetic Energy
1. Draw a diagram of the situation.
2. Add a coordinate system to the diagram, showing the positive direction(s). Doing so helps remind us that force and displacement are vector quantities.
3. Organize what you know, perhaps by drawing a free-body diagram of the object, or drawing a graph of the net force as a function of position.
4. Apply Equation 6.8 $\left(F_{net}\,\Delta r \cos\theta = \Delta K \right)$ to solve the problem.

We now have the tools needed to investigate some intriguing ideas about energy.

EXPLORATION 6.6A – Making gravity work
Step 1 - *Take a ball of weight mg = 10 N and move it through a distance of 2 m. How much work does gravity do on the ball during the motion?* It is tempting to multiply 10 N by 2 m to get 20 J and say that's the work, but the work depends on the angle between the force and the displacement (see Equation 6.7, $W = F_{net}\Delta r \cos\theta$). The direction of the displacement was not given, so we can't say how much work is done.

Let's consider the extreme cases. If we move the ball up 2 m, the force of gravity and the displacement are in opposite directions, so the work done is –20 J. If we move it down 2 m, the force and displacement are in the same direction, so the work done is +20 J. So, the work done is somewhere between –20 J and + 20 J. Work can even be zero, if the displacement is horizontal.

Step 2 – *What is the work done by gravity, if we give our 10 N ball a displacement of 2 m down at the same time we displace it 4 m horizontally?* Gravity still does +20 J of work. All we have to worry about is the vertical motion. There is no work done by gravity for the horizontal motion.

Step 3 - *Does the path followed make any difference?* In Figure 6.15, point B is 2 m below, and 4 m horizontally, from A. For any path starting at A and ending at B, the work done by gravity in moving a 10-N ball is +20 J. The horizontal motion does not matter. What matters is that every path involves the same net 2 m vertical downward displacement.

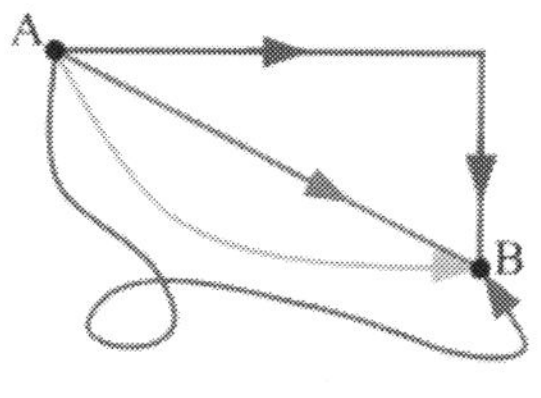

Figure 6.15: The work done by gravity, when an object is moved from point A to point B, is the same no matter what path the object is moved along.

Key idea: The work done by gravity on an object is **path-independent.** All that matters is the position of the initial point and the position of the final point. It doesn't matter how the object gets from the initial point to the final point.   **Related End-of-Chapter Exercise: 43.**

When the work done by a force is path-independent, we say the force is **conservative**. Gravity is a conservative force, and we will discuss other examples later in the book. Other conservative forces include the spring force (chapter 12) and the electrostatic force (chapter 16).

Instead of talking about the work done by a conservative force, we usually do something equivalent and talk about the change in **potential energy** associated with the force. Potential energy can, in general, be thought of as energy something has because of its position.

At the surface of the Earth, where we take the force of gravity to be constant, the work done by gravity is $W_g = -mg\Delta y$. The change in gravitational potential energy, ΔU_g, has the opposite sign: $\Delta U_g = -W_g = mg\Delta y$.   (Eq. 6.9: **Change in gravitational potential energy**)

When the force of gravity is constant, we define **gravitational potential energy** as

$U_g = mgh$,   (Equation 6.10: **Gravitational potential energy**)

where h is the height that the object is above some reference level. We can choose any convenient level to be the reference level.

EXPLORATION 6.6B – Talking about potential energy

A 10-N ball is moved by some path from A to B, where B is 2 m lower than A. What is the ball's initial gravitational potential energy? What is its final gravitational potential energy? What is the change in the gravitational potential energy? Analyze the following conversation.

Bob: "We can use Equation 6.9 to find the change in gravitational potential energy. Because B is 2 meters lower than A, the Δy in the equation is –2 meters. Multiplying this by an mg of 10 newtons gives a change in gravitational potential energy of –20 joules."

Andrea: "If I define B as the level where the potential energy equals zero, then the ball's potential energy at A is +20 joules. The ball's potential energy changes from +20 joules to zero for a change of –20 joules."

Bob: "I agree with what you get for the change but we have to define the zero for potential energy at A. That gives the object a potential energy of –20 joules at B."

Christy: "We can each pick our own zero. It doesn't make any difference. No matter where you put the zero you get –20 joules for the change in potential energy."

Which student is correct?

Bob's first statement is correct. Andrea is correct, and so is Christy. Christy makes an important point – everyone agrees on the value of the change in potential energy, no matter which level they choose as the zero. In his second statement, Bob is incorrect about having to set the potential energy to be zero at A. You can do that, but, as Christy points out, you don't have to.

Key idea: The change in potential energy, which everyone agrees on, is far more important than the actual value of the potential energy.   **Related End-of-Chapter Exercise: 12.**

Essential Question 6.6: We often use terminology like "the ball's gravitational potential energy." Does the ball really have gravitational potential energy all by itself?

Answer to Essential Question 6.6: No, gravitational potential energy really does not belong to an object. Rather, it is associated with the interaction between two objects, such as the interaction between the ball and the Earth in Exploration 6.6A. We will explore this idea further when we discuss gravity in more detail in Chapter 8.

6-7 Power

Let's say you are buying a new vehicle. While you are searching the Internet to compare the latest models, an advertisement for a fancy sports car catches your eye. You read that the car can go from rest to 100 km/h in under five seconds, considerably less time than it takes a base-model Honda Civic, for instance, to do the same. Then, when you tell your friend about what you're planning, he encourages you to buy a pickup truck. The truck and the Civic have similar accelerations, but the truck can achieve that acceleration while loaded down with bikes and kayaks. What is the difference between these vehicles? Their engines can all do work, but an important difference between them is the rate at which they do work.

The ability to do work quickly is something that we celebrate. For instance, in many Olympic events, the gold medal goes to the individual who can do more work, and/or do work in less time, than the other athletes. Once again, we should name this important concept.

Power is the rate at which work is done. The unit of power is the watt, and 1 W = 1 J/s.

$$P = \frac{Work}{\Delta t} = \frac{F \Delta r \cos\theta}{\Delta t} = Fv\cos\theta ,$$ (Equation 6.11: **Power**)

where θ is the angle between the force and the velocity.

EXAMPLE 6.7 – Climbing the hill
A car with a weight of $mg = 16000$ N is climbing a hill that is 120 m long and rises 30 m vertically. The car is traveling at a constant velocity of 72 km/h. In addition to having to contend with the component of the force of gravity that acts down the slope, the car also has to deal with a constant 1000 N in resistive forces as it climbs.

(a) What is the power provided to the drive wheels by the car's engine?

(b) The power unit the horsepower was first used by James Watt in 1782 to compare steam engines and horses. What is the car's power in units of horsepower, where 1 hp = 746 W?

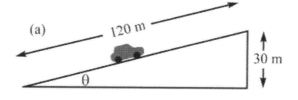

SOLUTION
Let's begin by sketching a diagram of the situation (see Figure 6.16), along with a free-body diagram. If we use a coordinate system aligned with the slope, with the positive *x*-direction up the slope, we can re-draw the free-body diagram with all the forces parallel to the coordinate axes. Doing so involves breaking the force of gravity into components.

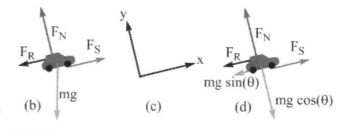

Figure 6.16: (a) A diagram for the car climbing the hill, along with (b) a free-body diagram, (c) an appropriate coordinate system, aligned with the slope, and (d) a revised free-body diagram, with all forces aligned with the coordinate system.

(a) Let's assume that this case is typical and the tires do not slip on the road surface as the car climbs the hill. If so, the force propelling the car up the slope is a static force of friction, much like the force propelling you forward when you walk is a static force of friction. This force of static friction, directed up the hill, must balance the sum of the 1000 N resistive force and the component of the force of gravity acting down the hill, which is $mg \sin\theta$. The value of $\sin\theta$ can be found from the geometry of the hill:

$$\sin\theta = \frac{\text{opposite}}{\text{hypotenuse}} = \frac{30 \text{ m}}{120 \text{ m}} = \frac{1}{4} .$$

The net force directed up the hill is:

$$F_{up} = 1000 \text{ N} + mg \sin\theta = 1000 \text{ N} + 16000 \times \frac{1}{4} = 1000 \text{ N} + 4000 \text{ N} = 5000 \text{ N} .$$

The car's velocity is also directed up, so, if we multiply the force by the speed, we get the power. The speed has to be expressed in units of m/s, however. So, we perform the conversion:

$$72 \frac{\text{km}}{\text{h}} \times \frac{1000 \text{ m}}{1 \text{ km}} \times \frac{1 \text{ h}}{3600 \text{ s}} = 20 \text{ m/s} .$$

The power associated with the drive wheels is:

$$P = F v \cos\theta = F v = 5000 \text{ N} \times 20 \text{ m/s} = 100000 \text{ W} .$$

(b) Converting watts to horsepower gives:

$$100000 \text{ W} \times \frac{1 \text{ hp}}{746 \text{ W}} = 134 \text{ hp} .$$

Not every car is capable of putting out that much power, but many cars are, so that's a reasonable value. The car is probably working at close to its maximum power output, however.

Related End-of-Chapter Exercise: 57, 58.

Question: A typical adult takes in about 2500 nutritional Calories of food energy in a day. Using the fact that 1 Calorie is equivalent to 1000 calories, and that 1 calorie is equivalent to 4.186 J, show that a typical adult takes in about 1×10^7 J worth of food energy in a day.

Answer: Much like converting from watts to horsepower, this is an exercise in unit conversion.

$$2500 \text{ Cal} \times \frac{1000 \text{ cal}}{1 \text{ Cal}} \times \frac{4.186 \text{ J}}{1 \text{ cal}} = 1.05 \times 10^7 \text{ J} .$$

Essential Question 6.7: As we have just shown, a typical adult takes in about 1×10^7 J of food energy in a day. Assuming this energy equals the work done by the person in a day, what average power output does this correspond to? Compare this power to the power output of a world-class cyclist, who can sustain a power output of 500 W for several hours.

Answer to Essential Question 6.7: The average power output can be found by dividing the energy by the time. The energy in question is the energy for one day, so we use one day as the time period. To obtain power in watts, we should express time in seconds. There are 86400 s in one day. Thus, the average power is:

$$P_{av} = \frac{Energy}{time} = \frac{1 \times 10^7 \text{ J}}{86400 \text{ s}} \approx \frac{1 \times 10^7 \text{ J}}{1 \times 10^5 \text{ s}} = 100 \text{ W} .$$

On average, we're about as bright as your average light bulb, and our power output is generally considerably less than what the world-class athlete is capable of sustaining. In our defense, this 100 W is averaged over the entire day, including the time we are sleeping.

Chapter Summary

Essential Idea about the Link between Force, Momentum, and Energy
In this chapter, we extended our understanding of force by connecting force to both momentum and energy. A net force applied over a particular time interval is directly connected to a change in momentum. A net force applied over a particular distance is directly connected to a change in kinetic energy.

Momentum
There are four key points to remember about momentum:

1. Momentum equals mass multiplied by velocity: $\vec{p} = m\vec{v}$. (Equation 6.2)
2. Momentum is a vector.
3. If no net force acts on a system, the system's momentum is conserved--the system's net momentum maintains its value.
4. If there is a net force acting on a system, the system's momentum changes according to equation 6.3: $\Delta\vec{p} = \vec{F}_{net}\Delta t$, which is known as an ***impulse.***

Momentum conservation is a consequence of Newton's third law. Whenever two objects collide, the force that one object exerts on the second object is always equal in magnitude, and opposite in direction, to the force that the second object exerts back on the first object. If we combine the two objects into one system, the forces between the objects cancel out, and the system's momentum is conserved as long as no net external force acts on the system.

Solving a Problem Involving Impulse and Momentum
A general method for solving a problem that relates a net force acting on an object over some time interval to the change in momentum (or velocity) of that object is:

1. Draw a diagram of the situation.
2. Add a coordinate system to the diagram, showing the positive direction(s). The coordinate system helps to remind us that force and momentum are vector quantities.
3. Organize what is known, perhaps by drawing a free-body diagram of the object, or drawing a graph of the net force as a function of time.
4. Apply equation 6.3 $\left(\vec{F}_{net}\Delta t = \Delta(m\vec{v}) \right)$ to solve the problem.

Center of Mass

The center of mass of an object is the point that moves as though the entire mass of the object was concentrated there. For a uniform object, the center of mass is located at the geometric center. When no net external force acts on a system, the system's momentum is conserved, and the motion of the center of mass continues unchanged.

The x-coordinate of the center of mass can be found using the equation:

$$X_{CM} = \frac{x_1 m_1 + x_2 m_2 + x_3 m_3 + \ldots}{m_1 + m_2 + m_3 + \ldots}$$

(Equation 6.4)

The y-coordinate of the center of mass is given by an equivalent equation.

Work and Energy

Energy is a scalar. The MKS unit of energy is the joule (J).

The kinetic energy of an object is given by: $K = \frac{1}{2}mv^2$.

(Equation 6.6: **Kinetic energy**)

The work done by a force is the displacement, multiplied by the component of the force in the direction of the displacement:

$W = F \Delta r \cos\theta$,

(Equation 6.7: **Work**)

where θ is the angle between the force and the displacement.

Solving a Problem Involving Work and Kinetic Energy

A general method for solving a problem that relates a net force acting on an object over some distance to the change in kinetic energy of that object is:

1. Draw a diagram of the situation.
2. Add a coordinate system to the diagram, showing the positive direction(s). Doing so helps remind us that force and displacement are vector quantities.
3. Organize what is known, perhaps by drawing a free-body diagram of the object, or drawing a graph of the net force as a function of position.
4. Apply equation 6.8 $\left(F_{net} \Delta r \cos\theta = \Delta K \right)$ to solve the problem.

Gravitational Potential Energy

Near the surface of the Earth (or at any location where the acceleration due to gravity is constant), gravitational potential energy is defined as:

$U_g = mgh$,

(Equation 6.10: **Gravitational potential energy**)

where h is the height that the object is above some reference level. We can choose any convenient level to be the reference level. The change in potential energy is more important than the value of the potential energy.

Power

Power is the rate at which work is done: $P = \frac{Work}{\Delta t} = Fv \cos\theta$.

(Eq. 6.11: **Power**)

The unit of power is the watt (W). 1 W = 1 J/s.

End-of-Chapter Exercises

Exercises 1 – 12 are conceptual questions designed to see whether you understand the main concepts of the chapter.

1. Three identical objects are traveling north with identical speeds v. Each object experiences a collision, after which the states of motion are: Object A is at rest; object B is traveling south with a speed v; and object C is traveling east with a speed v. Rank these objects, from largest to smallest, based on the magnitude of the impulse they experienced during their collision.

2. Case 1: you run with speed v toward a wall and then stick to it because you and the wall are covered with Velcro. Case 2: you run with speed v toward a wall and bounce straight back at speed v because the wall is covering with an elastic material. If you assume that the time during which you experience an acceleration, because of the force applied by the wall, is the same in both cases, in which case do you experience a larger force?

3. You are driving down the road at high speed. All of a sudden, you see your evil twin, driving an identical car with an equal-and-opposite velocity to you. You both apply the brakes, but it is too late and a collision is imminent. At the last instant, you see a large immovable (completely rigid) object on the side of the road. Considering only the likelihood that you will survive the crash, is it better for you to hit your evil twin or to hit the immovable object? Briefly explain your answer.

4. As you are driving along the road, you hit a mosquito, squashing it on the windshield of your car. During the collision, which object (a) exerts a force of a larger magnitude on the other? (b) experiences a change in momentum of larger magnitude? (c) experiences a change in velocity of larger magnitude?

5. Must the center of mass of an object always be located at a point where the object has some mass? If it must, explain why. If not, give an example (or two) of objects where the center of mass is located at a point where none of the mass of the object is located.

6. (a) Give an example of an object (or system of objects) such that when you make a single straight cut through the center of mass the object (or system) is split into two parts with the same mass. (b) If possible, give an example of an object or system such that when you make a single straight cut through the center of mass the two parts have different masses.

7. Consider the following four cases, in which a net force is applied to an object initially moving in the $+x$ direction with a velocity of 5 m/s. Case 1: the object's mass is 1 kg, and a force of 10 N in the $+x$ direction is applied for 1 s. Case 2: the same as case 1 except the object's mass is 2 kg. Case 3: the same as case 1 except the force is in the $-x$ direction. Case 4: the same as case 1 except the magnitude of the force is 5 N. Rank the four cases from largest to smallest based on (a) the magnitude of the change in momentum the object experiences; (b) the magnitude of the object's final momentum; (c) the object's final speed; (d) the final kinetic energy of the object.

8. Cars have crumple zones that are designed to crumple and compress when your car is in a collision. In many cases after a collision, this crumpling means that the car is ruined and you have to buy a new one (preferably with the aid of a payment from your insurance company). Is the crumple zone a huge conspiracy on the part of the auto industry, or is it

an important safety feature? Briefly explain your answer, using concepts of impulse and momentum, or work and kinetic energy.

9. Two objects, *A* and *B*, are initially at rest. The mass of object *B* is two times larger than that of object *A*. Identical net forces are then applied to the two objects, making them accelerate. Each net force is removed once the object that it is applied to has moved through a distance *d*. After both net forces are removed, how do: (a) the kinetic energies compare? (b) the speeds compare? (c) the momenta compare?

10. Repeat Exercise 9, assuming that both net forces are removed after the same amount of time instead.

11. (a) Is it possible to apply a force to an object so that the object's momentum changes but its kinetic energy remains the same? If so, give an example. (b) Is it possible to apply a force to an object so that the object's kinetic energy changes but its momentum remains the same? If so, give an example.

12. Consider Exploration 6.6B, in which Andrea and Bob chose different points as the zero point of the ball's gravitational potential energy. Do Andrea and Bob agree or disagree about the following? The value of (a) the ball's initial gravitational potential energy? (b) the ball's final gravitational potential energy? (c) the ball's change in gravitational potential energy? (d) the work done by gravity on the ball?

Exercises 13 – 18 deal with momentum and impulse.

13. Three identical objects are traveling north with identical speeds *v*. Each object experiences a collision, after which the states of motion are: Object A is at rest; object B is traveling south with a speed *v*; and object C is traveling east with a speed *v*. If the mass of each object is 40 kg and *v* = 12 m/s, find the magnitude and direction of the impulse experienced by (a) object A, (b) object B, and (c) object C.

14. Just before hitting the boards of a hockey rink, a puck is sliding along the ice at a constant velocity. As shown in Figure 6.17, the components of this velocity are 3 m/s in the direction perpendicular to the boards and 4 m/s parallel to the boards. Immediately after bouncing off the boards, the puck's velocity component parallel to the boards is unchanged at 4 m/s, and its velocity component perpendicular to the boards is 1 m/s in case A, 2 m/s in case B, and 3 m/s in case C. Without doing any calculations, rank the three cases based on the impulse the puck experienced because of its collision with the boards.

15. Return to the situation described in Exercise 14, and shown in Figure 6.17. If the puck's mass is 160 g, find the impulse applied by the boards in (a) case C; (b) case A.

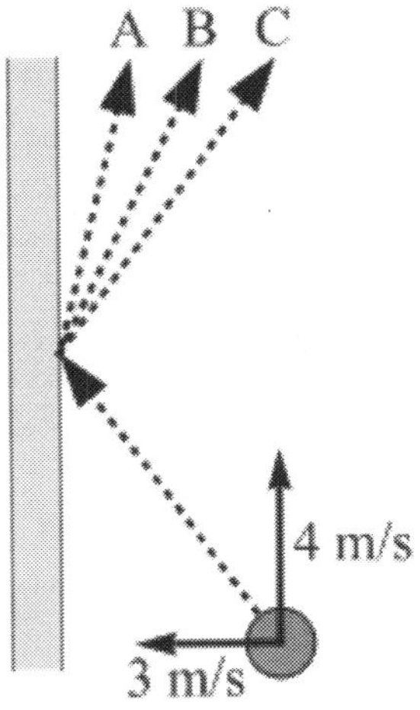

16. An object with a mass of 5.00 kg is traveling east at 4.00 m/s. It is then subjected to a constant net force for a period of 2.00 s. In which direction should the force be applied if you want the object (a) to be moving fastest once the force is removed? (b) to experience the largest-magnitude change in momentum over the time period during which the force is applied?

Figure 6.17: Three situations involving a hockey puck colliding with the boards, for Exercises 14 and 15.

17. Return to the situation described in Exercise 16. What are the magnitude and direction of the applied force if the object's velocity after the force is removed is (a) 12.0 m/s east? (b) zero? (c) 12.0 m/s west?

18. Return to the situation described in Exercise 16. What are the magnitude and direction of the applied force if the object's velocity after the force is removed is (a) 4.00 m/s north? (b) $4\sqrt{2}$ m/s northeast? (c) 8.00 m/s south?

Exercises 19 – 23 are designed to give you some practice in applying the general method for solving a problem involving impulse and momentum. For each exercise, begin with the following parts: (a) Sketch a diagram of the situation. (b) Choose a coordinate system, and show it on the diagram. (c) Organize what you know, such as by drawing a free-body diagram or a graph of the net force as a function of time.

19. You throw a 200-gram ball straight up into the air, releasing it with a speed of 20 m/s. The goal here is to use impulse and momentum concepts to determine the time it takes the ball to reach its maximum height, assuming $g = 10$ m/s² down. Parts (a) – (c) as described above, where you should draw a free-body diagram of the ball after it leaves your hand for part (c). (d) What is the ball's momentum at the instant you let go of it? (e) What is the ball's momentum at the maximum-height point? (f) What is the change in the ball's momentum between the time you release it and the time it reaches its maximum height? (g) What is the force acting on the ball over this time interval? (h) Using equation 6.3, determine the time the ball takes to reach its maximum height.

20. You launch a 200-gram ball horizontally, with a speed of 30 m/s, from the top of a tall building, 80 m above the ground. The goal in this exercise is to use impulse and momentum concepts to determine the ball's momentum just before it hits the ground, assuming $g = 10$ m/s² down and air resistance is negligible. Parts (a) – (c) as described above, where you should draw a free-body diagram of the ball after it leaves your hand for part (c). (d) Using one or more constant-acceleration equations, determine the time it takes the ball to reach the ground. (e) What is the ball's momentum at the instant you let go of it? (f) Using equation 6.3, what is the change in the ball's momentum during the time it is in flight? (g) Use your answers to parts (e) and (f), noting that they are vectors, to find the ball's momentum just before it reaches the ground.

21. The Williams sisters are playing one another in the semi-finals at Wimbledon. At the instant Venus' racket makes contact with one of Serena's serves, the ball is traveling horizontally at 20 m/s, and it has no vertical velocity. The racket is in contact with the ball (which has a mass of 100 g) for 0.030 s, and the ball leaves the racket traveling at 40 m/s horizontally, in a direction exactly opposite to the path it was traveling just as her racket made contact with it. Parts (a) – (c) as described above, where you should sketch the x and y components of the average force exerted on the ball by the racket for the free-body diagram of the ball in part (c). (d) What is the x-component of the average force exerted by the racket on the ball in this case? (e) Does the average force exerted by the racket on the ball also have a non-zero y-component? Briefly explain your answer.

22. A box, with a weight of 40 N, is initially at rest on a horizontal surface. The coefficients of friction between the box and the surface are $\mu_S = 0.40$ and $\mu_K = 0.20$. You then exert a horizontal force on the box that increases linearly from 0 to 40 N over a 1.0-second period. Assume $g = 10$ m/s². Parts (a) – (c) as described above, where you should sketch a graph of the net force acting on the box, as a function of time, in part (c). (d) When does the box start to move? (e) What is the area under the net force versus time graph over the 1.0-second period? (f) Determine the speed of the box at the end of the 1.0-second period.

23. While you are out for a run, you see a patch of smooth ice ahead of you. You decide to slide (on your running shoes) across the ice. Your initial speed is 6.0 m/s. When you reach the end of the horizontal ice patch, after sliding for 2.0 s, your speed is 4.0 m/s. Your goal here is to determine the coefficient of kinetic friction between your running shoes and the ice, assuming that $g = 10$ m/s². Parts (a) – (c) as described above, where you should sketch a free-body diagram for the period you are sliding, in part (c). (d) Write an expression for the net force acting on you while you are sliding. This should involve the coefficient of kinetic friction and g. (e) Write an expression representing your change in momentum while you are sliding. (f) Use equation 6.3 to relate the expressions you wrote down in parts (d) and (e). (g) Solve for the coefficient of kinetic friction.

Exercises 24 – 30 deal with working with graphs.

24. At a time $t = 0$, a wheeled cart with a mass of 2.00 kg has an initial velocity of 5.00 m/s in the +x-direction. For the next 8.00 seconds, the cart then experiences a net force. As shown in the graph in Figure 6.18, the x-component of the applied force is +1.00 N for 2.00 seconds, then –4.00 N for 5.00 seconds, then +2.00 N for 1.00 seconds. (a) Sketch a graph of the x-component of the cart's momentum as a function of time. (b) What is the cart's maximum speed during the 8.00-second interval when the varying force is being applied? At what time does the cart reach this maximum speed? (c) What is the cart's minimum speed during the 8.00-second interval when the varying force is being applied? At what time does the cart reach this minimum speed?

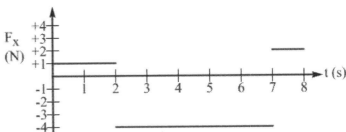

25. Return to the situation described in Exercise 24. (a) How much work is done on the cart during the 8.00-second interval over which the force acts? (b) If the cart starts at the origin at $t = 0$, where is it at $t = 8$ s?

Figure 6.18: A plot of the force applied to a cart as a function of time, for Exercises 24 and 25.

26. A spaceship of mass 4000 kg is drifting at constant velocity through outer space, unaffected by any gravitational interactions. Figure 6.19 shows the trajectory followed by the spaceship in a particular x-y coordinate system during a 2.00 second interval. At $t = 2.00$ seconds, the spaceship fires its engine, producing a net force on the spaceship of 8000 N in the $+y$ direction. The engine is turned off again after 2.00 seconds, at $t = 4.00$ seconds. Assume the mass of the spaceship does not change. The square boxes in the Figure 6.19 measure 1.00 m by 1.00 m. (a) Carefully plot the trajectory followed by the spaceship after $t = 2.00$ seconds. Note in particular where the spaceship is at $t = 3.00$ s, $t = 4.00$ s, and $t = 5.00$ s. (b) What is the speed of the spaceship at $t = 5.00$ seconds?

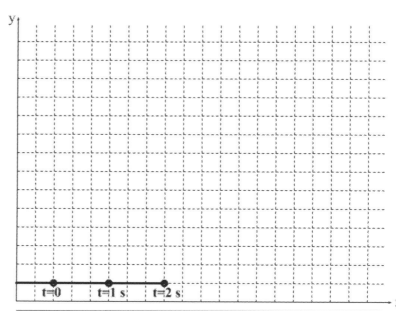

Figure 6.19: A plot of a spaceship's position as a function of time between $t = 0$ and $t = 2$ seconds, for Exercise 26.

27. An object of mass 2.0 kg is at rest at the origin, at $t = 0$, when it is subjected to a net force in the x-direction that varies in magnitude and direction as shown by the graph in Figure 6.20. (a) When does the object reach its maximum speed? (b) What is the maximum speed reached by the object? (c) What is the object's velocity at $t = 8$ s?

28. Repeat Exercise 27, with the only change being that the object has an initial velocity of 2.0 m/s in the negative x direction at $t = 0$.

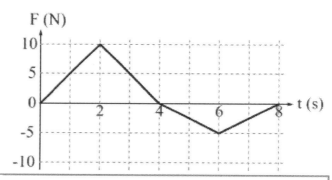

Figure 6.20: A graph of the net force in the x-direction that an object experiences, for Exercises 27 and 28.

29. After the time $t = 0$, an object of mass $m = 1.0$ kg is moving in the positive x direction at a constant speed of 8.0 m/s. The object is on a frictionless horizontal surface. Before $t = 0$, however, the object experienced a net force in the positive x-direction as shown in Figure 6.21. Determine the object's velocity at a time of (a) $t = -1.0$ s (b) $t = -2.0$ s (c) $t = -4.0$ s.

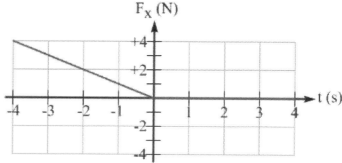

30. Repeat Exercise 29, but this time use a mass of $m = 0.25$ kg.

Figure 6.21: A graph of the net force applied to an object as a function of time, for Exercises 29 and 30.

Exercises 31 – 34 deal with the concept of center of mass.

31. A system consists of three balls. (a) Find the center of mass of the system, given that: Ball 1 has a mass of 2.0 kg and is located at $x = +3$ m, $y = 0$; ball 2 has a mass of 3.0 kg and is located at $x = -1$ m, $y = -1$ m; and ball 3 has a mass of 5.0 kg and is located at $x = 0$, $y = +2$ m. (b) If the mass of ball 3 is increased, the position of the center of mass shifts. In which direction does it shift?

32. A system consists of three balls at different locations on the x-axis. Ball 1 has a mass of 6.0 kg and is located at $x = +3$ m; ball 2 has a mass of 2.0 kg and is located at $x = -1$ m; ball 3 has an unknown mass and is located at $x = -4$ m. (a) If the center of mass of this system is located at $x = -2$ m, what is the mass of ball 3? (b) Let's say that you can make ball 3 as light or as heavy as you like. By adjusting the mass of ball 3, what range of positions on the x-axis can the center of mass of this system occupy?

33. A man with a mass of 120 kg is out fishing with his daughter, who has a mass of 40 kg. They are initially sitting at opposite ends of their 3.0-m boat, which has a mass of 80 kg and is at rest in the middle of a calm lake. If the man and the daughter then carefully trade places, how far does the boat move?

34. A uniform sheet of plywood measuring 4L by 4L is centered on the origin, as shown in Figure 6.22. One quarter of the sheet (the part in the first quadrant) is removed. Where is the center of mass of the remaining piece?

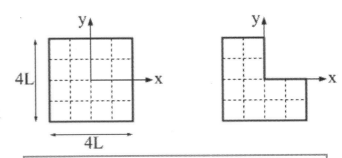

Figure 6.22: One quarter of a sheet of plywood is removed, for Exercise 34.

Exercises 35 – 39 are designed to give you some practice in applying the general method for solving a problem involving work and kinetic energy. For each exercise, begin with the following parts: (a) Sketch a diagram of the situation. (b) Choose a coordinate system, and show it on the diagram. (c) Organize what you know, such as by drawing a free-body diagram or a graph of the net force as a function of position.

35. You throw a 200-gram ball straight up into the air, releasing it with a speed of 20 m/s. The goal here is to use work and energy concepts to determine the ball's maximum height, assuming $g = 10$ m/s² down. Parts (a) – (c) as described above, where you should draw a free-body diagram of the ball after it leaves your hand for part (c). (d) What is the ball's kinetic energy at the instant you let go of it? (e) What is the ball's kinetic energy at the maximum-height point? (f) What is the change in the ball's kinetic energy between the point at which you release it and the point of maximum height? (g) What is the force acting on the ball over this distance? (h) Using equation 6.8, determine the distance between the point from which you released the ball and the point of maximum height.

36. You launch a 200-gram ball horizontally, with a speed of 30 m/s, from the top of a tall building, 80 m above the ground. The goal in this exercise is to use work and kinetic energy concepts to determine the ball's speed just before it hits the ground, assuming $g = 10$ m/s² down and air resistance is negligible. Parts (a) – (c) as described above, where you should draw a free-body diagram of the ball after it leaves your hand for part (c). (d) What is the ball's kinetic energy when you release it? (e) What is the net work done on the ball over its path from your hand to just above the ground? Hint: you can find

the net work by multiplying the net force acting on the ball by the ball's displacement in the direction of the net force. (f) Using equation 6.8, what is the ball's kinetic energy just before the ball reaches the ground? (g) What is the ball's speed just before it reaches the ground?

37. An object with a mass of 2.0 kg is following a straight path, along a line we can call the x axis. When it passes $x = 4.0$ m, it is traveling with a velocity of 8.0 m/s in the positive x direction. The object experiences a net force of 4.0 N in the positive x-direction at all locations where $x < 2.0$ m and a net force of 10 N in the positive x-direction at all locations where $x > 2.0$ m. Parts (a) – (c) as described above, where you should draw a graph of the net force acting on the ball as a function of position in part (c). The goal of this exercise is to determine, with the help of the graph, at what location the object has a velocity of 10 m/s in the positive x-direction, and at what location the object has a velocity of 2.0 m/s in the positive x-direction. (d) What is the object's kinetic energy at $x = 4.0$ m? (e) What is the object's kinetic energy when its speed is 10 m/s? (f) How much work is required to change the object's kinetic energy from what the object has at $x = 4.0$ m to what it has at the point at which the velocity is 10 m/s in the positive x-direction? Shade in the corresponding area on the force-vs.-position graph. (g) At what location will the object have a velocity of $+10$ m/s? (h) Follow a similar procedure to determine at what location the object will have a velocity of $+2.0$ m/s.

38. You are traveling in a car at 54 km/h when the car is involved in an accident. Assume that your mass is 60 kg. (a) What is your kinetic energy? (b) You are wearing your seat belt, and you come to rest after you and the car move through a distance of 2.0 m. What is the average force exerted on you by the seat belt? (c) What is your average acceleration? Express this in units of g, assuming $g = 10$ m/s^2. (d) If you are not wearing your seat belt, you may come to rest after striking the windshield and moving through a distance of 10 cm. What is your average acceleration in this case? Again, express this in units of g.

39. While you are out for a run you see a long patch of smooth ice ahead of you. You decide to slide (on your running shoes) across the ice. When you begin sliding, your speed is 6.0 m/s. When you reach the end of the horizontal ice patch, after sliding for a distance of 5.0 m, your speed is 4.0 m/s. Your goal here is to determine the coefficient of kinetic friction between your running shoes and the ice, assuming that $g = 10$ m/s^2. Parts (a) – (c) as described above, where you should sketch a free-body diagram for the period in which you are sliding, in part (c). (d) Write an expression for the net force acting on you while you are sliding. This expression should involve the coefficient of kinetic friction and g. (e) Write an expression representing your change in kinetic energy while you are sliding. (f) Use equation 6.8 to relate the expressions you wrote down in parts (d) and (e). (g) Solve for the coefficient of kinetic friction.

General Exercises and Conceptual Questions

40. A hose is used to spray water horizontally at a wall. The water has a speed of 4 m/s, and the flow rate is 5 liters per second. (a) Assuming that the water stops completely when it hits the wall, how much force does the water exert on the wall? (b) Rather than stopping completely, the water rebounds when it hits the wall. Does this change the force exerted by the water on the wall? If so, how?

41. An object has a momentum with a magnitude of 20 kg m/s and a speed of 4 m/s. It is then subjected to an impulse of 15 kg m/s in the $+x$ direction. What is the object's final velocity if the initial momentum is in the (a) $+x$ direction? (b) $-x$ direction? (c) $+y$ direction?

42. A hockey puck is sliding east at a constant velocity v over some ice. A net force F is then applied to the puck for 5 seconds. In case 1, the net force is directed west. In case 2, the net force is directed south. In case 3, the net force is directed east. The magnitude of the applied force is the same in each case. Rank the cases from largest to smallest, based on: (a) the magnitude of the change in momentum experienced by the puck, (b) the magnitude of the puck's final momentum, and (c) the work done on the puck.

43. You are shooting a free throw in basketball. If the center of the basket is 1.0 m higher, and 4.0 m horizontally, from the point at which the ball loses contact with your hands, what momentum (magnitude and direction) must the ball have when you release it, if the ball takes exactly 1.0 s to reach the center of the basket? The basketball has a mass of 0.50 kg. Use $g = 9.8$ m/s^2 for this exercise.

44. A firework of mass $10M$ is launched from the ground and follows a parabolic trajectory (assume air resistance is negligible) as shown in Figure 6.23. Its initial velocity has components $v_{ix} = 30$ m/s to the right and $v_{iy} = 20$ m/s up. It follows the parabolic trajectory shown at right. When the firework reaches its maximum height, it explodes into four pieces, A, B, C, and D (not shown on the diagram). The masses and velocities of the four pieces immediately after the explosion are:
$m_A = 1M$, $v_{Af} = 24$ m/s vertically up;
$m_B = 2M$, $v_{Bf} = 50$ m/s horizontally to the right;
$m_C = 3M$, $v_{Cf} =$ an unknown speed vertically down;
$m_D = 4M$, $v_{Df} =$ an unknown speed horizontally right or left.
(a) What is the speed of piece C after the collision? (b) What is the velocity (magnitude and direction) of piece D after the collision? (c) Before the explosion, the firework follows the typical parabolic path of an object moving under the influence of gravity alone. What path will the center of mass follow after the collision? Qualitatively, when will the center of mass divert from this path?

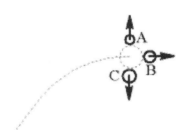

45. Repeat Exercise 44, parts (a) and (b), with the firework exploding not at the top of its trajectory, but 2.3 s after launch instead. Use $g = 10$ m/s^2, so you can do the calculations without a calculator.

Figure 6.23: An exploding firework, for Exercises 44 and 45.

46. How much work do you do on a box with a weight of 10 N in the following situations? (a) You hold the box motionless over your head for 2.0 s. (b) You move the box 2.0 m horizontally at constant velocity. (c) Starting and ending with the box at rest, you move the box 2.0 m straight up.

47. A box with a weight of 20.0 N is initially at rest on a horizontal surface, when a force is applied to it for 6.00 seconds. As shown in Figure 6.24, in case 1, the force is 5.00 N to the right, while in case 2, the force is 10.0 N at an angle of 60° above the horizontal. (a) If there is no friction between the box and the surface, in which case is more work done on the object? (b) What is the net work done in the two cases? (c) If, instead, the coefficients of friction are $\mu_s = 0.400$ and $\mu_k = 0.300$, in which case is more work done on the object? What is the net work done in the two cases now? Use $g = 10.0$ m/s^2 to simplify the calculations.

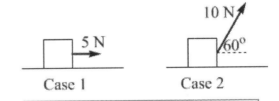

Figure 6.24: Two situations of a box subjected to a force, for Exercise 47.

48. A wheeled cart, which is free to move along the x-axis, is initially at rest at the origin. As the graph in Figure 6.25 shows, if the cart is between $x = -1$ m and $x = +2$ m, the net force is 1.00 N in the positive x-direction. If the cart is between $x = +2$ m and $x = +7$ m, the net force is 4.00 N in the negative x-direction. If the cart is between $x = +7$ m and $x = +8$ m, the net force is 2.00 N in the positive x-direction. The net force is zero at all other locations. (a) Describe, qualitatively, the resulting motion of the cart. (b) What is the maximum distance the cart gets from the origin? (c) Graph the cart's kinetic energy as a function of position as it moves. (d) If you wanted the cart to travel at least as far as $x = +8$ m, what is the minimum kinetic energy the cart needs to have at the origin?

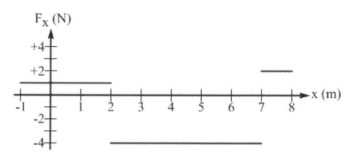

49. Consider again the situation in Exercise 48. Assume the cart has a mass of 0.250 kg and is released from rest at the origin. (a) How long after the cart is released does it first pass $x = +2$ m? (b) What is the cart's maximum speed during its motion? (c) How long after it is released does the cart first return to the origin? (d) Graph the cart's velocity, as a function of time, for the first 10 seconds after its release.

Figure 6.25: A graph showing the force applied to an object as a function of position, for Exercises 48 and 49.

50. We'll deal with springs in detail in chapter 12, but consider the situation shown in Figure 6.26. A block of mass $m = 0.25$ kg is traveling with a velocity $v = 4.0$ m/s to the left on a frictionless horizontal surface. When it reaches $x = 0$, the block encounters a spring, which exerts a force directed right on the block that depends on how much the spring is compressed. The graph shows the force the spring exerts on the block as a function of position, x. (a) How far will the block compress the spring in this case? (b) How far is the spring compressed when the block has a speed of $v = 2.0$ m/s?

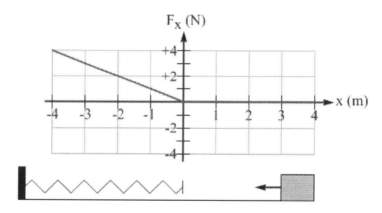

51. Consider the situation described in Exercise 50. How far will the block compress the spring (a) if the mass of the block is doubled? (b) if, instead, the initial velocity of the block is doubled?

Figure 6.26: The graph shows the force a spring exerts on a block as a function of position. The diagram below the graph shows the block moving left on a frictionless surface before encountering the spring. For Exercises 50 and 51.

52. Two identical boxes of mass m are sliding along a horizontal floor, but both eventually come to rest because of friction. Box A has an initial speed of v, while box B has an initial speed of $2v$. The coefficient of kinetic friction between each box and the floor is μ_K, and the acceleration due to gravity is g. (a) If it takes box A a time T to come to a stop, how much time does it take for box B to come to a stop? (b) Find an expression for T in terms of the variables specified in the exercise. (c) If box A travels a distance D before coming to rest, how far does box B travel before coming to rest? (d) Find an expression for D in terms of the variables specified in the exercise.

53. Return to the situation described in Exercise 52. How does T, the stopping time for box A, change if (a) m is doubled? (b) v is doubled? (c) μ_K is doubled?

54. Return to the situation described in Exercise 52. How does D, the stopping distance for box A, change if (a) m is doubled? (b) v is doubled? (c) μ_K is doubled? (d) g is doubled?

55. A car traveling 50 km/h can be brought to a stop in a distance of 40 m under controlled braking conditions. (a) Assuming the force used to bring the car to rest is the same, how much distance is required to bring the car to a stop if the car is traveling 100 km/h, twice as fast as it was originally? (b) How do the stopping times compare? (Ignore the reaction time of the driver and find the distance and time after the brakes are applied.)

56. A box, with a weight of $mg = 25$ N, is placed at the top of a ramp and released from rest. The ramp is in the shape of a 3-4-5 triangle, measuring 4 meters horizontally and 3 meters vertically. The box accelerates down the incline, attaining a kinetic energy at the bottom of the ramp of 55 J. There is a force of kinetic friction acting on the box as it slides down the incline. (a) Sketch a free-body diagram of the box, showing all the forces acting on it. (b) How much work does the normal force do on the box as the box slides down the incline? (c) Calculate the change in gravitational potential energy that the box experiences in this process. (d) How much work does the force of friction do on the box as the box slides down the incline? (e) What is the coefficient of kinetic friction between the box and ramp?

57. A car is accelerating from rest and takes a time T to reach speed v. (a) Assuming the force accelerating the car is constant, what is the total time (measured from the starting point) needed to reach a speed of $2v$? (b) Assuming instead that the power associated with accelerating the car is constant, what is the total time needed to reach a speed of $2v$?

58. You are cycling at a constant speed of 10 m/s. (a) If the net resistive force acting against you from things like air resistance is 35 N, what is your power output as you pedal? (b) How much additional power is required to maintain this speed up a hill inclined at 8.0° with the horizontal? Assume the combined mass of you and your bicycle is 50 kg.

59. On a monthly electricity bill, the power companies charge you for the number of kilowatt-hours you consume. (a) What kind of unit is the kilowatt-hour? Is it power? Momentum? Something else? (b) Convert 1 kW-h to MKS units. (c) 1 kilowatt-hour typically costs about 20 cents. If you were somehow able to obtain your daily intake of 2500 Cal by plugging yourself into a wall socket (don't try this, of course!), how much would it cost you?

60. An energy bar contains about 200 Cal. If your brain consumes about 20 W under typical conditions, for how long does one energy bar keep the brain functioning? 1 Cal = 1000 calories, and 1 calorie is approximately 4 J.

61. Consider the Earth, with a mass of 6.0 x 10^{24} kg, in its orbit around the Sun, with a mass of 2.0 x 10^{30} kg. Assume the orbit is circular, with a radius of 1.5 x 10^{11} m. The Earth, traveling at 30 km/s, takes six months to travel halfway around the orbit. (a) What is the magnitude of the Earth's change in momentum over this six-month period? (b) How much work does the Sun do on the Earth over this six-month period?

62. Comment on the statements made by three students who are working together to solve the following problem, and state the answer to the problem. A cart with a mass of 2.0 kg has an initial velocity of 4.0 m/s in the positive x-direction. A constant net force of 8.0 N, in the positive x-direction, is then applied to the cart for 0.50 s. What is the cart's kinetic energy at the end of this 0.50 second interval?

Christina: I think we should use impulse here. Using impulse, we can figure out the change in velocity, and then the final velocity. Once we get that, we can use the mass and velocity to get the kinetic energy.

Sandy: Don't we need to find the acceleration? That's just 4.0 meters per second squared. Then we can use one of the constant-acceleration equations to find the final speed, and get the kinetic energy that way.

Phil: I like getting the acceleration first, but then we can find the displacement using one of the constant-acceleration equations. After that, we can get the work, which is the change in kinetic energy, and then get the final kinetic energy. They basically give us the initial kinetic energy.

The device in the photo is called Newton's cradle, an excellent example of conservation ideas. When a ball swings down at the right and collides with another ball, a series of collisions takes place, in which momentum is conserved and in which the first ball's momentum in transferred to the ball at the left end. The ball on the left then swings up and comes back down, as its energy is transformed from gravitational potential energy to kinetic energy and back again, and the process happens again in the other direction. Photo credit: Ali Ender Birer / iStockphoto.

Chapter 7 – Conservation of Energy and Conservation of Momentum

CHAPTER CONTENTS

When a quantity is conserved, its value is constant. In many situations, momentum and/or energy remain constant, so we say that they are conserved. The corresponding conservation laws, in fact, are among the most important ideas in physics. Conservation laws are the fundamental laws of the universe, the guiding principles upon which systems interact and evolve. We will continue to apply the lessons we learn in this chapter to help us to understand a wide variety of situations throughout the rest of the book.

Whenever we analyze collisions, or explosions, we apply momentum conservation. This idea applies to everyday situations, such as the collisions of vehicles on the street, or of football players on the field. However, momentum conservation also applies to the collisions of entire galaxies, as well as to events at the atomic level, such as radioactive decay processes.

Energy conservation will also be extremely helpful to us in understanding a wide variety of situations. Make a list of all the situations in which we apply energy conservation as we work our way through the rest of this book – you will probably be amazed at how long the list is by the time we are finished.

7-1 The Law of Conservation of Energy

In Chapter 6, we looked at the work-energy relationship: $W_{net} = \Delta K$. If we break the net work up into two pieces, W_{con}, the work done by conservative forces (such as gravity), and W_{nc}, the work done by non-conservative forces (such as tension or friction), then we can write the equation as: $W_{con} + W_{nc} = \Delta K$. Recall that the work done by a conservative force does not depend on the path taken between points. The work done by non-conservative forces is path dependent, however.

As we did in Chapter 6 with the work done by gravity, the work done by any conservative force can be expressed in terms of potential energy, using $W_{con} = -\Delta U$. We can now write the work-energy equation as

$$-\Delta U + W_{nc} = \Delta K .\qquad \text{(Equation A)}.$$

Let's use i to denote the initial state and f to denote the final state. The change in a quantity is its final value minus its initial value, so we can use $\Delta K = K_f - K_i$ and $\Delta U = U_f - U_i$. Substituting these expressions into Equation A gives $-\left(U_f - U_i\right) + W_{nc} = K_f - K_i$. With a bit of re-arranging to make everything positive, we get Equation 7.1, below.

Equation 7.1 expresses a basic statement of the **Law of Conservation of Energy**: "Energy can neither be created nor destroyed, it can only be changed from one form to another."

$$K_i + U_i + W_{nc} = K_f + U_f \qquad \text{(Equation 7.1: \textbf{Conservation of energy})}$$

The law of conservation of energy is so important that we will use it in Chapters 8, 9, and 10, as well as in many chapters after that. With equation 7.1, we have the only equation we need to solve virtually any energy problem. Let's discuss its five different components.

K_i and K_f are the initial and final values of the kinetic energy, respectively.

U_i and U_f are the initial and final values of the potential energy, respectively.

W_{nc} is the work done by non-conservative forces (such as by the force of friction).

The conservation of energy equation is very flexible. So far, we have discussed one form of kinetic energy, the translational kinetic energy given by $K = (1/2)mv^2$. When we get to Chapter 11, we will be able to build rotational kinetic energy into energy conservation without needing to modify equation 7.1. Similarly, no change in the equation will be necessary when we define a general form of gravitational potential energy in Chapter 8, and define spring potential energy in Chapter 12. It will only be necessary to expand our definitions of potential and kinetic energy.

Mechanical energy is the sum of the potential and the kinetic energies. If no net work is done by non-conservative forces (if $W_{nc} = 0$), then mechanical energy is conserved. This is the **principle of the conservation of mechanical energy**.

EXAMPLE 7.1 – A frictional best-seller
    A popular book, with a mass of 1.2 kg, is pushed across a table. The book has an initial speed of 2.0 m/s, and it comes to rest after sliding through a distance of 0.80 m.
        (a) What is the work done by friction in this situation?
        (b) What is the average force of friction acting on the book as it slides?

SOLUTION
        (a) As usual, we should draw a diagram of the situation and a free-body diagram. These diagrams are shown in Figure 7.1. Three forces act on the book as it slides. The normal force is directed up, at 90° to the displacement, so the normal force does no work. The effect of the force of gravity is accounted for via the potential energy terms in equation 7.1, but the gravitational potential energy does not change, because the book does not move up or down. The only force that affects the energy is the force of friction. The work done by friction is accounted for by the W_{nc} term in the conservation of energy equation.

Figure 7.1: A diagram of the sliding book and a free-body diagram showing the forces acting on it as it slides.

        So, the five-term conservation of energy equation, $K_i + U_i + W_{nc} = K_f + U_f$, can be reduced to two terms, because $U_i = U_f$ and the final kinetic energy $K_f = 0$. We are left with:

$$K_i + W_{nc} = 0 \quad \text{so, } W_{nc} = -K_i = -\frac{1}{2}mv_i^2 = -\frac{1}{2}(1.2 \text{ kg})(2.0 \text{ m/s})^2 = -2.4 \text{ J}.$$

        The work done by the non-conservative force, which is kinetic friction in this case, is negative because the force of friction is opposite in direction to the displacement.

        (b) To find the force of kinetic friction, F_K, use the definition of work. In this case, we get: $W_{nc} = F_K \Delta r \cos\theta = -F_K \Delta r$.

        Solving for the force of friction gives $F_K = -\dfrac{W_{nc}}{\Delta r} = -\dfrac{-2.4 \text{ J}}{0.80 \text{ m}} = 3.0 \text{ N}$.

Related End-of-Chapter Exercises: 37, 40, 42.

A general method for solving a problem involving energy conservation
        This general method can be applied to a wide variety of situations.
        1. Draw a diagram of the situation. Usually, we use energy to relate a system at one point, or instant in time, to the system at a different point, or a different instant.

        2. Write out equation 7.1, $K_i + U_i + W_{nc} = K_f + U_f$.

        3. Choose a level to be the zero for gravitational potential energy. Defining the zero level so that either U_i or U_f (or both) is zero is often best.

        4. Identify the terms in the equation that are zero.

        5. Take the remaining terms and solve.

Essential Question 7.1: Did we have to solve Example 7.1 using energy ideas, or could we have used forces and Newton's second law instead?

Answer to Essential Question 7.1: In the case of the book sliding on the table, we can apply either an energy analysis or a force analysis. Let's now compare these different methods.

7-2 Comparing the Energy and Force Approaches

In Example 7.1, there is no real advantage in using an energy analysis over a force analysis. In some cases, however, the energy approach is much easier than the force approach.

EXPLORATION 7.2A – Which ball has the higher speed?
Three identical balls are launched with equal speeds v from a height h above level ground. Ball A is launched horizontally, while the initial velocity of ball B is at 30° above the horizontal, and the initial velocity of ball C is at 45° below the horizontal. Rank the three balls, based on their speeds when they reach the ground, from largest to smallest. Neglect air resistance.

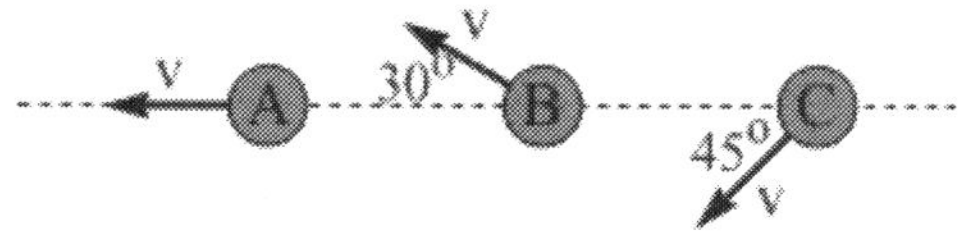

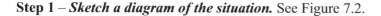

Step 1 – *Sketch a diagram of the situation.* See Figure 7.2.

Figure 7.2: A diagram showing the directions of the initial velocities of the three balls in Exploration 7.2A.

Step 2 – *Briefly describe how to solve this problem using methods applied in earlier chapters.* Consider the projectile-motion analysis we applied in chapter 4. For each ball, we would break the initial velocity into components, determine the y-component of the final velocity using one of the constant-acceleration equations, and then find the magnitude of the final velocity by using the Pythagorean theorem. We would have to go through the process three times, once for each ball.

Step 3 – *Instead, solve the problem using an energy approach.* Our starting point for energy is always the conservation of energy equation, $K_i + U_i + W_{nc} = K_f + U_f$. There is no air resistance, so $W_{nc} = 0$. If we define the zero for gravitational potential energy as the ground level, then $U_f = 0$, and $U_i = mgh$, where m is the mass of a ball (each ball has the same mass). Substituting this expression into the conservation of energy equation gives: $K_i + mgh = K_f$.

Both terms on the left are the same for all three balls. The balls have the same initial kinetic energy and they experience the same change in potential energy. Thus, all three balls have identical final kinetic energies. Because $K_f = (1/2)mv_f^2$, and the balls have equal masses, the final speeds are equal. Based on one energy analysis that works for all three balls, instead of three separate projectile-motion analyses, the ranking of the balls based on final speed is A=B=C.

Step 4 – *Would the answer be different if the balls had unequal masses?* Starting from $K_i + mgh = K_f$, and using the definition of kinetic energy, we can show that mass is irrelevant:

$$\frac{1}{2}mv_i^2 + mgh = \frac{1}{2}mv_f^2 .$$

Factors of mass cancel, giving: $v_f = \sqrt{2gh + v_i^2}$.

So, the balls have the same final speed even if their masses are different.

> **Key idea for solving problems**: We now have two powerful ways of analyzing physical situations. We can either apply force ideas, or apply energy ideas. In certain situations the energy approach is simpler than the force approach.   **Related End-of-Chapter Exercises: 6, 38.**

EXPLORATION 7.2B – Which cart wins the race?

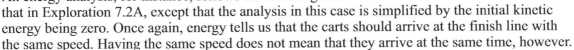

As shown in Figure 7.3, two identical carts have a race on separate tracks. Cart A's track follows a straight path sloping down, while cart B's track dips down below A's just after the start and rises up to meet A's again just before the finish line. If the carts are released at the same time, which cart reaches the end of the track first? Make a prediction, and justify your answer.

Figure 7.3: A race between two identical carts. Cart A's track has a constant slope, while cart B's track dips down below A's before rising up to meet A's again at the end.

After considering the three balls of Exploration 7.2A, it is tempting to predict that the race will end in a tie. An energy analysis, for instance, follows the same logic as that in Exploration 7.2A, except that the analysis in this case is simplified by the initial kinetic energy being zero. Once again, energy tells us that the carts should arrive at the finish line with the same speed. Having the same speed does not mean that they arrive at the same time, however.

Another popular answer is that A wins the race because B travels farther. In actuality, for most tracks, cart B wins the race. Cart A gradually picks up speed as it loses potential energy. In contrast, cart B immediately drops below A, transforming potential energy into kinetic energy, and reaching a speed larger than that of cart A. The carts then travel along parallel paths, with B always moving faster than A. Even while B slows as it climbs the hill near the end, it is traveling faster than A. The larger distance B travels is more than made up for by B's larger average speed.

Key idea about energy and time: Energy can be a powerful concept, but energy generally gives us no direct information about time. **Related End-of-Chapter Exercises: 4, 35.**

Let's compare and contrast the energy approach with the force approach. Energy can be a very effective tool, because in many cases we only have to consider the initial and final states and we don't have to worry about how the system gets from one state to the other. On the other hand, energy tells us nothing about the time it takes to get from one state to another. In Exploration 7.2B, for instance, the three balls reach the ground at different times, but we would have to use forces, and the constant-acceleration equations, to find those times. Energy is also a scalar, so it tells us nothing about direction. Energy is perfect, however, for connecting speed and position.

If we want to learn about time, or about the direction of a vector, analyzing forces is a better approach. So far, though, we are limited to applying force concepts to situations in which the net force is constant, when we can apply the constant-acceleration equations. We will go beyond this in Chapter 8 but, at the level of physics we are concerned with in this book, we will always be limited in how far we can go with forces. A good example of the limitations of force is shown in Figure 7.4, where an object slides from point A to point B along various paths. If the object comes down path 1, the straight line connecting A and B, we can use forces or energy to analyze the motion, even if friction acts as the object slides. If the object slides along a path other than path 1, then we can't get far with forces. The difficulty is that the force approach is *path dependent* – the forces applied to the object depend on the path, and the forces change if the object moves along a different path.

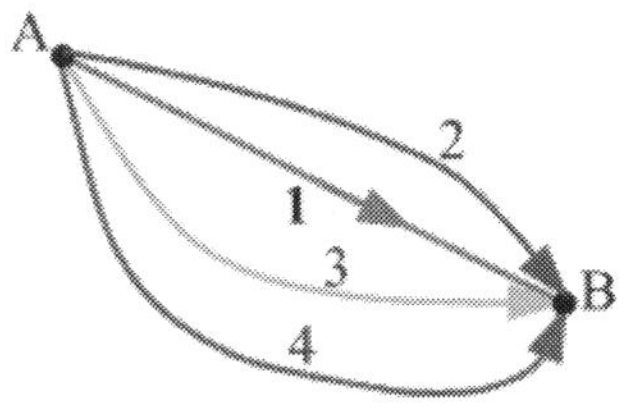

Figure 7.4: Various paths for an object to slide along in traveling from A to B.

Essential Question 7.2: Could we use energy to analyze the motion along paths 2, 3, or 4 in Figure 7.4?

Answer to Essential Question 7.2: If there is no friction, the energy analysis is path independent, so we treat all paths the same. If friction does act on the object, however, even energy is hard to apply, because then the work done by friction depends on the path.

7-3 Energy Bar Graphs: Visualizing Energy Transfer

Energy conservation is a powerful tool. To make it easier for us to use this tool, it can be useful to use a visual aid to keep track of the various types of energy. One way to visualize how energy in a system is transformed from one type of energy to another is to use energy bar graphs. Consider the following example.

EXAMPLE 7.3A – Learning to use energy bar graphs

Consider a ball that you throw straight up into the air, and neglect air resistance. Define the gravitational potential energy to be zero at the point from which you release it. Draw energy bar graphs to show how the ball's mechanical energy is divided between kinetic energy and gravitational potential energy at the following points: (a) the point you release it; (b) the point halfway between the release point and the maximum-height point, on the way up; (c) the maximum-height point; (d) the point halfway between the release point and the maximum height, on the way down; and (e) the release point, on the way down.

SOLUTION

The five sets of bar graphs are shown in Figure 7.5. The vertical position of the ball is indicated above each set of graphs, making it clearer why the second and fourth sets of graphs are the same, and why the first and last sets are the same. As the ball rises, its kinetic energy is transformed into potential energy, reaching 100% potential at the maximum height point. As the ball falls, the potential energy transforms back to kinetic energy.

Related End-of-Chapter Exercises: 21, 22.

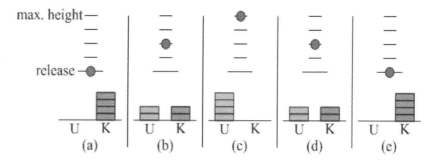

Figure 7.5: Energy bar graphs for a ball going straight up and down. The ball's corresponding vertical position is shown above each of the bar

In this case, the total mechanical energy is conserved. Let's do another example to see how bar graphs are used to depict a situation in which energy is not conserved.

EXAMPLE 7.3B – Extending the use of energy bar graphs

A string is tied to a block that has a mass of 2.0 kg. As shown in Figure 7.6, the string passes over a pulley, and you hang onto the end of the string to prevent the block from moving. Initially, the block is 1.0 m above the ground. You then pull down on the string so the block accelerates upward at a constant rate of 4.0 m/s². Use $g = 10$ m/s², and define the zero for gravitational potential energy to be at ground level. Draw bar graphs to show the block's gravitational potential energy, kinetic energy, total mechanical energy, and the work you have done on the block (a) at the instant the block starts to move, and (b) 0.50 s after the block starts to move.

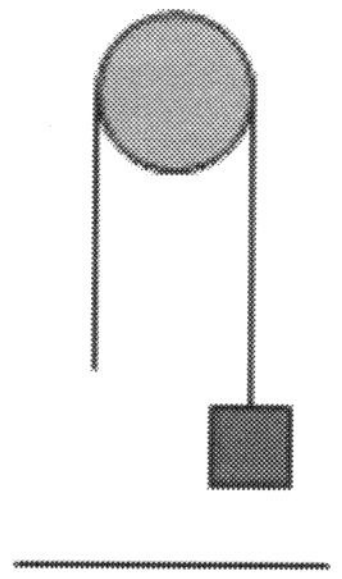

Figure 7.6: The system of the string, pulley, and block. You hold the string.

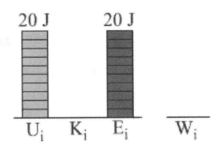

SOLUTION

(a) Let's define up to be the positive y-direction. Initially, the block is at rest and you have not done any work, so the energy is all gravitational potential energy. This relation is written $U_i = mgy_i = (2.0 \text{ kg}) \times (10 \text{ m/s}^2) \times (1.0 \text{ m}) = 20 \text{ J}$. Because there is no kinetic energy, the block's total mechanical energy is 20 J. The bar graphs for the initial energies are shown in Figure 7.7.

Figure 7.7: Energy bar graphs showing the energy at the start of the motion. Each small rectangle represents an energy of 2 J. E is the total mechanical energy.

(b) To find the block's gravitational potential energy at t = 0.50 s, we must determine how far off the ground the block is at that time. Because the acceleration is constant, we can use a constant-acceleration equation:

$$\vec{y}_{t=0.5} = \vec{y}_i + \vec{v}_{iy}t + \frac{1}{2}\vec{a}_y t^2 = +1.0 \text{ m} + 0 + \frac{1}{2}(+4.0 \text{ m/s}^2)\times(0.50 \text{ s})^2 = +1.5 \text{ m}.$$

This y-position corresponds to a gravitational potential energy of:

$$U_{t=0.5} = mgy_{t=0.5} = (2.0 \text{ kg})\times(10 \text{ m/s}^2)\times(+1.5 \text{ m}) = +30 \text{ J}.$$

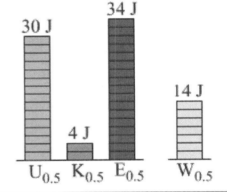

To find the kinetic energy at t = 0.50 s, let's find the speed of the block. We can use a constant-acceleration equation to find the block's velocity at t = 0.50 s and then take the magnitude to get the speed.

$$\vec{v}_{y,t=0.5} = \vec{v}_{iy} + \vec{a}_y t = 0 + (+4.0 \text{ m/s}^2)\times(0.50 \text{ s}) = +2.0 \text{ m/s}.$$

The kinetic energy is then

$$K_{t=0.5} = \frac{1}{2}mv^2 = \frac{1}{2}(2.0 \text{ kg})\times(2.0 \text{ m/s})^2 = 4.0 \text{ J}.$$

Figure 7.8: Energy bar graphs showing the energy of the block, and the work (W) you have done on the block up to that point, at t = 0.50 s.

The total mechanical energy is the sum of the potential energy and the kinetic energy, for a total of 34 J. This is 14 J larger than the initial mechanical energy, meaning that you must have done 14 J of work on the block. All the energies are represented by the bar graphs in Figure 7.8.

To summarize this section, the visualization technique of drawing energy bar graphs can be applied to systems in which mechanical energy is conserved, or even to systems when it is not conserved. Drawing these bar graphs is a good way to keep track of the different types of energy in a system.

Related End-of-Chapter Exercises: 23, 24.

Essential Question 7.3: In Example 7.3B, the bar graphs representing the various energies and the work are all either positive or zero. Could any of these be negative in some circumstances?

Answer to Essential Question 7.3: All of them can be negative except for the kinetic energy, which can't be negative because $K = (1/2)mv^2$ and neither the square of the speed nor the mass can be negative. What is key is how the energies change, not what the values of the energies are.

7-4 Momentum and Collisions

Let's extend our understanding of momentum by analyzing a **collision**, which is an event in which two objects interact. As we learned in Chapter 6, Newton's third law tells us that, when no net external force acts on a system, the total momentum of the system is conserved. The momenta of the individual objects can change, but the total momentum of the system does not.

Generally, when we analyze a collision, we look at the situation immediately before the collision and compare it to the situation immediately after the collision. What happens during the collision itself can be interesting, and complicated. Fortunately, by using momentum we don't have to worry about such complications. The usual starting point in analyzing a collision is to write down a conservation of momentum equation reflecting the following relation:

Total momentum before the collision = total momentum after the collision.

$$m_1\vec{v}_{1i} + m_2\vec{v}_{2i} = m_1\vec{v}_{1f} + m_2\vec{v}_{2f} \qquad \text{(Equation 7.2: Momentum conservation)}$$

where the subscripts i and f stand for initial and final, and the two colliding objects are denoted by 1 and 2.

EXPLORATION 7.4 – Two carts collide…again

Two identical carts experience a collision on a horizontal track. Before the collision, cart 1 is moving at speed v to the right, directly toward cart 2, which is at rest. Immediately after the collision, cart 2 is moving with a velocity of $v/2$ to the right.

Step 1 - *What is the velocity of cart 1 immediately after the collision?* Let's begin, as usual, by drawing a diagram of the situation in Figure 7.9, showing the carts before and after the collision.

Figure 7.9: Two carts, immediately before and immediately after their collision.

The first step in applying equation 7.2 is to remember that momentum is a vector. Let's define right as the positive x-direction. We can say that each cart has a mass m, and we are given that $\vec{v}_{1i} = +v$, $\vec{v}_{2i} = 0$, and $\vec{v}_{2f} = +v/2$. Substituting all these terms into the conservation of momentum equation gives:

$$+mv = m\vec{v}_{1f} + m\frac{v}{2}$$

Dividing out a factor of m and solving for the velocity of cart 1 after the collision gives:

$$\vec{v}_{1f} = +\frac{v}{2}.$$

The two carts have the same velocity, and thus move together, after the collision. We could arrange this special case by attaching Velcro to both carts so they stick together. When the objects move together afterwards, we say that the collision is ***completely inelastic***.

Step 2 - *Is kinetic energy conserved in this collision?* Kinetic energy does not have to be conserved in a collision, although in certain special cases it is. Let's see what happens to the kinetic energy in this case.

Before the collision: $K_i = \frac{1}{2}mv_{1i}^2 + \frac{1}{2}mv_{2i}^2 = \frac{1}{2}mv^2$.

After the collision: $K_f = \frac{1}{2}mv_{1f}^2 + \frac{1}{2}mv_{2f}^2 = \frac{1}{2}m\left(\frac{v}{2}\right)^2 + \frac{1}{2}m\left(\frac{v}{2}\right)^2 = \frac{1}{4}mv^2 = \frac{K_i}{2}$.

In this case, only 50% of the kinetic energy from before the collision is in the system as kinetic energy after the collision. The total energy has to be conserved, but in this case, some of the kinetic energy of cart 1 before the collision is transformed to other forms of energy (such as thermal energy, which is energy associated with the motion of molecules, and sound energy) in the collision process.

Step 3 - *What is the velocity of the system's center of mass before the collision?* By dividing both sides of Equation 6.4, for the position of the center of mass, by a time interval, Δt, and using the definition of velocity, we can obtain an equation for the velocity of the center of mass:

$$\vec{v}_{CM} = \frac{m_1\vec{v}_1 + m_2\vec{v}_2 + m_3\vec{v}_3 + \ldots}{m_1 + m_2 + m_3 + \ldots}$$ (Equation 7.3: **Velocity of the center of mass**)

The *m*'s represent the masses of the various pieces of the object or system. The terms in the numerator on the right represent the momenta of the individual parts of the system, so the equation really says that the total momentum of the system is the vector sum of the momenta of its parts, which seems sensible.

Applying the equation to the two-cart system before the collision gives:

$$\vec{v}_{CM,i} = \frac{+mv + m \times 0}{m + m} = +\frac{v}{2}.$$

This result makes sense because the center of mass is halfway between the carts, so the center of mass covers half the distance as cart 1 does in the same time.

Step 4 - *What is the velocity of the system's center of mass after the collision?* Applying equation 7.3 to the system after the collision gives:

$$\vec{v}_{CM,f} = \frac{+m\dfrac{v}{2} + m\dfrac{v}{2}}{m + m} = +\frac{v}{2}.$$

It should come as no surprise that the velocity of the center of mass after the collision is the same as the velocity of the center of mass before the collision. Rather, this result is expected as a consequence of momentum conservation. In short, the center of mass does not even register that a collision has taken place.

Key ideas: In a collision, in general, the system's momentum is conserved while the system's kinetic energy is not necessarily conserved. In addition, in general, the motion of the system's center of mass is unaffected by the collision.   **Related End-of-Chapter Exercises: 25 – 28.**

Essential Question 7.4: Under what condition is the momentum of a system conserved in a collision?

Answer to Essential Question 7.4: For momentum to be conserved, either no net force is acting on the system, or the net force must act over such a small time interval that it has a negligible effect on the momentum of the system.

7-5 Classifying Collisions

If we attach Velcro to our colliding carts, and the carts stick together after the collision, as in Exploration 7.4, the collision is **completely inelastic**. If we remove the Velcro, so the carts do not stick together, we can set up a collision with the same initial conditions (cart 1 moving toward cart 2, which is stationary) and get a variety of outcomes. We generally classify these outcomes into four categories, depending on what happens to the kinetic energy in the collision.

We can also define a parameter k called the *elasticity*. Elasticity is the ratio of the relative velocity of the two colliding objects after the collision to the negative of their relative velocity before the collision. By this definition, the elasticity should always be positive:

$$k = \frac{\vec{v}_{2f} - \vec{v}_{1f}}{\vec{v}_{1i} - \vec{v}_{2i}}$$

(Equation 7.4: **Elasticity**)

The four categories of collisions can also be defined in terms of the elasticity.

Type of Collision	Kinetic Energy	Elasticity	Example
Super-elastic	$K_f > K_i$	$k > 1$	Carts are initially stationary, then pushed apart by a spring-loaded piston, as in Exploration 6.4. An explosion.
Elastic	$K_f = K_i$	$k = 1$	Carts with repelling magnets.
Inelastic	$K_f < K_i$	$k < 1$	Describes most collisions, such as two cars that make contact when colliding but that don't stick together.
Completely inelastic	$K_f < K_i$, and the objects stick together	$k = 0$	Carts with Velcro, as in Exploration 7-3, or chewing gum hitting the sidewalk.

Table 7.1: Collisions can be classified in terms of what happens to the kinetic energy or in terms of the elasticity. Note that, in an elastic collision, the fact that $k = 1$ can be obtained by combining the momentum conservation equation with the conservation of kinetic energy equation.

EXAMPLE 7.5 – An assist from gravity

Sending a space probe from Earth to another planet requires a great deal of energy. In many cases, a significant fraction of the probe's kinetic energy can be provided by a third planet, through a process known as a *gravitational assist*. For instance, the Cassini-Huygens space probe launched on October 15, 1997, used four gravitational assists, two from Venus, one from Earth, and one from Jupiter, to speed it on its more than 1 billion km trip to Saturn, arriving there on July 1, 2004. We can treat a gravitational assist as an elastic collision, because the long-range interaction of the probe and the planet provides no mechanism for a loss of mechanical energy.

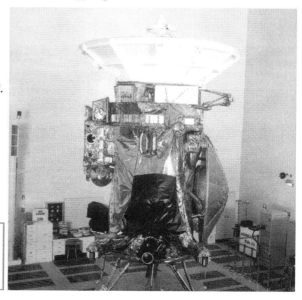

Figure 7.10: The Cassini-Huygens space probe while it was being assembled. The desk and chair at the lower left give a sense of the scale. Photo courtesy NASA/JPL-Caltech.

A space probe with a speed v is approaching Venus, which is traveling at a velocity V in the opposite direction. The probe's trajectory around the planet reverses the direction of the probe's velocity. (a) How fast does the probe travel away from Venus? (b) If $v = 1 \times 10^5$ m/s, and $V = 3.5 \times 10^5$ m/s, what is the ratio of the probe's final kinetic energy to its initial kinetic energy?

SOLUTION

Let's begin with a diagram of the situation, shown in Figure 7.11. Although we will analyze this situation as a collision, the objects do not make contact with one another.

(a) The probe's speed depends on how far away it is from Venus. Because no distances were given, let's assume the probe has speed v when it is so far from Venus that the gravitational pull of Venus is negligible. We will work out the final velocity under the same assumption. This is an elastic collision. We could apply conservation of momentum and conservation of energy, but we were not given any masses and the resulting equations can be challenging to combine to find the final velocity of the probe. Let's try working with the elasticity k instead.

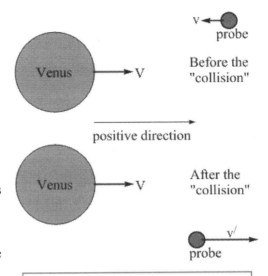

Because this collision is elastic, $k = 1$. The elasticity is the ratio of the final relative speed to the initial relative speed, so those two relative speeds must be equal for k to equal 1. Also, we can reasonably assume that the probe's mass is much smaller than the planet's mass and that the planet's motion is unaffected by its interaction with the probe. Thus, the planet's velocity is V in the direction shown both before and after the collision.

Figure 7.11: Before and After situations for the interaction between Venus and the space probe.

Defining the direction of the planet's velocity as the positive direction, plugging everything into the elasticity equation gives:

$$1 = \frac{\vec{v}_{2f} - \vec{v}_{1f}}{\vec{v}_{1i} - \vec{v}_{2i}} = \frac{V - v_{1f}}{-v - V} \quad,$$

where both the numerator and denominator are negative.

Solving for the final speed of the probe gives $v_{1f} = 2V + v$.

(b) Substituting in the numbers gives a final speed of 8×10^5 m/s, an increase by a factor of 8 in speed. Kinetic energy is proportional to the square of the speed, so the probe's kinetic energy is increased by a factor of 64. A more formal way to show this relation is the following:

$$\frac{K_f}{K_i} = \frac{\frac{1}{2}mv_f^2}{\frac{1}{2}mv_i^2} = \frac{v_f^2}{v_i^2} = \frac{\left(8 \times 10^5 \text{ m/s}\right)^2}{\left(1 \times 10^5 \text{ m/s}\right)^2} = 64 \, .$$

The probe gains an enormous amount of energy, and it does so without requiring massive amounts of fuel to be burned. This is why probes to the outer planets are often first sent toward Venus, because the large increase in speed more than makes up for the extra distance traveled.

Related End-of-Chapter Exercises: 11, 54, 55.

Essential Question 7.5: In our analysis in example 7.5, we assumed that the planet's speed is constant. Is this absolutely correct? Is it a reasonable assumption?

Answer to Essential Question 7.5: To conserve energy of the planet-probe system, the planet's speed must decrease when the probe's speed increases. Because the planet's mass is so much larger than the probe's, this decrease in speed is negligible. Thus, the assumption is reasonable.

7-6 Collisions in Two Dimensions

Momentum conservation also applies in two and three dimensions. The standard approach to a two-dimensional (or even three-dimensional) problem is to break the momentum into components and conserve momentum in both the x and y directions separately. For colliding objects, the conservation of momentum equation in the x-direction, for instance, is:

$$\vec{p}_{1ix} + \vec{p}_{2ix} = \vec{p}_{1fx} + \vec{p}_{2fx}. \quad \text{(Eq. 7.5: Conserving momentum in the } x\text{-direction)}$$

This can be written in an equivalent form:

$$m_1\vec{v}_{1ix} + m_2\vec{v}_{2ix} = m_1\vec{v}_{1fx} + m_2\vec{v}_{2fx} \quad \text{(Eq. 7.6: Momentum conservation, } x\text{-direction)}$$

Similar equations apply in the y-direction.

EXPLORATION 7.6 – A two-dimensional collision

An object of mass m, moving in the $+x$-direction with a velocity of 5.0 m/s, collides with an object of mass $2m$. Before the collision, the second object has a velocity given by $\vec{v}_{2i} = -3.0 \text{ m/s } \hat{x} + 4.0 \text{ m/s } \hat{y}$, while, after the collision, its velocity is 3.0 m/s in the $+y$-direction. What is the velocity of the first object after the collision?

Step 1 – Draw a diagram of the situation. This is shown in Figure 7.12.

Step 2 - Set up a table showing the momentum components of each object before and after the collision. Organizing components into a table helps us keep the x-direction information separate from the y-direction information. We can combine the components into one vector at the end.

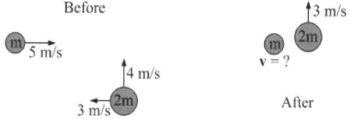

Figure 7.12: A diagram of the objects before and after they collide.

	x-direction	**y-direction**
Before the collision	$\vec{p}_{1ix} = m\vec{v}_{1ix} = +(5 \text{ m/s})m$	$\vec{p}_{1iy} = m\vec{v}_{1iy} = 0$
	$\vec{p}_{2ix} = 2m\vec{v}_{2ix} = -(6 \text{ m/s})m$	$\vec{p}_{2iy} = m\vec{v}_{2iy} = +(8 \text{ m/s})m$
After the collision	$\vec{p}_{1fx} = m\vec{v}_{1fx} = ?$	$\vec{p}_{1fy} = m\vec{v}_{1fy} = ?$
	$\vec{p}_{2fx} = 2m\vec{v}_{2fx} = 0$	$\vec{p}_{2fy} = 2m\vec{v}_{2fy} = +(6 \text{ m/s})m$

Table 7.2: Organizing the collision data in a table helps to keep the x-direction information separate from the y-direction information, and doing so can also help us solve the problem.

Step 3 – Apply conservation of momentum in the x-direction, and find the x-component of the first object's final velocity. Applying momentum conservation in the x-direction involves writing down the equation $\vec{p}_{1ix} + \vec{p}_{2ix} = \vec{p}_{1fx} + \vec{p}_{2fx}$.

This gives $\vec{p}_{1fx} = \vec{p}_{1ix} + \vec{p}_{2ix} - \vec{p}_{2fx} = +(5 \text{ m/s})m - (6 \text{ m/s}) m - 0 = -(1 \text{ m/s})m$.

To get the velocity component in the x-direction we just divide by the mass, m.

$$\vec{v}_{1fx} = \frac{\vec{p}_{1fx}}{m} = \frac{-(1\ \text{m/s})m}{m} = -1\ \text{m/s} .$$

Step 4 – Use a similar process in the y-direction to find the y-component of the first object's final velocity. Applying momentum conservation in the y-direction involves writing down the equation $\vec{p}_{1iy} + \vec{p}_{2iy} = \vec{p}_{1fy} + \vec{p}_{2fy}$.

This equation gives $\vec{p}_{1fy} = \vec{p}_{1iy} + \vec{p}_{2iy} - \vec{p}_{2fy} = 0 + (8\ \text{m/s})m - (6\ \text{m/s})m = +(2\ \text{m/s})m$.

To get the velocity component in the y-direction, we divide by the mass, m.

$$\vec{v}_{1fy} = \frac{\vec{p}_{1fy}}{m} = \frac{+(2\ \text{m/s})m}{m} = +2\ \text{m/s} .$$

Step 5 – Combine the x and y components to find the first object's final speed. Also, write down an expression for the first object's final velocity. We can use the Pythagorean theorem to find the final speed of the first object:

$$v_{1f} = \sqrt{v_{1fx}^2 + v_{1fy}^2} = \sqrt{1^2 + 2^2} = \sqrt{5}\ \text{m/s} .$$

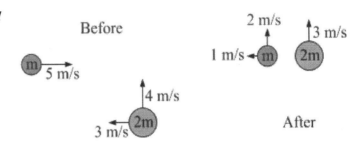

The velocity can be written in terms of components as $\vec{v}_{1f} = -1\ \text{m/s}\ \hat{x} + 2\ \text{m/s}\ \hat{y}$. The first ball's final velocity is shown in Figure 7.13.

Figure 7.13: The situation Before and After the collision.

Key idea for momentum problems: We can solve a momentum problem in two dimensions with a strategy based on the independence of x and y, breaking a two-dimensional problem into two independent one-dimensional problems. **Related End-of-Chapter Exercises: 29, 57.**

Now that we've looked at a few examples, let's summarize a general method for solving a problem in which there is a collision.

A General Method for Solving a Problem That Involves a Collision
1. Draw a diagram of the situation, showing the velocity of the objects immediately before and immediately after the collision.
2. In a two-dimensional situation, set up a table showing the components of the momentum before and after the collision for each object.
3. Use momentum conservation: $m_1\vec{v}_{1i} + m_2\vec{v}_{2i} = m_1\vec{v}_{1f} + m_2\vec{v}_{2f}$. (Apply this twice, once for each direction, in a two-dimensional situation.) Account for the fact that momentum is a vector by using appropriate $+$ and $-$ signs.
4. If you need an additional relationship (such as in the case of an elastic collision), use the elasticity relationship or write an energy-conservation equation.

Essential Question 7.6: The strategy outlined above, which we applied in Exploration 7.6, relies on breaking vectors into components. Is there another method that we could use to solve the problem without using components?

Answer to Essential Question 7.6: A whole-vector approach, not splitting the velocity and momentum vectors into components, would also work (see End-of-Chapter Exercise 58).

7-7 Combining Energy and Momentum

To analyze some situations, we apply both energy conservation and momentum conservation in the same problem. The trick is to know when to apply energy conservation (and when not to!) and when to apply momentum conservation. Consider the following Exploration.

EXPLORATION 7.7 – Bringing the concepts together
Two balls hang from strings of the same length. Ball A, with a mass of 4.0 kg, is swung back to a point 0.80 m above its equilibrium position. Ball A is released from rest and swings down and hits ball B. After the collision, ball A rebounds to a height of 0.20 m above its equilibrium position and ball B swings up to a height of 0.050 m. Let's use $g = 10$ m/s^2 to simplify the calculations.

Step 1 – *Sketch a diagram of the situation.* This is shown in Figure 7.14.

Step 2 – *Our goal is to find the mass of ball B. Can we find the mass by setting the initial gravitational potential energy of ball A equal to the sum of the final potential energy of ball A and the final potential energy of ball B? Explain why or why not.* The answer to the question is no. We can use energy conservation to help solve the problem, but setting the mechanical energy before the collision equal to the mechanical energy after the collision is assuming too much. The balls make contact in the collision, so it is likely that some of the mechanical energy is transformed to thermal energy, for instance.

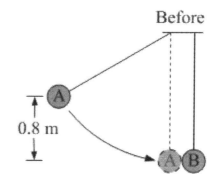

Figure 7.14: A diagram of the two balls on strings. Ball A is swung back until it is 0.80 m higher than its equilibrium point and released from rest.

Step 3 – *Apply energy conservation to find the speed of ball A just before the collision.* The gravitational potential energy of ball A is transformed into kinetic energy just before the collision. We will neglect the work done by air resistance, so we can apply energy conservation before the collision. Let's start with the conservation of energy equation:
$K_i + U_i + W_{nc} = K_f + U_f$.

For ball A's swing before the collision, we know that the initial kinetic energy is zero. We are assuming that non-conservative forces do no work. We can also define the zero level for gravitational potential energy to be the lowest point in the swing, just before A hits B, so $U_f = 0$. The five-term equation reduces to:
$$U_i = K_f ;$$
$$mgh = \frac{1}{2}mv_f^2 ;$$
So, $v_f = \sqrt{2gh} = \sqrt{2 \times 10 \times 0.8} = \sqrt{16} = 4.0$ m/s .

Step 4 – *Apply conservation of energy again to find the speed of ball A just after the collision.* We could try applying conservation of momentum here, but there are too many unknowns. Instead, we can follow the conservation of energy method we used above. Note that we will not state that the kinetic energy immediately before the collision is equal to the kinetic energy after the collision, because that is not true. We can apply energy conservation, however, if we confine ourselves to the mechanical energy before the collision (as in step 3) or to the mechanical energy after the collision (this step). If we focus on the upswing, we have the kinetic energy of ball A, immediately after the collision, being transformed into gravitational potential energy. The conservation of energy equation reduces to:

$$K_i = U_f \, ;$$

$$\frac{1}{2}mv_i^2 = mgh \, ;$$

So, $v_i = \sqrt{2gh} = \sqrt{2 \times 10 \times 0.2} = \sqrt{4.0} = 2.0 \text{ m/s}$.

Let's be clear on what we have calculated in parts 3 and 4, because the notation can be confusing. We are analyzing the motion in three separate parts. The first part of ball A's motion is the downswing, which we analyzed in step 3. The third part is the upswing, which we analyzed in step 4. The second part is the collision, which we still have to analyze. The velocity of ball A immediately before the collision, at the end of the downswing, is $\vec{v}_{Ai} = 4.0 \text{ m/s}$ to the right, while A's velocity just after the collision, at the start of the upswing, is $\vec{v}_{Af} = 2.0 \text{ m/s}$ to the left.

These are the values we will use in the conservation of momentum equation in step 5.

Step 5 – First, apply energy conservation to find the speed of ball B after the collision. Then, apply momentum conservation to find the mass of ball B. We still have to find the velocity of ball B, after the collision, before we use the conservation of momentum equation to find ball B's mass. We can find B's speed immediately after the collision by following the same process we used for ball A in step 3. We get:

$$K_i = U_f$$

$$\frac{1}{2}mv_i^2 = mgh$$

So, $v_{Bf} = \sqrt{2gh_B} = \sqrt{2 \times (10 \text{ m/s}^2) \times 0.050 \text{ m}} = \sqrt{1.0 \text{ m}^2/\text{s}^2} = 1.0 \text{ m/s}$.

The velocity of ball B immediately after the collision is $\vec{v}_{Bf} = 1.0 \text{ m/s}$ to the right.

Now, we can write out a conservation of momentum equation to solve for the mass of ball B. It is critical to account for the fact that momentum is a vector. In this case, we account for the vector nature of momentum by using a minus sign for the velocity of ball A after the collision to reflect that it is moving to the left, when we chose right to be the positive direction. This gives:

$m_A\vec{v}_{Ai} + m_B\vec{v}_{Bi} = m_A\vec{v}_{Af} + m_B\vec{v}_{Bf}$, where $\vec{v}_{Bi} = 0$. Solving for the mass of ball B gives:

$$m_B = \frac{m_A\vec{v}_{Ai} - m_A\vec{v}_{Af}}{\vec{v}_{Bf}} = \frac{(4.0 \text{ kg}) \times (+4.0 \text{ m/s}) - (4.0 \text{ kg}) \times (-2.0 \text{ m/s})}{+1.0 \text{ m/s}} = 24 \text{ kg} \, .$$

Key idea: In some situations, we can apply conservation of energy and conservation of momentum ideas together. In general, we apply conservation of momentum to connect the situation before the collision to the situation after the collision. We use energy conservation to learn something about the situation before the collision and/or the situation afterwards.
Related End-of-Chapter Exercises: 30 – 32.

Essential Question 7.7: Is the collision in Exploration 7.7 super-elastic, elastic, inelastic, or completely inelastic? Justify your answer in two different ways.

Answer to Essential Question 7.7: The balls don't stick together, so we know the collision is not completely inelastic. One way to classify the collision is to find the elasticity, k (see equation 7.4).

$$k = \frac{\vec{v}_{Bf} - \vec{v}_{Af}}{\vec{v}_{Ai} - \vec{v}_{Bi}} = \frac{1.0 \text{ m/s} - (-2.0 \text{ m/s})}{4.0 \text{ m/s} - 0} = 0.75.$$

The fact that k is less than 1 means the collision is inelastic. We can confirm this result by looking at the kinetic energy before and after the collision.

$$K_i = \frac{1}{2} m_A v_{Ai}^2 + \frac{1}{2} m_B v_{Bi}^2 = \frac{1}{2} \times (4.0 \text{ kg}) \times (4.0 \text{ m/s})^2 + 0 = 32 \text{ J}.$$

$$K_f = \frac{1}{2} m_A v_{Af}^2 + \frac{1}{2} m_B v_{Bf}^2 = \frac{1}{2} \times (4.0 \text{ kg}) \times (2.0 \text{ m/s})^2 + \frac{1}{2} \times (24.0 \text{ kg}) \times (1.0 \text{ m/s})^2;$$

$$K_f = 8.0 \text{ J} + 12 \text{ J} = 20 \text{ J}.$$

The kinetic energy in the system after the collision is less than it is before the collision, so we have an inelastic collision.

Chapter Summary

Essential Idea about Conservation Laws
Many physical situations can be analyzed using forces, which we learned about in previous chapters, and/or by applying the fundamental concepts of conservation of momentum and conservation of energy, which we learned about in this chapter.

Comparing the Energy and Force Methods
The primary methods we use to analyze situations are to use forces and Newton's Laws, or to use energy conservation. Let's compare these two methods.
- The energy approach can be very effective, because we often just have to deal with the initial and final states and we don't have to account for the path taken by the system in going from one state to another, as we do with the force approach.
- The energy approach, by itself, does not give us any information about time, such as about how long it takes a system to move from one state to another. If you need to know about time, use a force analysis.
- Energy is a scalar. Thus energy, by itself, tells us nothing about direction. Force is a vector, and this it can give us information about direction.
- If W_{nc}, the work done by non-conservative forces, is zero, then the total ***mechanical energy*** (the sum of the kinetic and potential energies) is conserved.

A General Method for Solving a Problem Involving Energy Conservation

1. Draw a diagram of the situation. Usually, we use energy to relate a system at one point, or instant in time, to the system at a different point, or a different instant.
2. Apply energy conservation: $K_i + U_i + W_{nc} = K_f + U_f$. (Eq. 7.1)
3. Choose a level to be the zero for gravitational potential energy. Setting the zero level so that either U_i or U_f (or both) is zero is often best.
4. Identify the terms in the equation that are zero.
5. Take the remaining terms and solve.

Collisions and Momentum Conservation

In general, the momentum of a system is conserved in a collision, but the system's kinetic energy is often not conserved in a collision. In fact, one of the two ways in which we classify collisions is based on how the kinetic energy before the collision compares to that afterwards. The second way collisions can be classified is in terms of the *elasticity*, k, which is the ratio of the relative speed of the colliding objects after the collision to their relative speed after the collision:

$$k = \frac{\vec{v}_{2f} - \vec{v}_{1f}}{\vec{v}_{1i} - \vec{v}_{2i}} \quad .$$
(Equation 7.4: **Elasticity**)

This equation is particularly useful when the collision is elastic and the relative velocity of the objects has the same magnitude before and after the collision.

The four collision categories are:

Type of Collision	Kinetic Energy	Elasticity
Super-elastic	$K_f > K_i$	$k > 1$
Elastic	$K_f = K_i$	$k = 1$
Inelastic	$K_f < K_i$	$k < 1$
Completely inelastic	$K_f < K_i$, and the objects stick together	$k = 0$

A General Method for Solving a Problem Involving a Collision

1. Draw a diagram of the situation, showing the velocity of the objects immediately before and immediately after the collision.
2. In a two-dimensional situation, set up a table showing the components of the momentum before and after the collision for each object.
3. Use momentum conservation: $m_1 \vec{v}_{1i} + m_2 \vec{v}_{2i} = m_1 \vec{v}_{1f} + m_2 \vec{v}_{2f}$. (Eq. 7.2)

    Apply equation 7.2 twice, once for each direction, in a two-dimensional situation. Account for the fact that momentum is a vector with $+$ and $-$ signs.
4. If you require an additional relationship (such as in the case of an elastic collision) use the elasticity relationship or write an energy-conservation equation.

End-of-Chapter Exercises

Several of these exercises can be answered without a calculator, if you use $g = 10$ m/s^2.

Exercises 1 – 12 are conceptual questions designed to see whether you understand the main concepts of the chapter.

1. Why is it more tiring to walk for an hour up a hill than it is to walk for an hour on level ground?

2. (a) Is it possible for the gravitational potential energy of a system to be negative? (b) Is it possible for the kinetic energy of a system to be negative? (c) Can the total mechanical energy of a system be negative?

3. Given the right (or wrong, depending on your perspective) conditions, a mudslide or avalanche can occur, in which a section of earth or snow that has been at rest slides down a steep slope, reaching impressive speeds. Where does all the kinetic energy that the mud or snow has at the bottom of the slope come from?

4. Three identical blocks (see Figure 7.15) are released simultaneously from rest from the same height h above the floor. Block A falls straight down, while blocks B and C slide down frictionless ramps. B's ramp is steeper than C's. (a) Rank the blocks according to their speed, from largest to smallest, when they reach the floor. (b) Rank the blocks according to the time it takes them to reach the floor, from greatest to least. (c) If the two ramps are not frictionless, and the coefficient of friction between the block and ramp is identical for the ramps, do any of your rankings above change? If so, how?

    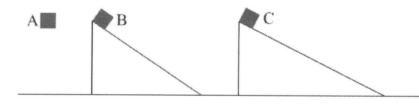

    Figure 7.15: Three identical blocks are simultaneously released from rest from the same height above the floor, for Exercise 4.

5. You are on a diving platform 3.0 m above the surface of a swimming pool. Compare the speed you have when you hit the water if you: A, drop almost straight down from rest; B, run horizontally at 4.0 m/s off the platform; C, leap almost straight up, with an initial speed of 4.0 m/s, from the end of the platform.

6. Consider the following situations. For each, state whether or not you would apply energy methods, force/projectile motion methods, or either to solve the exercise. You don't have to solve the exercise, but you can if you wish. (a) Find the maximum height reached by a ball fired straight up from level ground with a speed of 8.0 m/s. (b) Find the maximum height reached by a ball launched from level ground at a 45° angle above the horizontal if its launch speed is 8.0 m/s. (c) Find the time taken by the ball in part (b) to reach maximum height. (d) Determine which of the balls, the one in (a) or the one in (b), returns to ground level with the higher speed. (e) Determine the horizontal distance traveled by the ball in (b) before it returns to ground level.

7. You drop a large rock on an empty soda can, crushing the can. (a) Is mechanical energy conserved in this process? Explain. (b) Is energy conserved in this process? Explain.

8. A block of mass m is released from rest at a height h above the base of a frictionless loop-the-loop track, as shown in Figure 7.16. The loop has a radius R. In this situation, $h = 3R$, and, defining the block's gravitational potential energy to be zero at point a, the block's gravitational potential energy at point b is twice the size of the block's kinetic energy at point b. Sketch energy bar graphs showing the block's gravitational potential energy, kinetic energy, and total mechanical energy at (a) the starting point; (b) point a; (c) point b.

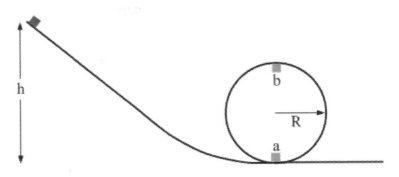

Figure 7.16: A block released from rest from a height h above the bottom of a loop-the-loop track, for Exercise 8.

9. Two boxes, A and B, are released simultaneously from rest from the top of ramps that have the same shape. Box A slides without friction down its ramp, while a kinetic friction force acts on box B as it slides down its ramp. The two boxes have the same mass. For the two boxes, plot the following as a function of time: (a) the kinetic energy; (b) the gravitational potential energy, taking the bottom of the ramp to be zero; (c) the total mechanical energy. There are no numbers here, so just show the general trend on each graph.

10. Repeat Exercise 9, but now plot the graphs as a function of distance traveled along the ramp instead of as a function of time.

11. A block is sliding along a frictionless horizontal surface with a speed v when it encounters a spring. The spring compresses, bringing the spring momentarily to rest, and then the spring returns to its original length, reversing the direction of the block's motion. If the block moves away from the spring at speed v, how can we explain what the spring has done in terms of conservation of energy? Note: this is a preview of how we will handle energy conservation for springs in Chapter 12. Hint: is there a parallel between what the spring does to the block and what the force of gravity does to the block if we toss the block straight up in the air?

12. Comment on the applicability of conservation of energy, conservation of mechanical energy, and momentum conservation in each of the following situations. (a) A car accelerates from rest. (b) In six months, the Earth goes halfway around the Sun. (c) Two football players collide and come to rest on the ground. (d) A diver leaps from a cliff and plunges toward the ocean below.

Exercises 13 – 16 deal with various aspects of the same situation.

13. A ball with a mass of 200 g is tied to a light string with a length of 2.4 m. The end of the string is tied to a hook, and the ball hangs motionless below the hook. Keeping the string taut, you move the ball back and up until it is a vertical distance of 1.25 m above its equilibrium point. You then release the ball from rest. (a) What is the highest speed the ball achieves in its subsequent motion? (b) Where does the ball achieve this maximum speed? (c) What is the maximum height reached by the ball in its subsequent motion? (d) Of the three numerical values stated in this exercise, which one(s) do you actually require to solve the problem?

14. Take a ball with a mass of 200 g and drop it from rest. (a) When the ball has fallen a distance of 1.25 m, how fast is it going? (b) How does this speed compare to the maximum speed of the identical ball in Exercise 13? Briefly explain this result. (c) Which ball takes longer to drop through a distance of 1.25 m? Justify your answer.

15. Consider the ball in Exercise 13. (a) Is it reasonable to assume that the work done by non-conservative forces is negligible over the time during which the ball swings down through the equilibrium position and up to its maximum height point on the other side? Why or why not? (b) If we watch this ball for a long time, it will eventually stop and hang motionless below the hook. Explain, in terms of energy conservation, why the ball eventually comes to rest. (c) In part (b), how much work is done by resistive forces in bringing the ball to rest?

16. Consider the ball in Exercise 13. Assuming that mechanical energy is conserved (that friction and air resistance are negligible), graph the ball's potential energy, kinetic energy, and total energy as a function of height above the equilibrium position. Take the zero of potential energy to be the equilibrium position.

Exercises 17 – 20 are designed to give you some practice in applying the general method of solving a problem involving energy conservation. For each exercise, begin with the following: (a) Sketch a diagram of the situation, showing the system in at least two states that you will relate by using energy conservation. (b) Write out equation 7.1, and define a zero level for gravitational potential energy. It is usually most convenient to define a zero level so that the initial and/or final gravitational potential energy terms are zero. (c) Identify which, if any of the terms in the equation equal zero, and explain why they are zero.

17. You drop your keys, releasing them from rest from a height of 1.2 m above the floor. The goal of this exercise is to use energy conservation to determine the speed of the keys just before they reach the floor. Assume g = 9.8 m/s². Parts (a) – (c) as above. (d) Use the remaining terms in the equation to find the speed of the keys before impact.

18. During a tennis match, you mis-hit the ball, making the ball go straight up in the air. The ball, which has a mass of 57 g, reaches a maximum height of 7.0 m above the point at which you hit it, and the ball's velocity just before you hit it was 12 m/s directed horizontally. The goal of the exercise is to determine how much work your racket did on the ball. Parts (a) – (c) as described above. (d) Determine the work the racket did on the ball. (e) Would your answer to part (d) change if the initial velocity was not horizontal but had the same magnitude?

19. You and your bike have a combined mass of 65 kg. Starting from rest, you pedal to the top of a hill, arriving there with a speed of 6.0 m/s. The net work done on you and the bike by non-conservative forces during the ride is 1.5 x 10⁴ J. The goal of the exercise is to determine the height difference between your starting point and the top of the hill. Parts (a) – (c) as described above. (d) Determine the height difference between your start and end points.

20. A block slides back and forth, inside a frictionless hemispherical bowl. The block's speed is 20 cm/s when it is halfway (vertically) between the lowest point in the bowl and the point where it reaches its maximum height. The goal of the exercise is to determine the maximum height of the block, relative to the bottom of the bowl. Parts (a) – (c) as described above. (d) Determine the block's maximum height.

Exercises 21 – 25 involve energy bar graphs.

21. You throw a ball to your friend, launching it at an angle of 45° from the horizontal. Neglect air resistance, define the zero of gravitational potential energy to be the height from which you release the ball, and assume your friend catches the ball at the same height from which you released it. Draw a set of energy bar graphs, showing the ball's gravitational potential energy and the kinetic energy, for each of the following points: (a) the launch point; (b) the point at which the ball is halfway, vertically, between the launch point and the maximum height; (c) the point where it reaches maximum height.

22. You are on your bicycle at the top of an incline that has a constant slope. You release your brakes and coast down the incline with constant acceleration, taking a time T to reach the bottom. Neglecting all resistive forces, and taking the zero of gravitational potential energy to be at the bottom of the incline, sketch a set of energy bar graphs, showing your gravitational potential energy and kinetic energy for the following points: (a) your starting point (b) at a time of $T/2$ after you start to coast (c) halfway down the incline (d) at the bottom of the incline.

23. Repeat Exercise 22, but this time make it more realistic by accounting for a resistive force. The bar graphs should show your gravitational potential energy, kinetic energy, and total mechanical energy, with a separate bar graph for the work done by the resistive force. Assume the resistive force is constant, and that it causes your kinetic energy at the bottom of the incline to be half of what it would be if the resistive force were not present. If the total time it takes you to come down the incline is now T', in part (b) the energy bar graphs should represent the energies at a time of $T'/2$ after you start to coast.

24. You show three of your friends a set of energy bar graphs. The bar graphs represent the energy, at the release point, of a ball hanging down from a string that you have pulled up and back and released from rest, so it swings with a pendulum motion. These bar graphs are the "Initial" set in Figure 7.17. You ask your three friends to draw the bar graphs representing the ball's energy as it passes through the lowest point in its swing. Margot draws the set of bar graphs shown at the upper right, Jean the set on the lower left, and Wei the set on the lower right. (a) Are the sets of bar graphs, drawn by your friends, consistent with the idea of energy conservation? Justify your answer. (b) Which (if any) of your friends has the right answer? (b) If Jean has the right answer, from what height above the lowest point was the ball released? Assume each of the small rectangles making up the bar graphs represents 1 J, that $g = 10$ m/s², and that the ball's mass is 1.0 kg.

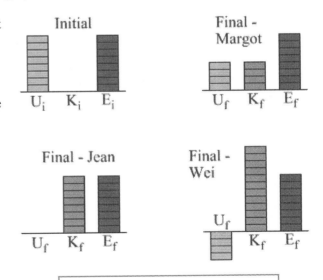

Figure 7.17: Energy bar graphs, for Exercise 24.

Exercises 25 – 29 are designed to give you some practice in applying the general method for solving a problem that involves a collision. For each exercise, start with the following parts: (a) Draw a diagram showing the objects immediately before and immediately after the collision. (b) Apply equation 7.2, the momentum-conservation equation. Choose a positive direction, and account for the fact that momentum is a vector with appropriate + and – signs.

25. A car with a mass of 2000 kg is traveling at a speed of 50 km/h on an icy road when it collides with a stationary truck. The two vehicles stick together after the collision, and their speed after the collision is 10 km/h. The goal of this exercise is to find the mass of the truck. Parts (a) – (b) as described above. (c) Solve for the mass of the truck.

26. Repeat Exercise 25, except that, in this case, the truck is moving at 20 km/h in the opposite direction of the car before the collision, and, after the collision, the two vehicles move together at 10 km/h in the same direction the truck was traveling initially.

27. Two identical air-hockey pucks experience a one-dimensional elastic collision on a frictionless air-hockey table. Before the collision, puck A is moving at a velocity of v to the right, while puck B has a velocity of $2v$ to the left. The goal of the exercise is to determine the velocity of each puck after the collision. Parts (a) – (b) as described above. (c) Use the elasticity relationship to get a second connection between the two final velocities. (d) Find the two final velocities.

28. Repeat Exercise 27, except that in this case puck B has a mass twice as large as the mass of puck A.

29. While shooting pool, you propel the cue ball at a speed of 1.0 m/s. It collides with the 8-ball (initially stationary), propelling the 8-ball into a corner pocket. The cue ball is deflected by 42° from its original path by the collision, and it moves away from the collision with a speed of 0.70 m/s. The goal of this exercise is to determine the magnitude and direction of the 8-ball's velocity after the collision. The cue ball has a little more mass than the 8-ball, but assume for this exercise that the masses are equal. Parts (a) – (b) as described above. For part (b), set up a table to keep track of the x and y components of the momenta of the two balls before and after the collision. (c) Use the information in the table to determine the velocity of the 8-ball after the collision.

Exercises 30 – 32 involve combining energy conservation and momentum conservation.

30. As shown in Figure 7.18, a wooden ball with a mass of 250 g swings back and forth on a string, pendulum style, reaching a maximum speed of 4.00 m/s when it passes through its equilibrium position. Use $g = 10.0$ m/s². (a) What is the maximum height above the equilibrium position reached by the ball in its motion? (b) At one instant, when the ball is at its equilibrium position and moving left at 4.00 m/s, it is struck by a bullet with a mass of 10.0 g. Before the collision, the bullet has a velocity of 300 m/s to the right. The bullet passes through the ball and emerges with a velocity of 100 m/s to the right. What is the magnitude and direction of the ball's velocity immediately after the collision? Neglect any change in mass for the ball.

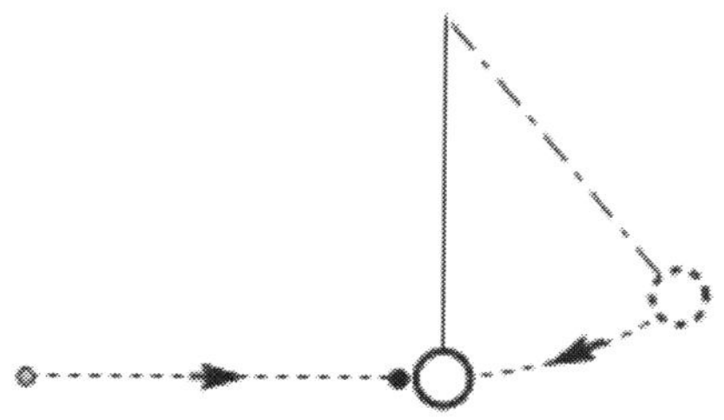

Figure 7.18: A bullet colliding with a ball on a string, for Exercise 30.

31. A pendulum, consisting of a ball of mass m on a light string of length 1.0 m, is swung back to a 45° angle and released from rest. The ball swings down and, at its lowest point, collides with a block of mass $2m$ that is on a frictionless horizontal surface. After the collision, the block slides 1.0 m across the frictionless surface and an additional 0.50 m across a horizontal surface where the coefficient of friction between the block and the surface is 0.10. (a) What is the block's speed after the collision? (b) What is the velocity of the ball after the collision? (c) Is the ball-block collision elastic, inelastic, or completely inelastic? Justify your answer. Use $g = 10$ m/s^2 to simplify the calculations.

32. Two balls hang from strings of the same length. Ball A, with a mass m, is swung back to a height h above its equilibrium position. Ball A is released from rest and swings down and hits ball B, which has a mass of $3m$. Assuming that all collisions between the balls are elastic, describe the subsequent motion of the two balls.

General Problems and Conceptual Questions

33. A Boeing 747 has a mass of about 3×10^5 kg, a cruising speed of 965 km/h, and cruises at an altitude of about 10 km. (a) Assuming the plane starts from rest at an airport at sea level, how much energy is required to reach its cruising height and altitude? Neglect air resistance in this calculation. (b) Comment on the validity of neglecting air resistance.

34. One way to estimate your power is to time yourself as you run up a flight of stairs. (a) In terms of simplifying the analysis, should you start from rest at the bottom of the stairs or should you give yourself a running start and try to keep your speed as constant as possible? (b) Which of the following distance(s) is/are most important for the power calculation, the magnitude of the straight-line displacement along the staircase or the vertical or horizontal components of this displacement? (c) Find a staircase and a stopwatch and estimate your average power.

35. A toy car rolls along a track. Starting from rest, the car drops gradually to a level 2.0 m below its starting point and then gradually rises to a level 1.0 m below its starting point, where it is traveling at a speed v_f. The goal of the exercise is to find v_f. Assume that mechanical energy is conserved, and use $g = 9.8$ m/s^2. (a) Should you first use energy conservation to relate the initial point to the lowest point, and then apply energy conservation to relate the lowest point to the final point, or can you relate the initial point directly to the final point using energy conservation? Justify your answer. (b) Find v_f.

36. Ball A is released from rest at a height h above the floor and has a speed v when it reaches the floor. (a) If ball B, which has half the mass of ball A, is released from rest at a height of $4h$ above the floor, what is its speed when it reaches the floor? Neglect air resistance. (b) What if ball B had double the mass of A instead?

37. A box with a mass of 2.0 kg slides at a constant speed of 3.0 m/s down a ramp. The ramp is in the shape of a 3-4-5 triangle, as shown in Figure 7.19. (a) Does friction act on the box? Briefly justify your answer. (b) If you decide that friction does act on the box, calculate the coefficient of kinetic friction between the box and the ramp. (c) The mass and speed of the box are given, but could you solve this exercise without them? Briefly explain.

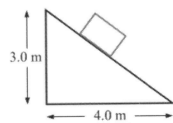

Figure 7.19: A box sliding on an incline, for Exercise 37.

3.0 m

4.0 m

38. A ball is launched with an initial velocity of 28.3 m/s, at a 45° angle, from the top of a cliff that is 10.0 m above the water below. Use $g = 10.0$ m/s² to simplify the calculations. (a) What is the ball's speed when it hits the water? (b) What is the ball's speed when it reaches its maximum height? (c) What is the maximum height (measured from the water) reached by the ball in its flight? Note: you could answer these questions using projectile motion methods, but try using an energy conservation approach instead.

39. You drop a 50-gram Styrofoam ball from rest. After falling 80 cm, the ball hits the ground with a speed of 3.0 m/s. Use $g = 10$ m/s². (a) With what speed would the ball have hit the ground if there had been no air resistance? (b) How much work did air resistance do on the ball during its fall? (c) Is your answer to (b) positive, negative, or zero? Explain.

40. As shown in Figure 7.20, two frictionless ramps are joined by a rough horizontal section that is 4.0 m long. A block is placed at a height of 124 cm up the ramp on the left and released from rest, reaching a maximum height of 108 cm on the ramp on the right before sliding back down again. (a) How far up the ramp on the left does the block get in its subsequent motion? (b) What is the coefficient of kinetic friction between the block and the rough surface? (c) At what location does the block eventually come to a permanent stop?

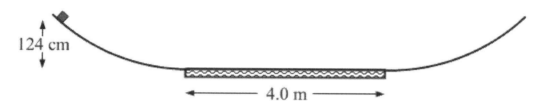

124 cm

4.0 m

Figure 7.20: A block released from rest 124 cm above the bottom of a track. The curved parts of the track are frictionless, while there is some friction between the track and the block on the 4.0-meter long horizontal section of the track. For Exercises 40 – 42.

41. Consider again the situation described in Exercise 40. If you took this apparatus to the Moon, where the acceleration due to gravity is one-sixth of what it is on Earth, and released the block from rest from the same point, what (if anything) would change about the motion?

42. Consider again the situation described in Exercise 40. Now, a different block is released from the point shown, 124 cm above the flat part of the track. This block does not reach the other side at all, but instead it stops somewhere in the rough section of the track. (a) What could be different about this block compared to the block in exercise 35? (b) What, if anything, can you say about the coefficient of kinetic friction between this block and the rough surface based on the information given here?

43. Two ramps have the same length, height, and angle of incline. One of the ramps is frictionless, while for the second ramp the coefficient of kinetic friction between the ramp and a particular block is $\mu_K = 0.25$. You release the block from rest at the top of the frictionless ramp, and when it reaches the bottom of the incline its kinetic energy is a particular value K_1. When you repeat the process with the second ramp, you find that the block's kinetic energy at the bottom of the ramp is 80% of K_1. At what angle with respect to the horizontal are the ramps inclined?

44. Two blocks are connected by a string that passes over a massless, frictionless pulley, as shown in Figure 7.21. Block A, with a mass $m_A = 2.0$ kg, rests on a ramp measuring 3.0 m vertically and 4.0 m horizontally. Block B hangs vertically below the pulley. Note that you can solve this exercise entirely using forces and the constant-acceleration equations, but see if you can apply energy ideas instead. Use $g = 10$ m/s². When the system is released from rest, block A accelerates up the slope and block B accelerates straight down. When block B has fallen through a height $h = 2.0$ m, its speed is $v = 6.0$ m/s. (a) At any instant in time, how does the speed of block A compare to that of block B? (b) Assuming that no friction is acting on block A, what is the mass of block B?

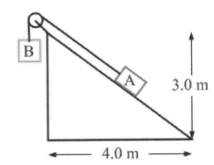

45. Repeat Exercise 44, this time accounting for friction. If the coefficient of kinetic friction for the block A – ramp interaction is 0.625, what is the mass of block B?

Figure 7.21: Two blocks connected by a string passing over a pulley, for Exercises 44 and 45.

46. Tarzan, with a mass of 80 kg, wants to swing across a ravine on a vine, but the cliff on the far side of the ravine is 1.0 m higher than the cliff where Tarzan is now and 2.0 m higher than Tarzan's lowest point in his swing. Use $g = 10$ m/s² to simplify the calculations. (a) If Tarzan wants to reach the cliff on the far side, how much kinetic energy must he have when he jumps off the cliff where he starts? (b) How fast is Tarzan going at the bottom of his swing? (c) If Tarzan swings along a circular arc of radius 10 m, what is the tension in the vine when Tarzan reaches the lowest point in his swing?

47. A block of mass m is released from rest at a height h above the base of a frictionless loop-the-loop track, as shown in Figure 7.22. The loop has a radius R. When the block is at point b, at the top of the loop, the normal force exerted on the block by the track is equal to mg. (a) What is h, in terms of R? (b) What is the normal force acting on the block at point a, at the bottom of the loop?

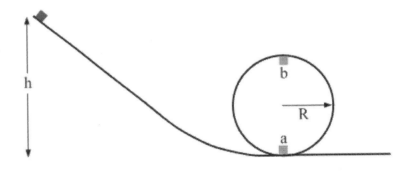

48. Consider again the situation described in Exercise 47, and shown in Figure 7.22. What is the block's speed at (a) point a (b) point b? Your answers should be given in terms of m, g, and/or R only.

Figure 7.22: A block released from rest from a height h above the bottom of a loop-the-loop track, for Exercises 47 – 49.

49. A block of mass m is released from rest, at a height h above the base of a frictionless loop-the-loop track, as shown in Figure 7.22. The loop has a radius R. What is the minimum value of h necessary for the block to make it all the way around the loop without losing contact with the track? Express your answer in terms of R.

50. On an incline, set up a race between a low-friction block that slides easily down the incline and a ball that rolls down the incline. A good approximation of a low-friction block is a toy car, or a wheeled cart, with low-friction bearings in its wheels. (a) Predict the winner of the race if you release both objects from rest. Run the race to check your prediction. (b) If we assume that mechanical energy is conserved for both objects over the course of the race, how can you explain the result? Note: this is a preview of how we will handle energy conservation for rolling objects in chapter 11.

51. Two air-hockey pucks collide on a frictionless air-hockey table, as shown in Figure 7.23. Before the collision puck A, with a mass of m, is traveling at 20 m/s to the right, while puck B, with a mass of $4m$, is stationary. After the collision puck A is traveling to the left at 4.0 m/s. (a) What is the velocity of puck B after the collision? (b) Is this collision super-elastic, elastic, or inelastic? Justify your answer.

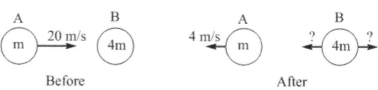

Before   After

Figure 7.23: Two air-hockey pucks just before and just after they collide, for Exercise 51.

52. Two identical carts experience a collision on a horizontal track. Immediately before the collision, cart 1 is moving at speed v to the right, directly toward cart 2, which is moving at speed v to the left. If the collision is completely inelastic then: (a) What is the velocity of cart 1 immediately after the collision? (b) Is kinetic energy, or momentum, conserved in this collision? (c) What is the velocity of the system's center of mass before the collision? (d) What is the velocity of the system's center of mass after the collision?

53. Two carts experience a collision on a horizontal track. Immediately before the collision, cart 1 is moving at speed v to the right, directly toward cart 2, which is moving at speed v to the left. If cart 2's mass is three times larger than cart 1's mass, and the collision is completely inelastic, what is the velocity of cart 1 immediately after the collision?

54. A one-dimensional collision takes place between object 1, which has a mass m_1 and a velocity $\vec{v}_{1i}$ that is directed toward object 2, which has mass m_2 and is initially stationary.

   (a) If the collision is completely inelastic, what is the velocity of the two objects immediately after the collision? (b) If the collision is completely elastic, what are the velocities of the two objects after the collision? Hint: for part (b) make use of the elasticity, k, defined in equation 7.4. Making use of the result of part (b), (c) under what condition is object 1 stationary after the collision? (d) Under what condition does object 1 reverse its direction because of the collision?

55. A one-dimensional elastic collision between an object of mass m and velocity $+\vec{v}$, and a second object of mass $3m$ and velocity $-\vec{v}$, is a special case. (a) Find the velocities of the two objects after the collision to see why. Note that you can arrange such a collision by placing a baseball or tennis ball on top of a basketball and letting the balls fall straight down from rest. (b) Assuming the masses of the basketball and baseball are in the special 3:1 ratio, that all collisions are elastic, and that the balls are dropped from a height h above the floor, how high up should the baseball go after the collision?

56. Two cars of the same mass collide at an intersection. Just before the collision, one car is traveling east at 30 km/h and the other car is traveling south at 40 km/h. If the collision is completely inelastic, so that the two cars move as one object after the collision, what is the speed of the cars immediately after the collision?

57. Because you are an accident reconstruction expert, working with the local police department, you are called to the scene of an accident at a local parking lot. The speed limit posted in the parking lot is 20 miles per hour. Although nobody was hurt in the accident, the police officer in charge would like to determine whether or not anyone was at fault, for insurance purposes. When you reconstruct the accident, you find that the cars, an Acura MDX and a Volkswagen Jetta, were approaching one another at a 90° angle. After the collision, the cars locked together and slid for 3.3 m, traveling along a path at 45° to their original paths, before coming to rest. You also determine that the Acura has a mass of 2000 kg, the Jetta's mass is 1500 kg, and the coefficient of kinetic friction for the car tires sliding on the dry pavement is somewhere between 0.75 and 0.85. (a) Which car was traveling faster before the collision? (b) Should either one of the drivers be given a speeding ticket and be determined to be at fault for the accident? Justify your answer.

58. A wooden block with a mass of 200 g rests on two supports. A piece of sticky chewing gum with a mass of 50 g is fired straight up at the block, colliding with the block when the gum's speed is 10 m/s. The gum sticks to the block, and we want to find the maximum height reached by the block and gum in its subsequent motion. (a) To solve for this maximum height, should we set the gum's kinetic energy before the collision equal to the gravitational potential energy of the gum-block system after the collision? Why or why not? (b) What is the maximum height reached by the gum-block system?

59. Re-do Exploration 7.6, but solve it another way, using a whole-vector approach by adding vectors graphically. First, add the momentum of the first object before the collision to that of the second object before the collision. That resultant vector is the total momentum before the collision, and because momentum is conserved, it is also the total momentum after the collision. Using this fact and the known momentum of the second object after the collision, you should be able to use the cosine law to find the momentum of the first object after the collision. Does the result match what we found using the component method in Exploration 7.6?

60. You release a rubber ball, from rest, at a point 1.00 m above the floor, and you observe that the ball bounces back to a height of 87.0 cm. (a) What is the net impulse experienced by the ball, which has a mass of 50.0 g, while it is in contact with the floor? (b) What is the elasticity, k, characterizing the collision between the ball and the floor? (c) Assuming the elasticity is the same for each collision, how many times will the ball bounce off the floor before losing half its mechanical energy?

61. Two different collisions take place in a large level parking lot, which is otherwise empty of vehicles. In collision A, a car with mass M traveling at a speed of v_i, runs into a stationary truck of mass $4M$. In collision B, a truck of mass $4M$, traveling at the same speed v_i, runs into a stationary car of mass M. In both collisions, the two vehicles stick together and the combined object skids to a halt because of friction. Assume that the force of friction is constant and the same for both collisions. (a) What is the speed of the combined object immediately after (i) collision A? (ii) collision B? (b) If, in collision A, the combined object slides for a time T and a distance D after the collision, for how long and through what distance does the combined object slide in collision B?

62. Comment on the statements made by two students who are working together to solve the following problem, and state the answer to the problem. A cart with a mass of 2.0 kg has a velocity of 4.0 m/s in the positive *x*-direction. The first cart collides with a second cart, which is identical to the first and has a velocity of 2.0 m/s in the negative *x*-direction. After the collision, the first cart has a velocity of 1.0 m/s in the positive *x*-direction. What is the velocity of the second cart after the collision?

Martha: This is pretty easy. We can use momentum conservation, and we don't even have to worry about the masses, because the masses are the same. So, we have a total of 4 plus 2 equals 6 meters per second before the collision, so we must have a total of 6 meters per second afterwards, too. The first cart has 1 meter per second afterwards, so the second cart must have 5 meters per second afterwards.

George: But, what direction is it going afterwards? We need to give the velocity, so is it in the plus x-direction or the minus x-direction?

Martha: It can't be minus x, because that would mean the two carts would pass through each other. It must bounce back, and go in the plus x-direction.

This image, from the Hubble Space Telescope, shows two spiral galaxies. In each galaxy, the gravitational pull of the central core keeps stars and dust clouds in orbit around the core. In addition, the gravitational interaction between the galaxy on the left and the smaller one on the right is tearing the smaller one apart. After billions of years, they will have merged to become a single galaxy. Photo courtesy of NASA.

Chapter 8 – Gravity

CHAPTER CONTENTS

Our goals in this chapter are to understand how ideas from previous chapters (such as forces and energy) apply to situations involving gravitational interactions between objects, and to lay a foundation for our discussion in Chapter 16 of interactions between charged objects. There are many parallels between how objects with mass interact and how objects with charge interact.

Although the force of gravity is relatively weak, we are aware of the influence of gravity at almost all times because we live on an object, the Earth, which has such an enormous mass (about 6×10^{24} kg). We celebrate athletes, such as high jumpers and pole vaulters, not to mention basketball players, who excel at overcoming gravity, however briefly. Some people get their thrills by flirting with death-by-gravity, jumping out of airplanes with parachutes and leaping off bridges with bungee cords tied to their ankles. As a species, we have also invested considerable sums of money into breaking our gravitational bond to mother Earth, via various space programs. One of the neat things to keep in mind, however, is that, even though each of us is tiny compared to the Earth, we all exert the same magnitude force on the Earth that the Earth exerts on us.

8-1 Newton's Law of Universal Gravitation

One of the most famous stories of all time is the story of Isaac Newton sitting under an apple tree and being hit on the head by a falling apple. It was this event, so the story goes, that led Newton to realize that the same force that brought the apple down on his head was also responsible for keeping the Moon in its orbit around the Earth, and for keeping all the planets of the solar system, including our own planet Earth, in orbit around the Sun. This force is the force of gravity.

It is hard to over-state the impact of Newton's work on gravity. Prior to Newton, it was widely thought that there was one set of physical laws that explained how things worked on Earth (explaining why apples fall down, for instance), and a completely different set of physical laws that explained the motion of the stars in the heavens. Armed with the insight that events on Earth, as well as the behavior of stars, can be explained by a relatively simple equation (see the box below), humankind awoke to the understanding that our fates are not determined by the whims of gods, but depend, in fact, on the way we interact with the Earth, and in the way the Earth interacts with the Moon and the Sun. This simple, yet powerful idea, that we have some control over our own lives, helped trigger a real enlightenment in many areas of arts and sciences.

The force of gravity does not require the interacting objects to be in contact with one another. The force of gravity is an attractive force that is proportional to the product of the masses of the interacting objects, and inversely proportional to the square of the distance between them.

A gravitational interaction involves the attractive force that any object with mass exerts on any other object with mass. The general equation to determine the gravitational force an object of mass M exerts on an object of mass m when the distance between their centers-of-mass is r is:

$$\vec{F}_G = -\frac{GmM}{r^2}\hat{r} \qquad \text{(Equation 8.1: \textbf{Newton's Law of Universal Gravitation})}$$

where $G = 6.67\times10^{-11}\text{N m}^2/\text{kg}^2$ is known as the universal gravitational constant. The magnitude of the force is equal to GmM/r^2 while the direction is given by $-\hat{r}$, which means that the force is attractive, directed back toward the object exerting the force.

At the surface of the Earth, should we use $\vec{F}_G = m\vec{g}$ or Newton's Law of Universal Gravitation instead? Why is g equal to 9.8 N/kg at the surface of the Earth, anyway? The two equations must be equivalent to one another, at least at the surface of the Earth, because they represent the same gravitational interaction. If we set the expressions equal to one another we get:

$$mg = \frac{GmM}{r^2} \quad \text{which gives} \quad g = \frac{GM}{r^2}.$$

At the surface of the Earth M is the mass of the Earth, $M_E = 5.98\times10^{24}$ kg, and r is the radius of the Earth, $R_E = 6.37\times10^6$ m. So, the magnitude of g at the Earth's surface is:

$$g_E = \frac{GM_E}{R_E^2} = \frac{(6.67\times10^{-11} \text{ N m}^2/\text{kg}^2)(5.98\times10^{24} \text{ kg})}{\left(6.37\times10^6 \text{ m}\right)^2} = 9.83 \text{ N/kg.}$$

For any object at the surface of the Earth, when we use Newton's Law of Universal Gravitation, the factors G, M_E, and R_E are all constants, so, until this point in the book, we have simply been replacing the constant value of GM_E/R_E^2 by $g = 9.8$ N/kg.

EXAMPLE 8.1 – A two-dimensional situation

Three balls, of mass m, $2m$, and $3m$, are placed at the corners of a square measuring L on each side, as shown in Figure 8.1. Assume this set of three balls is not interacting with anything else in the universe. What is the magnitude and direction of the net gravitational force on the ball of mass m?

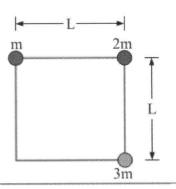

Figure 8.1: Three balls placed at the corners of a square.

SOLUTION

Let's begin by attaching force vectors to the ball of mass m. In Figure 8.2, each vector points toward the object exerting the force. The length of each vector is proportional to the magnitude of the force it represents.

We can find the two individual forces acting on the ball of mass m using Newton's Law of Universal Gravitation. Let's define $+x$ to the right and $+y$ up.

From the ball of mass $2m$: $\vec{F}_{21} = \dfrac{Gm(2m)}{L^2}$ to the right.

From the ball of mass $3m$: $\vec{F}_{31} = \dfrac{Gm(3m)}{L^2 + L^2}$ at $45°$ below the x-axis.

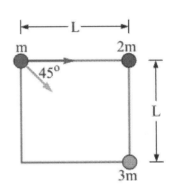

Figure 8.2: Attaching force vectors to the ball of mass m.

Finding the net force is a vector-addition problem.

In the x-direction, we get:

$$\vec{F}_{1x} = \vec{F}_{21x} + \vec{F}_{31x} = +\frac{2Gm^2}{L^2} + \frac{3Gm^2}{2L^2}\cos 45° = \left(2 + \frac{3}{2\sqrt{2}}\right)\frac{Gm^2}{L^2}.$$

In the y-direction, we get: $\vec{F}_{1y} = \vec{F}_{21y} + \vec{F}_{31y} = 0 - \dfrac{3Gm^2}{2L^2}\sin 45° = \left(-\dfrac{3}{2\sqrt{2}}\right)\dfrac{Gm^2}{L^2}.$

The Pythagorean theorem gives the magnitude of the net force on the ball of mass m:

$$F_1 = \sqrt{F_{1x}^2 + F_{1y}^2} = \sqrt{\left(4 + \frac{6}{\sqrt{2}} + \frac{9}{8} + \frac{9}{8}\right)}\frac{Gm^2}{L^2} = 3.24\frac{Gm^2}{L^2}.$$

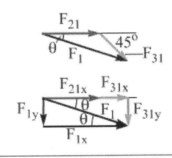

The angle is given by: $\tan\theta = \dfrac{F_{1y}}{F_{1x}} = \dfrac{\frac{3}{2\sqrt{2}}}{\frac{4\sqrt{2}+3}{2\sqrt{2}}} = \dfrac{3}{4\sqrt{2}+3}.$

Figure 8.3: The triangle representing the vector addition problem being solved in Example 8.1.

So, the angle is $19.1°$ below the x-axis.

Related End-of-Chapter Exercises: 16, 56 – 58.

Essential Question 8.1: The Sun has a much larger mass than the Earth. Which object exerts a larger gravitational force on the other, the Sun or the Earth?

Answer to Essential Question 8.1: Newton's third law tells us that the gravitational force the Sun exerts on the Earth is equal in magnitude (and opposite in direction) to the gravitational force the Earth exerts on the Sun. This follows from Equation 8.1, because, whether we look at the force exerted by the Sun or the Earth, the factors going into the equation are the same.

8-2 The Principle of Superposition

EXPLORATION 8.2 – Three objects in a line

Three balls, of mass m, $2m$, and $3m$, are equally spaced along a line. The spacing between the balls is r. We can arrange the balls in three different ways, as shown in Figure 8.4. In each case the balls are in an isolated region of space very far from anything else.

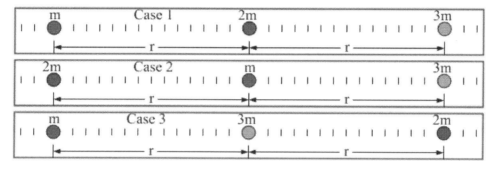

Figure 8.4: Three different arrangements of three balls of mass m, $2m$, and $3m$ placed on a line with a distance r between neighboring balls.

Step 1 – *How many forces does each ball experience in each case?* Each ball experiences two gravitational forces, one from each of the other balls. We can neglect any other interactions.

Step 2 – *Consider Case 1. Is the force that the ball of mass m exerts on the ball of mass 3m affected by the fact that the ball of mass 2m lies between the other two balls?* Interestingly, no. To find the net force on any object, we simply add the individual forces acting on an object as vectors. This is known as **the principle of superposition**, and it applies to many different physical situations. In case 1, for instance, we find the force the ball of mass m applies to the ball of mass $3m$ as if the ball of mass $2m$ is not present. The net force on the ball of mass $3m$ is the vector addition of that force and the force on the $3m$ ball from the ball of mass $2m$.

Step 3 – *In which case does the ball of mass 2m experience the largest-magnitude net force? Argue qualitatively.* Let's attach arrows to the ball of mass $2m$, as in Figure 8.5, to represent the two forces the ball experiences in each case. The length of each arrow is proportional to the force.

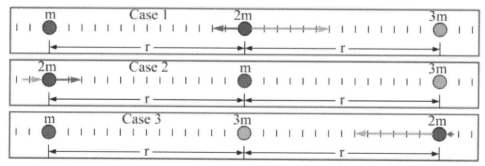

Figure 8.5: Attaching force vectors to the ball of mass $2m$. The vectors point toward the object exerting the force. The length of each vector is drawn in units of Gm^2/r^2.

In case 1, the two forces partly cancel, and, in case 2, the forces add but give a smaller net force than that in case 3. Thus, the ball of mass $2m$ experiences the largest-magnitude net force in Case 3.

Step 4 – *Calculate the force experienced by the ball of mass 2m in each case.*
To do this, we will make extensive use of Newton's universal law of gravitation. Let's define right to be the positive direction, and use the notation $\vec{F}_{21}$ for the force that the ball of mass $2m$ experiences from the ball of mass m. In each case, $\vec{F}_{2,net} = \vec{F}_{21} + \vec{F}_{23}$.

$$\text{Case 1: } \vec{F}_{2,net} = \vec{F}_{21} + \vec{F}_{23} = -\frac{Gm(2m)}{r^2} + \frac{G(2m)(3m)}{r^2} = -\frac{2Gm^2}{r^2} + \frac{6Gm^2}{r^2} = +\frac{4Gm^2}{r^2}$$

$$\text{Case 2: } \vec{F}_{2,net} = \vec{F}_{21} + \vec{F}_{23} = +\frac{Gm(2m)}{r^2} + \frac{G(2m)(3m)}{(2r)^2} = +\frac{2Gm^2}{r^2} + \frac{3Gm^2}{2r^2} = +\frac{7Gm^2}{2r^2}$$

$$\text{Case 3: } \vec{F}_{2,net} = \vec{F}_{21} + \vec{F}_{23} = -\frac{Gm(2m)}{(2r)^2} - \frac{G(2m)(3m)}{r^2} = -\frac{Gm^2}{2r^2} - \frac{6Gm^2}{r^2} = -\frac{13Gm^2}{2r^2}$$

This approach confirms that the ball of mass $2m$ experiences the largest-magnitude net force in case 3.

Step 5 - *Rank the three cases, from largest to smallest, based on the magnitude of the net force exerted on the ball in the middle of the set of three balls.* Let's extend our pictorial method by attaching force vectors to each ball in each case, as in Figure 8.6.

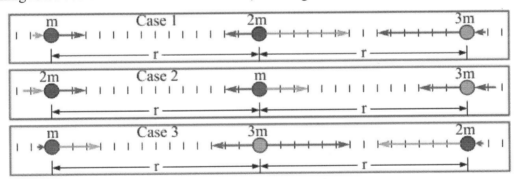

Figure 8.6: Attaching force vectors to the balls in each case. The force vectors point toward the ball applying the force. The length of each vector is drawn in units of Gm^2/r^2.

Again, when considering the net force on the middle ball, we need to add the individual forces as vectors. Referring to Figure 8.6, ranking the cases based on the magnitude of the net force exerted on the middle ball gives **Case 1 > Case 3 > Case 2**.

Key idea about the principle of superposition: The net force acting on an object can be found using the principle of superposition, adding all the individual forces together as vectors and remembering that each individual force is unaffected by the presence of other forces.
Related End of Chapter Exercises: 15 and 27.

Essential Question 8.2: In the Exploration above, which ball experiences the largest-magnitude net force in (i) Case 1 (ii) Case 2 (iii) Case 3?

Answer to Essential Question 8.2: We could determine the net force on each object quantitatively, but Figure 8.6 shows that the object experiencing the largest-magnitude net force is the object of mass $3m$ in cases 1 and 2, and the object of mass $2m$ in case 3.

In general, in the case of three objects of different mass arranged in a line the object experiencing the largest net force will be one of the objects at the end of the line, the one with the larger mass. The object in the middle will not have the largest net force because the two forces it experiences are in opposite directions.

8-3 Gravitational Field

Let's discuss the concept of a gravitational field, which is represented by $\vec{g}$. So far, we have referred to $\vec{g}$ as "the acceleration due to gravity", but a more appropriate name is "the strength of the local gravitational field."

A field is something that has a magnitude and direction at all points in space. One way to define the gravitational field at a particular point is in terms of the gravitational force that an object of mass m would experience if it were placed at that point:

$$\vec{g} = \frac{\vec{F}_G}{m}.$$ (Equation 8.2: **Gravitational field**)

The units for gravitational field are N/kg, or m/s².

A special case is the gravitational field outside an object of mass M, such as the Earth, that is produced by that object:

$$\vec{g} = -\frac{GM}{r^2}\hat{r},$$ (Equation 8.3: **Gravitational field from a point mass**)

where r is the distance from the center of the object to the point. The magnitude of the field is GM/r^2, while the direction is given by $-\hat{r}$, which means that the field is directed back toward the object producing the field.

One way to think about a gravitational field is the following: it is a measure of how an object, or a set of objects, with mass influences the space around it.

Visualizing the gravitational field

It can be useful to draw a picture that represents the gravitational field near an object, or a set of objects, so we can see at a glance what the field in the region is like. In general there are two ways to do this, by using either field lines or field vectors. The field-line representation is shown in Figure 8.7. If Figure 8.7 (a) represents the field at the surface of the Earth, Figure 8.7 (b) could represent the field at the surface of another planet where g is twice as large as it is at the surface of the Earth. In both these cases we have a **uniform field**, because the field lines are equally spaced and parallel. In Figure 8.7 (c) we have zoomed out far from a planet to get a wider perspective on how the planet affects the space around it, while in Figure 8.7 (d) we have done the same thing for a different planet with half the mass, but the same radius, as the planet in (c).

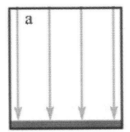 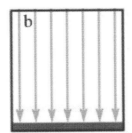 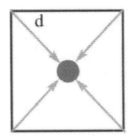 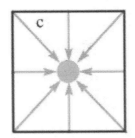

Figure 8.7: Field-line diagrams for various situations. Diagrams *a* and *b* represent uniform gravitational fields, with the field in *b* two times larger than that in *a*. Diagrams *c* and *d* represent non-uniform fields, such as the fields near a planet. The field at the surface of the planet in *c* is two times larger than that at the surface of the planet in *d*.

    Question: How is the direction of the gravitational field at a particular point shown on a field-line diagram? What indicates the relative strength of the gravitational field at a particular point on the field-line diagram?

    Answer: Each field line has a direction marked on it with an arrow that shows the direction of the gravitational field at all points along the field line. The relative strength of the gravitational field is indicated by the density of the field lines (i.e., by how close the lines are). The more lines there are in a given area the larger the field.

    A second method of representing a field is to use field vectors. A field vector diagram has the nice feature of reinforcing the idea that every point in space has a gravitational field associated with it, because a grid made up of equally spaced dots is superimposed on the picture and a vector is attached to each of these grid points. All the vectors are the same length. The situations represented by the field-line patterns in Figure 8.7 are now re-drawn in Figure 8.8 using the field-vector representation.

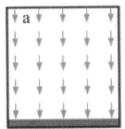 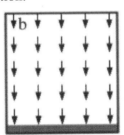 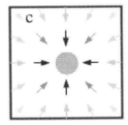 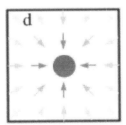

Figure 8.8: Field-vector diagrams for various situations. In figures *a* and *b* the field is uniform and directed down. The field vectors are darker in figure *b*, reflecting the fact that the field has a larger magnitude in figure *b* than in figure *a*. Figures *c* and *d* represent non-uniform fields, such as those found near a planet. Again, the fact that each field vector in figure *c* is darker than its counterpart in figure *d* tells us that the field at any point in figure *c* has a larger magnitude than the field at an equivalent point in figure *d*.

Related End of Chapter Exercises: 18, 36.

Essential Question 8.3: How is the direction of the gravitational field at a particular point shown on a field-vector diagram? What indicates the relative strength of the gravitational field at a particular point on the field-vector diagram?

Answer to Essential Question 8.3: The direction of the gravitational field at a particular point is represented by the direction of the field vector at that point (or the ones near it if the point does not correspond exactly to the location of a field vector). The relative strength of the field is indicated by the darkness of the arrow. The larger the field's magnitude, the darker the arrow.

8-4 Gravitational Potential Energy

The expression we have been using for gravitational potential energy up to this point, $U_G = mgh$, applies when the gravitational field is uniform. In general, the equation for gravitational potential energy is:

$$U_G = -\frac{GmM}{r}.$$ (Equation 8.4: **Gravitational potential energy, in general**)

This gives the energy associated with the gravitational interaction between two objects, of mass m and M, separated by a distance r. The minus sign tells us the objects attract one another.

Consider the differences between the *mgh* equation for gravitational potential energy and the more general form. First, when using Equation 8.4 we are no longer free to define the potential energy to be zero at some convenient point. Instead, the gravitational potential energy is zero when the two objects are infinitely far apart. Second, when using Equation 8.4 we find that the gravitational potential energy is always negative, which is certainly not what we found with *mgh*. That should not worry us, however, because **what is critical is how potential energy changes** as objects move with respect to one another. If you drop your pen and it falls to the floor, for instance, both forms of the gravitational potential energy equation give consistent results for the change in the pen's gravitational potential energy.

Equation 8.4 also reinforces the idea that, when two objects are interacting via gravity, neither object has its own gravitational potential energy. Instead, gravitational potential energy is associated with the interaction between the objects.

EXPLORATION 8.4 – Calculate the total potential energy in a system

Three balls, of mass m, $2m$, and $3m$, are placed in a line, as shown in Figure 8.9. What is the total gravitational potential energy of this system?

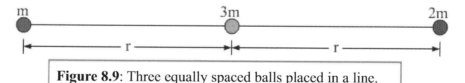

Figure 8.9: Three equally spaced balls placed in a line.

To determine the total potential energy of the system, consider the number of interacting pairs. In this case there are three ways to pair up the objects, so there are three terms to add together to find the total potential energy. Because energy is a scalar, we do not have to worry about direction. Using a subscript of 1 for the ball of mass m, 2 for the ball of mass $2m$, and 3 for the ball of mass $3m$, we get:

$$U_{Total} = U_{13} + U_{23} + U_{12} = -\frac{Gm(3m)}{r} - \frac{G(2m)(3m)}{r} - \frac{Gm(2m)}{2r} = -\frac{10Gm^2}{r}.$$

Key ideas for gravitational potential energy: Potential energy is a scalar. The total gravitational potential energy of a system of objects can be found by adding up the energy associated with each interacting pair of objects.   **Related End-of-Chapter Exercises: 25, 29, 40.**

EXAMPLE 8.4 – Applying conservation ideas

A ball of mass 1.0 kg and a ball of mass 3.0 kg are initially separated by 4.0 m in a region of space in which they interact only with one another. When the balls are released from rest, they accelerate toward one another. When they are separated by 2.0 m, how fast is each ball going?

SOLUTION

Figure 8.10 shows the balls at the beginning and when they are separated by 2.0 m. Analyzing forces, we find that the force on each ball increases as the distance between the balls decreases. This makes it difficult to apply a force analysis. Energy conservation is a simpler approach. Our energy equation is:

$$U_i + K_i + W_{nc} = U_f + K_f.$$

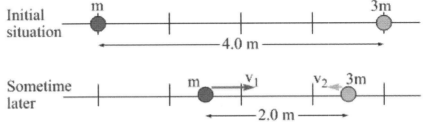

Figure 8.10: The initial situation shows the balls at rest. The force of gravity causes them to accelerate toward one another.

In this case, there are no non-conservative forces acting, and in the initial state the kinetic energy is zero because both objects are at rest. This gives $U_i = U_f + K_f$. The final kinetic energy represents the kinetic energy of the system, the sum of the kinetic energies of the two objects.

Let's solve this generally, using a mass of m and a final speed of v_1 for the 1.0 kg ball, and a mass of $3m$ and a final speed of v_2 for the 3.0 kg ball. The energy equation becomes:

$$-\frac{Gm(3m)}{4.0 \text{ m}} = -\frac{Gm(3m)}{2.0 \text{ m}} + \frac{1}{2}mv_1^2 + \frac{1}{2}(3m)v_2^2.$$

Canceling factors of m gives: $-\dfrac{3Gm}{4.0 \text{ m}} = -\dfrac{3Gm}{2.0 \text{ m}} + \dfrac{1}{2}v_1^2 + \dfrac{3}{2}v_2^2.$

Multiplying through by 2, and combining terms, gives: $+\dfrac{3Gm}{2.0 \text{ m}} = v_1^2 + 3v_2^2.$

Because there is no net external force, the system's momentum is conserved. There is no initial momentum. For the net momentum to remain zero, the two momenta must always be equal-and-opposite. Defining right to be positive, momentum conservation gives: $0 = +mv_1 - 3mv_2$, which we can simplify to $v_1 = 3v_2$.

Substituting this into the expression we obtained from applying energy conservation:

$$+\frac{3Gm}{2.0 \text{ m}} = (3v_2)^2 + 3v_2^2 = 12v_2^2$$

This gives $v_2 = \sqrt{\dfrac{Gm}{8.0 \text{ m}}}$, and $v_1 = 3v_2 = 3\sqrt{\dfrac{Gm}{8.0 \text{ m}}}$.

Using $m = 1.0$ kg, we get $v_2 = 2.9 \times 10^{-6}$ m/s and $v_1 = 8.7 \times 10^{-6}$ m/s.

Related End-of-Chapter Exercises: Problems 43 – 45.

Essential Question 8.4: Return to Example 8.4. If you repeat the experiment with balls of mass 2.0 kg and 6.0 kg instead, would the final speeds change? If so, how?

Answer to Essential Question 8.4: If we double each mass, the analysis above still works. Plugging $m = 2.0$ kg into our speed equations shows that the speeds increase by a factor of $\sqrt{2}$.

8-5 Example Problems

EXAMPLE 8.5A – Where is the field zero?

Locations where the net gravitational field is zero are special, because an object placed where the field is zero experiences no net gravitational force. Let's place a ball of mass m at the origin, and place a second ball of mass $9m$ on the x-axis at $x = +4a$. Find all the locations near the balls where the net gravitational field associated with these balls is zero.

SOLUTION

A diagram of the situation is shown in Figure 8.11. Let's now approach the problem conceptually. At every point near the balls there are two gravitational fields, one from each ball. The net field is zero only where the two fields are equal-and-opposite. These fields are in exactly opposite directions only at locations on the x-axis between the balls. If we get too close to the first ball it dominates, and if we get too close to the second ball it dominates; there is just one location between the balls where the fields exactly balance.

An equivalent approach is to use forces. Imagine having a third ball (we generally call this a **test mass**) and placing it near the other two balls. The third ball experiences two forces, one from each of the original balls, and these forces have to exactly balance. This happens at one location between the original two balls.

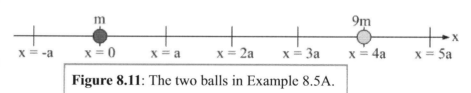

Figure 8.11: The two balls in Example 8.5A.

Whether we think about fields or forces, the approach is equivalent. The special place where the net field is zero is closer to the ball with the smaller mass. To make up for a factor of 9, representing the ratio of the two masses, we need to have a factor of 3 (which gets squared to 9) in the distances. In other words, we need to be three times farther from the ball with a mass of $9m$ than we are from the ball of mass m for the fields to be of equal magnitude. This occurs at $x = +a$.

We can also get this answer using a quantitative approach. Using the subscript 1 for the ball of mass m, and 2 for the ball of mass $9m$, we can express the net field as:

$$\vec{g}_{net} = \vec{g}_1 + \vec{g}_2 = 0.$$

Define right to be positive. If the point we're looking for is between the balls a distance x from the ball of mass m, it is $(4a - x)$ from the ball of mass $9m$. Using the definition of $\vec{g}$ gives:

$$+\frac{Gm}{x^2} - \frac{G(9m)}{(4a - x)^2} = 0.$$

Canceling factors of G and m, and re-arranging gives: $\dfrac{1}{x^2} = \dfrac{9}{(4a - x)^2}.$

Cross-multiplying leads to: $(4a - x)^2 = 9x^2.$

We could use the quadratic equation to solve for x, but let's instead take the square root of both sides of the equation. When we take a square root the result can be either plus or minus:

$$4a - x = \pm 3x.$$

Using the positive sign, we get $4a = +4x$, so $x = +a$. This is the correct solution, lying between the balls and closer to the ball with the smaller mass. Because it is three times farther from the ball of mass $9m$ than the ball of mass m, and because the distance is squared in the equation for field, this exactly balances the factor of 9 in the masses.

Using a minus sign gives a second solution, $x = -2a$. This location is three times farther ($6a$) from the ball of mass $9m$ than from the ball of mass m ($2a$). Thus at $x = -2a$ the two fields have the same magnitude, but they point in the same direction so they add rather than canceling.

Related End-of-Chapter Exercises: 13, 14, 20.

EXAMPLE 8.5B – Escape from Earth
When you throw a ball up into the air, it comes back down. How fast would you have to launch a ball so that it never came back down, but instead it escaped from the Earth? The minimum speed required to do this is known as the escape speed.

SOLUTION
A diagram is shown in Figure 8.12. Let's assume the ball starts at the surface of the Earth and that we can neglect air resistance (this would be fine if we were escaping from the Moon, but it is a poor assumption if we're escaping from Earth - let's not worry about that, however). We'll also assume the Earth is the only object in the Universe. So, this is an interesting calculation but the result will only be a rough approximation of reality.

Let's apply the energy conservation equation:
$$U_i + K_i + W_{nc} = U_f + K_f.$$

We're neglecting any work done by non-conservative forces, so $W_{nc} = 0$. The final gravitational potential energy is negligible, because the distance between the ball and Earth is very large (we can assume it to be infinite). What about the final kinetic energy? Because we're looking for the minimum initial speed let's use the minimum possible speed of the ball when it is very far from Earth, which we can assume to be zero. This leads to an equation in which everything on the right-hand side is zero:
$$U_i + K_i = 0.$$
$$-\frac{GmM_E}{R_E} + \frac{1}{2}mv_{escape}^2 = 0.$$

The mass of the ball does not matter, because it cancels out. This gives:

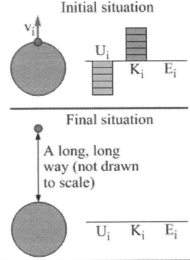

Figure 8.12: Energy bar graphs are shown in addition to the pictures showing the initial and final situations.

$$v_{escape} = \sqrt{\frac{2GM_E}{R_E}} = \sqrt{\frac{2(6.67\times10^{-11}\,\mathrm{N\,m^2/kg^2})(5.97\times10^{24}\,\mathrm{kg})}{6.37\times10^6\,\mathrm{m}}} = 11.2\,\mathrm{km/s}.$$

This is rather fast, and explains why objects we throw up in the air come down again!

Related End-of-Chapter Exercises: 41, 42.

Essential Question 8.5: Let's say we were on a different planet that had the same mass as Earth but twice Earth's radius. How would the escape speed compare to that on Earth?

Answer to Essential Question 8.5: Since $v_{escape} = \sqrt{\dfrac{2GM_E}{R_E}}$, keeping the mass the same while doubling the radius reduces the escape speed by a factor of $\sqrt{2}$.

8-6 Orbits

Imagine that we have an object of mass m in a circular orbit around an object of mass M. An example could be a satellite orbiting the Earth. What is the total energy associated with this object in its circular orbit?

The total energy is the sum of the potential energy plus the kinetic energy:
$$E = U + K = -\frac{GmM}{r} + \frac{1}{2}mv^2 .$$

This is a lovely equation, but it doesn't tell us much. Let's consider forces to see if we can shed more light on what's going on. For the object of mass m to experience uniform circular motion about the larger mass, it must experience a net force directed toward the center of the circle (i.e., toward the object of mass M). This is the gravitational force exerted by the object of mass M. Applying Newton's second law gives:

$$\Sigma \vec{F} = m\vec{a} = \frac{mv^2}{r} , \text{ directed toward the center.}$$

$$\frac{GmM}{r^2} = \frac{mv^2}{r} , \text{ which tells us that } mv^2 = \frac{GmM}{r} .$$

Substituting this result into the energy expression gives:

$$E = -\frac{GmM}{r} + \frac{GmM}{2r} = -\frac{GmM}{2r} .$$

This result is generally true for the case of a lighter object traveling in a circular orbit around a more massive object. We can make a few observations about this. First, the magnitude of the total energy equals the kinetic energy; the kinetic energy has half the magnitude of the gravitational potential energy; and the total energy is half of the gravitational potential energy. All this is true when the orbit is circular. Second, the total energy is negative, which is true for a **bound system** (a system in which the components remain together). Systems in which the total energy is positive tend to fly apart.

What happens when an object has a velocity other than that necessary to travel in a circular orbit? One way to think of this is to start the orbiting object off at the same place, with a velocity directed perpendicular to the line connecting the two objects, and simply vary the speed. If the speed necessary to maintain a circular orbit is denoted by v_O, let's consider what happens if the speed is 20% less than v_O; 20% larger than v_O; the special case of $\sqrt{2}v_O$; and $1.5v_O$. The orbits followed by the object in these cases are shown in Figure 8.13.

Unless the object's initial speed is too small, causing it to eventually collide with the more massive object, an initial speed that is less than v_O will produce an elliptical orbit where the initial point turns out to be the farthest the object ever gets from the more massive object. The initial point is special because at that point the object's velocity is perpendicular to the gravitational force the object experiences.

If the initial speed is larger than v_O, the result depends on how much larger it is. When the initial speed is $\sqrt{2}v_O$ that is the escape speed, and is thus a special case. The shape of the orbit is parabolic, and this path marks the boundary between the elliptical paths in which the object remains in orbit and the higher-speed hyperbolic paths in which the object escapes from the gravitational pull of the massive object.

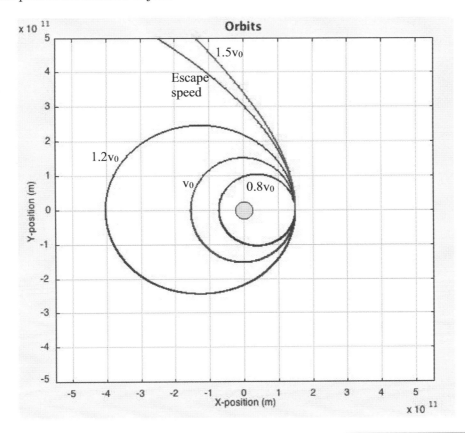

Figure 8.13: The orbits resulting from starting at a particular spot, the right-most point on each orbit, with initial velocities directed the same way (up in the figure) but with different initial speeds. The dark circular orbit represents the almost-circular orbit of the Earth, where the distances on each axis are in units of meters and the Sun is not shown but is located at the intersection of the axes. If the Earth's speed was suddenly reduced by 20%, the Earth would instead follow the smallest orbit, coming rather close to the Sun. If, instead, the Earth's speed was increased by 20%, the resulting elliptical orbit would take us quite a long way from the Sun before coming back again. Increasing the Earth's speed to $\sqrt{2}$ times its current speed (an increase of a little more than 40%) the Earth would be moving at the escape speed and we would follow the parabolic orbit to infinity (and beyond). Any initial speed larger than this would result in a hyperbolic orbit to infinity. Note that the speeds given in the picture represent initial speeds, the speed the Earth would have at the right-most point in the orbit to follow the corresponding path.

Related End-of-Chapter Exercises: 47, 59, and 60.

Essential Question 8.6: Is linear momentum conserved for any of these orbits? If so, which?

Answer to Essential Question 8.6: Linear momentum is not conserved for any orbit, because linear momentum is a vector and the direction of the momentum changes. The magnitude of the linear momentum is constant for the circular orbit, but not for any of the others. Linear momentum is not conserved because the Sun exerts a net force on the orbiting object.

Chapter Summary

Essential Idea
Gravity is one of the four fundamental forces in the universe, and it strongly influences each of us all the time. In addition, because the way objects with mass interact with each other is similar to the way objects with charge interact with one another, the material covered in this chapter lays a foundation for our understanding of charged particles in Chapters 16 and 17.

Newton's Law of Universal Gravitation
The gravitational force an object of mass M exerts on an object of mass m when the distance between their centers-of-mass is r is known as Newton's Law of Universal Gravitation:

$$\vec{F}_G = -\frac{GmM}{r^2}\hat{r} \qquad \text{(Equation 8.1: \textbf{Gravitational force between two objects})}$$

where $G = 6.67 \times 10^{-11} N\ m^2/kg^2$ is the universal gravitational constant. The magnitude of the force is equal to GmM/r^2 while the direction is given by $-\hat{r}$, which means that the force is attractive, directed back toward the object exerting the force.

The Gravitational Field, $\vec{g}$

In previous chapters we have referred to $\vec{g}$ as "the acceleration due to gravity", but a more appropriate name is "the strength of the local gravitational field". A field is something that has a magnitude and direction at all points in space. It is also one way to examine how an object with mass influences the space around it. The gravitational field can be defined as the gravitational force per unit mass:

$$\vec{g} = \frac{\vec{F}_G}{m}. \qquad \text{(Equation 8.2: \textbf{Gravitational field})}$$

For an object of mass M, such as the Earth, the gravitational field outside of the object that is produced by that object is:

$$\vec{g} = -\frac{GM}{r^2}\hat{r}, \qquad \text{(Equation 8.3: \textbf{Gravitational field from a point mass})}$$

where r is the distance from the center of the object to the point. The magnitude of the field is GM/r^2, while the direction is given by $-\hat{r}$, which simply means that the field is directed back toward the object producing the field.

Gravitational Potential Energy

Previously we have defined gravitational potential energy as *mgh*, but that applies only in a uniform gravitational field. More generally the gravitational potential energy associated with the interaction between objects of mass *m* and *M*, separated by a distance *r*, is given by:

$$U_G = -\frac{GmM}{r}.$$ (Equation 8.4: **Gravitational potential energy**)

The negative sign is associated with the fact that gravitational interactions are always attractive. In other words, the force of gravity always causes objects to attract one another, rather then repel one another.

Orbits and Energy

When an object is held in orbit around another object via the force of gravity, the total energy is always negative, indicating that the system is bound. If the total energy in the system is positive then the system is not bound, and the objects tend to fly apart from one another.

In the special case of a circular orbit, the total energy is half the value of the gravitational potential energy, as well as equal in magnitude, but opposite in sign, to the kinetic energy of the orbiting object.

End-of-Chapter Exercises

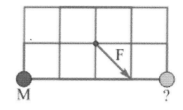

Exercises 1 – 12 are primarily conceptual questions that are designed to see if you have understood the main concepts of the chapter. Treat all balls with mass as point masses.

1. Figure 8.14 shows the force a small object feels when it is placed at the location shown, near two balls. The ball on the left has a mass *M*, while the ball on the right has an unknown mass. Based on the force experienced by the small object, state whether the mass of the ball on the right is more than, less than, or equal to *M*. Justify your answer.

Figure 8.14: A small object experiences the net gravitational force shown. The ball on the right has an unknown mass. For Exercises 1 and 2.

2. Return to the situation described in the previous problem, and shown in Figure 8.14. Determine the mass of the ball on the right, in terms of *M*.

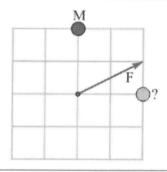

3. Figure 8.15 shows the net gravitational force experienced by a small object located at the center of the diagram. The force comes from two nearby balls, one with a charge of *M* and one with an unknown mass. (a) Is the mass of the second ball more than, less than, or equal to *M*? (b) Find the mass of the second ball.

Figure 8.15: The two balls produce a net gravitational force directed up and to the right, as shown, on the object at the center of the diagram. For Exercise 3.

4. Figure 8.16 shows the net gravitational force experienced by a small object located at the center of the diagram. The force comes from two nearby balls, one with a mass of *M* and one with an unknown mass. (a) Is the mass of the second ball more than, less than, or equal to *M*? (b) Find the mass of the second ball.

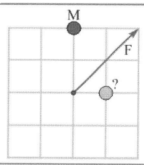

Figure 8.16: The two balls produce a net gravitational force at a 45° angle directed up and to the right, as shown, on the object at the center of the diagram. For Exercise 4.

5. Figure 8.17 shows the net gravitational force experienced by a small object located at the center of the diagram. The force comes from two nearby balls, one with a charge of *M* and one with an unknown mass. Find the mass of the second ball.

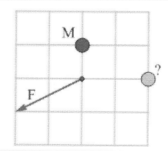

Figure 8.17: The two balls produce a net force directed down and to the left, as shown, on the object at the center of the diagram. For Exercise 5.

6. Two identical objects are located at different positions, as shown in Figure 8.18. The interaction between the objects themselves can be neglected. They experience forces of the same magnitude, and in the directions shown. Could these forces be produced by a single nearby object? If so, state where that object would be. If not, explain why not.

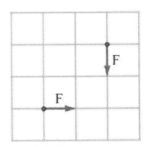

Figure 8.18: Two identical objects experience the forces shown in the diagram. The interaction between the objects themselves can be neglected. For Exercise 6.

7. Five balls, two of mass m and three of mass $2m$, are arranged as shown in Figure 8.19. What is the magnitude and direction of the net gravitational force on the ball of mass m that is located at the origin?

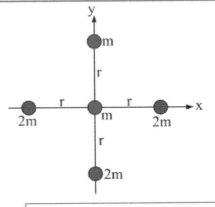

8. Consider a region of space in which there is at least one object with mass. Can there be a location in this region where the net gravitational field is zero? Briefly justify your answer and, if you answer Yes, draw an arrangement of one or more objects that produces a zero gravitational field at a particular location.

Figure 8.19: An arrangement of five balls, for Exercise 7.

9. You have four balls, two of mass m and two of mass $2m$, and you will place one at each corner of a square. Show how you can arrange the balls so that the net gravitational field at the center of the square is (a) zero (b) directed to the right.

10. Let's say you were able to tunnel into the center of the Earth (and were somehow able to withstand the tremendous pressure and temperature). (a) What would be the magnitude of the gravitational force you would experience at the center of the Earth? (b) What does Equation 8.1 predict for the magnitude of this force? (c) Are your answers to (a) and (b) consistent with one another? Explain.

11. Which of the following is/are conserved for an object that is held in a circular orbit around another more massive object by the gravitational force between the objects? Justify your answers. (a) Kinetic energy? (b) Gravitational potential energy? (c) Total mechanical energy? (d) Linear momentum?

12. Repeat Exercise 11 in the situation where the orbit is elliptical rather than circular.

Exercises 13 – 17 deal with gravitational force.

13. Two balls are placed on the x-axis, as shown in Figure 8.20. The first ball has a mass *m* and is located at the origin, while the second ball has a mass 2*m* and is located at *x* = +4*a*. A third ball, with a mass of 4*m*, is then brought in and placed somewhere on the x-axis. Assume that each ball is influenced only by the other two balls. (a) Could the third ball be placed so that all three balls simultaneously experience no net force due to the other two? (b) Could the third ball be placed so that at least one of the three balls experiences no net force due to the other two? Briefly justify your answers.

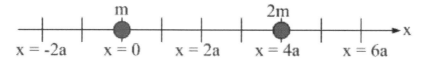

Figure 8.20: Two balls on the x-axis. These could represent planets you are placing in a solar system, in your role as designer of the cosmos. For Exercises 13 and 14.

14. Return to the situation described in Exercise 13, and find all the possible locations where the third ball could be placed so that at least one of the three balls experiences no net force due to the other two.

15. Refer back to Figure 8.9, showing a ball of mass 3*m* located halfway between a ball of mass *m* and a ball of mass 2*m*. Rank the three balls based on the magnitude of the net force they experience, from largest to smallest.

16. Three identical balls are arranged so there is one ball at each corner of an equilateral triangle. Each side of the triangle is exactly 1 meter long. If each ball experiences a net force of 5.00 x 10^{-6} N because of the other two balls, what is the mass of each ball?

17. Rank the four inner planets of the solar system (Mercury, Venus, Earth, and Mars) based on the magnitude of the gravitational force they each experience from the Sun, from largest to smallest.

Problems 18 – 23 deal with gravitational field.

18. Using the fact that the gravitational field at the surface of the Earth is about six times larger than that at the surface of the Moon, and the fact that the Earth's radius is about four times the Moon's radius, determine how the mass of the Earth compares to the mass of the Moon.

19. Five identical balls of mass M are placed so there is one ball at each corner of a regular pentagon. (a) If each ball is a distance R from the geometrical center of the pentagon, what is the magnitude of the gravitational field at the center of the pentagon due to the balls? (b) If the ball at the top of the pentagon is completely removed from the system, as shown in Figure 8.21, what is the magnitude and direction of the gravitational field at the center of the pentagon?

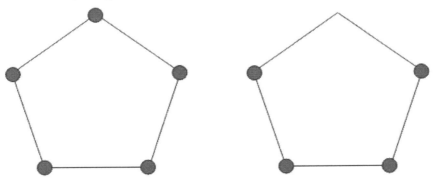

Figure 8.21: Initially, there are five identical balls, each placed at one vertex of a regular pentagon. The ball at the top is then removed, as shown at right. For Exercise 19.

20. At some point on the line connecting the center of the Earth to the center of the Moon the net gravitational field is zero. How far is this point from the center of the Earth?

21. A ball of mass $6m$ is placed on the x-axis at $x = -2a$. There is a second ball of unknown mass at $x = +a$. If the net gravitational field at the origin due to the two balls has a magnitude of $\dfrac{Gm}{a^2}$, what is the mass of the second ball? Find all possible solutions.

22. Repeat the previous exercise if the net gravitational field at the origin has a magnitude of $\dfrac{3Gm}{a^2}$.

23. A ball of mass $2m$ is placed on the x-axis at $x = -a$. There is a second ball with a mass of m that is placed on the x-axis at an unknown location. If the net gravitational field at the origin due to the two balls has a magnitude of $\dfrac{6Gm}{a^2}$, what is the location of the second ball? Find all possible solutions.

Exercises 24 – 31 deal with gravitational force, field, and potential energy.

24. A ball of mass $2m$ is placed on the x-axis at $x = -2a$. A second ball of mass m is placed nearby so that the net gravitational field at the origin because of the two balls is $\dfrac{Gm}{2a^2}$ in the negative x direction. Where is the second ball?

25. Consider the three cases shown in Exploration 8.2. (a) Rank these cases based on their gravitational potential energy, from most positive to most negative. (b) Determine the gravitational potential energy of the system shown in case 2.

26. Consider the three cases shown in Figure 8.22. Rank these cases, from largest to smallest, based on the (a) magnitude of the gravitational force experienced by the ball of mass m; (b) magnitude of the gravitational field at the origin; (c) gravitational potential energy of the system (do this ranking from most positive to most negative).

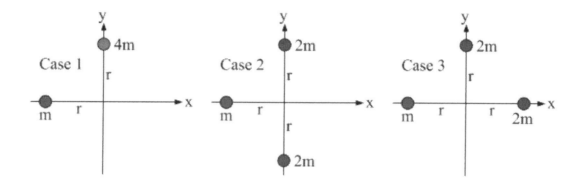

Figure 8.22: Three configurations of balls with mass, for Exercises 26 – 29.

27. Consider the three cases shown in Figure 8.22. Determine the magnitude and direction of the gravitational force experienced by the ball of mass m in (a) case 1; (b) case 2; (c) case 3.

28. Consider the three cases shown in Figure 8.22. Determine the magnitude and direction of the gravitational field at the origin in (a) case 1; (b) case 2; (c) case 3.

29. Consider the three cases shown in Figure 8.22. Determine the gravitational potential energy of the system in (a) case 1; (b) case 2; (c) case 3.

30. A ball of mass $4m$ is placed on the x-axis at $x = +3a$ and a second ball of mass m is placed on the x-axis at $x = +6a$, as shown in Figure 8.23. (a) What is the gravitational potential energy associated with this system? (b) If you bring in a third ball of with a mass of $3m$ and place it at $x = +4a$, what is the gravitational potential energy of the three-ball system? (c) If the third ball had a mass of $2m$ instead what is the gravitational potential energy of the three-ball system?

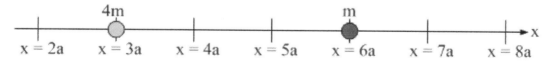

Figure 8.23: Two balls on the x-axis, a ball of mass $4m$ at $x = +3a$ and a ball of mass m at $x = +6a$, for Exercises 30 – 31.

31. Consider again the system shown in Figure 8.23. (a) At how many locations near the two balls is the net gravitational field equal to zero? (b) Specify the locations of all such points.

General exercises and conceptual questions

32. (a) Which object, the Sun or the Moon, exerts a larger gravitational force on the Earth? (b) By approximately what factor do these forces differ? (c) Is the Sun or the Moon primarily responsible for tides on the Earth? How do you explain this, given the answers to parts (a) and (b)?

33. Why do we need to have a leap year almost every four years? Sometimes we skip a leap year. Why would this be? What is the rule that determines which years are leap years and which years are skipped?

34. (a) If time zones were all one hour apart (this is not always the case), how many time zones would there be? Using the same assumption, what would the average width of a time zone be at (b) the equator? (c) a latitude equal to the latitude of Paris, France, which is 48.8° north?

35. (a) What is the speed of the Earth in its orbit around the Sun? (b) What is the acceleration of the Earth because of the gravitational force exerted on it by the Sun? (c) What is the acceleration of the Sun because of the gravitational force exerted on it by the Earth?

36. Two identical balls are placed some distance apart from one another. (a) Sketch a field vector diagram for this situation, assuming only the two balls contribute to the field. (b) Sketch a field line diagram for this situation.

37. Three balls, of mass m, $2m$, and $3m$, are arranged so there is one ball at each corner of an equilateral triangle. Each side of the triangle is exactly 1 meter long. (a) Rank the balls based on the magnitude of the net force they each experience, from largest to smallest. (b) Find the magnitude of the net force acting on the ball of mass $2m$.

38. Return to the situation described in Exercise 37. What is the magnitude of the gravitational field at the center of the triangle?

39. (a) Referring to Figure 8.24, which of the two balls of mass m experiences the larger net gravitational force? Justify your answer. (b) What is the magnitude of the net gravitational force experienced by the three different balls of mass $2m$?

40. Referring to Figure 8.24, what is the gravitational potential energy of this arrangement of five balls?

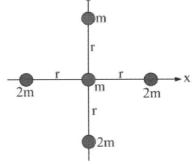

Figure 8.24: An arrangement of five balls, for Exercises 39 – 40.

41. (a) What is the escape speed for projectiles launched from the surface of the Moon? (b) Some people think that if we launch a manned mission to Mars it makes more sense to launch the spacecraft from the Moon rather than from the Earth. Comment on whether this makes sense, from an energy perspective.

42. If a projectile is launched straight up from the surface of the Moon with 90% of the speed necessary to escape from the Moon's gravity, what is the maximum distance it gets from the surface of the Moon before turning around? Assume the Moon is the only object influencing the projectile after launch.

43. Two identical objects of mass m are completely isolated from anything else and interact only with one another via gravity. If the two objects are both at rest when they are separated by a distance L, how fast are they each traveling when the distance between them is $L/4$?

44. Repeat the previous problem in the case when one object has a mass m and the second has a mass of $2m$.

45. Two identical objects of mass 5.0 kg are initially 20 cm apart. They are then each given initial velocities of 0.10 m/s directed away from the other object. (a) Assuming they are completely isolated from anything else and interact only with one another via gravity, will they eventually come back together? (b) If so, determine their maximum separation distance; if not, determine how fast each object is traveling when they are very far apart.

46. Do some research on Johannes Kepler, and write two or three paragraphs explaining his contributions to our understanding of planetary orbits.

47. There is a law known as Kepler's third law that states that when an object of mass m is held in a circular orbit around another object of mass M because of the gravitational interaction between them, the square of the period of the orbiting object is proportional to the cube of the orbital radius. Expressed as an equation, this is $T^2 = \dfrac{4\pi^2}{GM}r^3$. Let's derive the equation. (a) First, express the speed of the object of mass m in terms of the radius and period of the orbit. (b) Second, apply Newton's second law, using the fact that we're dealing with uniform circular motion. (c) Third, substitute your result from (a) into your result from (b) and re-arrange to get the result stated above.

48. Knowing that the Earth's orbit around the Sun is approximately circular with a radius of 150 million km, determine the mass of the Sun (see Exercise 47).

49. The speed of light is $3.0 \times 10^8 \ m/s$. When Neil Armstrong and Buzz Aldrin landed on the Moon in July 1969, they left reflectors that would reflect a laser beam fired at the Moon from the Earth back to the Earth (see Figure 8.25). These reflectors are still used today. By measuring the round-trip time for the light scientists can determine the distance from the surface of the Earth to the surface of the Moon to within about 1 mm. (a) Assuming the laser is fired along the line connecting the centers of the Earth and Moon, and the round-trip time for the laser beam is measured to be 2.53 s, determine the center-to-center distance from the Earth to the Moon. The radius of the Earth is $6.38 \times 10^6 \ m$, and the radius of the Moon is $1.74 \times 10^6 \ m$. (b) Using the previous result, and knowing that the Moon takes 27.3 days to orbit the Earth, determine the mass of the Earth.

Figure 8.25: A photo of the set of 100 reflectors left by Neil Armstrong and Buzz Aldrin on the Moon in 1969, and described in Exercise 49. Photo courtesy of NASA's Marshall Space Flight Center and Science @ NASA.

50. (a) Neglecting air resistance and the fact that the Earth is spinning, and assuming the ball does not hit anything in its travels, how fast would you have to launch a ball horizontally near the surface of the Earth so that it traveled in a circular path (with a radius equal to the radius of the Earth) around the Earth? (b) How long would the ball take to complete one orbit?

51. Some satellites are located in what is called a *geosynchronous* orbit around the Earth, in which they maintain their position over a particular location on the equator as the Earth spins on its axis (e.g., one might be over Ecuador at all times). How far from the center of the Earth is such a satellite (see Exercise 47), assuming it is over the equator?

52. The space shuttle orbits the Earth at an altitude that is typically 360 km above the Earth's surface. (a) What is the magnitude of the Earth's gravitational field at that altitude? (b) Explain why astronauts in the spaceship feel weightless.

53. A ball of mass $2m$ is placed on the x-axis at $x = -a$. There is a second ball with an unknown mass that is placed on the x-axis at an unknown location. If the force the second ball exerts on the first ball has a magnitude of $\dfrac{2Gm^2}{3a^2}$ and the gravitational potential energy associated with the interacting balls is $-\dfrac{2Gm^2}{a}$, what is the mass and location of the second ball? Find all possible solutions.

54. A ball of mass $3m$ is placed on the x-axis at $x = -a$. There is a second ball with an unknown mass that is placed on the x-axis at an unknown location. If the force the second ball exerts on the first ball has a magnitude of $\dfrac{Gm^2}{2a^2}$ and the net gravitational field at $x = 0$ due to these balls is $\dfrac{69Gm}{25a^2}$ in the positive x-direction, what is the mass and location of the second ball? Find all possible solutions.

55. A ball of mass $2m$ is placed on the x-axis at $x = -2a$. There is a second ball with an unknown mass that is placed on the x-axis at an unknown location. If the gravitational potential energy associated with the interacting balls is $-\dfrac{2Gm^2}{a}$ and the net gravitational field at $x = 0$ due to these balls has a magnitude of $\dfrac{Gm}{2a^2}$, what is the mass and location of the second ball? Find all possible solutions.

56. Four small balls are arranged at the corners of a square that measures L on each side, as shown in Figure 8.26. (a) Which ball experiences the largest-magnitude force due to the other three balls? (b) What is the direction of the net force acting on the ball with the mass of $4m$? (c) If you reduced the length of each side of the square by a factor of two, so neighboring balls were separated by a distance of $L/2$ instead, what would happen to the magnitude of the force experienced by each ball?

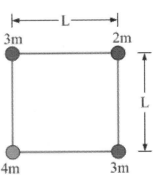

Figure 8.26: Four balls at the corners of a square, for Exercises 56 – 58.

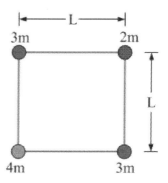

57. Four small balls are arranged at the corners of a square that measures L on each side, as shown in Figure 8.26. Calculate the magnitude and direction of the force experienced by (a) the ball with the mass of 2*m*, and (b) the ball with the mass of 3*m* in the lower right corner.

58. Four small balls are arranged at the corners of a square that measures L on each side, as shown in Figure 8.26. (a) Calculate the magnitude and direction of the gravitational field at the center of the square that is produced by these balls. (b) Could you change the mass of just one of the balls to produce a net electric field at the center that is directed horizontally to the right? If so, which ball would you change the mass of and what would you change it to? If not, explain why not.

Figure 8.26: Four balls at the corners of a square, for Exercises 56 – 58.

59. (a) What is the speed of the Earth in its approximately circular orbit around the Sun? (b) If the Earth's mass was suddenly reduced by a factor of 4, what would its speed have to be to maintain its orbit?

60. The four inner planets of the Solar System have orbits that are approximately circular. (a) Find the orbital speed of each of these four planets. (b) Rank the planets based on their orbital speed. Why is the ranking in this order?

61. What is the minimum amount of work required to move a satellite from a circular orbit around the Earth at an altitude of 350 km to a circular orbit at an altitude of 500 km?

62. (a) Look up the radius of the Sun, the distance from the Sun to the Earth, and the radius of the Earth. Use those numbers to determine the angle the Sun's diameter subtends if you were to look at the Sun. (b) Repeat for the Moon, to determine the angle the Moon's diameter subtends when you look at the Moon. (c) Comment on the relative size of these angles.

63. Let's say you are an elf and are standing exactly at the North Pole looking due south at Santa's workshop. If you turn around 180° to face the other way, in which direction are you looking now?

64. Three students are having a conversation. Explain what you think is correct about what they say, and what you think is incorrect.

Anna: This question says that we have two objects, one with a mass of m and the other with a mass of 2m. It asks us for which one experiences a larger magnitude force because of the other object. That should be the smaller one, I think – shouldn't it feel twice as much force as the bigger one?

Mark: Well, the field created by the larger one should be twice as big as the field from the smaller one. Does that tell us anything?

Suzanne: But what about Newton's third law? Doesn't that say that any two objects, no matter how big, always exert equal-and-opposite forces on one another?

Mark: I just kind of feel like the smaller one should feel more force.

Anna: I do, too, but what if we look at Newton's gravitation law? To get the force, you actually multiply the two masses together. So, it has to work out the same for each object.

Flowing fluids, such as the water flowing down a waterfall, can make interesting patterns. In this chapter, we will investigate the basic physics behind such flow. Photo credit: by Jiri Hodan, from http://www.publicdomainpictures.net.

It is interesting to think about how much physics is involved in the situation shown in the photograph. First of all, there is the influence of gravity, which makes the water flow down. Second, you can see the parabolic trajectories followed by the water as it is in free-fall at a number of different locations in the photograph. Third, there is the somewhat ghostly appearance of the water. This is caused by the photographer keeping the camera shutter open for an extended period as the picture was taken, with the motion of the water during this period causing a blurring.

Chapter 9 – Fluids

CHAPTER CONTENTS

In this chapter on fluids, we will introduce some new concepts, but the main focus will be on how to incorporate fluids into the framework of forces and energy that we have examined in the earlier chapters.

Although we will address basic issues related to flowing fluids, in the first half of this chapter our focus will be on static fluids. What determines whether something floats or sinks in a fluid? How can an object sink in one liquid, and yet float in a different liquid? How can a huge ocean liner float, when its mass is so huge, and being made from raw materials that would sink in water? Our study of fluids begins with us addressing questions such as these.

9-1 The Buoyant Force

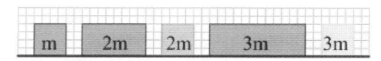

We should begin by defining what a fluid is. Many people think of a fluid as a liquid, but **a fluid is anything that can flow**. By this definition, a fluid can be a liquid or a gas. Flowing fluids can be rather complicated, so let's start with static fluids – fluids that are at rest.

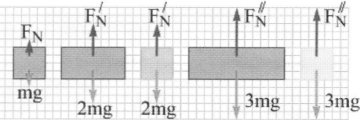

Let's consider some experiments involving various blocks that float in a container of water. The blocks are represented in Figure 9.1, which shows how the masses of the blocks compare, and also shows the free-body diagrams of the blocks as they sit in equilibrium on a table. Starting from the left, the first, second, and fourth blocks are all made from the same material. The other two blocks are both made from different material.

Figure 9.1: A diagram of the blocks we will place in a beaker of water, and the free-body diagram for each block as it sits on a table.

Our first goal is to look at the similarities between the normal force (a force arising from contact between solid objects) and the force arising from the interaction between an object and a fluid that the object is completely or partly submerged in. Figure 9.2 illustrates how the blocks float when they are placed in the container of water. We have

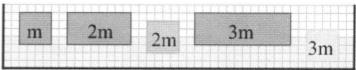

Figure 9.2: A diagram of the blocks floating in the beaker of water.

taken some liberties here, because in reality some of the blocks would tilt 45° and float as shown in Figure 9.3. Neglecting this rotation simplifies the analysis without affecting the conclusions.

Figure 9.3: We will ignore the fact that blocks that are submerged more than 50% tend to float rotated by 45° from the way they are drawn in Figure 9.2. Neglecting this fact will simplify the analysis without affecting the conclusions.

EXPLORATION 9.1 – Free-body diagrams for floating objects

Sketch the free-body diagram of the blocks in Figure 9.2 as they float in the container of water. Note that each block is in equilibrium – what does that imply about the net force acting on each block? Because each block is in equilibrium, the net force acting on each block must be zero.

What forces act on each block? As usual there is a downward force of gravity. Because each block is in equilibrium, however, the net force acting on each block is zero. For now, let's keep things simple and show, on each block, one upward force that balances the force of gravity. The free-body diagrams are shown in Figure 9.4. Note that there is no normal force, because the blocks are not in contact with a solid object. Instead, they are supported by the fluid. We call the upward force applied by a fluid to an object in that fluid **the buoyant force**, which we symbolize as $\vec{F}_B$.

Because the objects are only in contact with the fluid, the fluid must be applying the upward buoyant force to each block. Compare the free-body diagrams in Figure 9.1, when the blocks are in equilibrium on the table, with the free-body diagrams in Figure 9.4, when the blocks

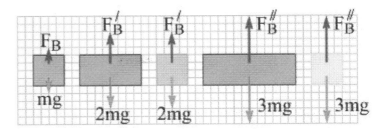

are in equilibrium while floating in the fluid. For a floating object, at least, there are a lot of similarities between the buoyant force exerted by a fluid and the normal force exerted by a solid surface.

Figure 9.4: Free-body diagrams for the blocks floating in equilibrium in the beaker of water. F_B represents the buoyant force, an upward force applied on each block by the fluid.

Examine Figures 9.2 and 9.4 closely. Even though the two blocks of mass 2*m* are immersed to different levels in the fluid, they displace the same volume of fluid, so they experience equal buoyant forces. The 3*m* blocks displace 50% more volume than do the blocks of mass 2*m*, and they experience a buoyant force that is 50% larger. The block of mass *m*, on the other hand, displaces half the volume of fluid that the blocks of mass 2*m* do, and experiences a buoyant force that is half as large. We can conclude that ***the buoyant force exerted on an object by a fluid is proportional to V*dip*, the volume of fluid displaced by that object***. We can express this as an equation (where ∝ means "is proportional to"),

$$F_B \propto V_{disp}. \qquad \text{(Eq. 9.1: Buoyant force is proportional to volume of fluid displaced)}$$

Key idea about the buoyant force: An object in a fluid experiences a net upward force we call the buoyant force, $\vec{F}_B$. The magnitude of the buoyant force is proportional to the volume of fluid displaced by the object.　　　**Related End-of-Chapter Exercise: 2.**

The conclusion above is supported by the fact that if we push a block farther down into the water and let go, the block bobs up. The buoyant force increases when we push the block down because the volume of fluid displaced increases, so, when we let go, the block experiences a net upward force. Conversely, when a block is raised, it displaces less fluid, reducing the buoyant force and giving rise to a net downward force when we let go. Figure 9.5 shows these situations and the corresponding free-body diagrams.

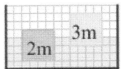

Figure 9.5: In this case, the blocks are not at equilibrium. The block on the left has been pushed down into the water and released. Because it displaces more water than it does at equilibrium, the buoyant force applied to it by the water is larger than the force of gravity applied to it by the Earth and it experiences a net upward force. The reverse is true for the block on the right, which has been lifted up and released. Displacing less water causes the buoyant force to decrease, giving rise to a net downward force.

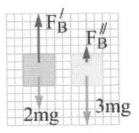

Figure 9.6: To be able to float, this large ship needs to displace a very large volume of fluid. This large volume of fluid is displaced by the part of the ship that is below the water surface, and which, therefore, is not visible to us in this photograph. Photo credit: by Peter Griffin, from http://www.publicdomainpictures.net..

Essential Question 9.1: Two objects float in equilibrium in the same fluid. Object A displaces more fluid than object B. Which object has a larger mass?

Answer to Essential Question 9.1: Object A. When an object floats in equilibrium, the buoyant force exactly balances the force of gravity. Object A displaces more fluid, so it experiences a larger buoyant force. This must be because object A weighs more than object B.

9-2 Using Force Methods with Fluids

EXAMPLE 9.2 – A block on a string

A block of weight $mg = 45$ N has part of its volume submerged in a beaker of water. The block is partially supported by a string of fixed length that is tied to a support above the beaker. When 80% of the block's volume is submerged, the tension in the string is 5.0 N.

    (a) What is the magnitude of the buoyant force acting on the block?

    (b) Water is steadily removed from the beaker, causing the block to become less submerged. The string breaks when its tension exceeds 35 N. What percent of the block's volume is submerged at the moment the string breaks?

    (c) After the string breaks, the block comes to a new equilibrium position in the beaker. At equilibrium, what percent of the block's volume is submerged?

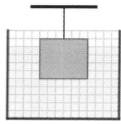

SOLUTION

As usual, we should begin with a diagram of the situation. A free-body diagram is also very helpful. These are shown in Figure 9.7.

(a) On the block's free-body diagram, we draw a downward force of gravity, applied by the Earth. We also draw an upward force of tension (applied by the string), and, because the block displaces some fluid, an upward buoyant force (applied by the fluid). The block is in equilibrium, so there must be no net force acting on the block.

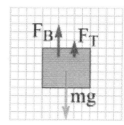

Figure 9.7: A diagram and a free-body diagram for the 45 N block floating in the beaker of water while partly supported by a string.

Taking up to be positive, applying Newton's Second Law gives:
$$\sum \vec{F} = 0.$$

Evaluating the left-hand side with the aid of the free-body diagram gives:
$+F_T + F_B - mg = 0.$

Solving for the buoyant force gives: $F_B = mg - F_T = +45 \text{ N} - 5.0 \text{ N} = +40 \text{ N}.$

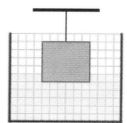

(b) As shown in Figure 9.8, removing water from the beaker causes the block to displace less fluid, so the magnitude of the buoyant force decreases. The magnitude of the tension increases to compensate for this. Applying Newton's Second Law again gives us essentially the same equation as in part (a). We can use this to find the new buoyant force, F_B'. Just before the string breaks we have:

$F_B' = mg - F_T' = +45 \text{ N} - 35 \text{ N} = +10 \text{ N}.$

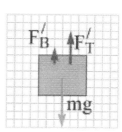

Figure 9.8: A diagram and free-body diagram of the situation just before the string breaks.

Now, we can apply the idea that the buoyant force is proportional to the volume of fluid displaced. If a buoyant force of 40 N corresponds to a displaced volume equal to 80% of the block's volume, a buoyant force of 10 N (1/4 of the original force) must correspond to a displaced volume equal to 20% of the block's volume (1/4 of the original displaced volume).

(c) After the string breaks and the block comes to a new equilibrium position, we have a simpler free-body diagram, as shown in Figure 9.9. The buoyant force now, F_B'', applied to the block by the fluid, must balance the force of gravity applied to the block by the Earth. This comes from applying Newton's Second Law:

$$\sum \vec{F} = 0.$$

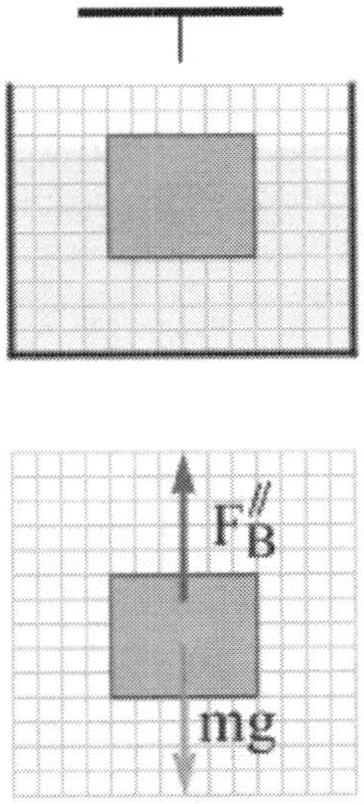

Taking up to be positive, evaluating the left-hand side with the aid of the free-body diagram gives:

$$F_B'' - mg = 0, \text{ so } F_B'' = mg = 45\,\text{N}.$$

Using the same logic as in (b), if a buoyant force of 40 N corresponds to a displaced volume equal to 80% of the block's volume, a buoyant force of 45 N must correspond to a displaced volume equal to 90% of the block's volume.

Figure 9.9: A diagram and free-body diagram for the situation after the string breaks, when the block has come to a new equilibrium position in the beaker.

Related End-of-Chapter Exercises: 21, 36.

Let's now extend our analysis to objects that sink. First, hang a block from a spring scale (a device that measures force) to measure the force of gravity acting on the block. With the block hanging from the spring scale, the scale reads 10 N, so there is a 10 N force of gravity acting on the block. A diagram and two free-body diagrams (one for the spring scale and one for the block) are shown in Figure 9.10.

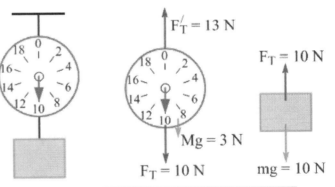

Figure 9.10: A diagram showing a block hanging from a spring scale, as well as free-body diagrams for the spring scale (which itself has a force of gravity of 3 N acting on it) and the block.

Question: With the block still suspended from the spring scale, let's dip the block into a beaker of water until it is exactly half submerged. Make a prediction. As we lower the block into the water, will the reading on the spring scale increase, decrease, or stay the same? Briefly justify your prediction.

Answer: The reading on the spring scale should decrease. This is because the spring scale no longer has to support the entire weight of the block. The more the block is submerged, the larger the buoyant force, and the smaller the spring-scale reading.

Essential Question 9.2: The spring scale reads 10 N when the block is out of the water. Let's say it reads 6.0 N when exactly 50% of the block's volume is below the water surface. What will the scale read when the entire block is below the water surface? Why?

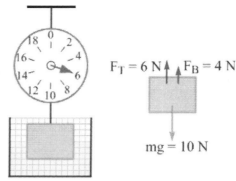

$F_T = 6\,N \uparrow \quad \uparrow F_B = 4\,N$

$mg = 10\,N$

Answer to Essential Question 9.2: We can apply the idea that the buoyant force acting on an object is proportional to the volume of fluid displaced by that object. When the block is half submerged, the buoyant force is 4.0 N up because the buoyant force and the spring scale, which exerts a force of 6.0 N up, must balance the downward 10 N force of gravity acting on the block. When the block is completely submerged, it displaces twice as much fluid, doubling the buoyant force to 8.0 N up. The spring scale only has to apply 2.0 N of force up on the block to make the forces balance. Diagrams and free-body diagrams for these situations are shown in Figure 9.11.

Figure 9.11: The top diagrams show the situation and free-body diagram for a block suspended from a spring scale when the block is half submerged in water. The bottom diagrams are similar, except that the block is completely submerged.

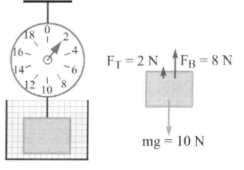

$F_T = 2\,N \quad \uparrow F_B = 8\,N$

$mg = 10\,N$

9-3 Archimedes' Principle

EXPLORATION 9.3 – What does the buoyant force depend on?
We know that the buoyant force acting on an object is proportional to the volume of fluid displaced by the object. What else does it depend on? Let's experiment to figure this out. We'll use a special beaker with a spout, as shown in Figure 9.12. In each case, we will fill the beaker to a level just below the spout, so that when we add a block to the beaker any fluid displaced by the block will flow down the spout into a second catch beaker. The catch beaker sits on a scale, so we can measure the weight of the displaced fluid. The fluid in the beaker will be either water or a second liquid, so we can figure out whether the fluid in the beaker makes any difference.

Figure 9.12: The beaker with the spout, and the catch beaker sitting on the scale. The scale is tared so it will read directly the weight of fluid in the catch beaker.

The blocks we will use have equal volumes but different masses. The weights of the blocks are 8 N, 16 N, and 24 N. Before we add a block to the beaker, we will make sure the beaker is filled to just below the level of the spout, and that the catch beaker is empty. If a block sinks in the fluid, we will hang it from a spring scale before completely submerging the block, so we can find the buoyant force from the difference between the force of gravity acting on the block and the reading on the spring scale. Also, the scale under the catch beaker is tared, which means that with the empty catch beaker sitting on it the scale reads zero and will read directly the weight of any fluid in the catch beaker.

The results of the experiments with water are shown in Figure 9.13, along with the corresponding free-body diagrams. In every case, ***the magnitude of the buoyant force acting on the block is equal to the weight of the fluid displaced by the block.***

Does this only work with water? Let's try it with the second fluid. The results are shown in Figure 9.13. Here we notice some differences - the 8 N block still floats buts displaces twice the volume of fluid it did in the water; the 16 N block now sinks; and the 24 N block still sinks but has half the buoyant force it had when it was in the water. Once again, however, the magnitude of the buoyant force on the block is equal to the weight of fluid displaced by the block.

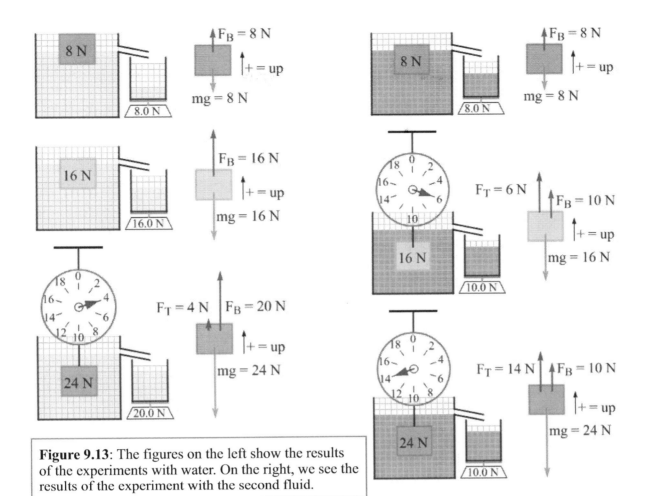

Figure 9.13: The figures on the left show the results of the experiments with water. On the right, we see the results of the experiment with the second fluid.

With the second fluid, we see that the buoyant force the fluid exerts on an object is still proportional to the volume of fluid displaced. However, we can also conclude that displacing a particular volume of water gives a different buoyant force than displacing exactly the same amount of the other fluid. Some property of the fluid is involved here.

To determine which property of the fluid is associated with the buoyant force, let's focus on the fact that the buoyant force is equal to the weight of the fluid displaced by the object:

$$F_B = m_{disp} g \, .$$

If we bring in mass density, for which we use the symbol ρ , we can write this equation in terms of the volume of fluid displaced. The relationship between mass, density, and volume is:

$$m = \rho V \, . \qquad \text{(Equation 9.2: \textbf{Mass density})}$$

Using this relationship in the equation for buoyant force gives:

$$F_B = m_{disp} g = \rho_{fluid} V_{disp} \, g \, . \qquad \text{(Equation 9.3: \textbf{Archimedes' Principle})}$$

Key Idea regarding Archimedes' Principle: The magnitude of the buoyant force exerted on an object by a fluid is equal to the weight of the fluid displaced by the object. This is known as Archimedes' principle.   **Related End-of-Chapter Exercises: 4, 7.**

Essential Question 9.3: How does the mass density of the second fluid in Exploration 9.3 compare to the mass density of water?

Answer to Essential Question 9.3: The second fluid has a density that is half that of water. We can see that because a particular volume of water has a mass that is twice as much as the mass of an equal volume of the second fluid.

9-4 Solving Buoyancy Problems

Archimedes was a Greek scientist who, legend has it, discovered the concept while taking a bath, whereupon he leapt out and ran naked through the streets shouting "Eureka!" Archimedes was thinking about this because the king at the time wanted Archimedes to come up with some way to make sure that the king's crown was made out of solid gold, and was not gold mixed with silver. Archimedes realized that he could use his principle to determine the density of the crown, and he could then compare it to the known density of gold.

Using Equation 9.3, we can now explain the results of the block-and-two-fluid experiment above. The differences we observe between when we place the blocks in water and when we place them in the second fluid can all be explained in terms of the difference between the density of water and the density of the second fluid. In fact, to explain the results of Exploration 9.2 the second fluid must have half the density of water. The 10-N block, for instance, floats in both fluids and therefore the buoyant force is the same in both cases, exactly equal-and-opposite to the 10 N force of gravity acting on the block. Because the density of the second fluid is half the density of water, the block needs to displace twice the volume of fluid in the second fluid to achieve the same buoyant force. The 30-N block, on the other hand, displaces the same amount of fluid in each case. However, it experiences twice the buoyant force from the water as it does from the second fluid because of the factor of two difference in the densities.

What happens with the 20-N block is particularly interesting, because it floats in water and yet sinks in the second fluid. This raises the question, what determines whether an object floats or sinks when it is placed in a fluid?

EXPLORATION 9.4 – Float or sink?

How can we tell whether an object will float or sink in a particular fluid? As we have considered before, when an object floats in a fluid the upward buoyant force exactly balances the downward force of gravity. This gives: $F_B = mg$.

Using Archimedes' principle, we can write the left-hand side as: $\rho_{fluid} V_{disp} g = mg$.

The factors of g cancel (this tells us that it doesn't matter which planet we're on, or where on the planet we are), giving: $\rho_{fluid} V_{disp} = m$.

If we write the right-hand side in terms of the density of the object, we get, for a floating object:

$$\rho_{fluid} V_{disp} = \rho_{object} V_{object} .$$

Re-arranging this equation leads to the interesting result (that applies for floating objects only):

$$\frac{\rho_{object}}{\rho_{fluid}} = \frac{V_{disp}}{V_{object}} . \qquad \text{(Equation 9.4: \textbf{For floating objects})}$$

Equation 9.4 answers the question of what determines whether an object floats or sinks in a fluid – the density. *If an object is less dense than the fluid it is in then it floats.* An object that is less dense than the fluid it is in floats because the object displaces a volume of fluid smaller than its own volume – in other words, the object floats with part of it sticking out above the surface of the fluid. On the other hand, an object more dense than the fluid it is in must displace a volume of fluid larger than its own volume in order to float. This is certainly not possible for the solid blocks we have considered above. Thus, we can conclude that *an object with a density larger than the density of the fluid it is in will sink in that fluid*.

Key Ideas: Whether an object floats or sinks in a fluid depends on its density. An object with a density less than that of a fluid floats in that fluid, while an object with a larger density than that of a fluid will tend to sink in that fluid. **Related End-of-Chapter Exercises: 1, 13.**

Equation 9.4 tells us that we can determine the density of a floating object by observing what fraction of its volume is submerged. For instance, if an object is 30% submerged in a fluid its density is 30% of the density of the fluid. Table 9.1 shows the density of various materials.

Material	Density (kg/m³)	Material	Density (kg/m³)
Interstellar space	10^{-20}	Planet Earth (average)	5500
Air (at 1 atmosphere)	1.2	Iron	7900
Water (at 4°C)	1000	Mercury (the metal)	13600
Sun (average)	1400	Black hole	10^{+19}

Table 9.1 The density of various materials.

What about an object that has the same density as the fluid it is in? This is known as **neutral buoyancy**, because the upward buoyant force on the object balances the downward force of gravity on the object when the object is 100% submerged. Because the net force acting on the object is zero it is in equilibrium at any of the positions shown in Figure 9.14. This is true as long as the fluid density does not change with depth, which is something of an idealization. Again we are using a model, with an assumption of the model being that a fluid is incompressible – its density is constant.

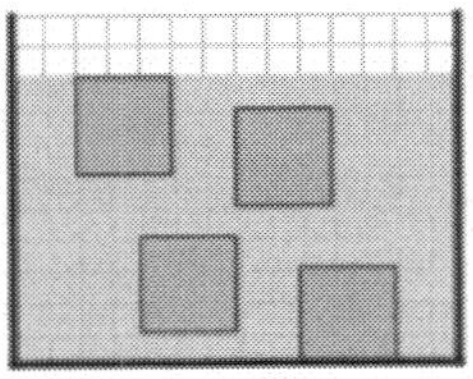

Figure 9.14: A neutrally buoyant object (an object with the same density as the surrounding fluid) will be at equilibrium at any of the positions shown. All other objects will either float at the surface, or sink to the bottom.

The general method for solving a typical buoyancy problem is based on the method we used in chapter 3 for solving a problem involving Newton's Laws. Now, we include Archimedes' principle. In general buoyancy problems are 1-dimensional, involving vertical forces, so that simplifies the method a little.

A General Method for Solving a Buoyancy Problem
1. Draw a diagram of the situation.
2. Draw one or more free-body diagrams, with each free-body diagram showing all the forces acting on an object as well as an appropriate coordinate system.
3. Apply Newton's Second Law to each free-body diagram.
4. If necessary, bring in Archimedes' principle, $F_B = \rho_{fluid} V_{disp} g$.
5. Put the resulting equations together and solve.

Essential Question 9.4: Let's say the four objects shown in Figure 9.14 have densities larger than that of the fluid. Can any of the objects be at equilibrium at the positions shown? Explain.

Answer to Essential Question 9.4: Objects that are denser than the fluid they are in tend to sink to the bottom of the container. One object in Figure 9.14 already rests at the bottom, so it is in equilibrium. For the three higher objects, the force of gravity, acting down, is larger than the buoyant force that acts up. These three objects are not at equilibrium, and will sink to the bottom.

9-5 An Example Buoyancy Problem

EXAMPLE 9.5 – Applying the general method

Let's now consider an object that sinks to the bottom of a beaker of liquid. The object is a block with a weight of 20 N, when weighed in air. The beaker it is to be placed in contains some water, as well as a waterproof scale that rests on the bottom of the beaker. This scale is tared to read zero, and let's assume the scale is unaffected by any changes in the level of the water above it. The beaker itself rests on a second scale that reads 50 N, the combined weight of the beaker, the water, and the scale inside the beaker. When the 20-N block is placed in the beaker, it sinks to the bottom and comes to rest on the scale in the beaker, which now reads 5.0 N. This is known as the **apparent weight** of the block. Let's assume $g = 10$ m/s² to simplify the calculations.

(a) What is the magnitude and direction of the buoyant force applied on the block by the water?
(b) With the block now completely immersed in the water, what is the reading on the scale under the beaker?
(c) What is the block's density and volume?

SOLUTION

Let's begin with the first two steps in the general method, by drawing a diagram of the situation and a free-body diagram of the block. These are shown in Figure 9.15, where up is taken to be the positive direction. Note that three forces act on the block, one of which is the downward force of gravity. The 5.0 N reading on the scale is the magnitude of the downward normal force applied by the block on the scale. By Newton's Third Law, the scale applies an upward 5.0 N normal force on the block. The third force acting on the block is the upward buoyant force applied on it by the water.

(a) The block is in equilibrium (at rest with no acceleration), so we can apply Newton's Second Law to determine the buoyant force acting on the block.

$$\Sigma \vec{F} = m\vec{a} = 0 .$$

Looking at the free-body diagram to evaluate the left-hand side gives:

$$+F_B + F_N - mg = 0 .$$

Solving for the buoyant force gives:

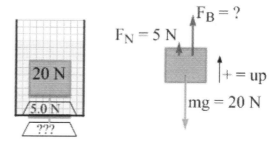

Figure 9.15: A diagram and free-body diagram for the block resting on the scale inside the beaker of fluid.

$$F_B = mg - F_N = 20 \text{ N} - 5.0 \text{ N} = 15 \text{ N} , \text{ directed up.}$$

(b) What is the reading on the scale under the beaker? The scale under the beaker supports everything on top of it, so with the block inside the beaker the scale under the beaker reads 70 N. This comes from adding the full 20-N weight of the block to the original 50 N, from the beaker, water, and scale inside the beaker.

Doesn't the water support 15 N of the block's weight, via the buoyant force? Yes, it does. However, if the water exerts a force of 15 N up on the block, then by Newton's third law the block exerts a 15 N force down on the water. The water passes this force along to the beaker, which passes it along to the scale under the beaker. Similarly, the block exerts a 5.0-N normal force down on the scale inside the beaker, and the scale passes this force along to the beaker, which passes it along to the scale under the beaker. Now matter how you look at it, adding a 20-N block to the beaker ends up increasing the reading on the scale under the beaker by 20 N.

(c) Let's derive a general equation that tells us how the density of a submerged object is related to its weight mg and apparent weight W_{app}. The apparent weight is numerically equal to the normal force experienced by the submerged object.

In part (a), we used Newton's Second Law to arrive at an expression for the buoyant force acting on our submerged object, obtaining: $F_B = mg - F_N$.

Writing this in terms of the apparent weight gives: $F_B = mg - W_{app}$.

For a submerged object, the apparent weight is less than the actual weight. If we call f the ratio of the apparent weight to the actual weight, $f = W_{app}/mg$, we can write the previous equation, using $W_{app} = f\,mg$, as:

$$F_B = mg - f\,mg = (1 - f)mg.$$

Now, use Archimedes' principle to transform the left-hand side of the equation:

$$\rho_{fluid} V_{disp}\, g = (1 - f)mg.$$

Finally, write the object's mass in terms of its density: $\rho_{fluid} V_{disp}\, g = (1 - f)\rho_{object} V_{object}\, g$.

The volume of fluid displaced by an object that is completely submerged is equal to its own volume, so we can cancel the factors of volume as well as the factors of g, leaving:

$$\rho_{fluid} = (1 - f)\rho_{object}.$$

Solving for the density of the object, we can write the equation in various ways:

$$\rho_{object} = \frac{\rho_{fluid}}{1 - f} = \frac{\rho_{fluid}}{1 - \dfrac{W_{app}}{mg}} = \frac{mg\,\rho_{fluid}}{mg - W_{app}}.$$

That applies generally to a completely submerged object. In our case, where we have $f = 1/4$, we find that the density of the block is:

$$\rho_{block} = \frac{4}{3}\rho_{water} = 1330\,\text{kg/m}^3.$$

Related End-of-Chapter Exercises: 9, 18.

Essential Question 9.5: It is possible for small metal objects, such as sewing needles or Japanese yen, to float on the surface of water, if carefully placed there. Can we explain this in terms of the buoyant force?

Answer to Essential Question 9.5: No. These metal objects are denser than water, so we expect them to sink in the water (which they will if they are not placed carefully at the surface). They are held up by the surface tension of the water. Surface tension is beyond the scope of this book but it is similar to how a gymnast is supported by a trampoline – the water surface can act like a stretchy membrane that can support an object that is not too massive.

9-6 Pressure

Where does the buoyant force come from? What is responsible, for instance, for the small upward buoyant force exerted on us by the air when we are surrounded by air? Let's use a model in which the fluid is considered to be made up of a large number of fast-moving particles that collide elastically with one another and with anything immersed in it. For simplicity, let's examine the effect of these collisions on a block of height h and area A that is suspended from a light string, as shown in Figure 9.16.

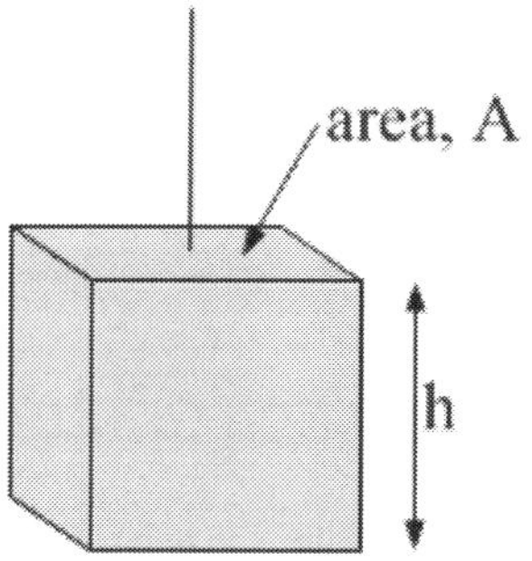

Figure 9.16: A block of height h and area A supported by a light string.

Consider a collision involving an air molecule bouncing off the left side of the block, as in Figure 9.17. Assuming the block remains at rest during the collision (the block's mass is much larger than that of the air molecule), then, because the collision is elastic, the magnitude of the molecule's momentum remains the same and only the direction changes: the component of the molecule's momentum that is directed right before the collision is directed left after the collision. All other momentum components remain the same. The block exerts a force to the left on the molecule during the collision, so the block experiences an equal-and-opposite force to the right.

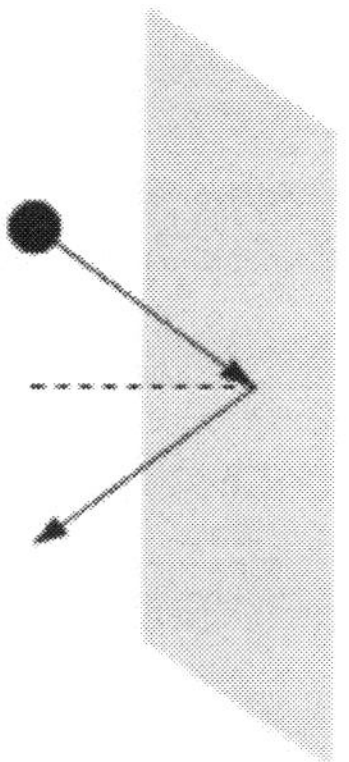

There are a many molecules bouncing off the left side of the block, producing a sizable force to the right on the block. The block does not accelerate to the right, however, because there is also a large number of molecules bouncing off the right side of the block, producing a force to the left on the block. Averaged over time, the rightward and leftward forces balance. Similarly, the forces on the front and back surfaces cancel one another.

If all the forces cancel out, how do these collisions give rise to the buoyant force? Consider the top and bottom surfaces of the cube. Because the buoyant force exerted on the cube by the air is directed vertically up, the upward force on the block associated with air molecules bouncing off the block's bottom surface must be larger than the downward force on the block from air molecules bouncing off the block's top surface. Expressing this as an equation, and taking up to be positive, we get:

Figure 9.17: A magnified view of a molecule bouncing off the left side of the block.

$$+F_{bottom} - F_{top} = +F_B = +\rho_{fluid} V_{disp} g .$$

The volume of air displaced by the block is the block's entire volume, which is its area multiplied by its height: $V_{disp} = Ah$. Substituting $V_{disp} = Ah$ into the expression above gives:

$$+F_{bottom} - F_{top} = +F_B = +\rho_{fluid} A h g .$$

This is the origin of the buoyant force – the net upward force applied to the block by molecules bouncing off the block's bottom surface is larger in magnitude than the net downward force applied by molecules bouncing off the block's upper surface. This is a gravitational effect – the buoyant force is proportional to g. One way to think about this is that if the molecules at the block's top surface have a particular average kinetic energy, to conserve energy those at the block's bottom surface should have a larger kinetic energy because their gravitational potential energy is less. Thus, molecules bouncing off the bottom surface are more energetic, and they impart a larger average force to the block than the molecules at the top surface.

Dividing both sides of the previous equation by the area A gives:

$$+\frac{F_{bottom}}{A}-\frac{F_{top}}{A}=+\rho_{fluid}\,h\,g\,.$$ (Equation 9.5)

> The name for the quantity of force per unit area is **pressure**.
>
> $$\text{Pressure}=\frac{\text{Force}}{\text{Area}} \qquad \text{or} \qquad P=\frac{F}{A}\,.$$ (Equation 9.6: **Pressure**)
>
> The MKS unit for pressure is the pascal (Pa). $1\,\text{Pa}=1\,\text{N}/\text{m}^2$.

Using the symbol P for pressure, we can write Equation 9.5 as:

$$P_{bottom}-P_{top}=\rho_{fluid}\,h\,g\,.$$

We can write this equation in a general way, so that it relates the pressures of any two points, points 1 and 2, in a static fluid, where point 2 is a vertical distance h below the level of point 1. This gives:

$$P_2=P_1+\rho gh\,.$$ (Equation 9.7: **Pressure in a static fluid**)

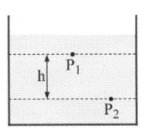

Figure 9.18: The pressure difference between two points is proportional to the vertical distance between them. Pressure increases with depth in a static fluid.

As represented by Figure 9.18, only the vertical level of the point matters. Any horizontal displacement in moving from point 1 to point 2 is irrelevant.

EXPLORATION 9.6 – Pressure in the L
Consider the L-shaped container in Figure 9.19. Rank points A, B, and C in terms of their pressure, from largest to smallest.

Because pressure in a static fluid depends only on vertical position, points B and C have equal pressures, and the pressure at that level in the fluid is higher than that at point A. The fact that there is a column of water of height d immediately above both points A and C is irrelevant. The fact that C is farthest from the opening is also irrelevant. Only the vertical position of the points matters.

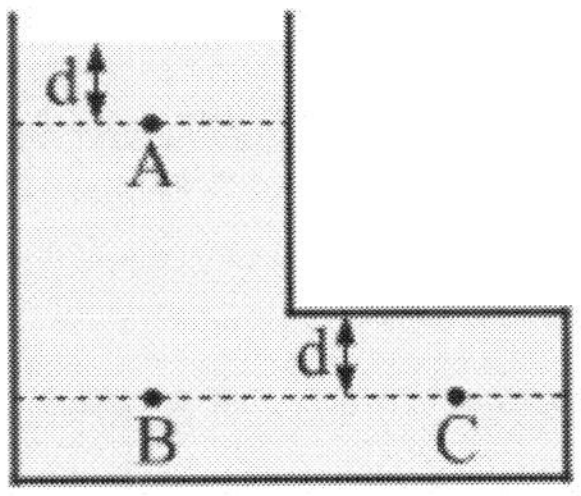

Figure 9.19: A container shaped like an L that is filled with fluid and open at the top.

> **Key ideas**: In a static fluid the pressure at any point is determined by that point's vertical position. All points at the same level have the same pressure, and points lower down have higher pressure than points higher up.
> **Related End-of-Chapter Exercises: 10, 51.**

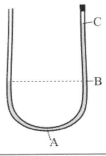

Essential Question 9.6: Water is placed in a U-shaped tube, as shown in Figure 9.20. The tube's left arm is open to the atmosphere, but the tube's right arm is sealed with a rubber stopper. Rank points A, B, and C based on their pressure, from largest to smallest.

Figure 9.20: A U-shaped water-filled tube that is sealed at the top right by a rubber stopper.

Answer to Essential Question 9.6: A>B>C. Point A, being the lowest of the three points, has the highest pressure. Point C, being the highest of the three points, has the lowest pressure.

9-7 Atmospheric Pressure

At sea level on Earth, standard atmospheric pressure is 101.3 kPa, or about 1.0×10^5 Pa, a substantial value. Atmospheric pressure is associated with the air molecules above sea level. Air is not very dense, but the atmosphere extends upward a long way so the cumulative effect is large. The reason we, and most things, don't collapse under atmospheric pressure is that in almost all situations there is pressure on both sides of an interface, so the forces balance. If you can create a pressure difference, however, you can get some interesting things to happen. This is how suction cups work, for instance – by removing air from one side the air pressure on the outside of the suction cup gives rise to a force that keeps the suction cup attached to a surface. It's also fairly easy to use atmospheric pressure to crush a soda can (see end-of-chapter Exercise 6).

In many situations what matters is the **gauge pressure**, which is the difference between the total pressure and atmospheric pressure. The total pressure is generally referred to as the **absolute pressure**. For instance, the absolute pressure at the surface of a lake near sea level is 1 atmosphere (1 atm), so the gauge pressure there would be 0. The gauge pressure 10 meters below the surface of a lake is about 1 atmosphere (1 atm), taking the density of water to be 1000 kg/m³, because:

$$\rho\, g\, h \approx 1000\,\frac{kg}{m^3} \times 10\,\frac{m}{s^2} \times 10\,m = 1 \times 10^5\,\frac{kg\,m}{s^2}\,\frac{1}{m^2} = 1 \times 10^5\,\frac{N}{m^2} = 1 \times 10^5\,Pa\ .$$

The absolute pressure 10 m below the surface is about 2 atm. This is particularly relevant for divers, who must keep in mind that every 10 m of depth in water is associated with an additional 1 atmosphere worth of pressure.

EXAMPLE 9.7 – Under pressure

A plastic box is in the shape of a cube measuring 20 cm on each side. The box is completely filled with water and remains at rest on a flat surface. The box is open to the atmosphere at the top. Assume atmospheric pressure is 1.0×10^5 Pa and use $g = 10$ m/s².

    (a) What is the gauge pressure at the bottom of the box?
    (b) What is the absolute pressure at the bottom of the box?
    (c) What is the force associated with this absolute pressure?
    (d) What is the force associated with the absolute pressure acting on the inside surface of one side of the box?
    (e) What is the net force associated with pressure acting on one side of the box?
    (f) What is the net force acting on one side of the box?

SOLUTION

As usual let's begin with a diagram of the situation, shown in Figure 9.21.

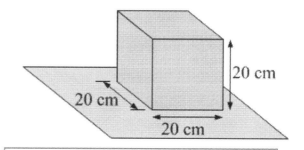

Figure 9.21: A box in the shape of a cube that is open at the top and filled with water.

(a) Because the pressure at the top surface is atmospheric pressure, the gauge pressure at the bottom is simply the pressure difference between the top of the box and the bottom. Applying Equation 9.7, regarding the pressure difference between two points in a static fluid, we get the gauge pressure at the bottom:

$$\Delta P = \rho\, g\, h = 1000\,\frac{\text{kg}}{\text{m}^3} \times 10\,\frac{\text{m}}{\text{s}^2} \times 0.20\,\text{m} = 2000\,\text{Pa}\,.$$

(b) The absolute pressure at the bottom is the gauge pressure plus atmospheric pressure. This gives: $P_{bottom} = P_{atm} + P_{gauge} = 1.0 \times 10^5\,\text{Pa} + 2000\,\text{Pa} = 1.02 \times 10^5\,\text{Pa}$. Stating this to three significant figures would violate the rules about significant figures when adding, so we should really round this off to $1.0 \times 10^5\,\text{Pa}$.

(c) To find the force from the pressure we use Equation 9.6, re-arranged to read Force = Pressure × Area. This gives a force of $F_{bottom} = \left(1.0 \times 10^5\,\text{Pa}\right)\left(0.2\,\text{m}\right)^2 = 4000\,\text{N}$, directed down at the bottom of the box.

(d) Finding the force associated with the side of the box is a little harder than finding it at the bottom, because the pressure increases with depth in the fluid. In other words, the pressure is different at points on the side that are at different depths. Because the pressure increases linearly with depth, however, we can take the average pressure to be the pressure halfway down the side of the box. The gauge pressure at a point inside the box that is halfway down the side is:

$$P_{gauge} = \rho gh = 1000\,\frac{\text{kg}}{\text{m}^3} \times 10\,\frac{\text{m}}{\text{s}^2} \times 0.1\,\text{m} = 1000\,\text{Pa}.$$

To find the force associated with the pressure we use absolute pressure, so we get:

$$F_{side} = \left(P_{atm} + P_{gauge}\right) \times \text{Area} = \left(1.0 \times 10^5\,\text{Pa} + 1000\,\text{Pa}\right) \times \left(0.2\,\text{m}\right)^2 = 4040\,\text{N}\,,\ \text{which we should}$$

round off to 4000 N directed out from the center of the box.

(e) In part (d) we were concerned with the fluid pressure applying an outward force on one side of the box. Now we need to account for the air outside the box exerting an inward force on the same side of the box. This force is simply atmospheric pressure multiplied by the area, and is thus 4000 N directed inward. The net force associated with pressure is thus the combination of the 4040 N force directed out and the 4000 N force directed in, and is thus 40 N directed out. The same result can be obtained from $F_{pressure} = P_{gauge} \times \text{Area}$.

(f) Because the box and all its sides remain at rest, the net force on any one side must be zero, so this 40 N outward force associated with the gauge pressure of the water must be balanced by forces applied to one side by the rest of the box.

Related End-of-Chapter Exercises: 27, 28.

Essential Question 9.7: In Example 9.7, we accounted for the change in water pressure with depth, but we did not account for the increase in air pressure with depth, which could affect our calculation of the inward force exerted by the air on a side of the box. Explain why we can neglect this change in air pressure.

Answer to Essential Question 9.7: The increase in pressure with depth is proportional to the product of the density multiplied by the vertical distance. Because the density of the water is on the order of 1000 times larger than that of air, we can neglect this effect for the air.

9-8 Fluid Dynamics

Let's turn now from analyzing fluids at rest to analyzing fluids in motion. The study of flowing fluids is known as fluid dynamics. Fluid dynamics can be rather complex, so we will make some simplifying assumptions. These include:
1. The flow is steady – flow patterns are maintained without turbulence.
2. The fluid is incompressible – its density is constant.
3. The fluid is non-viscous – there is no resistance to the flow.
4. The flow is irrotational – there are no swirls or eddies.

Under these assumptions we get what is called streamline flow, indicated by the blue streamlines in the pipe shown in Figure 9.22.

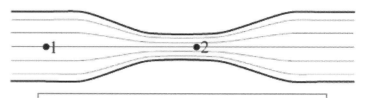

Figure 9.22: Streamline flow through a pipe.

Continuity

There are two main equations we will apply to analyze flowing fluids. The first of these is called the continuity equation, which comes from the fact that when an incompressible fluid flows through a tube of varying cross-section, the rate at which mass flows past any point in the tube is constant. If the flow rate varied, fluid would build up at points where the flow rate is low.

The mass flow rate is the total mass flowing past a point in a given time interval divided by that time interval. At a point where the flow is in the x-direction with a speed v and the tube has a cross-sectional area A, the mass flow rate is given by:

$$\text{mass flow rate} = \frac{\Delta m}{\Delta t} = \frac{\rho \, \Delta V}{\Delta t} = \frac{\rho \, A \, \Delta x}{\Delta t} = \rho \, A v \,.$$

Consider the streamline flow pattern in Figure 9.22. The mass flow rate is the same at two different points, 1 and 2, in the tube, so:

$$\rho_1 \, A_1 \, v_1 = \rho_2 \, A_2 \, v_2 \,.$$

One of our assumptions is that the density is the same at all points, so we can reduce the preceding equation to:

$$A_1 \, v_1 = A_2 \, v_2 \,. \qquad\qquad \text{(Equation 9.8: \textbf{The Continuity Equation})}$$

The main implication of the continuity equation is that the speed of the fluid increases as the cross-sectional area of the tube decreases, and vice versa. The streamlines in Figure 9.22 show the change in speed that corresponds to a change in area. Where the streamlines are farther apart, such as at point 1, the flow speed is less. Where the streamlines are close together, such as at point 2, the speed is higher.

The second equation we will apply to flowing fluids is an energy conservation equation, transformed to be particularly useful for fluids. Let's start by writing out our energy conservation equation from chapter 6:

$$U_1 + K_1 + W_{nc} = U_2 + K_2 \, .$$

The potential energy we're talking about here is gravitational potential energy, in the form $U = mgy$, and we can write the kinetic energy as $K = (1/2)mv^2$. The energy equation can thus be written as:

$$mgy_1 + \frac{1}{2}mv_1^2 + W_{nc} = mgy_2 + \frac{1}{2}mv_2^2 \, .$$

Let's apply this equation to a fluid flowing through a pipe. Figure 9.23 shows the two points we are considering, and two cylindrical regions of fluid are highlighted, one at each point. The cylindrical regions have equal volumes.

In the case of a flowing fluid, the work done by non-conservative forces is related to forces that arise because of pressure differences. We can write the W_{nc} term as:

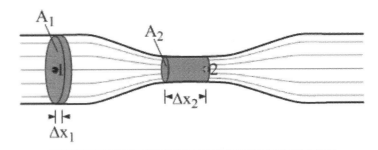

$$W_{nc} = F_{nc}\,\Delta x = F_1\,\Delta x_1 - F_2\,\Delta x_2 = P_1 A_1\,\Delta x_1 - P_2 A_2\,\Delta x_2 \, .$$

Substituting this expression for W_{nc} into the energy conservation relationship gives:

$$mgy_1 + \frac{1}{2}mv_1^2 + P_1 A_1 \Delta x_1 = mgy_2 + \frac{1}{2}mv_2^2 + P_2 A_1 \Delta x_1 \, .$$

Figure 9.23: The two cylindrical regions, one at point 1 and one at point 2, have the same volume.

The m here represents the mass of the fluid in one of the cylindrical regions in Figure 9.23 (the cylindrical regions have equal masses because of their equal volumes).

Let's simplify the equation by dividing both sides by the volume V of one of the cylindrical regions ($V = A_1\,\Delta x_1 = A_2\,\Delta x_2$). Because mass/volume = density, we get:

$$\rho g y_1 + \frac{1}{2}\rho v_1^2 + P_1 = \rho g y_2 + \frac{1}{2}\rho v_2^2 + P_2 \, . \quad \text{(Equation 9.9: \textbf{Bernoulli's Equation})}$$

Bernoulli's equation represents conservation of energy applied to fluids, although each term has units of energy density (energy per unit volume).

Related End-of-Chapter Exercises: 31, 56.

Essential Question 9.8: Consider a special case of Bernoulli's equation, when the fluid is at rest. Which of the equations we have examined previously in this chapter is equivalent to Bernoulli's equation with the two terms involving speed set to zero?

Answer to Essential Question 9.8: Eliminating the speed terms means that we can write Bernoulli's equation as: $\rho g y_1 + P_1 = \rho g y_2 + P_2$. Re-arranging this equation to solve for the pressure at point 2 gives: $P_2 = P_1 + \rho g y_1 - \rho g y_2 = P_1 + \rho g (y_1 - y_2)$. This equation is equivalent to Equation 9.7, the equation for pressure in a static fluid.

9-9 Examples Involving Bernoulli's Equation

EXPLORATION 9.9 – Pressure inside a pipe
Step 1 - *Make a prediction. In the pipe shown in Figure 9.24, is the pressure higher at point 2, where the fluid flows fastest, or at point 1? The fluid in the pipe flows from left to right.*

Many people predict that the pressure is higher at point 2, where the fluid is moving faster.

Figure 9.24: Fluid flowing through a pipe from left to right.

Step 2 - *Apply the continuity equation, and Bernoulli's equation, to rank points 1, 2, and 3 according to pressure, from largest to smallest.* Let's see if the common prediction, that the pressure is highest at point 2, is correct. First, apply the continuity equation: $A_1 v_1 = A_2 v_2 = A_3 v_3$. Looking at the tube, we know that $A_1 = A_3 > A_2$, which tells us that $v_2 > v_1 = v_3$.

Now, let's apply Bernoulli's Equation. Comparing points 1 and 2, we start with:

$$\rho g y_1 + \frac{1}{2} \rho v_1^2 + P_1 = \rho g y_2 + \frac{1}{2} \rho v_2^2 + P_2 .$$

The vertical positions of these two points are equal so the $\rho g y$ terms cancel out:

$$\frac{1}{2} \rho v_1^2 + P_1 = \frac{1}{2} \rho v_2^2 + P_2 .$$

Let's re-write this as: $\quad P_1 - P_2 = \frac{1}{2} \rho v_2^2 - \frac{1}{2} \rho v_1^2 .$

The continuity equation told us that $v_2 > v_1$, so the right-hand side of the above equation is positive. This means the left-hand side must also be positive, implying that $P_1 > P_2$. Thus, the pressure at point 2, where the fluid speed is highest, is less than the pressure at point 1. For points at the same height, higher speed corresponds to lower pressure. We can make sense of this by considering a parcel of fluid that moves from point 1 to point 2. Because this parcel of fluid speeds up as it travels from point 1 to point 2, there must be a net force acting on it that is directed right. This force must come from a difference in pressure between points 1 and 2. For the force to be directed right, the pressure must be larger on the left, at point 1.

We can also use Bernoulli's equation to show that the pressure at point 3 is equal to that at point 1. Thus we can conclude that $P_1 = P_3 > P_2$.

Key idea for an enclosed fluid: In general, in an enclosed fluid the pressure decreases as the speed of the fluid flow increases. **Related End-of-Chapter Exercises: 52, 53.**

EXAMPLE 9.9 – How fast?
A Styrofoam cylinder, filled with water, sits on a table. You then poke a small hole through the side of the cylinder, 20 cm below the top of the water surface. What is the speed of the fluid emerging from the hole?

SOLUTION
As usual, begin by drawing a diagram of the situation, as shown in Figure 9.25.

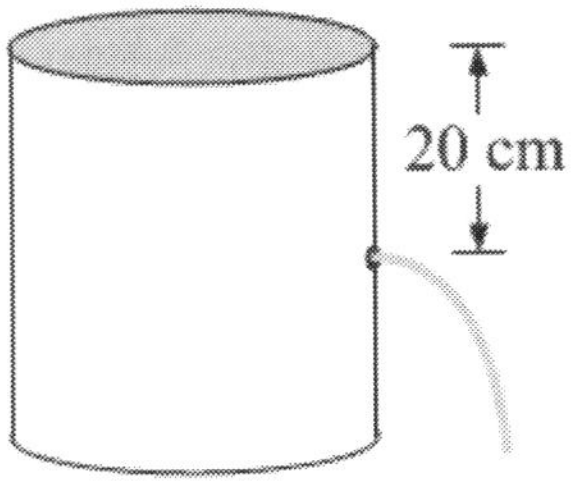

We're going to apply Bernoulli's equation, which means identifying two points that we can relate via the equation. Point 2 is outside the container where the hole is, because that is the place where we're trying to find the speed. Point 1 needs to be somewhere inside the container. Any point inside will do, although the most sensible places are either at the top of the container, where we know the pressure, or inside the container at the level of the hole. Let's choose a point at the very top, and apply Bernoulli's equation:

Figure 9.25: The Styrofoam container of water, with a small hole 20 cm from the top.

$$\rho g y_1 + \frac{1}{2}\rho v_1^2 + P_1 = \rho g y_2 + \frac{1}{2}\rho v_2^2 + P_2 .$$

First, we should recognize that, because both of our points are exposed to the atmosphere, we have $P_1 = P_2 = P_{atm}$. The pressure terms cancel in the equation, leaving:

$$\rho g y_1 + \frac{1}{2}\rho v_1^2 = \rho g y_2 + \frac{1}{2}\rho v_2^2 .$$

We can cancel factors of density. We are also free to define a zero for our y positions to be anywhere we find convenient. If we say $y = 0$ at the level of the hole we get $y_1 = +20$ cm and $y_2 = 0$, so the equation reduces to:

$$g y_1 + \frac{1}{2}v_1^2 = \frac{1}{2}v_2^2 .$$

If we knew the fluid speed at point 1 we could solve for the speed at point 2. This is a good time to bring in the continuity equation, which relates the speeds at the two points: $A_1 v_1 = A_2 v_2$. In this case we can say that, because A_1, the cross-sectional area of the cylinder, is so much larger than A_2, the cross-sectional area of the hole, then v_1 is much smaller than v_2. Thus, the $(1/2)v_1^2$ term is negligible compared to the $(1/2)v_2^2$ term. Our equation thus reduces to:

$$g y_1 = \frac{1}{2}v_2^2 .$$

Solving for the speed at which the fluid emerges from the hole gives:

$$v_2 = \sqrt{2 g y_1} = \sqrt{2 \times 9.8 \frac{m}{s^2} \times 0.20 \, m} = 2.0 \, m/s .$$

Related End-of-Chapter Exercises: 27, 28.

Essential Question 9.9: If you dropped an object from rest, what would its speed be after it had fallen through a distance of 20 cm? How does this compare to the result of the Example 9.9, where we found the speed of water emerging from a hole 20 cm below the water surface?

Answer to Essential Question 9.9: These two situations appear to be different, but the answers are the same. In both cases the speed is given by an equation of the form $v = \sqrt{2gh}$. The equations, and the speeds, are the same because, in both cases, we can apply conservation of energy, with gravitational potential energy being transformed to kinetic energy.

9-10 Viscosity and Surface Tension

Until this point, we have made a number of simplifying assumptions regarding the behavior of fluids, including assuming no viscosity (no resistance to flow). For any real fluid flowing through a pipe, there is some viscosity. In the case of air or water, the viscosity is generally low, but for applications such as blood flowing through blood vessels in the human body, the resistance to flow is an important factor.

Viscosity (a measure of a fluid's resistance to flow) is generally measured by means of a pair of horizontal parallel plates of area A with fluid filling the space between them. The bottom plate is at rest, as is the fluid right next to it, while the top plate has a horizontal speed v, with the fluid next to it moving with the same velocity as that plate. This situation is known as Couette flow. Viscosity arises because neighboring layers of fluid have different speeds, and there is some frictional resistance between the layers as they move past one another. In general, the speed of the fluid increases linearly as you move from the fixed plate to the moving plate. In a Newtonian fluid, the viscosity (η) is a constant equal to the force required to keep the moving plate moving with constant velocity v, multiplied by the length of the flow and divided by both the speed and the plate area. There are several classes of non-Newtonian fluids, in which the viscosity changes as the speed changes. Those are interesting, but beyond the scope of this book.

To give you some feel for typical numbers, the viscosity of water at 20°C is 0.001002 Pa s, while that of motor oil is about 0.250 Pa s.

Unlike the somewhat artificial situation of Couette flow, in a typical application fluid is flowing through a pipe in which the pipe is stationary. In this case, the fluid next to the pipe wall is at rest, and the fluid at the center of the pipe is moving fastest. If we use Q to denote the volume flow rate ($Q = Av$), then the volume flow rate of a viscous fluid is given by:

$$Q = \frac{\pi R^4 \Delta P}{8 \eta L}.$$
(Eq. 9.10: **The Hagen-Poiseuille equation**)

The equation is named for the German physicist and hydraulic engineer Gottlif Hagen and the French physician Jean Marie Louis Poiseuille, who independently arrived at the equation experimentally in 1839 and 1838, respectively.

The equation pertains to a fluid flowing through a pipe with a radius R and a length L, with a pressure difference of ΔP between the ends of the pipe. The viscosity is denoted by η, the Greek letter eta.

Surface Tension

In general, we can't walk on water (at least not the liquid variety), and we also can't float a quarter on water. However, certain insects (water striders, for instance) can support themselves perfectly well on the surface of a pond, and it is possible to float a Japanese yen, or a metal paper clip, in a glass of water. Based on what we learned earlier, the yen coin or the paper clip should sink - they are each made from material that is denser than water - but they float because of surface tension. In essence, for these light objects, the water surface acts as an elastic membrane, much like the surface of a trampoline does for us.

The elastic nature of the water surface comes from the attraction the water molecules have for one another. A water strider will dent the surface, but (unlike us) will not break the surface. Similarly, the paper clip floating on the water surface dents the surface, much like we do when standing on a trampoline. It can break the surface and sink to the bottom - it can take a few tried to get it to float at the surface, but if you place it gently and carefully on the water surface, the force of gravity acting on it can be balanced by the force associated with surface tension. Adding some liquid soap to the water reduces its surface tension, so you can make the floating paper clip sink just by adding a few drops of soap. In addition to a paper clip, you can also float a Japanese yen coin, as shown in Figure 9.26.

Figure 9.26: A Japanese yen coin, made from aluminum, which is supported by the surface tension at the surface of the water in a glass of water. Note how the water surface is indented, with the surface acting much like an elastic membrane. Photo credit: A. Duffy.

You have probably noticed that water often tends to form drops. This is because there is energy associated with the surface, so surfaces generally take on the smallest area possible, to minimize that surface energy. A sphere is the shape that minimizes the surface area, for a given volume. A similar effect is seen with soap films, which can take on interesting shapes when a frame is drawn out of soapy water - the shape of the film tends to minimize the film's surface area.

Surface tension and the lungs
Surface tension is actually quite important for our breathing. First, think about blowing up a balloon. You have to work to fill the balloon, but if you don't hold the neck of the balloon then the balloon will simply deflate by itself, because of the balloon's surface tension. Our lungs have a very large number of tiny balloons, essentially - these are called alveoli. We use muscles to breathe in, inflating the alveoli. However, just like the balloon, the alveoli deflate all by themselves, to minimize surface tension. That's a key part of the breathing process.

The alveoli have a mucus coating on their walls, which acts as a ***surfactant*** (a material that reduces surface tension, like dish soap does when you're doing dishes). Unlike dish soap, however, which has a fixed surface tension, the mucus has a surface tension that increases with the size of the alveoli. This is important for a number of reasons. Having a low surface tension when the alveoli are small prevents surface tension from collapsing the alveoli, and surface tension increasing as the alveoli expand prevents the alveoli from getting too large. This also explains why premature babies often have breathing issues - their lungs do not have the mucus coating the alveoli to reduce the surface tension, making it difficult to inflate the alveoli.

Essential Question 9.10: This question relates to viscosity. Let's say that Fred, who eats french fries on a daily basis, gradually experiences a narrowing of his blood vessels, because of plaque buildup in the vessel walls. All other things being equal, what would be the new flow rate through a blood vessel that experienced a 5% decrease in radius, compared to the original flow rate? In actuality, the blood pressure can change so that the flow rate does not drop quite so significantly. Would you expect the blood pressure to increase or decrease?

Answer to Essential Question 9.10: The new radius is 95% of the original radius. Taking a factor of 0.95 to the fourth power gives approximately 0.81, so the new flow rate would only be 81% of the original flow rate. The flow rate would not drop as much if the pressure difference between the ends of the blood vessel increased. This would generally be accomplished by increasing the blood pressure (which can lead to health issues, of course).

9-11 Drag and the Ultracentrifuge

When an object is falling through a viscous fluid, a drag force acts on it. Unlike the kinetic friction force we looked at earlier in the book, which has a magnitude that is independent of speed, the viscous drag force is generally proportional to the speed, and opposite in direction to the velocity. This is known as Stokes' drag, with the drag force on a spherical particle of radius r moving at speed v through a fluid of viscosity η being:

$$F_d = -6\pi\eta r v. \qquad \text{(Equation 9.11: \textbf{Stokes' drag force for a spherical particle})}$$

When an object falls through the fluid, it will reach a terminal velocity when the drag force plus the buoyant force is equal and opposite to the force of gravity. In general, the smaller the object, the smaller the magnitude of the terminal velocity. Very small objects fall very slowly. If your goal is to separate particles from the fluid the particles are in, this can be a problem - it can take a long time for the particles to settle out at the bottom.

This is where an ultracentrifuge comes in. The job of the ultracentrifuge is to spin the fluid very quickly in a circular path. In that case, the effect is just like increasing the value of g by a large factor. Effectively, as far as the particles are concerned, they are in a gravitational field with a strength given by the centripetal acceleration, $\omega^2 r$. Spinning very quickly gives very large values of the angular speed (ω), leading to very large "effective gravity" that separates out the particles quickly and efficiently. Let's consider an example.

EXAMPLE 9.11 – Analyzing a blood sample
You get a blood sample drawn while you're seeing your doctor, and the sample is sent to the lab for analysis. A key part of the analysis involves running the sample (contained in a cylindrical tube) through an ultracentrifuge to separate out the components, which have different densities (the red blood cells being most dense, at 1125 kg/m³, and the plasma being least dense, at 1025 kg/m³). The average density of blood is about 1060 kg/m³. What is the purpose of an ultracentrifuge, which, say, has a rotation rate of 5000 rpm and an acceleration 5000 times larger than the acceleration due to gravity? Why don't they just stand the tube of blood up vertically to let gravity separate it? Do a quantitative analysis, using the following values. The mass of a red blood cell is about 27×10^{-15} kg, the viscosity of blood is about 3.5×10^{-3} Pa s, and we will

model the cell as a sphere of radius 3.5×10^{-6} m.

SOLUTION
We'll start by determining what gravity can do by itself. A red blood cell in a vertical tube of blood will reach a terminal velocity (v_t) when the drag force plus the buoyant force is equal and opposite to the force of gravity.

$$mg = \rho_{fluid} V g + 6\pi\eta r v_t.$$

We can replace the volume of the cell by $V = m / \rho_{cell}$, which gives

$$mg = \frac{\rho_{fluid}}{\rho_{cell}} mg + 6\pi\eta r v_t.$$

This re-arranges to $v_t = \left(1 - \frac{\rho_{fluid}}{\rho_{cell}}\right) \frac{mg}{6\pi\eta r}$, solving for the terminal speed of a blood cell.

Now, we'll substitute the relevant values into our equation (recognizing that our model has some limitations, such as that red blood cells are not spheres, and issues with the fluid density not being constant).

$$v_t = \left(1 - \frac{\rho_{fluid}}{\rho_{cell}}\right) \frac{mg}{6\pi\eta r} = \left(1 - \frac{1060 \text{ kg/m}^3}{1125 \text{ kg/m}^3}\right) \frac{(27 \times 10^{-15} \text{ kg})(9.8 \text{ N/kg})}{6\pi(3.5 \times 10^{-3} \text{ Pa s})(3.5 \times 10^{-6} \text{ m})} = 6.6 \times 10^{-8} \text{ m/s}.$$

This is very slow, of course. If we needed to wait until the red blood cell traveled a distance of 3.3 cm through the tube of blood, say, we would have to wait for 500000 s, which is approximately 6 days.

In our ultracentrifuge, where the acceleration is 5000 g, what is the difference? We use the same equation we derived above for the terminal speed, but we replace the factor of g by 5000 g. That increases the terminal velocity by a factor of 5000, and reduces the time it takes the blood cell to travel 3.3 cm by a factor of 5000, so the time would be 100 s instead of 500000 s (about 2 minutes instead of 6 days). Despite the simplifying assumptions in our model, this is in keeping with the recommendations of ultracentrifuge manufacturers, that a spin time of five minutes at 5000 g is appropriate. Our analysis gives a value of the same order of magnitude.

Where does the 5000 g come from? This goes back to uniform circular motion, in which the acceleration is given by:

$$a_c = \frac{v^2}{R} = \omega^2 R.$$

R here is the radius of the circular path traveled by a blood cell inside the centrifuge (not to be confused with the r used above, for the radius of the blood cell itself). Expressing the angular speed in rad/s, and the acceleration in meters per second, we could solve for R, for instance.

Related End-of-Chapter Exercises: 67 - 71.

Essential Question 9.11: A manufacturer has two different centrifuges, one that spins at 5000 rpm, and another that spins at 10000 rpm. Everything else is the same. The manufacturer makes the following claim about the faster centrifuge - "It costs three times as much, but it is also three times as efficient - you can run samples through the faster centrifuge in one third the time!" Based on our analysis above, the factor of three seems somewhat surprising - what would we expect the difference to be between the two models? Accounting for the fact that it takes some time for the centrifuge to reach its maximum angular speed, and to slow down to a stop at the end of a run, might this factor of three actually be plausible?

Answer to Essential Question 9.11: If we double the angular velocity, the acceleration will increase by a factor of four (the acceleration goes as the square of the angular speed). That causes a corresponding decrease by a factor of 4 in the time to separate components of a sample - the manufacturer may be understating the case! However, the spin up and slow down time, which is probably longer for the faster centrifuge, means that the device is not consistently a factor of four better. That would decrease the ratio, although maybe not all the way down to a factor of three.

Chapter Summary

Essential Idea for Fluids

Even though situations involving fluids look quite different from those we examined in earlier chapters, we can apply the same methods we applied earlier. Forces are very useful for understanding situations in which an object floats or sinks in a static fluid, while energy-conservation ideas can help us to analyze situations involving moving fluids.

The Buoyant Force

An object in a fluid experiences a net upward force we call the buoyant force, $\vec{F}_B$. The magnitude of the buoyant force is proportional to the volume of fluid displaced by the object.

$$F_B \propto V_{disp}.$$ (Equation 9.1: **Buoyant force**)

Mass Density

Much of what happens with fluids involves mass density (often referred to simply as density). For instance, an object with a mass density larger than the mass density of the fluid it is in generally sinks in that fluid, while an object with a lower mass density than a fluid floats in the fluid. Using the symbol ρ for mass density, the relationship between mass, density, and volume is:

$$m = \rho V, \quad \text{or} \quad \rho = \frac{m}{V}$$ (Eq. 9.2: **Connection between mass and mass density**)

Archimedes' Principle

The magnitude of the buoyant force exerted on an object by a fluid is equal to the weight of the fluid displaced by the object. This is known as Archimedes' principle:

$$F_B = m_{disp}g = \rho_{fluid} V_{disp}\, g.$$ (Equation 9.3: **Archimedes' Principle**)

A General Method for Solving a Buoyancy Problem

1. Draw a diagram of the situation.

2. Draw one or more free-body diagrams, with each free-body diagram showing all the forces acting on an object as well as an appropriate coordinate system.

3. Apply Newton's Second Law to each free-body diagram.

4. If necessary, bring in Archimedes' principle, $F_B = \rho_{fluid} V_{disp}\, g$.

5. Put the resulting equations together and solve.

Pressure

$$\text{Pressure} = \frac{\text{Force}}{\text{Area}} \qquad \text{or} \qquad P = \frac{F}{A}. \qquad \text{(Equation 9.6: \textbf{Pressure})}$$

The MKS unit of pressure is the pascal (Pa). $1\,\text{Pa} = 1\,\text{N}/\text{m}^2$.

Standard atmospheric pressure is 101.3 kPa, or about $1.0 \times 10^5\,\text{Pa}$.

Pressure in a static fluid

$$P_2 = P_1 + \rho g h \quad . \qquad\qquad \text{(Equation 9.7: \textbf{Pressure in a static fluid})}$$

Fluid dynamics

$$A_1 v_1 = A_2 v_2. \qquad\qquad \text{(Equation 9.8: \textbf{The Continuity Equation})}$$

$$\rho g y_1 + \frac{1}{2}\rho v_1^2 + P_1 = \rho g y_2 + \frac{1}{2}\rho v_2^2 + P_2. \quad \text{(Equation 9.9: \textbf{Bernoulli's Equation})}$$

Bernoulli's Equation comes from applying energy-conservation ideas to fluids.

Viscosity, and the drag force

Viscosity is a measure of a fluid's resistance to flow, and arises because of friction between neighboring layers of fluid that are moving with different velocities.

If we use Q to denote the volume flow rate ($Q = Av$), then the volume flow rate of a viscous fluid is given by:

$$Q = \frac{\pi R^4 \Delta P}{8 \eta L}. \qquad\qquad \text{(Eq. 9.10: \textbf{The Hagen-Poiseuille equation})}$$

The equation pertains to a fluid flowing through a pipe with a radius R and a length L, with a pressure difference of ΔP between the ends of the pipe. The viscosity is denoted by η, the Greek letter eta.

The drag force on a spherical particle of radius r moving at speed v through a fluid of viscosity η is:

$$F_d = -6\pi \eta r v. \qquad\qquad \text{(Equation 9.11: \textbf{Stokes' drag force for a spherical particle})}$$

End-of-Chapter Exercises

Exercises 1 – 12 are primarily conceptual questions that are designed to see if you have understood the main concepts of the chapter.

1.   A particular block floats with 30% of its volume submerged in water, but with only 20% of its volume submerged in a second fluid. Which fluid exerts a larger buoyant force on the block? Briefly justify your answer.

2.   A solid brick and a solid wooden block have exactly the same dimensions. When they are placed in a bucket of water, the brick sinks to the bottom but the block floats. Which object experiences a larger buoyant force?

3.   An aluminum ball and a steel ball are exactly the same size, but the aluminum ball has less mass. Both balls sink to the bottom of a glass of water. (a) Briefly explain why the aluminum ball has less mass. (b) Which ball displaces more water in the glass?

4.   When a block suspended from a spring scale is half submerged in Fluid A, the spring scale reads 8 N. When the block is half submerged in Fluid B, the spring scale reads 6 N. Which fluid has a higher density? Explain.

5.   (a) Explain the physics behind a drinking straw. How is it possible to use a straw to drink through? (b) If you fill the straw partly with fluid and then seal the upper end of the straw with your thumb, you can get the column of fluid to remain in the straw. How does that work?

6.   To crush a soda can with atmospheric pressure, start by placing a small amount of water into an empty soda can. Carefully heat the can until steam comes out of the can. It's important to let this process continue long enough for water vapor to drive most of the air out of the can. If you then quickly remove the can from the heat source and invert it into a bowl of cold water so that the opening to the can is under water, the can should almost instantly collapse. Do this all very carefully to make sure you don't burn yourself. Explain why the can collapses.

7.   Three cubes of identical volume but different density are placed in a container of fluid. The blocks are in equilibrium when they are in the positions shown in Figure 9.27. If the strings connected to blocks A and C were cut, those blocks would not be in equilibrium. Rank the cubes based on the magnitude of (a) their densities; (b) the buoyant forces they experience.

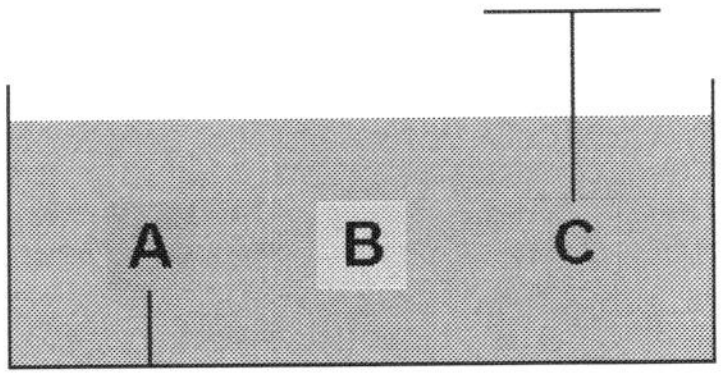

Figure 9.27: Three cubes of identical volume but different density at equilibrium in a container of fluid. For Exercise 7.

8.   The two strings in Figure 9.27 are now lengthened, giving the situation shown in Figure 9.28. The cubes still have identical volumes but different densities. Rank the cubes based on the magnitude of (a) the force applied by the fluid on the top surface of the cube; (b) the force applied by the fluid on the bottom surface of the cube; (c) the buoyant force acting on the cube.

Figure 9.28: The same three cubes of identical volume but different density at equilibrium in a container of fluid as in Figure 9.27, but with longer strings. For Exercise 8.

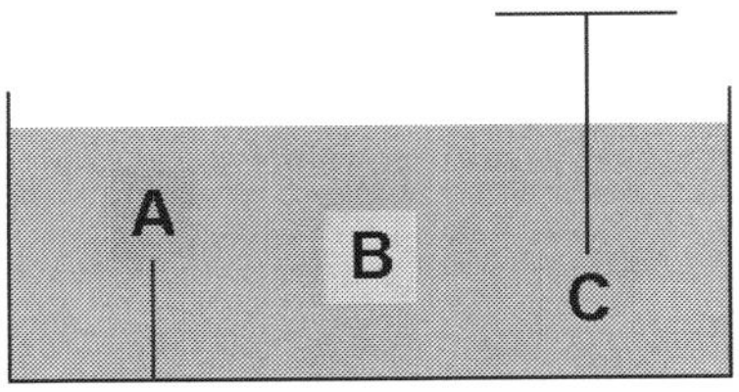

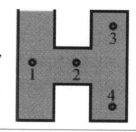

9. A beaker of water is placed on a scale. If you dip your little finger into the water, making sure that you don't touch the beaker itself, does the scale reading increase, decrease, or stay the same? Explain.

10. Four points are labeled in a container shaped like the letter H, as shown in Figure 9.29. The container is filled with fluid, and open to the atmosphere only at the top left of the container. Rank the points based on their pressure, from largest to smallest. Use only > and/or = signs in your ranking (e.g., 2>1=3>4).

Figure 9.29: Four points are labeled in an H-shaped container of fluid, that is open to the atmosphere only at the top left. For Exercise 10.

11. If you are careful when you lie down and get up again, it does not hurt to lie on a bed of nails. Supporting your weight on a single nail is a different story, however. Explain why you can lie comfortably on a bed of nails, but not with your weight supported by the point of a single nail.

12. Water is placed in a U-shaped tube, as shown in Figure 9.30. The tube is open to the atmosphere at the top left, but the tube is sealed with a rubber stopper at the top right. Can the water in the tube remain as shown, or must the level on the right drop and the level on the left rise? Explain.

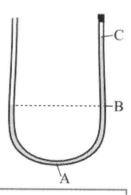

Figure 9.30: Fluid in a U-shaped tube that is open to the atmosphere on the left and sealed on the right. For Exercise 12.

Exercises 13 – 16 deal with buoyancy and/or density.

13. A particular block floats with 30% of its volume submerged in water, but with only 20% of its volume submerged in a second fluid. Taking the density of water to be 1000 kg/m³, determine the density of (a) the block; (b) the second fluid.

14. You put some water into a glass and carefully mark the level of the top of the water. You then pour some of the water into one section of an ice cube tray and place the tray in the freezer to form a single ice cube. (a) Keeping in mind that water expands in volume by about 10% when it freezes, what will happen when the ice cube is placed back in the glass? Will the top of the water be higher, lower, or the same as it was before? (b) As the ice melts, will the water level rise, fall, or stay the same? Briefly explain your answers.

15. You have a glass of water with one or more ice cubes in it. Test your friends by asking them what will happen to the ice when you pour some oil on top of the water and ice. The density of the oil must be less than that of ice so the ice does not float to the top of the oil. Will the ice cube(s) float higher or lower in the water when the oil is poured on top, or will the level be unchanged? How will you explain the result to your friends?

16. When a block suspended from a spring scale is half submerged in Fluid A, the spring scale reads 8 N. When the block is half submerged in Fluid B, the spring scale reads 6 N. If the density of one fluid is twice as large as the density of the other, determine the buoyant force acting on the block when the block is half submerged in (a) Fluid A; (b) Fluid B. (c) What is the weight of the block?

Exercises 17 – 26 are designed to give you some practice with applying the general method of solving a typical buoyancy problem. For each exercise begin with the following parts: (a) Draw a diagram of the situation. (b) Sketch one of more free-body diagrams, including appropriate coordinate systems for each. (c) Apply Newton's Second Law.

17. A wooden block with a weight of 8.0 N floats with 60% of its volume submerged in oil. Parts (a) – (c) as described above. (d) What is the magnitude and direction of the buoyant force exerted on the block by the oil?

18. A metal ball with a weight of 12.0 N hangs from a string tied to a spring scale. When the ball is half-submerged in a particular fluid, the spring scale reads 7.0 N. The goal of this exercise is to find the buoyant force exerted on the block by the fluid when the fluid is both half-submerged and completely submerged, and to find the reading on the spring scale when the ball is completely submerged. Parts (a) – (c) as described above, making sure that you draw two sets of diagrams, one when the ball is half submerged and one when the ball is completely submerged. Find the buoyant force exerted on the ball by the fluid when the ball is (d) half submerged; (e) completely submerged. (f) What is the reading on the spring scale when the ball is completely submerged?

19. A basketball floats in a large tub of water with $1/11^{th}$ of its volume submerged. The mass of the basketball is 500 grams. The goal of this exercise is to find the radius of the ball, assuming the water has a density of 1000 kg/m³. Parts (a) – (c) as described above. (d) What is the volume of water displaced by the ball? (e) What is the volume of the ball? (f) What is the equation for the volume of a sphere? (g) What is the radius of the basketball? (h) Did you use a particular value for g, the acceleration due to gravity? If so, what was it? Comment on the effect, if any, of you using a different value of g for the calculation.

20. Consider again the situation described in the previous exercise, with the basketball floating in the tub of water. Use $g = 9.80$ m/s². The goal of this exercise is to determine the force you need to apply to the ball to hold it below the surface of the water. Parts (a) – (c) as described above. You should have sketched diagrams for the floating ball in the previous exercise, so now draw a set of diagrams for the ball when you are holding it completely submerged below the water. What is the buoyant force applied to the ball by the water when the ball is (d) floating? (e) completely submerged? (f) What is the force you need to apply to the ball to hold it under the water?

21. A low-density block with a weight of 10 N is placed in a beaker of water and tied to the bottom of the beaker by a vertical string of fixed length. When the block is 25% submerged, the tension in the string is 15 N. The string will break if its tension exceeds 65 N. As water is steadily added to the beaker, the block becomes more and more submerged. Parts (a) – (c) as described above. (d) What fraction of the block is submerged at the instant the string breaks? (e) After the string breaks and the block comes to a new equilibrium position in the beaker, what fraction of the block's volume is submerged?

22. A large hot-air balloon has a mass of 300 kg, including the shell of the balloon, the basket, and the passengers, but not including the air inside the balloon itself. The goal of the exercise is to determine the volume of the balloon, assuming the air inside the balloon has a density that is 90% of the density of the air outside, that the density of the air outside the balloon is 1.30 kg/m³, and that the balloon is floating in equilibrium above the ground. Parts (a) – (c) as described above. (d) What is the volume of the balloon?

23. You are designing a pair of Styrofoam (density 1.30 kg/m³) shoes that you can wear to walk on water. Your mass is 50 kg, and you want the shoes to be 30% submerged in the water. Parts (a) – (c) as described above. (d) What volume of Styrofoam do you need for each shoe? (e) Estimate the volume of a typical shoebox, and compare the volume of one of the Styrofoam shoes to the volume of a shoebox.

24. You, Archimedes, suspect that the king's crown is not solid gold but is instead gold-plated lead. To test your theory, you weigh the crown, and find it to weigh 60.0 N and to have an apparent weight of 56.2 N when it is completely submerged in water. Parts (a) – (c) as described above. (d) What is the average density of the crown? (e) Is it solid gold? If not, find what fraction (by weight) is gold and what fraction is lead.

25. A square raft at the local beach is made from five 2.0-meter wooden boards that have a square cross-section measuring 40 cm x 40 cm. The goal of the exercise is to determine the largest number of children that can stand on the raft without the raft being completely submerged if the boards have a density of 500 kg/m³. Assume that each child has a mass of 35 kg. Parts (a) – (c) as described above. (d) How many children can stand on the raft without the raft being completely submerged, assuming each child is completely out of the water?

26. After heavy rains have stopped, you go out in a boat to check the level of the water in a reservoir behind a dam (your boat is in the reservoir). You notice that the water is dangerously close to spilling over the top of the dam, and when you look up at the sky you see more dark clouds approaching. Your boat has a very heavy anchor in it, so the goal of the exercise is to determine how throwing the anchor overboard would affect the level of the water in the reservoir (assuming the boat displaces a reasonable fraction of the water in the reservoir). Parts (a) – (c) as described above, where you should draw two sets of diagrams, one when the anchor is in the boat and the other when the anchor is resting at the bottom of the reservoir. (d) Using your diagrams to help you, determine whether the water level in the reservoir rises, falls, or stays the same when you toss the anchor overboard.

Exercises 27 – 30 deal with pressure.

27. A cylindrical barrel is completely full of water and sealed at the top except for a narrow tube extending vertically through the lid. The barrel has a diameter of 80 cm, while the tube has a diameter of 1 cm. You can actually cause the lid to pop off by pouring a relatively small amount of water into the tube. To what height do you need to add water to the tube to get the lid to pop off the barrel? The lid pops off when the vector sum of the force of the atmosphere pushing down on the top of the lid and the force of the water pushing up on the bottom of the lid is 250 N up.

28. Before the negative environmental impact of mercury was fully understood, many barometers utilized a column of mercury to measure atmospheric pressure. This was first done by Evangelista Torricelli in 1643. A design for a simple barometer is shown in Figure 9.31, where there is negligible pressure at the top of the inverted column. (a) What is the height of a column of mercury that produces a pressure at its base equal to standard atmospheric pressure? (b) If water was used instead, what is the height of the column of water required? (c) Does this help explain why mercury was chosen as the working fluid in many barometers? Are there any other advantages mercury offers over water in this application?

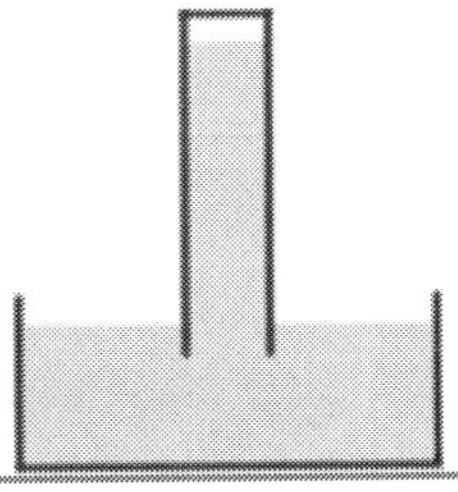

Figure 9.31: A simple liquid barometer, with an inverted column of fluid in a reservoir of that same fluid. The reservoir is open to the atmosphere, but the column is not. For Exercises 28 and 29.

29. Consider again the liquid barometer described in Exercise 28 and shown in Figure 9.31. Because of an approaching storm system, the local atmospheric pressure drops from 101.3 kPa to 99.7 kPa. (a) Does this cause the liquid to rise or fall in the tube? Determine the change in height of the column of liquid if the liquid is (b) mercury; (c) water.

30. In 2002, Tanya Streeter of the USA set a world record of 160 m for the deepest dive without breathing assistance (such as SCUBA gear). At that depth, what is the absolute pressure?

Exercises 31 – 35 address issues in fluid dynamics.

31. If you turn on a water faucet so that the water flows smoothly, you should observe that the cross-sectional area of the water stream decreases as the stream drops. (a) Explain why the water stream narrows. At a particular point, the flow speed is 10 cm/s and the stream has a cross-sectional area of 2.0 cm². At a point 20 cm below this point, determine (b) the flow speed, and (c) the cross-sectional area of the stream.

32. Take a bottle of water, filled to a depth of 25 cm, and carefully poke a small hole in the bottom of the bottle with a nail. (a) When you remove the cap from the bottle, what is the speed of the water emerging from the hole? (b) When you screw the cap back on the bottle, the water should stop coming out of the hole. Explain why.

33. A cylinder of height H sits on the floor. The cylinder is completely full of water, but a stream of water is emerging horizontally from the side of the cylinder at a distance h from the top. In terms of H, h, and g, determine: (a) the speed with which the water is emerging from the cylinder; (b) the time it takes the water to travel from the hole to the floor; (c) the horizontal distance traveled by the water as it falls.

34. Consider again the cylinder described in Exercise 33, but this time let's say there are three holes in the side of the cylinder. The holes are at distances of $H/4$, $H/2$, and $3H/4$ from the top of the cylinder. (a) Make a prediction – which stream of water travels furthest horizontally before reaching the floor? What do you base your prediction on? (b) Check your prediction using $H = 1.0$ m. Calculate the horizontal distance traveled, before reaching the floor, by the water from each hole.

35. While washing your hands at a sink, you determine that the water emerges from the faucet, which has a diameter of 1.0 cm, with a speed of 1.8 m/s. If the water comes from a pump located 1.5 m below the faucet, what is the absolute pressure at the pump if the pipe leading from the pump has a diameter of (a) 1.0 cm; (b) 8.0 cm?

General problems and conceptual questions

36. Consider the situation shown in Figure 9.32. (a) In figure A, what is the tension in the string? In figure B, what is the (b) buoyant force on the ball? (c) scale reading? In figure C, what is the (d) buoyant force on the ball? (e) tension in the string? (f) scale reading? In figure D, what is the (g) buoyant force on the ball? (h) scale reading?

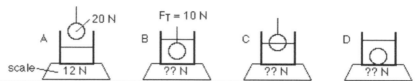

Figure 9.32: In figure A, a 20 N ball is supported by a string. It hangs over a beaker of fluid that sits on a scale. The scale reading is 12 N. In figure B the ball is completely submerged in the fluid. In figure C the ball is exactly half submerged. In figure D the string has been cut and the ball rests on the bottom of the beaker.

37. As shown in Figure 9.33, a wooden cube measuring 20.0 cm on each side floats in water with 80.0% of its volume submerged. Suspended by a string below the wooden cube is a metal cube. The metal cube measures 10.0 cm on each side and has a specific gravity of 5.00. (a) Which cube has a larger buoyant force acting on it? (b) Taking the density of water to be 1000 kg/m^3, what is the density of the wooden cube? (c) What is the tension in the string between the cubes? Assume the string itself has negligible mass and volume. (d) The pair of blocks is now placed in a different liquid. When the blocks are at equilibrium in this new liquid, the buoyant force acting on the wooden cube is exactly the same as the buoyant force acting on the metal cube. What is the density of this new liquid?

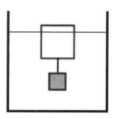

Figure 9.33: A metal cube suspended from a wooden cube, for Exercises 37 – 39.

38. Consider the situation shown in Figure 9.33, in which a wooden cube measuring 20.0 cm on each side floats in a fluid with 80.0% of its volume submerged. Suspended by a string below the wooden cube is a metal cube. The metal cube measures 10.0 cm on each side. If the wooden cube has a density of 800 kg/m^3 and the metal cube has a density of 1600 kg/m^3 what is the density of the fluid?

39. Consider the situation described in Exercise 38. (a) Describe qualitatively what will happen if the string is cut. (b) What is the magnitude and direction of the acceleration of each block immediately after the string is cut? (c) After a long time, where will the blocks be?

40. Three cubes of identical volume but different density are placed in a container of fluid. The blocks are in equilibrium when they are in the positions shown in Figure 9.34. If the strings connected to blocks A and C were cut, those blocks would not be in equilibrium. If the densities of the cubes have a 1:2:3 ratio and the magnitude of the tension in the string attached to block A is F_T, what is the magnitude of the tension in the string attached to block C? Express your answer in terms of F_T.

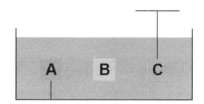

Figure 9.34: Three cubes of identical volume but different density at equilibrium in a container of fluid. For Exercise 40.

41. A toy balloon, which has a mass of 3.5 g before it is inflated, is filled with helium (with a density of 0.18 kg/m³) to a volume of 8000 cm³. What is the minimum mass that should be hung from the balloon to prevent it from rising up into the air?

42. Who was Archimedes? When did he live, and what else is he known for aside from buoyancy? Do some background reading and write a couple of paragraphs about him.

43. Use equation 9.7 to estimate the height of the atmosphere. Do you expect this to be a reasonably accurate measure of the height? Would your calculated value represent a lower limit or an upper limit of the true height of the atmosphere? Explain.

44. A popular demonstration about atmospheric pressure is called the Magdeburg hemispheres. Two hemispheres are held together while air is pumped out from between them. As long as there is a good seal between the hemispheres, it is extremely hard to separate them. (a) Explain why. (b) If the hemispheres are 20 cm in diameter, determine the force required to separate them, assuming all the air is evacuated from inside. This demonstration was first done by Otto von Guericke in the German town of Magdeburg around 1656, where two teams of horses tried unsuccessfully to pull the hemispheres apart.

45. Standard atmospheric pressure, which is 1 atm or 101.3 kPa, can be quoted in many different units. State atmospheric pressure in three other units, at least two of which are not SI units.

46. Pour your favorite carbonated beverage into a tall glass and watch the bubbles rise. What should happen to the size of the bubbles as they rise? Why? Can you observe the bubbles changing size as they rise?

47. A brick with a density of 4200 kg/m³ measures 8 cm by 15 cm by 30 cm. It can be placed with any of its six faces against the floor. (a) Find the maximum and minimum values of the normal force exerted by the floor on the block when the block is resting on the floor in its various orientations. (b) Find the maximum and minimum values of the pressure associated with the block resting on the floor in its various orientations. Neglect any contribution from atmospheric pressure.

48. What are "the bends", in reference to deep-sea diving? Write a paragraph or two on what causes the bends, how to prevent them, and how to treat a diver who has the bends.

49. Mountain climbers, and people who live at high altitudes, have difficulty making a good cup of tea or coffee, despite the fact that they follow the usual procedure of heating water until it boils, and bringing the tea or coffee together with the water. What is the problem?

50. As an engineer at a mine, you are in charge of pumping water out of a flooded mine shaft that extends 20 m down from the surface. You consider two different configurations, one in which the pump is placed at the surface and it essentially sucks water out of the shaft in the same manner that a drinking straw works, and another in which the pump is placed at the bottom of the shaft and pumps water up to the surface. Assuming the pump is fully submersible (i.e., that it will work when completely submerged in water at the bottom of the shaft), which of these configurations is more appropriate for this situation? Why?

51. Water is placed in a U-shaped tube, as shown in Figure 9.35. The tube is open to the atmosphere at the top left, but the tube is sealed with a rubber stopper at the top right. Point A is 20 cm below point B, and point C is 30 cm above point B. Determine the gauge pressure at (a) point A; (b) point B; (c) point C.

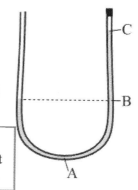

Figure 9.35: Fluid in a U-shaped tube that is open to the atmosphere on the left and sealed on the right. For Exercise 51.

52. A flexible tube can be used as a simple siphon to transfer fluid from one container to a lower container. This is shown in Figure 9.36. The fluid has a density of 800 kg/m³. If the tube has a cross-sectional area that is much smaller than the cross-sectional area of the higher container, what is the speed of the fluid at (a) point Z? (b) point Y? (c) What is the absolute pressure at point Y? See the dimensions given in Figure 9.35, and take atmospheric pressure to be 101.3 kPa.

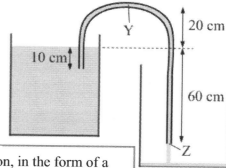

Figure 9.36: A siphon, in the form of a flexible tube, being used to transfer fluid from the container on the left to the container on the right.

53. A venturi tube is a tube with a constriction in it. Pressure in a venturi tube can be measured by attaching a U-shaped fluid-filled device to the venturi tube as shown in Figure 9.37. (a) If the fluid in the U is water, and there is a 10 cm difference between the water levels on the two sides, what is the pressure difference between points 1 and 2 in the venturi tube? (b) The venturi tube has air flowing through it. If the cross-sectional area of the venturi tube is 6 times larger at point 1 than it is at point 2, what is the air speed at point 2?

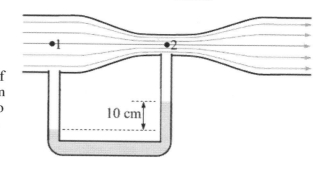

Figure 9.37: A venturi tube with a water-filled U-tube to measure pressure, for Exercise 53.

54. Your town supplies water with the aid of a tall water tower that is on top of the highest hill in town. The top surface of the water in the tank is open to the atmosphere. (a) Explain why it makes sense to use such a water tower in the water-distribution system. (b) Would you expect the water pressure to be highest in homes at higher elevations or at lower elevations in the town, all other factors being equal?

55. Consider the water tower described in the previous exercise. Opening a valve at the base of the tower allows water to flow out a pipe that bends up, projecting water straight up into the air. If the water reaches a maximum height of 8.5 m above the base of the tower, to what depth is the water tower filled with water?

56. As you wash your car, you are using an ordinary garden hose to spray water over the car. You notice that, if you cover most of the open end of the hose with your hand, the water sprays out with a higher speed. Explain how this works.

57. About once every 30 minutes, a geyser known as Old Faceful projects water 15 m straight up into the air. (a) What is the speed of the water when it emerges from the ground? (b) Assuming the water travels to the surface through a narrow crack that extends 10 m below the surface, and that the water comes from a chamber with a large cross-sectional area, what is the pressure in the chamber?

58. A drinking fountain projects water at a 45° angle with respect to the horizontal. The water reaches a maximum height of 10 cm above the opening of the fountain. With what speed was it projected into the air?

59. While taking a shower, you notice that the shower head is made up of 40 small round openings, each with a radius of 1.5 mm. You also determine that it takes 8.0 s for the shower to completely fill a 1-liter container you hold in the water stream. The water in your house is pumped by a pump in the basement, 7.2 m below the level of the shower head. If the pump maintains an absolute pressure of 1.5 atm., what is the cross-sectional area of the pipe connected to the pump?

60. There were many famous members of the Bernoulli family, and they did not all get along. Do some background reading on the Bernoullis and write a paragraph or two about them, making sure that you identify the member of the family credited with the Bernoulli equation we used in this chapter.

61. Three students are having a conversation. Explain what you think is correct about what they say, and what you think is incorrect.

Jenna: OK, so here's the question. "A block floats 60% submerged in Fluid A, and only 40% submerged in Fluid B. Which fluid applies a larger buoyant force to the block?" That should be easy, right? The buoyant force is proportional to the volume displaced, so Fluid A exerts the larger force.

Jaime: I think they give the same buoyant force. In both cases, the fluid has to support the full weight of the block. So, the buoyant force equals mg in both cases.

Jenna: Is it the full mg, though? Or is it 60% of mg in the first case and 40% in the second?

Michael: What if we think about densities? Fluid B has a bigger density, and the buoyant force is proportional to the fluid density, so it should really be B that exerts the bigger force.

Jaime: That's actually why it works out to be equal. The buoyant force depends on both the volume displaced and the density, so B has less volume displaced but more density, and it balances out.

62. In Essential Question 9.10, we analyzed what would happen to the flow rate if the radius of a blood vessel decreased by 5%. For that same situation, if the heart adjusted to maintain the original flow rate, by what factor would the pressure difference between the ends of the blood vessel increase?

63. By what factor would the cross-sectional area of a blood vessel have to change (all other factors being unchanged) for the flow rate to be reduced by 50%?

64. If the radius of a blood vessel drops to 80% of its original radius because of the buildup of plaque, and the body responds by increasing the pressure difference across the blood vessel by 10%, what will have happened to the flow rate?

65. A patient in the hospital is receiving a saline solution through a needle in their arm. The needle is 3.0 cm long, horizontal, with one end in a vein in the arm, and the other end attached to a wide tube that extends down from the bag of solution, which hangs from a pole so that the fluid level is 90 cm above the needle. The inner radius of the needle is 0.20 mm. The top of the fluid is exposed to the atmosphere, and the flow rate of the fluid (which has a density of 1025 kg/m^3 and a viscosity of 0.0010 Pa s) through the needle is 0.30 L/h. What is the average gauge pressure inside the vein where the needle is?

66. At 20°C, honey has a viscosity about 4000 times larger than that of water. For a particular tube, 400 mL/s of water will flow through for a particular pressure difference. (a) If the same tube, with the same pressure difference, is used for honey, what will the flow rate be? (b) If you want the flow rate for honey to be the same as that for water, with the same pressure difference, by what factor should you increase the radius of the tube?

67. In Example 9.11, we discussed an ultracentrifuge with an angular speed of 5000 rpm, producing a centripetal acceleration of 5000 g. (a) What is the angular speed, in rad/s? (b) What is the average radius of the circle through which the sample spins, to produce 5000 g?

68. A particular ultracentrifuge operates at two different angular speeds, one twice the other, and has two different places where sample tubes can be loaded into the device, one twice as far from the center as the other. If the smallest centripetal acceleration you can obtain with this centrifuge is 2000 g, what are the other values you can obtain for the centripetal acceleration?

69. Section 9.11, which discussed the terminal speed of an object falling through a viscous fluid, included the following statement: "In general, the smaller the object, the smaller the magnitude of the terminal velocity." In Example 9.11, we also derived an equation for the terminal speed, which was:

$$v_t = \left(1 - \frac{\rho_{fluid}}{\rho_{object}}\right)\frac{mg}{6\pi\eta r}$$

Note the factor of r in the denominator on the right side of the equation. You have two balls, made from identical material, so they have the same density, but one has twice the radius of the other. The smaller ball has a terminal speed of 4.0 mm/s in a particular viscous fluid. Predict the terminal speed of the larger ball in this same fluid. Is your prediction consistent with the statement that smaller objects have smaller terminal speeds? Explain why or why not.

70. Find a clear plastic shampoo bottle, almost full of shampoo. Make sure the bottle is tightly capped! Shake the bottle to get some air bubbles, with a good mixture of bubble sizes. When you invert the bottle, what do you observe? Which bubbles rise fastest? Is this consistent with what we learned about objects moving through a viscous fluid, or not?

71. You drop a steel ball bearing, with a radius of 2.0 mm, into a beaker of honey. Note that honey has a viscosity of 4.0 Pa s and a density of 1360 kg/m^3, and steel has a density of 7800 kg/m^3. (a) What is the terminal speed of the ball bearing? (b) Aluminum has a density of 2700 kg/m^3. What radius should an aluminum ball have to have the same terminal speed in honey that the steel ball has?

San Francisco - California - Powell & Mason Cable Car Turnaround - 14/08/1978

This photograph from 1978 shows the operator of a San Francisco cable car turning the car around at the end of the line. In this chapter, we will investigate how this man can push the very heavy car so that it rotates through 180°. Photo credit: public-domain image from Wikimedia Commons.

Chapter 10 – Rotation I:
Rotational Kinematics and Torque

CHAPTER CONTENTS

Our goal for Chapter 10 is to examine objects that either are rotating, or would rotate if we removed a force. To accomplish this, we will take many of the concepts we covered in the earlier chapters and apply them in this new context. Even though the situations we will consider look different from those we considered earlier, the methods and equations we apply will be closely based on those we have used previously.

10-1 Rotational Kinematics

Kinematics is the study of how things move. Rotational kinematics is the study of how rotating objects move. Let's start by looking at various points on a rotating disk, such as a compact disc in a CD player.

EXPLORATION 10.1 - A rotating disk

Step 1 – *Mark a few points on a rotating disk and look at their instantaneous velocities as the disk rotates.* Let's assume the disk rotates counterclockwise at a constant rate. Even though the rotation rate is constant, we observe that each point on the disk has a different velocity. The instantaneous velocities of five different points are shown in Figure 10.1. Points at the same radius have equal speeds, but their velocities are different because the directions of the velocities are different. We also observe that the speed of a point is proportional to its distance from the center of the disk.

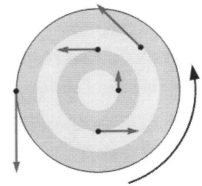

Figure 10.1: Instantaneous velocities, shown as red arrows, of various points on a disk rotating counterclockwise.

Step 2 – *Plot the paths followed by various points on the disk as the disk spins through 1/8th of a full rotation.* Figure 10.2 shows that each point travels on a circular arc, and the distance traveled by a particular point increases as the distance of that point from the center increases.

Step 3 – *What is the same for all the arcs shown in Figure 10.2?* One thing that is the same is the angle (45°, in this case) the points move through, measured from the center of the disk. This leads to an interesting conclusion. Maybe the parameters we used (position, velocity, and acceleration) to study projectile and one-dimensional motion are not the most natural parameters to use when describing rotational motion. For instance, in a given time interval every point on the rotating disk has a unique displacement, yet each point has the same angular displacement.

Step 4 – *Is there a connection between the distance traveled by a particular point in a given time interval (let's call this the arc length, s) and the corresponding angle, θ, of the arc the point moves along?* Absolutely. The connection between arc length and angle is:

$$s = r\theta \qquad \text{(Equation 10.1: \textbf{Arc length})}$$

where r is the radius of the arc (the distance from the point to the center). Note that the angle must be in units of radians in this equation, and that π radians $= 180°$.

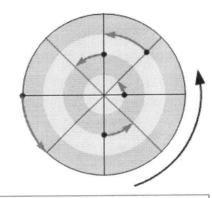

Figure 10.2: The circular arcs followed by five different points on the disk, as the disk moves through 1/8th of a full rotation.

There is an equivalent relationship for speed. Each point on the disk has a unique velocity, but each point moves through the same angle in a given time interval. Thus, every point has the same angular velocity, a quantity we symbolize using the Greek letter omega, $\vec{\omega}$. Angular velocity is related to angular displacement $\Delta\vec{\theta}$ in the same way that velocity is related to displacement. $\vec{v} = \Delta\vec{r} / \Delta t$, so for the angular variables we have:

$$\vec{\omega} = \frac{\Delta\vec{\theta}}{\Delta t}. \qquad \text{(Equation 10.2: \textbf{Angular velocity})}$$

Step 5 – *What is the connection between the instantaneous velocity of a point on the rotating disk and the disk's angular velocity?* The instantaneous velocity of a point is called the tangential velocity, $\vec{v}_T$, because the direction of the velocity is always tangential to the circular path followed by the point. At a given instant in time, every point on the disk has the same angular speed ω (this is the magnitude of the angular velocity, $\vec{\omega}$). As we noted in Step 1, however, the speed of a particular point is proportional to r, its distance from the center. The connection between the tangential speed v_T, and the angular speed ω is:

$$v_T = r\omega .$$ (Eq. 10.3: **Connecting tangential speed to angular speed**)

Step 6 – *How do we describe motion when the rotating object is not rotating at a constant rate?* If the rotating disk spins with constant angular velocity we can fully describe the motion of any point on the disk using the parameters described above (and time, t). If the angular velocity is changing, however, such as if the disk is speeding up or slowing down, we need an additional parameter to describe motion. This is the angular acceleration $\vec{\alpha}$, the rotational equivalent of the acceleration, defined as:

$$\vec{\alpha} = \frac{\Delta\vec{\omega}}{\Delta t} .$$ (Equation 10.4: **Angular acceleration**)

Key ideas for rotational motion: To describe rotational motion, we use the rotational variables $\theta, \omega,$ and α. These are more natural variables to use, instead of the more familiar r, v, and a, because every point on a rotating object has the same angular velocity $\vec{\omega}$ and angular acceleration $\vec{\alpha}$, while each point has unique values of position, velocity, and acceleration.
Related End-of-Chapter Exercises: 1, 43.

Figure 10.3: In this time exposure image of a rotating Ferris wheel, note how the tracks left by the parts farther away from the center are longer than those made by parts closer to the center. Comparing this picture to the diagram of the rotating disk in Figure 10.2, where we see the length of the tracks increasing as the distance from the center increases, we can understand why parts farther from the center of the Ferris wheel leave longer tracks on the photograph.
Photo credit: Gisele Wright / iStockPhoto.

Essential Question 10.1: If you travel a distance of 1.0 m as you walk around a circle that has a radius of 1.0 m, through what angle have you walked? Comment on the units here.

Answer to Essential Question 10.1: Let's re-arrange equation 10.1 to $\theta = s/r$. Thus, an arc length that is equal to the radius corresponds to an angle of 1.0 radian, which is about 57°. If the arc length and the radius have units of meters, the units cancel on the right side of the equation and we have units of radians on the left side. This violates the general rule that units have to match on two sides of an equation. We have two ways around this. One way is to treat the radian as dimensionless. Another way is to define the radius as having units of meters/radian.

10-2 Connecting Rotational Motion to Linear Motion

The angular variables we defined in Section 10-1 are vectors, so they have a direction. In which direction is the angular velocity of the disk shown in Figure 10.2? If we all observe the disk from the same perspective we can say that the direction is counterclockwise. In practice, we will generally use clockwise or counterclockwise to specify direction. In actuality, however, the direction is given by the **right-hand rule**. When you curl the fingers on your right hand in the direction of motion and stick out your thumb, your thumb points in the direction of the angular velocity. This is straight up out of the page for the disk in Figure 10.2.

EXPLORATION 10.2 – Connecting angular acceleration to acceleration

We can connect the magnitudes of the acceleration and angular acceleration in the same way that the distance traveled along an arc is connected to the angle ($s = r\theta$) and the speed is connected to the angular speed ($v = r\omega$). How?

Imagine yourself a distance r from the center of a rotating turntable, moving with the turntable. If the turntable has a constant angular velocity, you have no angular acceleration, but you have a centripetal acceleration, $\vec{a}_C = v^2/r$, directed toward the center of the turntable. The angular acceleration, α, cannot be connected to the centripetal acceleration by a factor of r, because $\alpha = 0$ in this case.

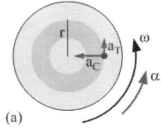
(a)

You have a non-zero angular acceleration if the turntable (and you) speeds up or slows down. If the turntable speeds up, the acceleration has two components (see Figure 10.4(a)), a centripetal acceleration $\vec{a}_C$ toward the center, and a component tangent to the circular path, which is called the tangential acceleration $\vec{a}_T$. If

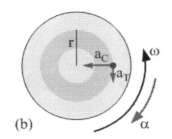
(b)

the turntable slows down, then the tangential acceleration reverses direction (see Figure 10.4(b)), as does the angular acceleration (because the angular velocity is decreasing instead of increasing). Thus, the magnitude of the tangential acceleration is directly related to the magnitude of the angular acceleration:

Figure 10.4: If you are rotating with a turntable as it speeds up (a) or slows down (b), your acceleration has two components, a centripetal component directed toward the center and a tangential component $\vec{a}_T$.

$a_T = r\alpha$ (Eq. 10.5: **Connecting tangential and angular accelerations**)

Key idea for angular acceleration: The angular acceleration $\vec{\alpha}$ is directly related to the tangential acceleration $\vec{a}_T$ (the component of acceleration tangent to the circular path), and is not related to the centripetal acceleration $\vec{a}_C$.   **Related End-of-Chapter Exercises: 44, 45.**

Equations for motion with constant angular acceleration

In Chapter 2, we considered one-dimensional motion with constant acceleration, and used three main equations to analyze motion. The analogous equations for rotational motion are summarized in Table 10.1. Note the parallels between the two sets of equations.

Straight-line motion equation		Analogous rotational motion equation	
$v = v_i + at$	(Equation 2.9)	$\omega = \omega_i + \alpha t$	(Equation 10.6)
$x = x_i + v_i t + \dfrac{1}{2}at^2$	(Equation 2.11)	$\theta = \theta_i + \omega_i t + \dfrac{1}{2}\alpha t^2$	(Equation 10.7)
$v^2 = v_i^2 + 2a\,\Delta x$	(Equation 2.12)	$\omega^2 = \omega_i^2 + 2\alpha\,\Delta\theta$	(Equation 10.8)

Table 10.1: Each kinematics equation has an analogous rotational-motion equation.

EXAMPLE 10.2 – Drawing a motion diagram for rotational motion

A turntable starts from rest, and has a counterclockwise angular acceleration of $(\pi/3)\,\text{rad}/\text{s}^2$. Sketch a motion diagram for an object 1.0 m from the center that rotates with the turntable, plotting its position at 0.50 s intervals for the first 3.0 s.

SOLUTION

Let's use equation 10.7 to find the object's angular position at 0.50-second intervals. The object starts at the position shown by the red circle in Figure 10.5 – the horizontal line will be the origin. Take counterclockwise to be positive, and then set up a table (see Table 10.2) summarizing what we know. This is similar to what we did for one-dimensional motion.

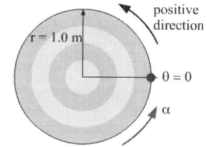

Parameter	Value
Positive direction	Counterclockwise
Initial position	$\theta_i = 0$
Initial angular velocity	$\omega_i = 0$
Angular acceleration	$\alpha = +(\pi/3)\,\text{rad}/\text{s}^2$

Table 10.2: Summarizing the initial information about the object.

Figure 10.5: The initial situation for the rotating object.

Using the values from Table 10.2, Equation 10.7 simplifies to: $\theta = \theta_i + \omega_i t + \dfrac{1}{2}\alpha t^2 = 0 + 0 + \dfrac{1}{2}\left(\dfrac{\pi}{3}\,\text{rad}/\text{s}^2\right)t^2 = +\left(\dfrac{\pi}{6}\,\text{rad}/\text{s}^2\right)t^2$.

Substituting different times into this equation gives the angular position of the object at the times of interest, as summarized in Table 10.3.

Time (s)	0	0.50	1.00	1.50	2.00	2.50	3.00
Angular position (radians)	0	$+\pi/24$	$+\pi/6$	$+3\pi/8$	$+2\pi/3$	$+25\pi/24$	$+3\pi/2$
Angular position (°)	0	+7.5	+30	+67.5	+120	+187.5	+270

Table 10.3: The angular position of the object at 0.50-second intervals.

Using the information in Table 10.3, we can sketch a motion diagram for the object. The motion diagram is shown in Figure 10.6.

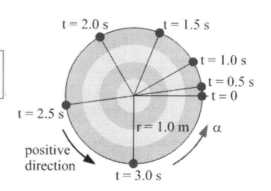

Figure 10.6: A motion diagram for an object moving with an accelerating turntable, showing the position at 0.5-second intervals.

Essential Question 10.2: If we repeated Example 10.2, for an object at a radius of 0.5 m from the center of the turntable, what would change in Table 10.3? Assume the object has an angular position of zero at $t = 0$.

Answer to Essential Question 10.2: Because Table 10.3 deals with angular variables, which are independent of the radius, nothing would change in the table.

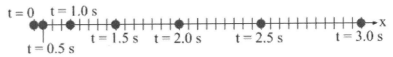

10-3 Solving Rotational Kinematics Problems

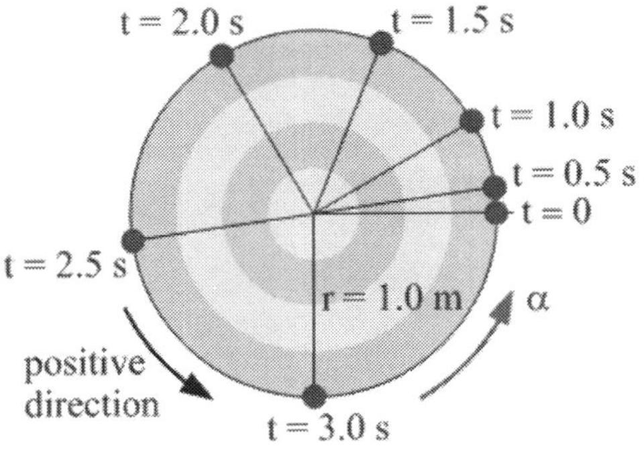

Figure 10.6: A motion diagram for an object moving with an accelerating turntable, showing the position at 0.5-second intervals.

EXPLORATION 10.3A – Unrolling the motion

Return to the situation from Example 10.2. Let's take the object's path, which is a circular arc (see Figure 10.6), and unroll it so it is a straight line. How would we analyze this straight-line motion? Unrolling the circular arc from Figure 10.6 gives the straight-line motion situation shown in Figure 10.7. Let's define the line the object moves along to be the *x*-axis. The origin is the object's initial position, and the positive direction is to the right. This situation should look familiar, because it is an excellent example of one-dimensional motion with constant acceleration, as we studied in Chapter 2.

For the rotational situation, the distance traveled is the length of the circular arc the object moves along. After unrolling the arc to get a straight line, we can use Equation 10.1, $s = r\theta$, to find the arc length corresponding to the distance traveled from the origin. When using this equation, use angles in radians. Table 10.4 builds on Table 10.3, bringing in a row for the arc length *s*, which is the same as the position, *x*, for the equivalent one-dimensional motion. Because we have the special case $r = 1.0$ m, *s* and θ are numerically equal, and differ only in their units.

Figure 10.7: The straight-line motion resulting from straightening the circular arc traveled by the object in Figure 10.6.

Time (s)	0	0.50	1.00	1.50	2.00	2.50	3.00
Angular position, θ (rad)	0	$+\pi/24$	$+\pi/6$	$+3\pi/8$	$+2\pi/3$	$+25\pi/24$	$+3\pi/2$
***s* or *x* (m)**	0	$+\pi/24$	$+\pi/6$	$+3\pi/8$	$+2\pi/3$	$+25\pi/24$	$+3\pi/2$

Table 10.4: Determining the arc length, and the displacement in the corresponding 1-dimensional motion situation, for the object on the turntable.

Key idea: Rotational motion with constant angular acceleration is analogous to one-dimensional motion with constant acceleration. **Related End-of-Chapter Exercises: 2, 39.**

Based on Example 10.2 and Exploration 10.3A, let's write down a general method for solving rotational kinematics problems. The method parallels the method used in Chapter 2 for solving one-dimensional kinematics problems.

A General Method for Solving a Rotational Kinematics Problem

1. Draw a diagram of the situation.
2. Choose an origin to measure positions from, and mark it on the diagram.
3. Choose a positive direction, and mark this on the diagram with an arrow.
4. Organize what you know, and what you're looking for. Making a date table is a useful way to organize the information.
5. Think about which of the constant-acceleration equations to apply, and then set up and solve the problem. The three main equations are:

$$\omega = \omega_i + \alpha\, t \,.$$ (Equation 10.6)

$$\theta = \theta_i + \omega_i\, t + \frac{1}{2}\alpha\, t^2 \,.$$ (Equation 10.7)

$$\omega^2 = \omega_i^2 + 2\alpha\, \Delta\theta \,.$$ (Equation 10.8)

EXPLORATION 10.3B – Graphs for rotational motion

Plot a set of graphs showing, as a function of time, the angular acceleration, the angular velocity, and the angular position of the object on the turntable we considered in Exploration 10.3A. How does this set of graphs compare to graphs showing, as a function of time, the acceleration, velocity, and position of the equivalent straight-line motion situation that we considered in Exploration 10.3A?

The angular acceleration is constant, with a value of $+(\pi/3)\,\mathrm{rad/s^2}$.

The graph of the angular acceleration is the horizontal line shown at the top of Figure 10.8.

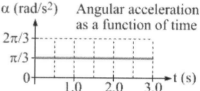

To graph angular velocity as a function of time, we can use Equation 10.6, $\omega = \omega_i + \alpha\, t$. Substituting values for the initial angular velocity and the angular acceleration gives: $\omega = 0 + (\pi/3)\,\mathrm{rad/s^2} \times t = (\pi/3)\,\mathrm{rad/s^2} \times t$.

This function is a straight line, starting from the origin, with a constant slope, as shown in the middle graph of Figure 10.8.

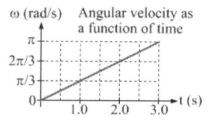

To graph the angular position as a function of time we can use Equation 10.7, as in Exploration 10.2, to get

$$\theta = \theta_i + \omega_i\, t + \frac{1}{2}\alpha\, t^2 = 0 + 0 + \left(\frac{\pi}{6}\,\mathrm{rad/s^2}\right)t^2 = +\left(\frac{\pi}{6}\,\mathrm{rad/s^2}\right)t^2 \,.$$

Recall that values of the angular position as a function of time are given in Table 10.3, and repeated in 10.4, so those points can be plotted on a graph and a smooth curve drawn through them. The result is the quadratic graph shown at the bottom of Figure 10.8.

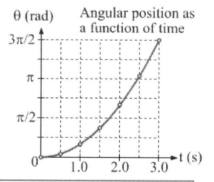

Note that, because $r = 1.0$ m, we can actually use these same graphs to represent the acceleration, velocity, and position of the equivalent straight-line motion that we considered in the previous Exploration. We would need to change the units and the labels on the three y-axes, but the graphs would otherwise look identical.

Figure 10.8: Graphs of the angular acceleration, angular velocity, and angular position for the object rotating with the turntable, all as a function of time.

Key idea: Plotting graphs of the angular acceleration, angular velocity, and angular position confirm the idea that rotational motion with constant angular acceleration is analogous to straight-line motion with constant acceleration, because the graphs in these two different situations have the same form. **Related End-of-Chapter Exercises: 40, 41.**

Essential Question 10.3: In Exploration 10.3B, we considered how to transform graphs for rotational motion into graphs for straight-line motion, but we did this with the special case of $r = 1.0$ m. What additional changes would be necessary if the radius r had a different value? Say, for example, that $r = 3.0$ m.

Answer to Essential Question 10.3: Because the straight-line motion variables are related to the equivalent rotational variables by a factor of r (e.g., $v = r\omega$), changing the value of r requires changing the graphs for the straight-line motion by a factor equal to the numerical value of r. One way to do this is to draw the lines on each graph exactly as before, but re-scale each y-axis. In the case of $r = 3.0$ m, for instance, each number on each y-axis would be multiplied by a factor of 3.0. A second approach is to keep the scales on the axes the same as before but move the graphs. For instance, the graph of velocity vs. time, which is given by the equation $v = +(\pi / 3)$ m/s$^2 \times t$ when $r = 1.0$ m, would be given by $v' = +(\pi$ m/s$^2) t$ when $r = 3.0$ m.

10-4 Torque

If an object is at rest, how can we get it to rotate? If an object is already rotating, how can we change its rotational motion? We answered equivalent questions about straight-line motion by saying "Apply a net force!" Let's now consider the rotational equivalent of force.

EXPLORATION 10.4 – Turning a revolving door

From an overhead view, a revolving door looks like a + sign mounted on a vertical axle. The door can spin freely, clockwise or counterclockwise, about its center.

Step 1 – Consider the three cases illustrated in Figure 10.9, in which a force (the red arrow) is applied to a revolving door. In each case, determine the direction the door will start to rotate, assuming it starts from rest.

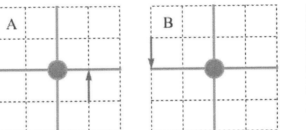 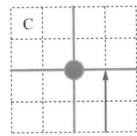

Figure 10.9: Three cases of forces applied to a revolving door, shown from an overhead perspective.

Although the direction of the force in case B is opposite to that in cases A and C, in each case the door will rotate counterclockwise. If you are ever confused about the direction an object will tend to rotate, place your pen or pencil on the diagram and hold it at the axis of the object, in this case at the center. Then push on the object in the direction, and at the location, of the applied force and see which way the object spins. Knowing the direction of a force applied to an object is not enough to determine the direction of rotation; we also need to know where the force is applied in relation to the axis of rotation.

Step 2 – Rank the three cases based on how quickly the revolving door spins, from largest to smallest, assuming the door is initially at rest. In case C, the door will rotate more quickly than in case A, because the applied force in C is twice as large as that in A while everything else (the point at which the force is applied, and the direction of the force) is equal. The door in case B also rotates faster than that in A because, even though the force has the same magnitude, in case B the force is applied further from the axis of rotation. Applying a force farther from the axis of rotation generally has a larger effect on the rotation of an object, which you have probably experienced. If you have ever come to a door where it was not obvious which side was connected to the hinges, and given the door a push on the edge where the hinges were, you most likely came close to running straight into the door as it opened very slowly in response to your push. Applying the same force at the edge of the door furthest from the hinges, however, is far more effective at opening the door.

The comparison that is hardest to rank is that between B and C. In case C the applied force is twice as large as that in B, but the force in B is applied twice as far from the axis of rotation as that in C. Which effect is more important? It turns out that these effects are equally important, so cases B and C are equivalent. The overall ranking is B=C>A.

The point of this discussion is that the angular acceleration of the door is proportional to both the applied force and the distance of the applied force from the axis of rotation. Let's now consider whether the direction at which the force is applied makes any difference.

Step 3 – *Consider the three cases shown in Figure 10.10. Rank these three cases based on the revolving door's angular acceleration, from largest to smallest.*

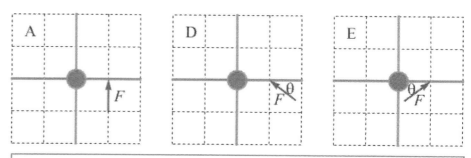

Let's split the forces in cases D and E into components, as shown in Figure 10.11. How do the components of the force influence the door in each case? If you've ever tried to open a door by exerting a force parallel to the door itself, you'll know that this is completely ineffective. Similarly, the parallel components in cases D and E do absolutely nothing to affect the door's rotation. Only the perpendicular components, which have a magnitude of $F \sin\theta$, affect the rotation. Because these components are smaller than F, the magnitude of the perpendicular force in case A, ranking the three cases gives A>D=E.

Figure 10.10: Three cases involving the same magnitude force applied at the same point on a revolving door, but applied in different directions.

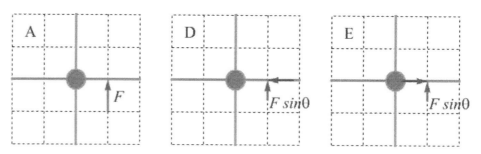

Figure 10.11: Splitting the force in case D, and case E, into components parallel to the door and perpendicular to the door.

Key ideas: The angular acceleration of a door depends on three factors: the magnitude of the applied force; the distance from the axis of rotation to where the force is applied; and the direction of the applied force. **Related End-of-Chapter Exercises: 48, 49.**

In Exploration 10.4, we learned about the rotational equivalent of force, which is torque.

The name for the rotational equivalent of force is **torque**, which we symbolize with the Greek letter tau (τ). Whereas a force is a push or a pull, a torque is a twist. A torque can result from applying a force. The torque resulting from applying a force F at a distance r from an axis of rotation is:

$$\tau = rF\sin\theta .$$ (Equation 10.9: **Magnitude of the torque**)

The angle θ represents the angle between the line of the force and the line the distance r is measured along.

Essential Question 10.4: Make a list of common household items or tools that exploit principles of torque.

Answer to Essential Question 10.4: Quite a number of tools and gadgets exploit torque, in the sense that they enable you to apply a small force at a relatively large distance from an axis, and the tool converts that into a large force acting at a relatively small distance from an axis. Examples include scissors, bottle openers, can openers, nutcrackers, screwdrivers, crowbars, wrenches, wheelbarrows, and bicycles.

10-5 Three Equivalent Methods of Finding Torque

EXPLORATION 10.5 – Three ways to find torque
A rod of length L is attached to a wall by a hinge. The rod is held in a horizontal position by a string that is tied to the wall and attached to the end of the rod, as shown in Figure 10.12.

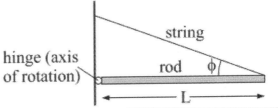

Step 1 – *In what direction is the torque applied by the string to the rod, about an axis that passes through the hinge and is perpendicular to the page?* As we did in previous chapters, it's a good idea to draw a free-body diagram of the rod (or at least part of a free-body diagram, as in Figure 10.13) to help visualize what is happening. For now the only force we'll include on the free-body diagram is the force of tension applied by the string (we'll go on to look at all the forces applied to the rod in Exploration 10.8). Try placing your pen over the picture of the rod. Hold the pen where the hinge is and push on the pen, at the point where the string is tied to the rod, in the direction of the force of tension. You should see the pen rotate counterclockwise. Thus, we can say that the torque applied by the string, about the axis through the hinge, is in a counterclockwise direction.

Figure 10.12: A rod attached to a wall at one end by a hinge, and held horizontal by a string.

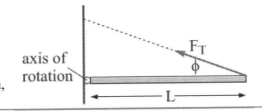

Figure 10.13: A partial free-body diagram for the rod, showing the force of tension applied to the rod by the string.

Note that we are dealing with direction for torque much as we did for angular velocity. The true direction of the torque can be found by curling your fingers on your right hand counterclockwise and placing your hand, little finger down, on the page. When you stick out your thumb it points up, out of the page. This is the true direction of the torque, but for simplicity we can state directions as either clockwise or, as in this case, counterclockwise.

Now we know the direction of the torque, relative to an axis through the hinge, applied by the string, let's focus on determining its magnitude.

Step 2 – *Measuring the distance r in Equation 10.9 along the bar, apply Equation 10.9 to find the magnitude of the torque applied by the string on the rod, with respect to the axis passing through the hinge perpendicular to the page.*
Finding the magnitude of the torque means identifying the three variables, r, F, and θ, in Equation 10.9. In this case we can see from Figure 10.13 that the distance r is the length of the rod, L; the force $\vec{F}$ is the force of tension, $\vec{F}_T$; and the angle θ is the angle between the line of the force (i.e., the string) and the line the distance r is measured along (the rod), so θ is the angle ϕ in Figure 10.13. In this case, then, applying Equation 10.9 tells us that the magnitude of the torque is $\tau = L F_T \sin \phi$.

Step 3 – *Now, determine the torque, about the axis through the hinge that is perpendicular to the page, by first splitting the force of tension into components, and then applying Equation 10.9.* Which set of axes should we use when splitting the force into components? The most sensible coordinate system is one aligned parallel to the rod and perpendicular to the rod, giving the two components shown in Figure 10.14. Because the force component that is parallel to the rod is directed at the hinge, where the axis goes through, that component gives a torque of zero (it's like trying to open a door by pushing on the door with a force directed at the line passing through the hinges). Another way to prove that the force is zero is to apply Equation 10.9 with an angle of 180°, which means multiplying by a factor of sin(180°), which is zero.

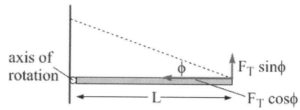

Figure 10.14: Splitting the force of tension into a component parallel to the rod, and a component perpendicular to the rod.

The torque from the force of tension is associated entirely with the perpendicular component of the force of tension. Now, identifying the three pieces of Equation 10.9 gives a force magnitude of $F = F_T \sin\phi$; a distance measured along the rod of $r = L$, and an angle of $\theta = 90°$ between the line of the perpendicular force component and the line we measured r along. Because $\sin(90°) = 1$, applying Equation 10.9 tells us that the magnitude of the torque from the tension, with respect to our axis through the hinge, is $\tau = L(F_T \sin\phi)\sin(90°) = L F_T \sin\phi$. This agrees with our calculation in Step 2.

Step 4 – *Instead of measuring r along the rod, draw a line from the hinge that meets the string (the line of the force of tension) at a 90° angle. Apply Equation 10.9 to find the magnitude of the torque applied by the string on the rod, with respect to the axis passing through the hinge, by measuring r along this line.*

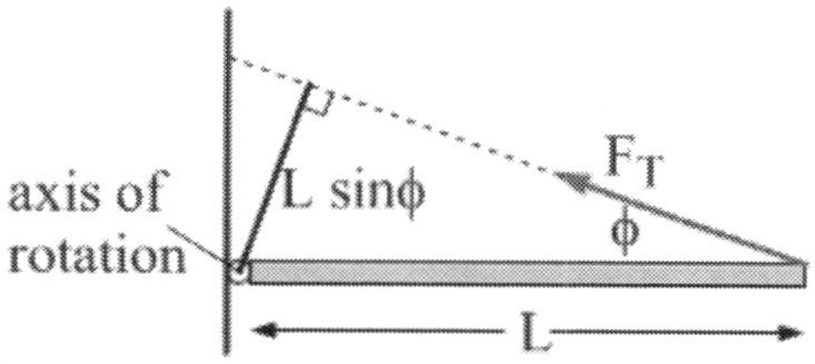

As we can see from Figure 10.15, the r in this case is not L, the length of the rod, but is instead $L\sin\phi$. This result comes from applying the geometry of right-angled triangles. The magnitude of the force, F, is F_T, the magnitude of the full force of tension, and the angle

Figure 10.15: A diagram showing the lever arm, in which the distance used to find torque is measured from the axis along a line perpendicular to the line of the force.

between the line we measure r along and the line of the force is 90°. This is known as the **lever-arm method** of calculating torque, where the lever-arm is the perpendicular distance from the axis of rotation to the force. Applying Equation 10.9 gives the magnitude of the torque as $\tau = (L\sin\phi)F_T \sin(90°) = L F_T \sin\phi$, agreeing with the other two methods discussed above.

Key idea for torque: We can find torque in three equivalent ways. It can be found using the whole force and the most obvious distance; after splitting the force into components; or by using the lever-arm method in which the distance from the axis is measured along the line perpendicular to the force. Use whichever method is most convenient in a particular situation.
Related End-of-Chapter Exercises: 7, 23, 50.

Essential Question 10.5: Torque can be calculated with respect to any axis. In Exploration 10.5, what is the torque, due to the force of tension, with respect to an axis passing through the point where the string is tied to the wall? In each case, assume the axis is perpendicular to the page.

The torque, from the tension, is zero with respect to any axis that passes through the string, because the line of the force (the string, in this case) passes through an axis that lies on the string. It is important to remember that the torque (both its direction and magnitude) associated with a force depends on the particular axis of rotation the torque is being measured with respect to.

10-6 Rotational Inertia

In Chapter 3, we found that an object's acceleration is proportional to the net force acting on the object:

$$\vec{a} = \frac{\sum \vec{F}}{m} .$$ (Equation 3.1: **Connecting acceleration to net force**)

A similar relationship connects the angular acceleration of an object to the net torque acting on it:

$$\vec{\alpha} = \frac{\sum \vec{\tau}}{I} .$$ (Eq. 10.10: **Connecting angular acceleration to net torque**)

Thus, the angular acceleration of an object is proportional to the net torque acting on the object. The I in the denominator of Equation 10.10 is known as the rotational inertia, which is the rotational equivalent of mass.

We have already looked at how the angular acceleration $\vec{\alpha}$ is the rotational equivalent of the acceleration $\vec{a}$, and how torque, $\vec{\tau}$, is the rotational equivalent of force, $\vec{F}$. The I in the denominator of Equation 10.10 must therefore be the rotational equivalent of the mass, m. I is known as the **rotational inertia**, or the **moment of inertia**. In the same way that mass is a measure of an object's tendency to maintain its state of straight-line motion, an object's rotational inertia is a measure of the object's tendency to maintain its rotational motion. Something with a large mass is hard to get moving, and it is also hard to stop if it is already moving. Similarly, if an object has a large rotational inertia it is difficult to start it rotating, and difficult to stop if it is already rotating.

One question to consider is, are rotational inertia and mass the same thing? In other words, does an object's mass, by itself, determine the rotational inertia? Let's check the units of rotational inertia. Re-arranging Equation 10.10, we find that rotational inertia has units of torque units (N m) divided by angular acceleration units (rad/s^2). Remembering that the newton is equivalent to kg m/s^2, and that we can treat the radian as being dimensionless, we find that rotational inertia has units of kg m^2. Rotational inertia depends on more than just mass, it depends on both mass and, somehow, length squared. Let's investigate this further.

EXPLORATION 10.6 – Rotational inertia
Consider a ball of mass M mounted at the end of a stick that has a negligible mass, and a length L (which is large compared to the ball's radius). The other end of the stick is pinned so the stick can rotate freely about the pin.

Step 1 – *If the ball and stick are held horizontal and then released from rest, what is the ball's initial acceleration?* The ball's initial acceleration is $\vec{g}$, the acceleration due to gravity. The force of the stick acting on the ball only becomes non-zero after the ball starts moving. We should also draw a diagram to help analyze the situation. The diagram is shown in Figure 10.16.

Step 2 – *What is the ball's initial angular acceleration?*
The angular acceleration can be found from the equation $a = r\alpha$. Here $r = L$, the length of the stick, so we have $\vec{\alpha} = g/L$, directed clockwise.

Figure 10.16: The initial position of the ball and stick. The system can rotate about an axis passing through the left end of the stick.

Step 3 – *What is the torque acting on the ball at the instant it is released?*
Here we can draw a free-body diagram of the ball, shown in Figure 10.17. Initially the only force acting on the ball is the force of gravity, Mg directed down. Considering an axis perpendicular to the page and passing through the pin, the torque is $\vec{\tau} = LMg$, directed clockwise.

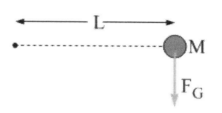

Step 4 – *Using Equation 10.10, and the results from steps 2 and 3, determine the rotational inertia of the ball relative to the axis passing through the pin.*

Figure 10.17: The free-body diagram of the ball immediately after the system is released from rest.

Re-arranging Equation 10.10 to solve for the rotational inertia gives:
$$I = \frac{\Sigma\vec{\tau}}{\vec{\alpha}}.$$

The torque and the angular acceleration are both clockwise, allowing us to divide the magnitude of the torque by the magnitude of the angular acceleration to determine the ball's rotational inertia about an axis through the pin.
$$I = \frac{LMg}{g/L} = ML^2.$$

Thus the rotational inertia of an object of mass M in which all the mass is at a particular distance L from the axis of rotation is $I = ML^2$.

Key ideas for rotational inertia: An object's rotational inertia is determined by three factors: the object's mass; how the object's mass is distributed; and the axis the object is rotating around.
Related End-of-Chapter Exercises: 10, 27.

Essential Question 10.6: Consider the three cases shown in Figure 10.18. In each case, a ball of a particular mass is placed on a light rod of a particular length. Each rod can rotate without friction about an axis through the left end. Rank the cases based on their rotational inertias, from largest to smallest.

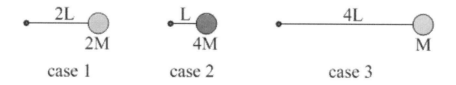

Figure 10.18: Three cases, each involving a ball on the end of a rod that can rotate about its left end.

Answer to Essential Question 10.6: The correct ranking is 3>1>2. In the rotational inertia equation, the distance from the axis to the ball (the length of the rod) is squared, while the mass is not. Thus, changing the length by a factor of 2 changes the rotational inertia by a factor of 4, whereas changing the mass by a factor of 2 changes the rotational inertia by only a factor of 2.

10-7 An Example Problem Involving Rotational Inertia

Our measure of inertia for rotational motion is somewhat more complicated than inertia for straight-line motion, which is just mass. Consider the following example.

EXAMPLE 10.7 – Spinning the system.
Three balls are connected by light rods. The mass and location of each ball are:
Ball 1 has a mass M and is located at $x = 0$, $y = 0$.
Ball 2 has a mass of $2M$ and is located at $x = +3.0$ m, $y = +3.0$ m.
Ball 3 has a mass of $3M$ and is located at $x = +2.0$ m, $y = -2.0$ m.
Assume the radius of each ball is much smaller than 1 meter.

(a) Find the location of the system's center-of-mass.
(b) Find the system's rotational inertia about an axis perpendicular to the page that passes through the system's center-of-mass.
(c) Find the system's rotational inertia about an axis parallel to, and 2.0 m from, the axis through the center-of-mass.

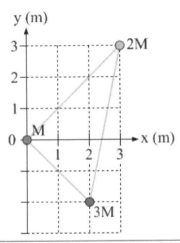

SOLUTION
Let's begin, as usual, by drawing a diagram of the situation. The diagram is shown in Figure 10.19.

(a) To find the location of the system's center-of-mass, let's apply Equation 6.3. To find the x-coordinate of the system's center-of-mass:

Figure 10.19: A diagram showing the location of the balls in the system described in Example 10.7.

$$X_{CM} = \frac{x_1 m_1 + x_2 m_2 + x_3 m_3}{m_1 + m_2 + m_3} = \frac{(0)M + (+3.0\,\text{m})(2M) + (+2.0\,\text{m})(3M)}{M + 2M + 3M} = \frac{(+12.0\,\text{m})M}{6M} = +2.0\,\text{m}$$

The y-coordinate of the system's center-of-mass is given by:

$$Y_{CM} = \frac{y_1 m_1 + y_2 m_2 + y_3 m_3}{m_1 + m_2 + m_3} = \frac{(0)M + (+3.0\,\text{m})(2M) + (-2.0\,\text{m})(3M)}{M + 2M + 3M} = \frac{(0)M}{6M} = 0.$$

(b) To find the system's rotational inertia about an axis through the center-of-mass we can find the rotational inertia for each ball separately, using $I = M\,L^2$, and then simply add them to find the total rotational inertia. Figure 10.20 is helpful for seeing where the different L values come from.

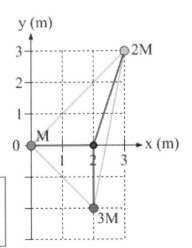

Figure 10.20: The center-of-mass of the system is marked at (+2 m, 0). The axis of rotation passes through that point. The dark lines show how far each ball is from the axis of rotation.

For ball 1, $L^2 = (2.0\,\text{m})^2 = 4.0\,\text{m}^2$ so $I_1 = M\,L^2 = (4.0\,\text{m}^2)M$.

For ball 2, $L^2 = 10\,\text{m}^2$ so $I_2 = 2M\,L^2 = (20\,\text{m}^2)M$.

For ball 3, $L^2 = 4.0\,\text{m}^2$ so $I_3 = 3M\,L^2 = (12\,\text{m}^2)M$.

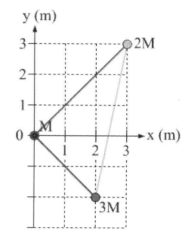

The total rotational inertia is the sum of these three values, $(36\,\text{m}^2)M$.

(c) To find the rotational inertia through an axis parallel to the first axis and 2.0 m away from it, let's choose a point for this second axis to pass through. A convenient point is the origin, $x = 0$, $y = 0$. Figure 10.21 shows where the L values come from in this case.

Repeating the process we followed in part (b) gives:

For ball 1, $L^2 = 0$ so $I_1' = 0$.

For ball 2, $L^2 = 18\,\text{m}^2$ so $I_2' = 2M\,L^2 = (36\,\text{m}^2)M$.

For ball 3, $L^2 = 8.0\,\text{m}^2$ so $I_3' = 3M\,L^2 = (24\,\text{m}^2)M$.

Figure 10.21: The axis of rotation now passes through the ball of mass M at the origin. The red lines show how far the other two balls are from the axis of rotation.

The total rotational inertia is the sum of these three values, $(60\,\text{m}^2)M$.

Related End-of-Chapter Exercises: 29, 31.

Does it matter which point the second axis passes through? What if we had used a different point, such as $x = +2.0$ m, $y = -2.0$ m, or any other point 2.0 m from the center-of-mass? Amazingly, it turns out that it doesn't matter. Any axis parallel to the axis through the center-of-mass and 2.0 m from it gives a rotational inertia of $(60\,\text{m}^2)M$. It turns out that the rotational inertia of a system is minimized when the axis goes through the center-of-mass, and the rotational inertia of the system about any parallel axis a distance h from the axis through the center-of-mass can be found from

$$I = I_{CM} + mh^2,$$ (Equation 10.11: **The parallel-axis theorem**)

where m is the total mass of the system.

Let's check the parallel-axis theorem using our results from (b) and (c). In part (b) we found that the rotational inertia about the axis through the center-of-mass is $I_{CM} = (36\,\text{m}^2)M$. The mass of the system is $m = 6M$ and the second axis is $h = 2.0$ m from the axis through the center-of-mass. This gives $I = (36\,\text{m}^2)M + 6M(2.0\,\text{m})^2 = (60\,\text{m}^2)M$, as we found above.

Essential Question 10.7: To find the total mass of a system of objects, we simply add up the masses of the individual objects. To find the total rotational inertia of a system of objects, can we follow a similar process, adding up the rotational inertias of the individual objects.

Answer to Essential Question 10.7: Yes, the rotational inertia of a system of objects can be found be adding up the rotational inertias of the various objects making up the system. This is precisely the process we followed in Example 10.7.

10-8 A Table of Rotational Inertias

Consider now what happens if we take an object that has its mass distributed over a length, area, or volume, rather then being concentrated in one place. Generally, the rotational inertia in such a case is calculated by breaking up an object into tiny pieces, finding the rotational inertia of each piece, and adding up the individual rotational inertias to determine the total rotational inertia.

Figure 10.22: A uniform rod of length L and mass M, divided into 10 equal pieces. The axis of rotation passes through the left end of the rod and is perpendicular to the page.

We can get a feel for the process by considering how we would find the rotational inertia of a uniform rod of length L and mass M, rotating about an axis through the end of the rod that is perpendicular to the rod itself. If all the mass were concentrated at the far end of the rod, a distance L from the axis, then the rotational inertia would be ML^2. Because most of the mass is closer than L to the axis of rotation, the rod's rotational inertia turns out to be less than ML^2. If we broke up the rod into ten equal pieces, with centers at 5%, 15%, 25%, 35%,…,95% of the length of the rod (see Figure 10.22), we would calculate a rotational inertia of 0.3325 ML^2. This is very close to the value we would get by doing the integration, $I_{rod,end} = ML^2/3$. The rotational inertia's of various shapes, and for various axes of rotation, are shown in Figure 10.23.

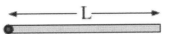

Rod rotating about an axis through one end, perpendicular to the rod.

$$I = \frac{1}{3}ML^2$$

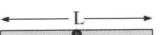

Rod rotating about an axis through the middle, perpendicular to the rod.

$$I = \frac{1}{12}ML^2$$

Solid disk or cylinder about an axis through the middle, perpendicular to the plane of the disk.

$$I = \frac{1}{2}MR^2$$

Solid sphere about an axis through the center.

$$I = \frac{2}{5}MR^2$$

Thin ring about an axis through the middle, perpendicular to the plane of the ring.

$$I = MR^2$$

Hollow sphere about an axis through the center.

$$I = \frac{2}{3}MR^2$$

Essential Question 10.8: In Figure 10.23, all the values for rotational inertia are of the form

$I = cMR^2$, or $I = cML^2$, where c is generally less than 1. The exception is the rotational inertia of a ring rotating about an axis through the center of the ring and perpendicular to the plane of the ring, where $c = 1$. Why do we expect to get $I = MR^2$ for the ring rotating about that central axis?

Figure 10.23: Expressions for the rotational inertia of various objects about a particular axis. In each case, the object has a mass M.

Answer to Essential Question 10.8: The expression for the rotational inertia of the ring has no factor less than 1 in front of the MR^2 because every bit of mass in the ring is a distance R from the center of the ring. In all the other cases shown in Figure 10.23, most of the mass of the given object is at a distance less than R (or less than L) from the axis in question.

10-9 Newton's Laws for Rotation

In Chapter 3 we considered Newton's three laws of motion. The first two of these laws have analogous statements for rotational motion.

> **Newton's First Law for Rotation**: an object at rest tends to remain at rest, and an object that is spinning tends to spin with a constant angular velocity, unless it is acted on by a nonzero net torque *or there is a change in the way the object's mass is distributed*.
>
> Recall that the net torque is the sum of all the forces acting on an object. Always remember to add torques as vectors. The net torque can be symbolized by $\sum \vec{\tau}$.

The first part of the statement of Newton's first law for rotation parallels Newton's first law for straight-line motion, but the phrase about how spinning motion can be affected by a change in mass distribution is something that only applies to rotation.

Newton's second law for rotation, on the other hand, is completely analogous to Newton's second law for straight-line motion, $\sum \vec{F} = m\vec{a}$. Replacing force by torque, mass by rotational inertia, and acceleration by angular acceleration, we get:

$$\sum \vec{\tau} = I\vec{\alpha}. \qquad \text{(Equation 10.12: Newton's Second Law for Rotation)}$$

We'll spend the rest of this chapter, and a good part of the next chapter, looking at how to apply Newton's second law in various situations. In Chapter 11, we will deal with rotational dynamics, involving motion and acceleration. For the remainder of this chapter, however, we will focus on situations involving static equilibrium.

Conditions for static equilibrium

An object is in static equilibrium when it remains at rest. Two conditions apply to objects in static equilibrium. These are:

$$\sum \vec{F} = 0 \qquad \text{and} \qquad \sum \vec{\tau} = 0.$$

Expressed in words, an object in static equilibrium experiences no net force and no net torque. Using these conditions, we will be able to analyze a variety of situations. Many excellent examples of static equilibrium involve the human body, such as when you hold your arm out; when you bend over; and when you stand on your toes. In each case, forces associated with muscles, bones, and tendons maintain the equilibrium situation.

Essential Question 10.9: Newton's first law for rotation includes a phrase that says spinning motion can be affected by a change in the way an object's mass is distributed. Can you think of a real-life example of this?

Answer to Essential Question 10.9: A familiar example is a figure skater who spins relatively slowly with her arms held out from her body, but then pulls her arms in and spins much faster.

10-10 Static Equilibrium

Let's first apply Newton's second law for rotation in a **static equilibrium** situation, in which an object remains at rest. Conditions for static equilibrium are given in section 10-9.

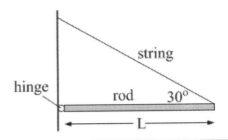

EXPLORATION 10.10 – A hinged rod.

Return to the hinged rod we looked at in Exploration 10.5. The rod's mass of 2.0 kg is uniformly distributed along its length L. The rod is attached to a wall by a hinge at one end. As shown in Figure 10.24, the angle between the rod and the string, which holds the rod in a horizontal position, is 30°. Use $g = 10$ N/kg to simplify the calculations.

Figure 10.24: A diagram of the rod, connected to the wall by a hinge and held horizontal by a string tied to the end of the rod.

Step 1 – *Sketch a free-body diagram of the rod.* The free-body diagram is in Figure 10.25. Start by drawing the force of tension applied to the rod by the string, which goes away from the rod along the string. Where should we draw the force of gravity? Until now, all we had to do was to show the direction of a force correctly on a free-body diagram. Now that we're dealing with torques, it is also critical to locate the force accurately. The force of gravity should be drawn at the **center-of-gravity** of the rod, which is at the rod's geometrical center because the rod is uniform. For now, we can assume that the center-of-mass and the center-of-gravity are the same point. We'll distinguish between the two later in the chapter.

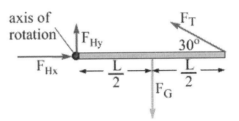

Figure 10.25: The free-body diagram of the rod.

What other forces act on the rod, in addition to gravity and tension? First, for the rod to remain in equilibrium, there must be a force directed right, to balance the component of the force of tension directed left. Second, because the hinge is in contact with the rod, the hinge very likely exerts a force on the rod. Generally, we draw this hinge force already split into components. The horizontal component of the hinge force, $\vec{F}_{Hx}$, is directed right to balance the horizontal

component of the force of tension. The vertical component of the hinge force, $\vec{F}_{Hy}$, is shown

directed up. This vertical component could, however, be directed down in some cases, or even be equal to zero. If you're not sure which direction a force is in, simply choose a direction. If the analysis gives a negative sign for a force, the force is opposite to the direction shown.

Step 2 – *Apply Newton's second law twice, once for the horizontal direction and once for the vertical direction, to come up with two force equations for this situation.* Figure 10.26 shows the x-y coordinate system, with positive x to the right and positive y up. The force of tension has been split into components parallel to the coordinate axes.

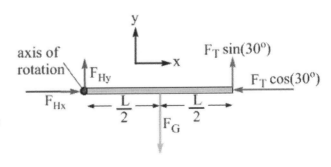

Figure 10.26: A free-body diagram showing the x-y coordinate system, and with all forces split into components.

Applying Newton's second law in the x-direction means:

1. Writing out Newton's second law: $\sum \vec{F}_x = m\vec{a}_x$.

2. Recognizing that the right-hand side equals zero, because the rod stays at rest.

3. Looking at the free-body diagram to evaluate the left-hand side of the equation:
   $+F_{Hx} - F_T \cos(30°) = 0$.

Using a similar process for the y-direction, we start with $\sum \vec{F}_y = m\vec{a}_y$, and end up with:

$+F_{Hy} - mg + F_T \sin(30°) = 0$

What can we solve for with these two force equations? We can't solve for anything! There are simply too many unknowns. If all we knew about were forces, we would be stuck.

Step 3 – *Choose an appropriate axis to take torques about. Then, apply Newton's second law for rotation to write a torque equation to solve for the tension.* What is an appropriate axis to use? Any axis can be used, but choosing an axis carefully can make a problem significantly easier to solve. ***The key is to choose an axis that one or more of the unknown forces pass through, because forces passing through an axis do not give any torque about that axis.*** In this case, we're trying to solve for the tension in the string, so we should pick an axis that eliminates the other unknown forces (the hinge forces), if possible. The most appropriate axis here is the axis perpendicular to the page that passes through the hinge. An axis through the hinge eliminates the two hinge forces, and the horizontal component of the force of tension, from the torque equation.

As with forces, we are free to choose a positive direction for torque. Let's use clockwise in this particular situation (although counterclockwise would be just as good). Applying Newton's second law for rotation means:

1. Writing out the equation: $\sum \vec{\tau} = I\vec{\alpha}$.

2. Recognizing that the right-hand side equals zero, because the rod stays at rest.

3. Looking at the free-body diagram to evaluate the left-hand side of the equation, and applying $\tau = rF\sin\theta$ to find the magnitude of the torque from each force. Recognizing that the torque due to the force of gravity is clockwise, while the torque due to the tension is counterclockwise, we get: $-L[F_T \sin(30°)]\sin(90°) + \dfrac{L}{2}(mg)\sin(90°) = 0$.

This equation demonstrates the power of using torque, because we can immediately solve for the tension. Note that the length of the rod, which is unknown, cancels out in the equation.

This gives: $F_T \sin(30°) = \dfrac{mg}{2}$.

Because $\sin(30°) = 0.5$, we find that $F_T = mg = (2.0\text{ kg})(10\text{ N}/\text{kg}) = 20\text{ N}$. Note that the fact that the force of tension has the same magnitude as the force of gravity in this case is highly coincidental, and happened only because the factor of two difference in the distances of these forces from the axis of rotation was exactly balanced by the factor of ½ we got from $\sin(30°)$.

Key ideas: By analyzing situations in terms of torque as well as force, we can solve problems that cannot be solved using force concepts alone. One of the keys to using torque is to choose an appropriate axis to take torques around. This is generally an axis that one or more of the unknown forces passes through. **Related End-of-Chapter Exercises: 8, 51 – 53.**

Essential Question 10.10: Return to Exploration 10.10, and solve for the x and y components of the force exerted on the rod by the hinge.

Answer to Essential Question 10.10: We solved for the force of tension in Exploration 10.10, so we can go on solve for the components of the hinge force. To find the *y*-component of the hinge force we could set up another torque equation, relative to an axis through the middle of the rod, for instance, or we could now make use of the force equation we worked out in Step 2. To find the *x*-component of the hinge force we can only use the force equation because, no matter which axis we choose, the torque from the *x*-component of the hinge force always exactly balances the torque from the *x*-component of the force of tension.

Making use of our *x*-component force equation, $+F_{Hx} - F_T \cos(30°) = 0$, we find that:

$$F_{Hx} = F_T \cos(30°) = (20 \text{ N}) \frac{\sqrt{3}}{2} = 10\sqrt{3} \text{ N} .$$

Using the *y*-component force equation, $+F_{Hy} - mg + F_T \sin(30°) = 0$, we get:

$$+F_{Hy} = mg - F_T \sin(30°) = 20 \text{ N} - 20 \text{ N} \left(\frac{1}{2} \right) = 10 \text{ N} .$$

Note that, if we combine the two components of the hinge force, we find that the hinge force is 20 N, with an angle of 30° between the hinge force and the rod. In other words, the hinge force is a mirror image of the force of tension, because of the symmetry of the situation (both forces are applied at the ends of the rod, while the force of gravity is applied at the exact center).

10-11 A General Method for Solving Static Equilibrium Problems

Now that we have explored the idea of applying the concept of torque to solve a static equilibrium problem, let's list the basic steps in the process.

A General Method for Solving a Static Equilibrium Problem

Objects in static equilibrium remain at rest, so both the acceleration and the angular acceleration are zero. This allows us to use special-case of Newton's second law and Newton's second law for rotation.

1. Draw a diagram of the situation.
2. Draw a free-body diagram showing all the forces acting on the object.
3. Choose a rotational coordinate system. Pick an appropriate axis to take torques about, and then apply Newton's second law for rotation ($\sum \vec{\tau} = 0$) to obtain one or more torque equations.
4. If necessary, choose an appropriate *x-y* coordinate system for forces. Apply Newton's second law ($\sum \vec{F} = 0$) to obtain one or more force equations.
5. Combine the resulting equations to solve the problem.

Let's apply the method in the following example.

EXAMPLE 10.11 – Supporting the board

A uniform board with a weight of 240 N and a length of 2.0 m rests horizontally on two supports. Support A is under the left end of the board, while Support B is 50 cm from the right end (150 cm from the left end, in other words).

(a) Which support exerts more force on the board? Without doing the calculations to find the two support forces, come up with a conceptual argument to justify your answer.

(b) Find the two support forces.

SOLUTION

As usual, let's begin by drawing a diagram of the situation. We should also sketch a free-body diagram to show all the forces acting on the board. The diagram and free-body diagram are shown in Figure 10.27.

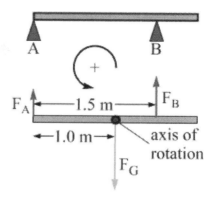

(a) Support B exerts a larger force on the board than support A. One way to see this is to sum torques about an axis through the center of the board, and perpendicular to the page. Taking counterclockwise to be the positive direction for torque, applying Newton's second law for rotation gives:

$$-(1.0\,\text{m})F_A + (0.5\,\text{m})F_B = 0.$$

Figure 10.27: A diagram and free-body diagram for the board on two supports.

The force of gravity does not appear in this torque equation because the force of gravity passes through the axis, and thus does not give rise to a torque about that axis. Because the torques from the two support forces must balance one another, and the distance from support B to the axis is half that of the distance from support A to the axis, the force exerted on the board by support B must be twice as large as that exerted by support A.

(b) To solve for the support forces, we could combine the torque equation above with the force equation we get by applying Newton's second law, or we could set up another torque equation by taking an axis perpendicular to the page through one of the supports. Let's do the latter, using Figure 10.28 to help us set up the new torque equation, summing torques about an axis through support A.

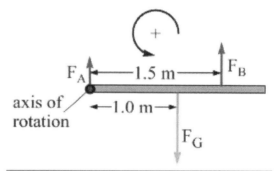

Applying Newton's second law for rotation, $\Sigma\vec{\tau} = I\vec{\alpha} = 0$, taking counterclockwise to be positive, gives:

$$-(1.0\,\text{m})mg + (1.5\,\text{m})F_B = 0.$$

Thus, $F_B = \dfrac{2}{3}mg = \dfrac{2}{3}(240\,\text{N}) = +160\,\text{N}.$

Figure 10.28: If we take torques about an axis through support A, the force applied to the board from support A does not give rise to a torque, because that force passes through the axis.

There are several ways to solve for the force applied by support A. Let's apply Newton's second law, $\Sigma\vec{F} = m\vec{a} = 0$, taking up to be the positive direction:

$$+F_A - mg + F_B = 0.$$

This gives: $+F_A = mg - F_B = mg - \dfrac{2}{3}mg = \dfrac{1}{3}mg = \dfrac{1}{3}(240\,\text{N}) = +80\,\text{N}.$

The fact that both support forces work out to be positive means they are in the direction shown in the diagrams, up.

Related End-of-Chapter Exercises: 33, 36.

Essential Question 10.11: In Example 10.11, would the support forces change if support B was moved a short distance to the right of its original position? If so, how would the forces change?

Answer to Essential Question 10.11: Moving a supporting force farther from the center-of-gravity generally reduces that support force. However, the two supports together support the weight of the board. Thus, moving support B to the right decreases the magnitude of force B, but support A's upward force increases to compensate. The simplest example is when support B is at the far right of the board. In that case, symmetry tells us that the supports share the load equally, with each support exerting an upward force of 120 N on the board.

10-12 Further Investigations of Static Equilibrium

Center-of-gravity

Previously, we discussed the importance of locating forces precisely on a free-body diagram. For instance, the force of gravity must be attached to the center-of-gravity of the system. If the acceleration due to gravity has the same direction at all points in a system, we can define the x-coordinate of the system's center-of-gravity as:

$$X_{CG} = \frac{x_1 m_1 g_1 + x_2 m_2 g_2 + x_3 m_3 g_3 + \dots}{m_1 g_1 + m_2 g_2 + m_3 g_3 + \dots}$$ (Eq. 10.13: *X-coordinate of the center-of-gravity*)

A similar equation gives the y-coordinate of the center-of-gravity. If the acceleration due to gravity is the same everywhere, g cancels out of the equation, giving the center-of-mass equation we used in Chapter 6. The center-of-gravity differs from the center-of-mass, therefore, only when the acceleration due to gravity is different for different parts of the object or system.

EXAMPLE 10.12 – Tipping the board

Let's continue from where we left off in Example 10.11, involving the 240 N board on two supports. Now you climb on the board and, starting at the left end of the board, you slowly walk along the board toward the right end. Your weight is 480 N.

(a) Defining up to be the positive direction, plot two graphs, on the same set of axes, of the support forces as a function of your distance *d* from the left end of the board. Use this graph to determine the value of *d* when the board tips over.

(b) Where is the center-of-gravity, of the system consisting of you and the board, when the board begins to tip?

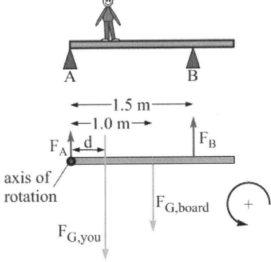

SOLUTION

(a) Again, we should draw a diagram to help us analyze the situation. Let's place you on the board at a distance *d* from the left end, and sketch a free-body diagram of the system. These diagrams are shown in Figure 10.29.

Let's define counterclockwise to be the positive direction for torques, and take torques about an axis perpendicular to the page that passes through the left end of the board. Choosing this axis eliminates, from the torque equation, the force exerted on the system by support A, and allows us to solve for the force exerted by support B. Let's use *M* to represent your mass and *m* to represent the mass of the board. Applying Newton's second law for rotation in this situation, $\sum \vec{\tau} = I\vec{\alpha} = 0$, gives:

Figure 10.29: A diagram and free-body diagram of the system consisting of you and the board. You are a distance *d* from the left end of the board.

$$-d\,(Mg)\sin(90°)-(1.0\,\text{m})(mg)\sin(90°)+(1.5\,\text{m})\,F_B\sin(90°)=0\,.$$

Solving for the force exerted by support B gives:

$$F_B=\frac{2}{3}mg+\frac{2d}{3.0\,\text{m}}Mg=160\,\text{N}+320\,d\,\text{N/m}\,.$$

We could follow a similar process to find an expression for the force exerted on the system by support A, taking torques about an axis through the board where support B is. In this case, however, it's probably easier to apply Newton's second law, $\Sigma\vec{F}=(M+m)\vec{a}=0$. Taking up to be positive gives: $+F_A-Mg-mg+F_B=0$.

Thus: $F_A=Mg+mg-F_B=480\,\text{N}+240\,\text{N}-160\,\text{N}-320d\,\text{N/m}=560\,\text{N}-320d\,\text{N/m}$.

Graphs of the two support forces, as a function of your position, are shown in Figure 10.30. Note that for values of $d>1.75$ m, the force from support A must be negative (directed down) to maintain the system's equilibrium. A's force could be negative if the board was bolted to the support, and the support either had a significant mass or it was fastened firmly to the ground. In this case, however, the board simply rests on the support, so the support can only provide an upward force.

Thus, when $d=1.75$ m, the board is on the verge of tipping, because the normal force between the board and support A goes to zero at that value of d. The board will tip if d exceeds 1.75 m. In this situation, then, Figure 10.30 shows the correct situation for $d \le 1.75$ m.

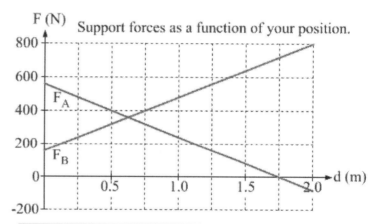

Figure 10.30: Graphs of the two support forces, as a function of your distance d from the left end of the board.

(b) What happens to the center-of-gravity of the system, which consists of you and the board, as you walk to the right? Because your weight is shifting right, the center-of-gravity of the system shifts right, also. The y-coordinate of the center-of-gravity has no bearing on whether the system tips, so let's simply determine the x-coordinate of the center-of-gravity when $d=1.75$ m:

$$X_{CG}=\frac{x_{board}\,mg+x_{you}\,Mg}{mg+Mg}=\frac{(1.0\,\text{m})(240\,\text{N})+(1.75\,\text{m})(480\,\text{N})}{240\,\text{N}+480\,\text{N}}=1.5\,\text{m}\,.$$

It is no coincidence that the position of the center-of-gravity corresponds to the location of support B. If the center-of-gravity of a system is between its supports, the system is stable. If the center-of-gravity moves out from the region bounded by the supports, the system tips over.

Related End-of-Chapter Exercises: 12, 34.

Essential Question 10.12: Return to the expressions we found for the support forces in part (a) of the Example 10.12. Add the two expressions. What is the significance of this result?

Answer to Essential Question 10.12: Adding the two expressions for the support forces gives:
$$F_A + F_B = 560 \text{ N} - 320d \text{ N/m} + 160 \text{ N} + 320d \text{ N/m} = 720 \text{ N} .$$

In other words, when the system is in equilibrium the sum of the support forces is always 720 N. This is expected because the supports combine to balance the weight of the system. Your weight of 480 N and the board's weight of 240 N add to 720 N.

Chapter Summary

Essential Idea for Rotational Motion
The methods we applied previously to solve straight-line motion problems, such as using constant-acceleration equations and Newton's laws of motion, can essentially be adapted to help us analyze situations involving rotational motion.

Rotational Kinematics
To help us understand how things move we defined the straight-line motion variables of position, displacement, velocity, and acceleration. The analogous rotational variables help us understand rotational motion.

Straight-line motion variable	Analogous rotational motion variable	Connection
Displacement, $\Delta \vec{s}$	Angular displacement, $\Delta \vec{\theta}$	$\Delta s = r \, \Delta \theta$
Velocity, $\vec{v} = \dfrac{\Delta \vec{s}}{\Delta t}$	Angular velocity, $\vec{\omega} = \dfrac{\Delta \vec{\theta}}{\Delta t}$	$v_T = r \omega$
Acceleration, $\vec{a} = \dfrac{\Delta \vec{v}}{\Delta t}$	Angular acceleration, $\vec{\alpha} = \dfrac{\Delta \vec{\omega}}{\Delta t}$	$a_T = r \alpha$

Table 10.5: Connecting straight-line motion variables to rotational variables. To prevent confusion with r, the radius, the variable $\vec{s}$ is used to represent position. The T subscripts denote tangential, for components that are tangential to the circular path.

In the special case of one-dimensional motion with constant acceleration, we derived a set of useful equations. An analogous set applies to rotation with constant angular acceleration.

Straight-line motion equation		Analogous rotational motion equation	
$v = v_i + at$	(Equation 2.9)	$\omega = \omega_i + \alpha t$	(Equation 10.6)
$x = x_i + v_i t + \dfrac{1}{2} a t^2$	(Equation 2.11)	$\theta = \theta_i + \omega_i t + \dfrac{1}{2} \alpha t^2$	(Equation 10.7)
$v^2 = v_i^2 + 2a \, \Delta x$	(Equation 2.12)	$\omega^2 = \omega_i^2 + 2\alpha \, \Delta \theta$	(Equation 10.8)

Table 10.1: Comparing the one-dimensional kinematics equations from chapter 2 to the rotational motion equations that can be applied to rotating objects.

Static Equilibrium

An object is in static equilibrium when it remains at rest. Two conditions apply to objects in static equilibrium. These are:

$$\sum \vec{F} = 0 \qquad \text{and} \qquad \sum \vec{\tau} = 0 .$$

Expressed in words, an object in static equilibrium experiences no net force and no net torque.

A General Method for Solving a Static Equilibrium Problem
1. Draw a diagram of the situation.
2. Draw a free-body diagram showing all the forces acting on the object.
3. Choose a rotational coordinate system. Pick an appropriate axis to take torques about, and then apply Newton's Second Law for Rotation ($\sum \vec{\tau} = 0$) to obtain one or more torque equations.
4. If necessary, choose an appropriate *x-y* coordinate system for forces. Apply Newton's Second Law ($\sum \vec{F} = 0$) to obtain one or more force equations.
5. Combine the resulting equations to solve the problem.

Rotational Dynamics

Mass is our measure of inertia for straight-line motion, while rotational inertia depends on the mass, the way the mass is distributed, and the axis about which rotation occurs. Torque is the rotational equivalent of force. The concepts of mass, force, and acceleration are linked by Newton's second law; an analogous law links the concepts of rotational inertia, torque, and angular acceleration.

Straight-line motion concept	Analogous rotational motion concept	Connection
Inertia: mass, m	Rotational Inertia, $I = cMR^2$ (c depends on axis and object's shape)	$I = \sum m_i r_i^2$
Can change motion: Force, $\vec{F}$	Can change rotation: Torque, $\vec{\tau}$	$\tau = r\,F\sin\theta$
Newton's Second Law, $\sum \vec{F} = m\vec{a}$	Second Law for Rotation, $\sum \vec{\tau} = I\vec{\alpha}$	Same form

Table 10.6: Rotational dynamics is governed by concepts that are similar to those that govern dynamics in straight-line motion.

End-of-Chapter Exercises

Exercises 1 – 12 are conceptual questions that are designed to see if you have understood the main concepts of the chapter.

1. As shown in the overhead view in Figure 10.31, four cylindrical objects (two red and two blue) are spinning with a turntable that is moving counterclockwise at a constant rate. The two red cylinders are the same distance from the center, and the two blue ones are also equally distant from the center, but farther from the center than the two red cylinders. Which cylinders have the same (a) speed? (b) velocity? (c) angular velocity? (d) acceleration? (e) angular acceleration?

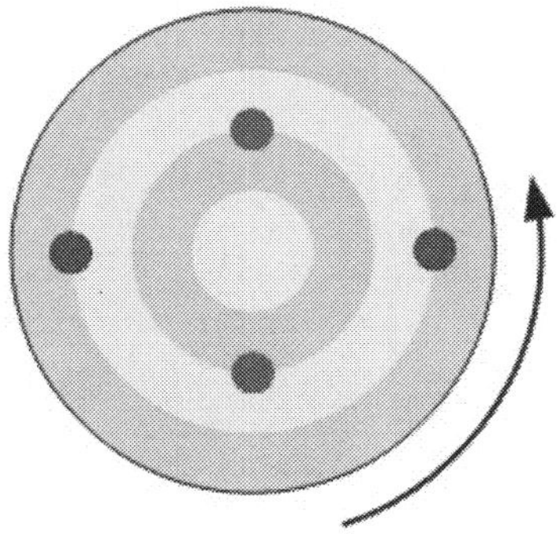

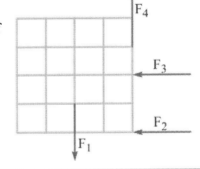

2. Return to the situation described in Exercise 2. (a) Draw a motion diagram for one of the red cylinders that corresponds to one complete rotation of the turntable. (b) Assuming the red cylinder is 2.0 m from the center of the turntable, construct a motion diagram that corresponds to the equivalent straight-line motion (as if you unrolled the motion diagram of the red cylinder you chose). Have we seen this kind of motion-diagram before? If so, what kind of motion did we classify it as?

Figure 10.31: An overhead view of a turntable that is spinning counterclockwise at a constant rate. Four cylinders are moving with the turntable, two red ones that are equally distant from the center, and two blue ones that are the same distance from the center as one another but farther out than the red ones. For Exercises 1 and 2.

3. A square sheet of plywood is subjected to four forces of equal magnitude, as shown in Figure 10.32. Relative to an axis that is perpendicular to the page and passes through the top left corner of the sheet, in which direction is the torque due to (a) $\vec{F}_1$; (b) $\vec{F}_2$;

   (c) $\vec{F}_3$; (d) $\vec{F}_4$?

4. Repeat Exercise 3, except this time use an axis that is perpendicular to the page and passes through the bottom left corner of the sheet.

5. A hockey puck is initially at rest on a frictionless ice rink. Two horizontal forces of equal magnitude are then simultaneously applied to the puck. For rotation, consider a vertical axis through the center of the puck. (a) Is it possible to apply the two forces so the puck has no acceleration and no angular acceleration? If so, sketch an example. (b) Is it possible to apply the forces so the puck's center-of-mass has no acceleration but the puck has a non-zero angular acceleration? If so, sketch an example. (c) Is it possible to apply the forces so the puck's center-of-mass has a non-zero acceleration but the puck has no angular acceleration? If so, sketch an example.

Figure 10.32: A square sheet of plywood subjected to four forces of equal magnitude, for Exercises 3 and 4.

6. Many common household tools (hand tools, as opposed to power tools) enable us to make use of torque to make it easier to do something. A can opener is a good example of such a device. (a) Briefly describe how torque is involved in the operation of a human-powered can opener. (b) Name two other tools or devices you would find in a typical house that involve torque in their operation and briefly describe them.

7. Figure 10.33 shows a side view of a uniform rod of length L and mass M that is pinned at its left end by a frictionless hinge. The rod is held horizontal by means of a force F that is applied at a distance $3L/4$ along the rod. The angle between the rod and this force is θ. Fill in the two blanks in the following statement using either "increase", "decrease", or "stay the same." As the angle θ decreases, the torque associated with the force F must _____ while the magnitude of the force F must _____ so that the rod remains in equilibrium.

Figure 10.33: A side view of a rod that is hinged at its left end, and which is held in a horizontal position by an applied force F.

8. As shown in Figure 10.34, a rod, with a length of 80 cm and a mass of 6.0 kg, is attached to a wall by means of a hinge at the left end. The rod's mass is uniformly distributed along its length. A string will hold the rod in a horizontal position; the string can be tied to one of three points, lettered A-C, spaced at 20 cm intervals along the rod, starting with point A which is 20 cm from the left end of the rod. The other end of the string can be tied to one of three hooks, numbered 1-3, in the ceiling 30 cm above the rod. Hook 1 is directly above point A, hook 2 is directly above B, etc. For each case below, draw a line (and only one line) from one lettered point to one numbered hook representing the string you would use to achieve the desired objective. If you think it is impossible to achieve the objective, explain why. (a) How would you attach a string so the rod is held in a horizontal position with the hinge exerting no force at all on the rod? (b) How would you attach a string so the rod is held in a horizontal position while the force exerted on the rod by the hinge has no horizontal component, but has a non-zero vertical component directed straight up? (c) How would you attach a string so the rod is held in a horizontal position while the force exerted on the rod by the hinge has no vertical component, but has a non-zero horizontal component?

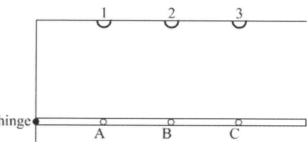

Figure 10.34: A hinged rod that you intend to hold horizontal by means of a string attached from one of the lettered points on the rod to one of the numbered hooks above the rod, for Exercise 8. This system represents a simple model of a broken arm you want to immobilize with a sling. The rod represents the lower arm, the hinge represents the elbow, and the string represents the sling.

9. You construct a mobile out of four objects, a sphere, a cube, a pyramid, and an ellipsoid. The mobile is in equilibrium in the configuration shown in Figure 10.35, where the vertical dashed lines are 20 cm apart. The mass of the strings (in blue) and rods (in red) can be neglected. If the pyramid has a mass of 400 g, what is the mass of each of the other objects?

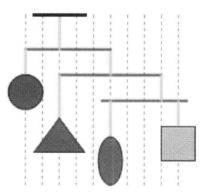

Figure 10.35: A mobile with four different objects, for Exercise 9.

10. Two cylinders have the same dimensions and mass, and the center-of-mass of each is at its geometric center. When you try to spin them, however, you notice that one cylinder is significantly more difficult to spin than the other. What is a good physical explanation for this? Assume you're trying to spin them about an axis through the center of the cylinder, perpendicular to the length of the cylinder, in each case.

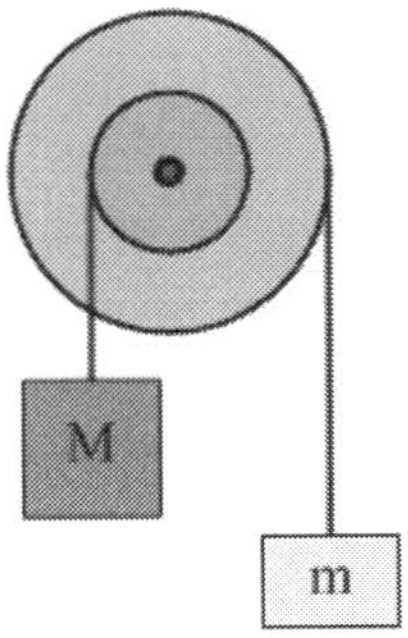

11. A pulley consists of a small uniform disk of radius $R/2$ mounted on a larger uniform disk of radius R. The pulley can rotate without friction about an axis through its center. As shown in Figure 10.36, a block with mass m hangs down from the larger disk while a block of mass M hangs down from the smaller disk. If the system remains in equilibrium, what is M in terms of m?

Figure 10.36: A dual-radius pulley system remains in equilibrium with two blocks hanging from it. For Exercise 11.

12. A particular type of leaning tower toy, as shown in Figure 10.37, remains upright until an extra piece is added to its top, at which point the tower falls over. Explain why this is.

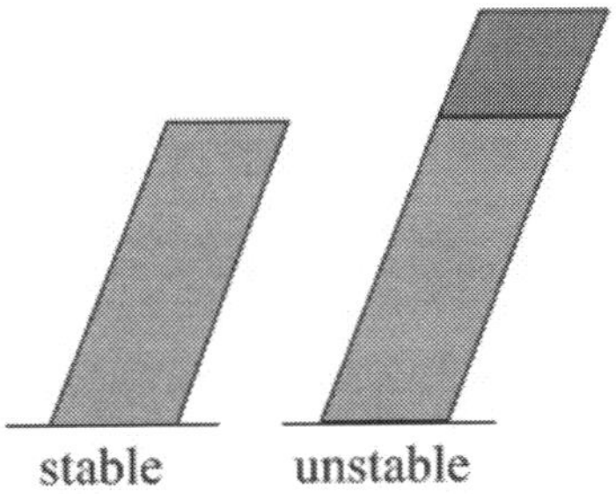

stable   unstable

Figure 10.37: This leaning-tower toy tips over when the extra piece is added at the top. For Exercise 12.

Exercises 13 – 21 are designed to give you practice in solving a typical rotational kinematics problem. For each exercise, start with the following parts: (a) Draw a diagram of the situation. (b) Choose an origin to measure displacements from and mark that on the diagram. (c) Choose a positive direction and indicate that with an arrow on the diagram. (d) Create a table summarizing everything you know, as well as the unknowns you want to solve for. Try to solve all exercises using a similar systematic approach. Compare your approach to those you used for Exercises 33 – 42 in Chapter 2.

13. While repairing your bicycle, you have your bicycle upside down so the front wheel is free to spin. You grab the front wheel by the edge and smoothly accelerate it from rest, giving it an angular acceleration of 5.0 rad/s² clockwise. You let go after the wheel has moved through one-quarter of a revolution. Your goals in this exercise are to determine the wheel's angular velocity at the instant you let go and the time it took to reach that angular velocity. Parts (a) – (d) as described above. (e) Which equation(s) will you use to determine the wheel's final angular velocity? (f) Find that angular velocity. (g) Which equation(s) will you use to determine the time the wheel was accelerating? (h) Solve for that time.

14. You release a ball from rest at the top of a ramp, and it experiences a constant angular acceleration of 1.2 rad/s². At the bottom of the ramp, the ball is rotating at 4.0 revolutions per second. The goal here is to determine how long it took the ball to reach that speed. Parts (a) – (d) as above. (e) Which equation(s) will you use to determine the time it takes to reach 4.0 rev/s? (f) What is that time?

15. You spin a disk, giving it an initial angular velocity of 2.4 rad/s clockwise. The disk has an angular acceleration of 1.2 rad/s² counterclockwise. Your goal in this exercise is to solve for the maximum angular displacement of the disk from its initial position before reversing direction. Parts (a) – (d) as described above. (e) Which equation(s) will you use to determine the maximum angular displacement of the disk? (f) Solve for that angular displacement.

16. In this exercise, you will analyze the method you used in Exercise 15, so all these questions pertain to what you did to solve Exercise 15. (a) Is there only one correct choice for the origin? Why did you make the choice you made? (b) Is there only one correct choice for the positive direction? Would your answer to (f) above change if you chose the opposite direction to be positive? (c) Find an alternative method to determine the maximum angular displacement, and show that it gives the same answer as the method you used.

17. A cylinder is rolling down a ramp. When it passes a particular point, you determine that it is traveling at an angular speed of 30 rad/s, and in the next 2.0 seconds it experiences an angular displacement of 80 radians. The goal of this exercise is to determine the cylinder's angular acceleration, which we will assume to be constant. Parts (a) – (d) as described above. (e) Which equation(s) will you use to determine the angular acceleration? (f) What is the angular acceleration?

18. Repeat parts (e) and (f) of the previous exercise, but do not use the equation(s) you used in the previous exercise.

19. A pulley with a radius of 0.20 m is mounted so its axis is horizontal. A block hangs down from a string wrapped around the pulley, as shown in Figure 10.38. You give the pulley an initial angular velocity of 0.50 rad/s directed counterclockwise. The block takes a total of 6.00 s to return to the level it was at when you released it. Assuming the acceleration is constant through the entire motion, the goal of the exercise is to determine the maximum distance the block rises above its initial point. Parts (a) – (d) as described above. (e) Which equation(s) will you use to determine the maximum distance the block rose above its initial point? (f) What is that maximum distance?

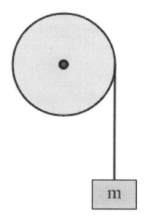

Figure 10.38: A block hanging down from a pulley, for Exercise 19.

20. An electric drill accelerates a drill bit, which has a radius of 3.0 mm, from rest to a maximum angular speed of 250 rpm (revolutions per minute) in 2.2 seconds. The goal of this exercise is to determine the drill bit's angular acceleration, assuming it to be constant. Parts (a) – (d) as described above. (e) What is the bit's angular acceleration?

21. With a quick flick of her wrist, an Ultimate Frisbee player can give a Frisbee an angular velocity of 8.0 revolutions per second. Assuming the player accelerates the Frisbee from rest through an angle of 75°, the goal of this exercise is to determine the Frisbee's angular acceleration and the time over which this acceleration occurs. Parts (a) – (d) as described above. (e) What is the Frisbee's angular acceleration? (f) What is the time over which the acceleration occurs?

Exercises 22 – 26 involve calculating torque in various situations.

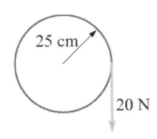

22. As shown in Figure 10.39, a disk mounted on an axle through its center is subjected to a 20 N force. The disk has a radius of 25 cm. What is the magnitude and direction of the torque associated with this force, measured with respect to an axis that is perpendicular to the page and (a) passes through the center of the disk? (b) passes through the point 50 cm above the center of the disk (i.e., move the axis up the page)? (c) is 50 cm to the right of the center of the disk?

Figure 10.39: A disk of radius 25 cm that is subjected to a 20 N force, for Exercise 22.

23. A box measuring 3.0 m high by 4.0 m wide is subjected to a 10 N force, as shown in Figure 10.40. Consider an axis that is perpendicular to the page and which passes through the bottom left corner of the box. (a) Follow the procedures outlined in Exploration 10.6 to first determine the direction of the torque due to this force. Now determine the magnitude of the torque due to this force by (b) applying Equation 10.9 directly; (c) breaking the force into horizontal and vertical components before applying Equation 10.9; (d) using the lever-arm method.

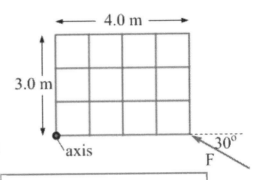

Figure 10.40: A box subjected to a 10 N force, for Exercise 23.

24. The plywood sheet shown in Figure 10.41 measures 2.0 m × 2.0 m, and each of the four forces the sheet is subjected to has a magnitude of 8.0 N. Relative to an axis that is perpendicular to the page and passes through the top left corner of the sheet, determine the magnitude of the torque due to (a) $\vec{F}_1$; (b) $\vec{F}_2$; (c) $\vec{F}_3$; (d) $\vec{F}_4$. (e) Find the magnitude and direction of the net torque, due to all four forces, about that axis.

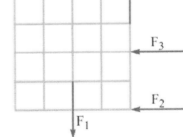

25. Repeat Exercise 24, except this time use an axis that is perpendicular to the page and passes through the bottom left corner of the sheet.

26. Consider again the plywood sheet shown in Figure 10.41. Is there an axis that is perpendicular to the page about which the four forces give a net torque of zero? If so, where would such an axis be located? If there is at least one such axis, is there only one, or are there more than one? Explain.

Figure 10.41: A square sheet of plywood subjected to four forces of equal magnitude, for Exercises 24 – 26.

Exercises 27 – 32 address issues associated with rotational inertia.

27. As shown in Figure 10.42, three identical balls, each with a mass $M = 1.0$ kg, are equally spaced along a rod of negligible mass. The distance between neighboring balls is 3.0 m, and you can assume that the radius of each ball is considerably less than the distance between them. If this system is spun about an axis that is perpendicular to the page, determine the system's total rotational inertia if the axis (a) passes through the center of the middle ball; (b) passes through the center of the ball at the left end.

Figure 10.42: A system of three balls, for Exercise 27.

28. A solid sphere has a mass of $M = 8.0$ kg and a radius of $R = 20$ cm. Determine the sphere's rotational inertia about an axis (a) passing through the center of the sphere; (b) tangent to the outer surface of the sphere.

29. Two balls of negligible radius are connected by a rod with a length of 1.2 m and a negligible mass. One ball has a mass M, while the other has a mass of $2M$. (a) If you spin this system about an axis that is perpendicular to the rod, where should you place the axis to minimize the system's rotational inertia? (b) If $M = 1.0$ kg, what is this minimum rotational inertia?

30. Repeat Exercise 29, except now the balls are joined by a 1.2-meter uniform rod with a mass of $3M$.

31. Four balls of equal mass M are placed so that there is one ball at each corner of a square measuring $d \times d$. The balls are joined by rods of negligible mass that run along the sides of the square. Assume the radius of each ball is small compared to d. What is the rotational inertia of the system about an axis that is perpendicular to the plane of the square and passes through (a) the center of the square? (b) one of the corners of the square? (c) a point a distance d from the center of the square, in any direction.

32. The rotational inertia of a uniform rod of length L and mass M, about an axis through the end of the rod and perpendicular to the rod, is $I = ML^2/3$. Use this expression, and the parallel-axis theorem (Equation 10.10), to show that the rotational inertia of the rod about a parallel axis through the center of the rod is $I = ML^2/12$.

Exercises 33 – 38 are designed to give you practice in solving a typical static equilibrium problem. For each of these exercises begin with the following: (a) Draw a diagram of the situation. (b) Draw a free-body diagram to show each of the forces acting on the object.

33. A board with a weight of 40 N and a length of 2.0 m is placed horizontally on a flat roof with 75 cm of the board hanging over the edge of the roof. The goal of this exercise is to determine the magnitude and direction of the normal force exerted on the board by the roof, and the exact location the normal force can be considered to be applied. Parts (a) and (b) as described above. (c) Apply Newton's second law to your free-body diagram, and solve for the magnitude and direction of the normal force exerted on the board by the roof. (d) Choose an axis to take torques about. Why did you select the axis you did? (e) Apply Newton's second law for rotation to determine the location the normal force can be considered to act on the board.

34. Return to the situation described in Exercise 33, except now we will add a 20 N bucket of nails to the end of the board that is hanging out over the edge of the roof. Repeat parts (a) – (e). (f) What is the maximum weight that could be placed on the end of the board without the board tipping over? (g) Where would the normal force act in that situation?

35. A particular door consists of a uniform piece of wood, with a weight of 20 N, measuring 2.0 m high by 1.0 m wide. The door is mounted on two hinges, one 20 cm down from the top of the door and the other 20 cm up from the bottom. The door also has an ornate handle, with a weight of 10 N, located halfway down the door and 10 cm from the edge of the door farthest from the hinges. The goal of this exercise is to determine the horizontal components of each hinge force. Parts (a) – (b) as described above. (c) Apply Newton's second law in the horizontal direction to obtain a relationship between the horizontal components of the hinge forces. (d) Now, choose an axis to take torques about so you can solve for the horizontal component of the force exerted on the door by the bottom hinge. Explain why you chose the axis you did. (e) Apply Newton's second law for rotation and solve for the horizontal component of the force exerted on the door by the bottom hinge. (f) Solve for the horizontal component of the force exerted on the door by the upper hinge.

36. Consider the following design for a one-person seesaw. The seesaw consists of a uniform board with a length of 5.0 m and a mass of 40 kg balanced on a support that is 2.0 m from one end. Julie, with a mass of 16 kg, sits on the board so that the system is in balance with the board horizontal. The goal is to determine how far from the support Julie is. Parts (a) – (b) as described above. (c) Which side of the board is Julie on, the side with 2.0 m or the side with 3.0 m of the board extending beyond the support? (d) Choose an axis to take torques about. Justify your choice of axis. (e) Choose a positive direction for rotation and apply Newton's second law for rotation to find Julie's position.

37. A ladder with a length of 5.0 m and a weight of 600 N is placed so its base is on the ground 4.0 m from a vertical frictionless wall, and its tip rests 3.0 m up the wall. The ladder remains in this position only because of the static friction force between the ladder and the ground. The goal of this exercise is to determine the magnitude of the normal force exerted by the wall on the ladder, and the minimum possible value of the coefficient of static friction for the ladder-ground interaction. Assume the mass of the ladder is uniformly distributed. Parts (a) – (b) as described above. (c) Apply Newton's second law twice, once for the horizontal forces and once for the vertical forces, to find relationships between the various forces applied to the ladder. (d) Choose an axis to take torques about so that you can solve for the normal force exerted by the wall on the ladder. Justify why you chose the axis you did. (e) Choose a positive direction for rotation and apply Newton's second law for rotation to find the normal force exerted by the wall on the ladder. (f) Solve for the minimum possible coefficient of static friction so the ladder remains in equilibrium.

38. Repeat Exercise 37, with the addition of you, with a weight of 500 N, standing on the ladder so that you are a horizontal distance of 1.0 m from the wall.

General Problems and Conceptual Questions

39. You drop a ball from rest from the top of a tall building. Let's assume the ball has an acceleration of 10 m/s² directed straight down. At the same time you drop the ball, you flick a switch that starts a motor, giving a disk that was initially at rest an angular acceleration of 10 rad/s² directed clockwise. Write a paragraph or two comparing and contrasting these two motions.

40. Return to the situation described in Exercise 39. (a) For the first four seconds of the motion, plot graphs of the ball's acceleration, velocity, and position as a function of time, taking down to be positive. (b) Over the same time period, plot graphs of the disk's angular acceleration, angular velocity, and angular position as a function of time, taking clockwise to be positive. (c) Comment on the similarities between the two sets of graphs.

41. Consider again the situation described in Exercise 39. The disk rotates about an axis through its center. It turns out that all points on the disk at a particular distance from the disk have a speed that matches the speed of the ball at all times (at least until the ball hits the ground!). What is this distance?

42. In the "old days", long before CD's and MP3's, people listened to music using vinyl records. Long-playing vinyl records spin at a constant rate of 33⅓ rpm. The music is encoded into a continuous spiral track on the record that starts at a radius of about 30 cm from the center and ends at a radius of about 10 cm from the center. If a record plays for 24 minutes, estimate how far apart the tracks are on the record.

43. Let's say that, during your last summer vacation, you drove your car across the United States from Boston to Seattle, staying on interstate I-90 the entire time. Estimate how many revolutions each tire made during this trip.

44. You are driving your car at a constant speed of 20 m/s around a highway exit ramp that is in the form of a circular arc of radius 100 m. What is the magnitude of your (a) angular velocity? (b) centripetal acceleration? (c) tangential acceleration? (d) angular acceleration?

45. Repeat Exercise 44, except now your speed is decreasing. At the instant we are interested in, your speed is 20 m/s, and you are planning to come to a complete stop at a red light in 5.0 s. Assume your acceleration is constant.

46. The rotational inertia of a uniform disk of mass M and radius R is found to be MR^2 about an axis perpendicular to the plane of the disk. How far is this axis from the center of the disk?

47. Archimedes once made a famous statement about a lever, which relies very much on the principle of torque, using words to the effect of *"Give me a lever long enough, and a fulcrum on which to place it, and I shall move the world."* Briefly explain what Archimedes was talking about.

48. Figure 10.43 shows four different cases involving a uniform rod of length L and mass M that is subjected to two forces of equal magnitude. The rod is free to rotate about an axis that either passes through one end of the rod, as in (a) and (b), or passes through the middle of the rod, as in (c) and (d). The axis is marked by the red and black circle, and is perpendicular to the page in each case. This is an overhead view, and we can neglect any effect of the force of gravity acting on the rod. Rank these four situations based on the magnitude of the net torque about the axis, from largest to smallest.

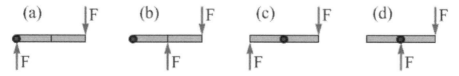

Figure 10.43: Four situations involving a uniform rod, which can rotate about an axis, being subjected to two forces of equal magnitude. For Exercises 48 and 49.

49. Return to the situation described in Exercise 48 and shown in Figure 10.43. If the rod has a length of 1.0 m, a mass of 3.0 kg, and each force has a magnitude of 5.0 N, determine the magnitude and direction of the net torque on the rod, relative to the axis in (a) Case (a); (b) Case (b); (c) Case (c); (d) Case (d).

50. Figure 10.44 shows a side view of a uniform rod of length L and mass M that is pinned at its left end by a frictionless hinge. The rod is held horizontal by means of a force F that is applied at a distance $3L/4$ along the rod.

Determine the angle θ between the rod and the force F, such that the magnitude of F is exactly equal to the magnitude of the force of gravity acting on the rod.

Figure 10.44: A side view of a rod that is hinged at its left end, and which is held in a horizontal position by an applied force F. For Exercise 50.

51. A rod, with a length of 80 cm and a mass of 6.0 kg, is attached to a wall by means of a hinge at the left end. The rod's mass is uniformly distributed along its length. A string will hold the rod in a horizontal position; the string can be tied to one of three points, lettered A-C, spaced at 20 cm intervals along the rod, starting with point A which is 20 cm from the left end of the rod. The other end of the string can be tied to one of three hooks, numbered 1-3, in the ceiling 30 cm above the rod. Hook 1 is directly above point A, hook 2 is directly above B, etc. **Use $g = 10$ N/kg.** As shown in Figure 10.44, a string is attached from point A to hook 3. Remember that point B is 40 cm from the hinge. (a) Calculate the tension in the string. (b) Determine the magnitude and direction of the horizontal component of the hinge force. (c) Determine the magnitude and direction of the vertical component of the hinge force.

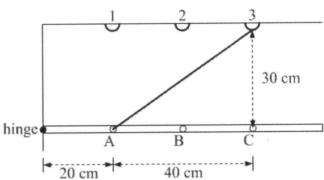

Figure 10.45: A diagram of a hinged rod, held horizontal by a string. Point B, in the middle of the uniform rod, is 40 cm from the hinge. For Exercises 51 – 53.

52. Repeat Exercise 51, with the string holding the rod horizontal attached from point A to hook 2 instead.

53. Repeat Exercise 51, with the string holding the rod horizontal attached from point C to hook 2 instead.

54. It is often useful to treat the lower arm as a uniform rod of length L and mass M that can rotate about the elbow. Let's say you are holding your arm so your upper arm is vertical (with your elbow below your shoulder), with a 90° bend at the elbow so the lower arm is horizontal. In this position we can say that three forces act on your lower arm: the force of gravity (Mg), the force exerted by the biceps, and the force exerted at the elbow joint by the humerus (the bone in the upper arm). Let's say the biceps muscle is attached to the lower arm at a distance of $L/10$ from the elbow, moving away from the elbow toward the hand. (a) Compare the force of gravity with the biceps force. Which has the larger magnitude? Briefly justify your answer. (b) Compare the force from the biceps with the force from the humerus. Which has the larger magnitude? Briefly justify your answer.

55. Return to the situation described in Exercise 54. If you now hold a 20 N object in your hand, at a distance L from the elbow joint, and your arm remains in the position described, the force from the biceps increases. (a) By how much does the force from the biceps increase? (b) Does the fact that you are holding a 20 N object in your hand change the force applied to your lower arm by the humerus at the elbow joint? If so, state both the magnitude and the direction of the change.

56. A uniform rod of mass M and length 1.0 m is attached to a wall by a hinge at one end. The rod is maintained in a horizontal position by a vertical string that can be attached to the rod at any point between 20 cm and 100 cm from the hinge. (a) Defining d to be the distance from the hinge to where the string is attached, plot a graph of the magnitude of the torque exerted on the rod by the string as a function of d, for $20\,\text{cm} \le d \le 100\,\text{cm}$. (b)

Over the same range of d values, plot a graph of the magnitude of the tension in the string as a function of d.

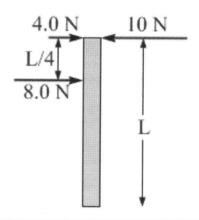

57. Figure 10.46 shows an overhead view of a piece of wood, with a mass of 2.0 kg, that is on a slippery ice rink. Three horizontal forces are shown on the rod, and there is a fourth force of unknown magnitude, direction, and location that is not shown. Determine the magnitude, direction, and location of the mystery force if the piece of wood remains at rest.

58. Consider again the situation described in the previous exercise and shown in Figure 10.46. Is it possible to apply a fourth force so the piece of wood does not spin but accelerates at 3.0 m/s² to the right? Justify your answer.

Figure 10.46: An overhead view of a piece of wood on a slippery ice rink. Four forces are applied to the wood but only three are shown. For Exercises 57 and 58.

59. Consider the mobile shown in Figure 10.47, in which all the balls have equal mass and in which the weight of the vertical strings and horizontal rods can be neglected. The vertical dashed lines in the figure are 10 cm apart. (a) Is this mobile in equilibrium in the configuration shown? How do you know? (b) If you add one additional ball to the configuration shown in the diagram can you get the mobile to be in equilibrium? If not, explain why not. If so, explain where you would place the additional ball. (c) If you have concluded that the mobile is not an equilibrium as shown, and that adding one additional ball will not achieve equilibrium, what is the minimum number of balls you can add to the system to achieve equilibrium, and where would you put them?

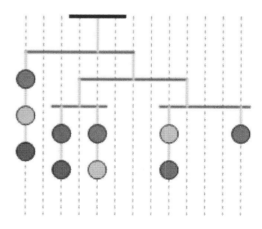

Figure 10.47: A mobile consisting of several balls of equal mass, for Exercise 59.

60. You are using a wheelbarrow to move a heavy rock, as shown in Figure 10.48. The diagram shows the location of the upward force you exert and the location of the force of gravity acting on the rock – wheelbarrow system. (a) Assuming the wheelbarrow is in equilibrium, how does the magnitude of the force you exert compare to the magnitude of the force of gravity acting on the system? (b) If the force of gravity has a magnitude of 420 N, solve for the magnitude of the force you exert on the wheelbarrow, and the magnitude and direction of any other force or forces necessary to maintain equilibrium.

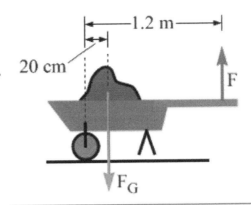

Figure 10.48: A side view of a wheelbarrow you are using to move a heavy rock, for Exercise 60.

61. Consider again the one-person seesaw described in Exercise 36. Would your answer for where Julie must sit to maintain equilibrium change if the system was on the Moon, as opposed to the Earth? Justify your answer.

62. Two of your friends, Latisha and Jorge, are carrying on a conversation about a physics problem. Comment on each of their statements, and state the answer to the problem the two of them are working on.

Latisha: In this problem, we have a uniform rod with a weight of 12 newtons, and it is supported at one end by a hinge. The question is, what is the smallest force that we can apply to keep the rod horizontal? Don't we need to apply a 12 newton force, at least, to hold it up?

Jorge: I think the idea is that the hinge can help support some of the weight, so, if we hold it in the right spot, we can apply a force that is less than 12 newtons.

Latisha: What if we start in the middle? If we hold it in the middle, then the rod is perfectly balanced, and we don't need the hinge at all. In that case, then we'd just need to apply a 12 newton force up to balance the 12 newton force of gravity, right?

Jorge: I think so. So, then, if we want to apply less force, should we move our force toward the hinge or away from the hinge? How do we figure that out if we don't know what the hinge force is?

Latisha: I guess this is why we're learning about torques. If we take torques about the hinge, then I think the hinge force cancels out – the distance is zero, for the torque from the hinge force. Then, if we apply our force farther from the hinge, can't we apply less force? But, how do we know how much less? We don't even know the length of the rod!

The photo shows a train wheel. Rolling is a concept we'll investigate in this chapter. Photo credit: by Leon Brooks, via http://www.public-domain-image.com.

Chapter 11 – Rotation II: Rotational Dynamics

CHAPTER CONTENTS

In this chapter, we continue to take concepts we examined previously, such as momentum and energy, and apply them to rotational situations. As with Chapter 10, understanding how these concepts apply in straight-line motion situations can give us considerable insight into how the concepts apply in a rotational setting.

There are a variety of practical applications of the ideas we will discuss here. These applications range from the rolling of wheels, which is relevant for driving cars and bicycles, to the workings of yo-yo's, to how gyroscopes work, which are important for maintaining the orientation of orbiting satellites.

11-1 Applying Newton's Second Law for Rotation

Let's learn how to apply Newton's second law for rotation to systems in which the angular acceleration $\vec{\alpha}$ is non-zero. The analysis of such systems is known as **rotational dynamics**.

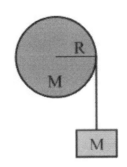

EXPLORATION 11.1 – A mass and a pulley

A pulley with a mass M and a radius R is mounted on a frictionless horizontal axle passing through the center of the pulley. A block with a mass M hangs down from a string that is wrapped around the outside of the pulley. Assume that the pulley is a uniform solid disk. The goal of this Exploration is to determine the acceleration of the block when the system is released from rest. Before we do anything else, we should draw a diagram of the situation based on the description above. The diagram is shown in Figure 11.1.

Figure 11.1: A diagram of the pulley and block system. The block hangs down from a string wrapped around the outside of the pulley.

Step 1 – Draw a free-body diagram of the block after the system is released from rest.
The block accelerates down when the system is released, which means that the block must have a net force acting on it that is directed down. There are only two forces acting on the block, an upward force of tension and a downward force of gravity. Thus, for the net force to be directed down, the force of tension must have a smaller magnitude than the force of gravity. Note that a common mistake is to assume that the force of tension is equal to the force of gravity acting on the block. Thinking about Newton's second law and how it applies to the block helps us to avoid making this error.

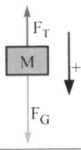

Figure 11.2: The free-body diagram of the block.

Step 2 – Draw a free-body diagram of the pulley.
A complete free-body diagram of the pulley, shown in Figure 11.3 (a), reflects that fact that the center-of-mass of the pulley remains at rest, so the net force must be zero. There is still a non-zero net torque, about an axis through the center of the pulley and perpendicular to the page, that gives rise to an angular acceleration. Generally, when we sum torques about an axis through the center, we draw a rotational free-body diagram, as in Figure 11.3 (b), including only forces that produce a torque. In this case, the only force producing a torque about the center of the pulley is the force of tension. As we discussed in Step 1, above, the force of tension is not equal to the weight of the block when the block has a non-zero acceleration.

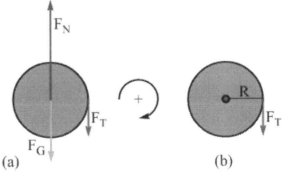

(a)   (b)

Step 3 – Apply Newton's Second Law to the block.
The block accelerates down, so let's define down to be the positive direction. Newton's second law is $\sum \vec{F} = M\vec{a}$. Using the free-body diagram in Figure 11.2 to evaluate the left-hand side gives:

$$+Mg - F_T = +Ma \, .$$

Figure 11.3: The full free-body diagram of the pulley, in (a), and the rotational free-body diagram in (b), showing the only force acting on the pulley that produces a torque about an axis perpendicular to the page through the center of the pulley.

Step 4 – *Apply Newton's second law for rotation to the pulley.* Here, let's define clockwise to be positive for rotation, both because this is the direction of the angular acceleration and because the pulley going clockwise is consistent with the block moving down, in the direction we defined as positive for the block's motion.

Newton's second law for rotation is $\sum \vec{\tau} = I\vec{\alpha}$. Using the free-body diagram in 11.4(b) to evaluate the left-hand side gives: $+R F_T \sin(90°) = +I\alpha$.

In the description of the problem we were told, "the pulley is a uniform solid disk." This tells us what to use for the rotational inertia, I, on the right-hand side of the equation. Looking up the expression for the rotational inertia of a solid disk in Figure 10.22, we get $I = MR^2/2$. Inserting this into the equation above, and using the fact that $\sin(90°) = 1$, gives:

$$+R F_T = +\frac{1}{2} MR^2 \alpha .$$

Canceling a factor of R gives: $+F_T = +\frac{1}{2} MR\alpha$.

Step 5 – *Put the resulting equations together to solve for the block's acceleration.* Looking at the force equation, $+Mg - F_T = +Ma$, and the equation we obtained from summing torques, $+F_T = +MR\alpha/2$, we have only two equations but three unknowns, F_T, a, and α. We could put the two equations together to eliminate the force of tension but then we'd be stuck.

Fortunately, we have another connection we can exploit, which is $\alpha = a/R$. The justification here is as follows. As the block accelerates down, every point on the string moves with the same magnitude acceleration as the block. We assume the string does not slip on the pulley, so the outer edge of the pulley (the part in contact with the string), moves with the string. Thus, the tangential acceleration of a point on the outer edge of the pulley is equal in magnitude to the acceleration of the string, which equals the magnitude of the block's acceleration. Finally, we connect the magnitude of the tangential acceleration of the outer edge of the pulley to the magnitude of the pulley's angular acceleration using $\alpha = a_T/R$. Putting everything together boils down to $\alpha = a/R$, which we can substitute into the equation that came from summing torques:

$$+F_T = +\frac{1}{2} MR\alpha = +\frac{1}{2} MR \frac{a}{R} = +\frac{1}{2} Ma .$$

Using this result in the force equation gives: $+Mg - F_T = +Ma$.

Substituting in for the force of tension gives: $+Mg - \frac{1}{2} Ma = +Ma$.

Thus, $a = 2g/3$, with the acceleration being directed down.

Key ideas: Applying Newton's second law for rotation helps us analyze situations that are purely rotational. Problems that involve both rotation and straight-line motion, as is the case in Exploration 11.1, can be analyzed by combining a torque analysis with a force analysis.
Related End-of-Chapter Exercises: 1, 13.

Essential Question 11.1: If the pulley in Exploration 11.1 is changed from a uniform solid disk to a uniform solid sphere of the same mass and radius as the disk, how does that affect the block's acceleration? How does that affect the tension in the string?

Answer to Essential Question 11.1: The effect of the change would be to decrease the rotational inertia of the pulley, because the rotational inertia of a solid sphere is $0.4MR^2$ compared with $0.5MR^2$ for the disk. The smaller the rotational inertia of the pulley, the less the pulley holds back the block, so the block's acceleration would increase. On the other hand, the force of tension would decrease. This is most easily seen by analyzing the block. If the block's acceleration increases, the net force on the block must increase. The force of gravity acting on the block remains constant, so the only way to increase the net force acting down on the block is to decrease the upward force of tension.

11-2 A General Method, and Rolling without Slipping

Let's begin by summarizing a general method for analyzing situations involving Newton's second law for rotation, such as the situation in Exploration 1.1. We will then explore rolling. We will tie together the two themes of this section in sections 11-3 and 11-4.

A General Method for Solving a Newton's Second Law for Rotation Problem
These problems generally involve both forces and torques.
1. Draw a diagram of the situation.
2. Draw a free-body diagram showing all the forces acting on the object.
3. Choose a rotational coordinate system. Pick an appropriate axis to take torques about, and then apply Newton's second law for rotation ($\sum \vec{\tau} = I\vec{\alpha}$) to obtain a torque equation.
4. Choose an appropriate x-y coordinate system for forces. Apply Newton's second law ($\sum \vec{F} = m\vec{a}$) to obtain one or more force equations. The positive directions for the rotational and x-y coordinate systems should be consistent with one another.
5. Combine the resulting equations to solve the problem.

Rolling without Slipping
Let's now examine a rolling wheel, which could be a bicycle wheel or a wheel on a car, truck, or bus. We will focus on a special kind of rolling, called **rolling without slipping**, in which the object rolls across a surface without slipping on that surface. This is actually what most rolling situations are, although our analysis would not apply to situations such as you spinning your car wheels on an icy road. Let's consider various aspects of rolling without slipping.

When we dealt with projectile motion in Chapter 4, we generally split the motion into two components, which were usually horizontal and vertical. To help understand rolling, we will follow a similar process. Rolling can be viewed as a combination, or superposition, of purely translational motion (moving a wheel from one place to another with no rotation) and purely rotational motion (only rotation with no movement of the center of the wheel). In the special case of rolling without slipping, there is a special connection between the translational component of the motion and the rotational component. Let's explore that connection.

EXPLORATION 11.2 – Rolling, rolling, rolling
We have a wheel of radius R that we will roll across a horizontal floor so that the wheel makes exactly one revolution. The wheel rolls without slipping on the floor.

Step 1 – *Consider the rotational part of the motion only (focus on the fact that the wheel spins around exactly once). What distance does a point on the outer edge of the wheel travel because of this spinning motion?*

Because we're ignoring the rotational motion, the distance traveled by a point on the outer edge of the wheel because of the spin is equal to the circumference of the wheel itself. This is a distance of $2\pi R$. See the top diagram in Figure 11.4.

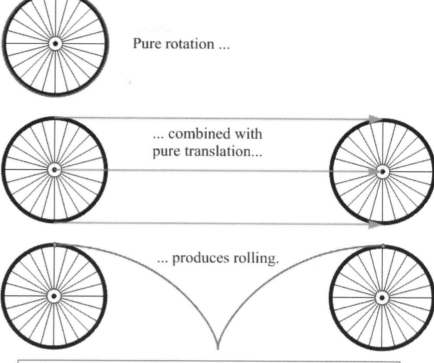

Pure rotation ...

... combined with pure translation...

... produces rolling.

Step 2 – *Now consider the translational part of the motion only (i.e., ignore the fact that the wheel is spinning, and imagine that we simply drag the wheel a particular distance without allowing the wheel to rotate). What is the distance that any point on the wheel moves if we drag the wheel a distance equal to that it would move if we rolled it so it rolled through exactly one revolution?* To determine what this distance is, imagine that we placed some double-sided tape around the wheel before we rolled it, and that the tape sticks to the floor. This is shown in the middle diagram in Figure 11.4. Rolling the wheel through one revolution lays down all the tape on the floor, covering a distance that is again equal to the

Figure 11.4: A pictorial representation of how the rotational component and the translational component of the motion combine to produce the interesting shape of the path traced out by a point on the outer edge of the wheel that is rolling without slipping. This shape is known as a cycloid.

circumference of the wheel. Thus, focusing on the translational distance only, the translational distance moved by every point on the wheel as the wheel rolls through one revolution is $2\pi R$.

Step 3 – *Assuming the rolling is done at constant speed, compare the speed of a point on the outer rim, associated only with the wheel's rotation, to the translational speed of the wheel's center of mass.* We can find these speeds by dividing the appropriate distances by the time during which the motion takes place. Because the distances associated with the two components of the motion are equal, and the time of the motion is the same for the two components, these two speeds are equal.

Key ideas for rolling: Rolling can be considered to be a superposition of a pure translational motion and a pure rotational motion. In the special case of rolling without slipping, the distance moved by a point on the outer edge of a wheel associated with the rotational component is equal to the translational distance of the wheel. The speed of a point on the outer edge because of the rotational component is also equal to the translational speed of the wheel.
Related End-of-Chapter Exercises: 4, 17.

Essential Question 11.2: Different points on a wheel that is rolling without slipping have different speeds. Considering one particular instant, which point on the wheel is moving slowest? Which point is moving the fastest?

Answer to Essential Question 11.2: As we will investigate in more detail in section 11-3, when a wheel rolls without slipping, the point at the bottom of the wheel has the smallest speed (the speed there is zero, in fact), while the point at the top of the wheel is moving fastest.

11-3 Further Investigations of Rolling

Let's continue our analysis of rolling, starting by thinking about the velocity of various points on a wheel that rolls without slipping. We will then go on to investigate rolling spools.

EXPLORATION 11.3 – Determining velocity
Let's turn now from thinking about speeds to thinking about velocities. Consider a wheel rolling without slipping with a constant translational velocity $\vec{v}$, directed to the right, across a level surface. For each point below, determine the point's net velocity by combining, as vectors, the point's translational velocity (the velocity associated with the translational component of the motion) with its velocity because of the rotational component of the motion.

Step 1 – Find the net velocity of the center of the wheel.
Our axis of rotation passes through the center of the wheel. The center of the wheel therefore has no rotational velocity (because $v_{rot} = r\omega$, and $r = 0$). Thus, the net velocity of the center of the wheel is its translational velocity, $\vec{v}$. Note that every point on the wheel has the same translational velocity. Two equal vectors are shown at the center of the wheel in Figure 11.5. One represents the translational velocity at that point, and the other represents the net velocity at that point.

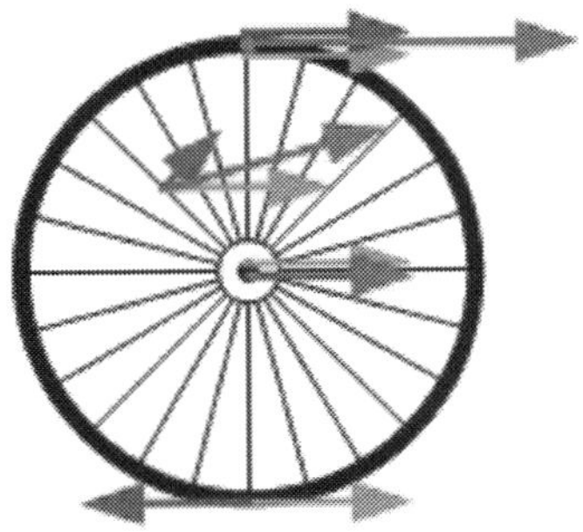

Step 2 – Find the net velocity of the point at the very top of the wheel.
Here, we use the fact that the rotational speed is equal to the translational speed, so we are adding two velocities of equal magnitude. At the top of the wheel, the velocities also point in the same direction, so the net velocity is $2\vec{v}$, as shown in Figure 11.5.

Figure 11.5: The translational (equal vectors all directed right), rotational (tangent to the circle), and net velocities of various points on the wheel. The net velocity at a point is a vector sum of the translational and rotational velocities.

Step 3 – Find the net velocity of the point at the very bottom of the wheel.
At the bottom of the wheel, the rotational velocity exactly cancels the translational velocity, because the vectors point in opposite directions and have equal magnitudes. The net velocity of that point is zero – the point is instantaneously at rest! This is a special condition that is characteristic of rolling without slipping. No slipping implies no relative motion between the surfaces in contact, which means the point at the bottom of the wheel that is in contact with the road surface is at rest.

Figure 11.5 also shows the net velocity at another point on the wheel, a point above and to the left of the center. As with all points, the translational velocity is a vector directed to the right. The velocity associated with the pure rotation is tangent to the circle that passes through the point (and centered at the center of the wheel) - this has a magnitude of $v/2$, because the point is halfway between the center and the rim. The net velocity is the longest of the three vectors at that point, the vector sum of the translational and rotational velocities.

Key ideas for rolling: The net velocity of a point on a rolling wheel can be found by adding, as vectors, the point's translational velocity and its rotational velocity. In the special case of a wheel rolling without slipping with a translational velocity $\vec{v}$, the net velocity of the center of the wheel is $\vec{v}$; while that of the point at the top of the wheel is $2\vec{v}$. A point on the outer edge of the wheel actually comes instantaneously to rest when it reaches the bottom of the wheel.
Related End-of-Chapter Exercises: 5, 6.

EXAMPLE 11.3 – Unrolling a ribbon from a spool

A long ribbon is wrapped around the outer edge of a spool. You pull horizontally on the end of the ribbon so the ribbon starts to unwind from the spool as the spool rolls without slipping across a level surface.

(a) When you have moved the end of the ribbon through a horizontal distance L, how far has the spool moved?

(b) Does your answer change if the ribbon is instead wrapped around the spool's axle, which has a radius equal to half the radius of the spool? If so, how does the answer change?

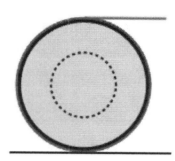

Figure 11.6: A spool is rolling without slipping to the right because you are pulling, to the right, on the red ribbon that is wrapped around the spool.

SOLUTION

(a) A diagram of the situation is shown in Figure 11.6. Once again, we can think of the spool's rolling motion as a combination of its translational motion and its rotational motion. We can thus say that the end of the ribbon moves because (a) the spool has a translational motion, and (b) the spool is rotating. The speed of the ribbon matches the speed of the top of the spool, because there is no slipping between the ribbon and the spool. Recalling the result from Exploration 11.3, the top of the spool has a velocity twice that of the center of the spool. Putting these facts together means that the center of the spool has a velocity half that of the end of the ribbon at any instant, and so the spool covers a distance of $L/2$, half the distance covered by the end of the ribbon.

(b) What if the ribbon is wrapped around the spool's axle and you move the end of the ribbon through a distance L? The answer changes because the rotational contribution to the net velocity changes. As shown in Figure 11.7, the ribbon now comes off the axle at the top of the axle, at a point halfway between the edge and the center of the spool. The net velocity at that point on the spool is 1.5 times the velocity of the center of the spool: the translational velocity is equal to the velocity of the center, while the rotational velocity is half that of the center, because at a radius of R/2 we have:

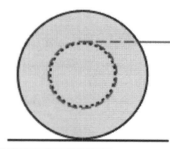

Figure 11.7: The ribbon is wrapped around the axle of the spool, which has a radius half that of the spool. The ribbon comes off the axle at the top.

$$v_{rot} = \frac{R}{2}\omega = \frac{1}{2}R\omega = \frac{1}{2}v_{trans}.$$

Putting it another way, the velocity of the center of the spool is now two-thirds of the velocity of the end of the ribbon. If the end of the ribbon travels a distance L, the spool translates through a distance of $2L/3$.

Related End-of-Chapter Exercises: 18, 19.

Essential Question 11.3: In a situation similar to that in Figure 11.7, you pull to the right on a ribbon wrapped around the axle of a spool. This time, however, the ribbon is wound so it comes away from the spool underneath the axle, as shown in Figure 11.8. When you pull to the right on the ribbon, the spool rolls without slipping. In which direction does it roll? Sketch a free-body diagram of the spool to help you think about this.

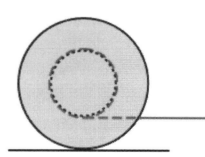

Figure 11.8: A ribbon is wrapped around the axle of the spool so the ribbon comes off the axle below the axle.

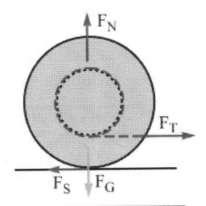

Answer to Essential Question 11.3: Many people focus on the counterclockwise torque, relative to an axis perpendicular to the page that passes through the center of the spool, exerted by the force of tension and conclude that the spool rolls to the left. Before jumping to conclusions, however, draw the free-body diagram (after drawing your own, see Figure 11.9). As usual there is a downward force of gravity and an upward normal force. Horizontally there is a force of tension, directed right, exerted by the ribbon. With no friction, the force of tension would cause the spool to move right and spin counterclockwise, so the bottom of the spool would move right with respect to the horizontal surface. Friction must therefore be directed left to oppose this, and, because we know the spool rolls without slipping, the force of friction must be static friction.

Figure 11.9: The free-body diagram of the spool.

Now we have the complete free-body diagram, we can see that the answer to the question is not obvious. There is one force left and one force right – which is larger? Relative to an axis through the center, there is one torque clockwise and one counterclockwise – which is larger? A quick way to get the answer is to consider an axis perpendicular to the page, passing through the point where the spool makes contact with the horizontal surface. Relative to this axis, three of the four forces give no torque, and the torque from the tension in the string is in a clockwise direction. Clockwise rotation of the spool, relative to the point where the spool touches the surface, is consistent with the spool rolling without slipping to the right. This is opposite to what you would conclude by focusing only on the torque about the center from the force of tension. The spool rolls to the right.

11-4 Combining Rolling and Newton's Second Law for Rotation

Let's now look at how we can combine torque ideas with rolling-without-slipping concepts.

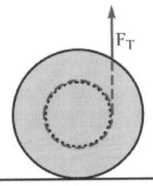

EXPLORATION 11.4 – A vertical force but a horizontal motion

A spool of mass M has a string wrapped around its axle. The radius of the axle is half that of the spool. An upward force of magnitude F_T is exerted on the end of the string, as shown in Figure 11.10. This causes the spool, which is initially at rest, to roll without slipping as it accelerates across the level surface.

Figure 11.10: An upward force is exerted on the string wrapped around the axle of the spool.

In which direction does the spool roll? Which horizontal force is responsible for the spool's horizontal acceleration? Let's begin by drawing a free-body diagram of the spool. Figure 11.11 shows a partial free-body diagram, showing only the vertical forces acting on the spool. There is a downward force of gravity acting on the spool, and an upward force of tension applied by the string (note that F_T must be less than or equal to Mg, so the spool has no vertical acceleration). There is also an upward normal force, required to balance the vertical forces.

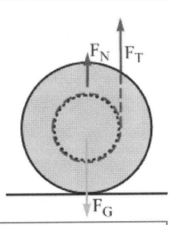

Is there a horizontal force? If there is, what could it be? Let's go back and think about what is interacting with the spool. The force of gravity accounts for the interaction between the Earth and the spool, and the force of tension accounts for the interaction between the string and the spool. The only interaction left is the interaction between the surface and the spool. The surface exerts a contact force on the spool. Remember that we generally

Figure 11.11: A partial free-body diagram of the spool, showing the vertical forces acting on it.

split the contact force into components, the normal force (which we have accounted for) and the force of friction (which we have not).

If there is a horizontal force acting, it can only be a force of friction. Do we need friction in this situation? Consider what would happen if the free-body diagram shown in Figure 11.11 was complete, and there was no friction. Taking an axis perpendicular to the page through the center of the spool, the tension force would give rise to a counterclockwise torque. Because the net force acting on the spool would be zero, however, the spool would simply spin counterclockwise without moving. This is inconsistent with the rolling-without-slipping motion we are told is occurring. There must be a force of friction acting on the spool to cause the horizontal motion.

Note that, without friction, the bottom of the spool rotates to the right relative to the surface. The force of friction must therefore be directed to the left, acting to oppose the relative motion that would occur without friction. Because the force of friction is the only horizontal force acting on the spool, the spool accelerates to the left.

To gain another perspective on this situation, we can follow the procedure discussed in the Answer to Essential Question 11.3, and consider the sum of the torques about the contact point (the point where the spool makes contact with the ground). Both the normal force and the force of gravity pass through the contact point, so they don't give rise to any torques about the contact point. If there is a force of friction, whether it is directed to the right or the left it would also pass through the contact point, giving rise to no torque about that point. Thus, the only force that produces a torque about the contact point is the tension force. Relative to the contact point, this torque is directed counter-clockwise, which is consistent with rolling without slipping to the left. Starting from rest, rolling to the left requires a horizontal force directed to the left, which can only be a friction force.

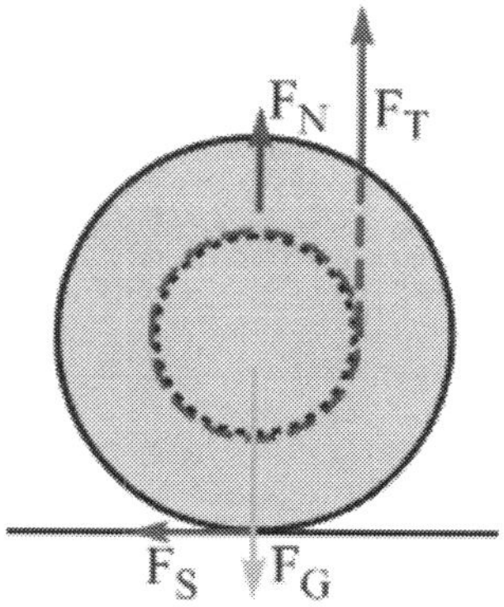

Figure 11.12: The complete free-body diagram of the spool.

Is the force of friction kinetic friction or static friction? Because the spool is rolling without slipping, and the bottom of the spool is instantaneously at rest relative to the surface it is in contact with, the force of friction is the static force of friction. This may sound counter-intuitive, since there is relative motion between the spool as a whole and the surface, but it is very similar to the walking (without slipping) situation that we thought about in Chapter 5. When walking, as long as our shoes do not slip on the floor, a force of static friction acts in the direction of motion. The same thing happens here - in this case, the force of static friction is the only horizontal force acting on the spool, so it is the force accelerating the wheel horizontally. The complete free-body diagram for the rolling-without-slipping situation is shown in Figure 11.12.

> **Key ideas for rolling without slipping**: Rolling without slipping often involves a force of friction, which must be a static force of friction. The static force of friction is often (although not always) in the direction of motion.
> **Related End-of-Chapter Exercises: 3, 50.**

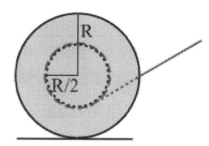

Essential Question 11.4: In the situation shown in Figure 11.13, you pull on the end of a ribbon wrapped around the axle of a spool. Your force is exerted in the direction shown. If the spool rolls without slipping, in which direction does the spool roll?

Figure 11.13: A ribbon is wrapped around the axle of the spool so the ribbon comes off the axle in the direction shown.

Answer to Essential Question 11.4: Once again, it is simplest to take torques about an axis perpendicular to the page, passing through the point at which the spool touches the ground. The only force giving rise to a torque about this point is the tension in the ribbon, which gives a clockwise torque. If the spool rotates clockwise with respect to its bottom point, the motion of the spool is to the right.

11-5 Analyzing the Motion of a Spool

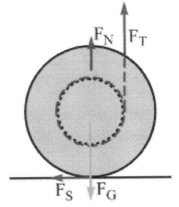

EXPLORATION 11.5 – Continuing the analysis of the rolling spool

Let's return to the situation described in Exploration 11.4, and focus in particular on the free-body diagram in Figure 11.12. Our goal is to determine the magnitude of the spool's acceleration in terms of F_T and M. The spool consists of two disks, each of mass $M/3$ and radius R, connected by an axle of mass $M/3$ and radius $R/2$.

Step 1 – *Apply Newton's Second Law for the horizontal forces.*

The spool accelerates left, so let's define left to be the positive direction.

$$\sum \vec{F}_x = M \vec{a}_x .$$

Figure 11.12: The complete free-body diagram of the spool.

Because the acceleration is entirely in the x-direction, we can replace $\vec{a}_x$ by $\vec{a}$.

Evaluating the left-hand side of this expression with the aid of Figure 11.12 gives:

$$+F_s = +M a .$$

Step 2 – *Find the expression for the spool's rotational inertia about an axis perpendicular to the page passing through the center of the spool.*

Why are we doing this? Well, we'll need to apply Newton's second law for rotation to solve this problem, and that involves the spool's rotational inertia. To find the spool's rotational inertia, we can use the expression for the rotational inertia of a solid disk or cylinder ($I = \frac{1}{2} mr^2$) about the center . Let's apply this equation to the three pieces of the spool and add them together to find the net rotational inertia.

Each of the two disks contributes $\dfrac{1}{2}\left(\dfrac{M}{3}\right)R^2 = \dfrac{1}{6}MR^2$ to the rotational inertia.

The axle contributes $\dfrac{1}{2}\left(\dfrac{M}{3}\right)\left(\dfrac{R}{2}\right)^2 = \dfrac{1}{24}MR^2$.

The total rotational inertia is $I = \dfrac{1}{6}MR^2 + \dfrac{1}{6}MR^2 + \dfrac{1}{24}MR^2 = \dfrac{9}{24}MR^2 = \dfrac{3}{8}MR^2$.

Step 3 – *Apply Newton's Second Law for Rotation to obtain a connection between the force of friction and the upward force $\vec{F}_T$ applied to the string.*

Taking torques about an axis perpendicular to the page and passing through the center of the spool is a good way to do this, because the force of gravity and the normal force pass through this axis and therefore give no torque about that axis. Since the spool has a counterclockwise angular acceleration let's take counterclockwise to be positive for torques. Applying Newton's second law for rotation gives:

$$\sum \vec{\tau} = I\vec{\alpha} .$$

Referring to Figure 11.12, and using the equation $\vec{\tau} = r F \sin\theta$, we have:

$$+\frac{R}{2}F_T \sin(90°) - R F_S \sin(90°) = +I\alpha .$$

Recognizing that $\sin(90°) = 1$, and substituting the expression for the spool's rotational inertia we found above, gives:

$$+\frac{R}{2}F_T - R F_S = +\frac{3}{8}MR^2\alpha .$$

Canceling a factor of R gives: $+\frac{1}{2}F_T - F_S = +\frac{3}{8}MR\alpha$.

Step 4 – *What is the connection between the spool's acceleration and its angular acceleration?*
For rolling without slipping, the connection between the acceleration and the angular acceleration is $a = R\alpha$, although it is always a good idea to check whether the positive direction for the straight-line motion is consistent with the positive direction for rotation. In our case they are consistent, since we chose them based on the motion. If we had reversed one of the positive directions, however, we would have had a negative sign in the equation.

Step 5 – *Combine the results above to determine the spool's acceleration in terms of F_T and M.*
Let's first substitute $a = R\alpha$ into our final expression from step 3, to get:

$$+\frac{1}{2}F_T - F_S = +\frac{3}{8}Ma .$$

In step 1, we determined that $F_s = M a$, so we get:

$$+\frac{1}{2}F_T - Ma = +\frac{3}{8}Ma ;$$

$$+\frac{1}{2}F_T = +\frac{11}{8}Ma ;$$

$$\vec{a} = \frac{4F_T}{11M} \text{ ,directed to the left.}$$

Key idea: Solving a rolling-without-slipping problem often involves analyzing the rotational motion, analyzing the one-dimensional motion, and combining the analyses.
Related End-of-Chapter Exercises: 51, 53.

Essential Question 11.5: Consider a hard ball that is rolling without slipping across a smooth level surface. If the ball maintains a constant velocity, in what direction is the static force of friction acting on the ball? Consider the three-possible free-body diagrams for the ball in Figure 11.14 below, and state which free-body diagram is appropriate for this situation.

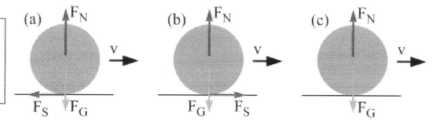

Figure 11.14: Possible free-body diagrams for a ball rolling, without slipping, at constant velocity to the right across a horizontal surface.

Answer to Essential Question 11.5: No friction force can act on the ball, so the correct free-body diagram is that shown in Figure 11.14 (c). A force of friction in the direction of motion would increase the ball's translational speed, and the counterclockwise torque from the force of friction would decrease the ball's angular speed. A force of static friction directed opposite to the ball's velocity would decrease the translational speed while increasing the rotational speed. The ball rolls horizontally at constant velocity only if no friction force acts.

11-6 Angular Momentum

By now, we have looked at enough analogies between straight-line motion and rotational motion that we can simply take a straight-line motion equation, replace the straight-line motion variables by their rotational counterparts, and write down the equivalent rotational equation. We could also derive the rotational equations following a derivation parallel to the one we used for the straight-line motion equation, but the end result would be the same.

Let's try this for angular momentum. In Chapter 6, we used the following expression for the linear momentum, $\vec{p}$, of an object of mass m moving with velocity $\vec{v}$: $\vec{p} = m\vec{v}$.

Using the symbol $\vec{L}$ to represent angular momentum, we can come up with the equivalent expression for angular momentum by replacing mass m by its rotational equivalent, rotational inertia I, and velocity $\vec{v}$ by its rotational equivalent $\vec{\omega}$:

$$\vec{L} = I\vec{\omega} .$$ (Equation 11.1: **Angular momentum**)

We made a number of statements about momentum in Chapter 6. Equivalent statements apply to angular momentum, including:
- Angular momentum is a vector, pointing in the direction of angular velocity.
- The angular momentum of a system can be changed by applying a net torque.
- If no net torque acts on a system, its angular momentum is conserved.

Let's explore this idea of angular momentum conservation.

EXPLORATION 11.6 – Jumping on the merry-go-round

A little red-haired girl named Sarah, with mass m, runs toward a playground merry-go-round, which is initially at rest, and jumps on at its edge. Sarah's velocity $\vec{v}$ is tangent to the circular merry-go-round. Sarah and the merry-go-round then spin together with a constant angular velocity $\vec{\omega}_f$. The merry-go-round has a mass M, a radius R, and has the form of a uniform solid disk. Assume that Sarah's "radius" is small compared to R. The goal of this Exploration is to determine an expression for $\vec{\omega}_f$. We can treat this as a collision.

Step 1 – *Sketch two diagrams, one showing Sarah running toward the merry-go-round and the other showing Sarah and the merry-go-round rotating together after Sarah has jumped on. Imagine that you're looking down on the situation from above.* These two diagrams are shown in Figure 11.15.

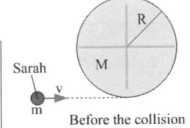

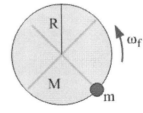

Figure 11.15: On the left is the situation before the collision, as Sarah runs toward the merry-go-round, while on the right is the situation after the collision, with Sarah and the merry-go-round rotating together with a constant angular velocity.

Before the collision   After the collision

Step 2 – *What kind of momentum does the Sarah/merry-go-round system have, if any, before Sarah jumps on the merry-go-round? What about after Sarah jumps on?* After the collision, when the system is rotating, the system clearly has a non-zero angular momentum. Before the collision, however, it is not obvious that the system has any angular momentum, because nothing is rotating. Sarah certainly has a linear momentum, however, because she has a non-zero velocity.

Step 3 – *Convert Sarah's linear momentum before the collision to an angular momentum, using a method modeled on the way we convert a force to a torque.* Although there is no rotation before the collision, we can say that the system has an angular momentum with respect to an axis perpendicular to the page that passes through the center of the merry-go-round. Consider how we get torque from force, where the magnitude of the torque is given by $\tau = r F \sin\phi$. Angular momentum is found from linear momentum in a similar fashion, with its magnitude given by:
$L = r\, p \sin\phi = r\,(mv)\sin\phi$,   (Eq. 11.2: **Connecting angular momentum to linear momentum**)

where ϕ is the angle between the line we measure distance along and the line of the linear momentum.

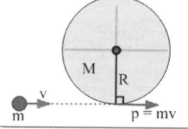

Relative to the axis through the center of the merry-go-round, the angular momentum is: $\vec{L}_i = R\,mv\sin(90°) = R\,mv$, in a counterclockwise direction.

Step 4 – *Apply angular momentum conservation to express $\vec{\omega}_f$, the angular velocity of the system after the collision, in terms of the variables above.* Angular momentum is conserved because there are no external torques acting on the Sarah/merry-go-round system, relative to a vertical axis passing through the center of the turntable. We will justify this further in section 11-7. Thus, we can say: Angular momentum before the collision = angular momentum afterwards.

Figure 11.16: The lever-arm method to determine Sarah's angular momentum, with respect to an axis passing through the center of the merry-go-round.

The angular momentum afterwards is $\vec{L}_f = I\vec{\omega}_f$. The system's rotational inertia after the collision is the sum of the rotational inertias of Sarah, and the ½ MR² of the merry-go-round. Sarah's "radius" is small compared to R, so we treat Sarah as a point, assuming that all her mass is the same distance, R, from the center of the turntable. Sarah's rotational inertia is thus mR^2.

Thus, the rotational inertia of the system after the collision is $I = \dfrac{1}{2}MR^2 + mR^2$.

Taking counterclockwise to be positive, angular momentum conservation gives: $\vec{L}_i = \vec{L}_f$.

$$+R\,mv = I\vec{\omega}_f = \left(\frac{1}{2}MR^2 + mR^2\right)\vec{\omega}_f .$$

Solving for the final angular velocity of the system gives:

$$\vec{\omega}_f = +\frac{mv}{\dfrac{1}{2}MR + mR} \quad \text{or,} \quad \vec{\omega}_f = \frac{mv}{\dfrac{1}{2}MR + mR} \quad \text{directed counterclockwise.}$$

Key ideas: Linear momentum converts to angular momentum in the same way force converts to torque. Also, we apply momentum conservation ideas to rotational collisions in the same way we analyze collisions in one and two dimensions.   **Related End-of-Chapter Exercises: 32, 34, 59.**

Essential Question 11.6: Is it possible for Sarah, with the same initial speed, to jump onto the merry-go-round at the same point, but not make it spin? If so, how could she do this?

Answer to Essential Question 11.6: One way for Sarah to jump onto the merry-go-round, without causing the merry-go-round to spin, is for Sarah to direct her velocity at the center of the merry-go-round, instead of tangent to it. If Sarah ran directly toward the center of the merry-go-round she would have no angular momentum before the collision and there would be no reason for the system to spin after the collision.

11-7 Considering Conservation, and Rotational Kinetic Energy

In step 4 of Exploration 11.6, we stated that the angular momentum of the system consisting of Sarah and the merry-go-round was conserved, because no external torques were acting on the system. Let's justify that statement. We do not have to concern ourselves with vertical forces, such as the force of gravity or the normal force applied to the merry-go-round by the ground, because vertical forces give no torque about a vertical axis of rotation. We also do not have to concern ourselves with the force that Sarah exerts on the merry-go-round, or the equal-and-opposite force the merry-go-round exerts on Sarah, because the system we're considering consists of the combination of Sarah and the merry-go-round, so those are internal forces and cancel one another. Still, let's examine those forces a little.

Individual free-body diagrams for Sarah and the merry-go-round when Sarah first jumps on the merry-go-round are shown in Figure 11.17. Through some combination of friction between her shoes and the merry-go-round, and a contact force between her hands and any handholds on the merry-go-round, there is a force component that acts to the left on Sarah from the merry-go-round (this reduces her speed), and an equal-and-opposite force component that acts to the right on the turntable by Sarah (providing the torque that gives the merry-go-round an angular acceleration). However, the turntable does not accelerate to the right. This is because there is a horizontal force applied on the turntable by whatever the turntable's axis is connected to, which we can consider to be the Earth. As shown in Figure 11.17, the Sarah/merry-go-round system has a net external force acting on it at this point, which is why the *linear* momentum of the system is not conserved. However, this net external force gives rise to no torque about an axis through the center of the merry-go-round, because the force passes through that axis. Because there is no net external torque acting on the system, the system's angular momentum is conserved.

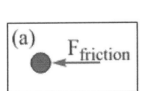
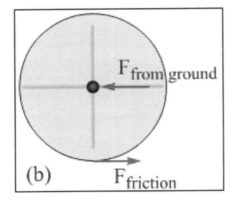
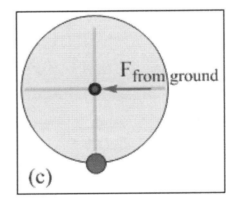

Figure 11.17: Free-body diagrams for Sarah, the merry-go-round, and the system consisting of Sarah and the merry-go-round together, when Sarah initially makes contact with the merry-go-round. Vertical forces are ignored in this overhead view.

Rotational Kinetic Energy

Let's now move from the rotational equivalent of linear momentum to the rotational equivalent of translational kinetic energy. The equation we used previously for kinetic energy is $K = \frac{1}{2}mv^2$. We can find the equivalent expression for kinetic energy in a rotational setting by replacing mass m by rotational inertia I, and speed v by angular speed ω. The kinetic energy of a purely rotating object is thus given by:

$$K = \frac{1}{2}I\omega^2 .$$ (Equation 11.3: **Rotational kinetic energy**)

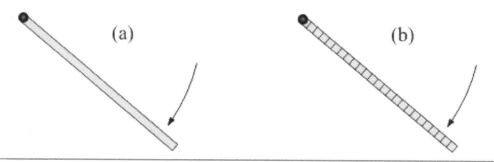

Figure 11.18: (a) A rod that has been released from rest when it was horizontal is now moving. We can find its kinetic energy by breaking the rod into small pieces, as shown in (b), finding the kinetic energy of each piece, and adding these kinetic energies together to find the net kinetic energy.

Let's make sure our substituting-the-equivalent-rotational-variables method of arriving at rotational equations makes sense. Consider, for instance, a uniform rod that can rotate about an axis through one end. If we hold the rod horizontal and then release it from rest, the rod swings down. What is the rod's kinetic energy at a particular instant, say at the instant shown in Figure 11.18 (a)? One thing we could do is, as shown in Figure 11.18 (b), break the rod into small pieces of mass m_i, determine the speed v_i of each piece, find the kinetic energy $\frac{1}{2}m_i v_i^2$ of each piece, and then add up all these kinetic energies to find the total kinetic energy:

$$K = \sum \frac{1}{2}m_i v_i^2 .$$

Because the speed of each piece is different, while the angular speed of each piece is the same, let's write the sum in terms of the rod's angular speed instead:

$$K = \sum \frac{1}{2}m_i (r_i \omega)^2 .$$

If we bring the constants of $\frac{1}{2}$ and ω^2 out in front of the sum, our expression becomes

$K = \frac{1}{2}\omega^2 \sum m_i r_i^2$, which we can write as $K = \frac{1}{2}I\omega^2$, because the definition of rotational inertia

is $I = \sum m_i r_i^2$. This expression for the kinetic energy agrees with what we came up with above (and it works for any rotating object, not just a rod!).

Essential Question 11.7: In Chapter 7 we used names such as "elastic collision" and "inelastic collision" to classify various collisions. Under what category would the Sarah/merry-go-round collision described in the previous Exploration fall?

Answer to Essential Question 11.7: Because Sarah and the merry-go-round stick together and move as one after the collision, the collision is completely inelastic.

11-8 Racing Shapes

Let's make use of the expression for rotational kinetic energy we derived in section 11-7, and apply it to analyze the motion of an object that rolls without slipping down a slope. The analysis can be done in terms of energy conservation (as we will do), or in terms of thinking about forces and torques and applying Newton's second law and Newton's second law for rotation. The analysis in those terms can be found on the accompanying web site.

EXPLORATION 11.8 – Racing shapes

You have various shapes, including a few different solid spheres, a few rings, and a few uniform disks and cylinders. The objects have various masses and radii. When you race the objects by releasing them from rest two at a time, they roll without slipping down an incline of constant angle. Our goal is to determine which object reaches the bottom of the incline in the shortest time. Let's analyze this for a generic object of mass M, radius R, and rotational inertia, about an axis through the center of mass, of cMR^2.

Step 1 – *Sketch a free-body diagram for the object as it rolls without slipping down the ramp.*
A diagram and a free-body diagram is shown in Figure 11.19. The Earth applies a downward force of gravity to the object, while the incline applies a contact force. We split the contact force into two forces, a normal force perpendicular to the incline and a force of friction directed up the slope. This is a static force of friction, because the object does not slip as it rolls. The force of static friction is directed up the slope, not because the motion of the object is down the slope, but because the object has a clockwise angular acceleration (its angular velocity is clockwise and increasing as it rolls down). Taking an axis through the center of the object, the static force of friction is the only force that can provide the torque associated with this angular acceleration – the other two forces pass through the center of the object and thus give no torque about that axis.

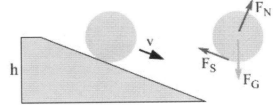

Figure 11.19: The diagram and free-body diagram of an object as it rolls without slipping down a ramp. A force of friction directed up the ramp provides the clockwise torque associated with the object's clockwise angular acceleration. The force of friction is static because the object does not slip as it rolls.

Step 2 – *Let's analyze this in terms of energy conservation, using the same conservation of energy equation we used in previous chapters. Start by eliminating the terms that are zero in the equation.* Recall that the energy conservation equation is: $K_i + U_i + W_{nc} = K_f + U_f$. The object is released from rest, so the initial kinetic energy K_i is zero. We can also define the bottom of the incline to be the zero level for gravitational potential energy, so the final potential energy is $U_f = 0$. We also have no work being done by non-conservative forces. This may seem somewhat counter-intuitive at first, because static friction acts on each object as it rolls down the hill, but it is kinetic friction that is associated with a loss of mechanical energy. Static friction, because it involves no relative motion (and therefore no displacement to use in the work equation), does not produce a loss of mechanical energy.

The conservation of energy equation can thus be written: $U_i = K_f$.

Let's say that each object starts from a height h above the bottom of the incline. Because the zero for potential energy is at the bottom, the initial gravitational potential energy can be written as: $U_i = Mgh$. Our energy conservation term can thus be written $Mgh = K_f$.

Step 3 – *Split the kinetic energy term into two pieces, one representing the translational kinetic energy and one representing the rotational kinetic energy. Express the rotational kinetic energy in terms of M and v_f (the speed at the bottom of the incline) and solve for v_f.* First, let's think

about why considering two types of kinetic energy is appropriate. When an object's center-of-mass is moving, the object has translational kinetic energy $KE_{trans} = \frac{1}{2}Mv^2$. When an object is only rotating, it has a rotational kinetic energy $KE_{rot} = \frac{1}{2}I\omega^2$. A rolling object, however, is both translating as well as rotating, and thus it has both these forms of kinetic energy.

Our energy equation now becomes: $Mgh = \frac{1}{2}Mv_f^2 + \frac{1}{2}I\omega_f^2$.

Let's make two substitutions to rewrite the rotational kinetic energy term. First, we can use our expression for rotational inertia, $I = cMR^2$. Then, we use the relationship between speed and angular speed that applies to rolling without slipping,: $\omega = v/R$. Our energy equation is now:

$$Mgh = \frac{1}{2}Mv_f^2 + \frac{1}{2}cMR^2\frac{v_f^2}{R^2}.$$

Note that all factors of mass M and radius R cancel, leaving: $gh = \frac{1}{2}v_f^2 + \frac{1}{2}cv_f^2$.

Solving for v_f, the object's speed at the bottom of the incline, gives: $v_f = \sqrt{\dfrac{2gh}{1+c}}$.

This result is consistent with the $v_f = \sqrt{2gh}$ result we obtained in previous chapters (for the speed of a ball dropped from rest through a height h, for instance), giving us some confidence that the answer is correct.

So, which object wins the race? The winner is the object with the highest speed at the bottom, which requires the smallest value of c. Recall that c is the numerical factor in the moment of inertia, $I = cMR^2$. For the various shapes we were racing we have $c = 2/5$ for solid spheres; $c = 1/2$ for uniform disks and cylinders; and $c = 1$ for rings. Thus, in the rolling races, a solid sphere beats any disk (or cylinder) and any ring, while any disk or cylinder beats any ring.

Key ideas: We can apply energy conservation in an analysis of rotating, or rolling, objects, just as we did in previous situations. Our energy conservation equation from Chapter 7 needs no modification. All we have to do is to use the expression for the kinetic energy of rotating objects: $KE_{rot} = \frac{1}{2}I\omega^2$. **Related End-of-Chapter Exercises: 7, 8, 10.**

Essential Question 11.8: In Exploration 11.8, we determined that, in the races of rolling objects, a solid sphere would beat a disk or cylinder, which would beat a ring. What if we raced two of the same kind of object against one another (such as a sphere versus a sphere)? Which object would win? The object with the larger mass, smaller mass, larger radius, or smaller radius?

Answer to Essential Question 11.8: Review the analysis in step 3 of Exploration 11.8. Both the mass and radius cancel out of the energy conservation equation. This tells us, surprisingly, that the mass and radius are irrelevant. In other words, all uniform solid spheres roll the same, all uniform solid disks (or cylinders) roll the same, and all rings roll the same – all the races involving two of the same kind of object end in a tie.

11-9 Rotational Impulse and Rotational Work

Let's continue our method of determining rotational equations from their straight-line motion counterparts by writing down expressions for rotational impulse and rotational work. In Chapter 6, the impulse relationship we came up with was: $\Delta \vec{p} = \vec{F}_{net}\Delta t$. In words, this equation tells us that the change in momentum an object experiences is equal to the product of the net force applied to the object multiplied by the time interval over which it is applied. Transforming this to a rotational setting, an object's change in angular momentum is equal to the net torque it experiences multiplied by the time interval over which that net torque is applied:

$$\Delta \vec{L} = \vec{\tau}_{net}\Delta t .$$ (Equation 11.4: **Rotational impulse**)

Similarly, we can consider the concept of work in a rotational setting. For straight-line motion, if we meld the work equation with the work-energy theorem we get:

$$\Delta K = W_{net} = \vec{F}_{net} \bullet \Delta \vec{r} = F_{net}\Delta r \cos\phi .$$ (Equation 6.8: **Work-kinetic energy theorem**)

In chapter 6, we used the variable θ to represent the angle between the net force $\vec{F}_{net}$ and the displacement $\Delta \vec{r}$. We'll use ϕ here instead because in this chapter we're using θ to represent the angular position of a rotating object.

To find the expression for work in a rotational setting, start with equation 6.8. Replace force $\vec{F}$ by its rotational equivalent, $\vec{\tau}$, and replace displacement $\Delta \vec{r}$ by its rotational equivalent $\Delta \vec{\theta}$. This gives:

$$\Delta K = W_{net} = \vec{\tau}_{net} \bullet \Delta \vec{\theta} = \tau_{net}\Delta\theta \cos\phi .$$ (Equation 11.5: **Rotational work**)

If the dot product notation confuses you, feel free to ignore it! Because we'll deal only with rotation about one axis (rotation in one dimension), we can make Equation 11.5 simpler:

$$\Delta K = W_{net} = \pm\tau_{net}\Delta\theta .$$ (Eq. 11.6: **Work-kinetic energy theorem, for rotation**)

We use the plus sign when the torque is in the same direction as the angular displacement, and the minus sign when the torque is opposite to the direction of the angular displacement.

EXAMPLE 11.9 – Comparing the motions
Note – compare this example to Example 6.3. The methods of analysis in that example and this one are virtually identical. Two objects, A and B, are initially at rest. The objects have the same mass and radius. Object A is a uniform solid disk, while object B is a bicycle wheel that can, for this purpose, be considered to be a ring. Each object rotates with no friction about an axis through its center, perpendicular to the plane of the disk/wheel. Identical net torques are then applied to the objects by pulling on strings wrapped around their outer rims. Each net torque is removed once the object it is applied to has accelerated through one complete rotation.
   (a) After the net torques are removed which object has more kinetic energy?
   (b) After the net torques are removed which object has more speed?
   (c) After the net torques are removed, which object has more momentum?

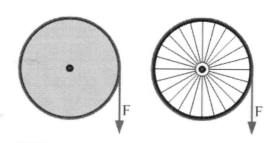

SOLUTION

(a) A diagram of this situation is shown in Figure 11.20. Because the objects start from rest, the angular displacement of each is in the same direction as the net torque (clockwise, in the case shown in Figure 11.20). Because the objects experience equal torques and equal angular displacements the work done on the objects is the same, by Equation 11.6. This means the change in kinetic energy is the same for each, and, because they both start with no kinetic energy, their final kinetic energies are equal.

(b) Unlike Example 6.3, in which the objects had different masses, these objects have the same mass M and the same radius R. This is a rotational situation, however, so what matters is how their rotational inertias compare. Object A, a uniform solid disk rotating about an axis through its center,

Figure 11.20: Diagrams of the disk and wheel. Each object starts from rest and rotates about an axis perpendicular to the page passing through the center of the object. The force exerted on the string wrapped around the object is removed once the object has accelerated through exactly one revolution.

has a rotational inertia of $I_A = \frac{1}{2}MR^2$. Object B, which we are treating as a ring, has a rotational

inertia of $I_B = MR^2$. Thus the relationship between the rotational inertias is $I_A = \frac{1}{2}I_B$. If the

objects have the same kinetic energy but B has a larger rotational inertia then A must have a larger angular speed. Setting the final kinetic energies equal, $K_A = K_B$, gives:

$$\frac{1}{2}I_A\omega_A^2 = \frac{1}{2}I_B\omega_B^2 .$$

Canceling factors of $\frac{1}{2}$ gives: $I_A\omega_A^2 = I_B\omega_B^2$

Bringing in the relationship between the rotational inertias gives: $\frac{1}{2}I_B\omega_A^2 = I_B\omega_B^2$.

This gives $\omega_A = \sqrt{2}\,\omega_B$, so object A has a larger angular speed than object B.

(c) One way to find the angular momenta is as follows:

$$\vec{L}_A = I_A\vec{\omega}_A = \frac{1}{2}I_B\vec{\omega}_A = \frac{1}{2}I_B\left(\sqrt{2}\,\vec{\omega}_B\right) = \frac{1}{\sqrt{2}}I_B\vec{\omega}_B = \frac{1}{\sqrt{2}}\vec{L}_B .$$

Thus, object B, the wheel, has a larger angular momentum than object A, the disk. As in Example 6.3, we can understand this result conceptually. The change in angular momentum is the net torque multiplied by the time over which the net torque acts. Both objects experience identical torques, but because B has a larger rotational inertia, B takes more time to spin through one revolution than A does. Because the torque is applied to B for a longer time, B's change in angular momentum, and final angular momentum, has a larger magnitude than A's.

Related End-of-Chapter Exercises: 22, 23.

Essential Question 11.9: Return to the situation described in Example 11.9, but now object B is replaced by object C, a bicycle wheel of the same mass as object A but with a different radius. Once again, we can treat the bicycle wheel as a ring. The situation described in Example 11.9 is repeated, but this time objects A and C end up with the same rotational kinetic energy and the same angular momentum. How is this possible? Be as quantitative about your answer as you can.

Answer to Essential Question 11.9: If the two objects have the same kinetic energy and angular momentum, they must have the same rotational inertia. This allows us to solve for the radius of object C:

$$I_A = I_C; \qquad \frac{1}{2}M R_A^2 = M R_C^2; \qquad R_C = \frac{1}{\sqrt{2}}R_A.$$

Chapter Summary

Essential Idea
Concepts that we found to be powerful for analyzing motion in previous chapters, such as Newton's Second Law, Conservation of Momentum, and Conservation of Energy, are equally powerful for analyzing motion in a rotational setting.

A General Method for Solving a Problem Involving Newton's Second Law for Rotation
1. Draw a diagram of the situation.
2. Draw a free-body diagram showing all the forces acting on the object.
3. Choose a rotational coordinate system. Pick an appropriate axis to take torques about, and apply Newton's Second Law for Rotation ($\sum \vec{\tau} = I\vec{\alpha}$) to obtain a torque equation.
4. Choose an appropriate *x-y* coordinate system for forces. Apply Newton's Second Law ($\sum \vec{F} = m\vec{a}$) to obtain one or more force equations. The positive directions for the rotational and *x-y* coordinate systems should be consistent with one another.
5. Combine the resulting equations to solve the problem.

Rolling
It can be very helpful to look at rolling as a combination of purely translational motion and purely rotational motion.

Angular Momentum
Angular momentum is a vector, pointing in the direction of angular velocity. The angular momentum of a system can be changed by applying a net torque. If no net torque acts on a system its angular momentum is conserved.

Straight-line motion concept	Analogous rotational motion concept	Connection
Newton's Second Law, $\sum \vec{F} = m\vec{a}$	Second Law for Rotation, $\sum \vec{\tau} = I\vec{\alpha}$	Same form
Momentum: $\vec{p} = m\vec{v}$	Angular momentum: $\vec{L} = I\vec{\omega}$	$L = r\,p\sin\theta$
Translational kinetic energy: $K = \frac{1}{2}mv^2$	Rotational kinetic energy: $K = \frac{1}{2}I\omega^2$	Same form
Impulse: $\vec{F}\Delta t = \Delta\vec{p}$	Rotational impulse: $\vec{\tau}\,\Delta t = \Delta\vec{L}$	Same form
Work: $\Delta K = W_{net} = F_{net}\,\Delta r\cos\phi$	Work: $\Delta K = W_{net} = \tau_{net}\,\Delta\theta\cos\phi = \pm\tau_{net}\,\Delta\theta$	Same form

Table 11.1: The equations we use in rotational situations are completely analogous to those we use in analyzing motion in one, two, or three dimensions.

End-of-Chapter Exercises

Exercises 1 – 12 are conceptual questions that are designed to see if you have understood the main concepts of the chapter.

1. Figure 11.21 shows four different cases involving a uniform rod of length L and mass M is subjected to two forces of equal magnitude. The rod is free to rotate about an axis that either passes through one end of the rod, as in (a) and (b), or passes through the middle of the rod, as in (c) and (d). The axis is marked by the red and black circle, and is perpendicular to the page in each case. This is an overhead view, and we can neglect any effect of the force of gravity acting on the rod. Rank these four situations based on the magnitude of their angular acceleration, from largest to smallest.

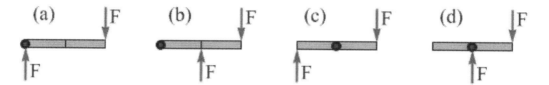

Figure 11.21: Four situations involving a uniform rod that can rotate about an axis being subjected to two forces of equal magnitude. For Exercise 1.

2. A pulley has a mass M, a radius R, and is in the form of a uniform solid disk. The pulley can rotate without friction about a horizontal axis through its center. As shown in Figure 11.22, the string wrapped around the outside edge of the pulley is subjected to an 8.0 N force in case 1, while, in case 2, a block with a weight of 8.0 N hangs down from the string. In which case is the angular acceleration of the pulley larger? Briefly justify your answer.

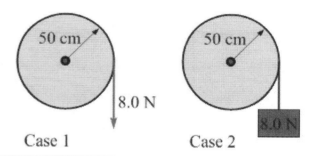

Case 1   Case 2

Figure 11.22: A frictionless pulley with a string wrapped around its outer edge. The string is subjected to an 8.0 N force in case 1, while in case 2 a block with a weight of 8.0 N hangs from the end of the string. For Exercise 2.

3. Consider again the spool shown in Figure 11.23, which we examined in Essential Question 11.2. In Essential Question 11.2, the spool rolled without slipping when a force to the right was exerted on the end of the string, but in this case let's say there is no friction between the spool and the horizontal surface. (a) Does the spool move? If so, which way does it move? (b) Does the spool rotate? If so, which way does it rotate?

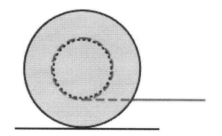

Figure 11.23: A spool with a ribbon wrapped around its axle. A force directed to the right is applied to the end of the ribbon. For Exercise 3.

4. In Exploration 11.2, we looked at how rolling motion can be viewed as a superposition of two simpler motions, pure translation and pure rotation. Have we done this before, broken down a more complicated motion into two simpler motions? If so, in what sort of situation? Comment on the similarities and differences between what we did previously and what we're doing in this chapter, for rolling.

5. A uniform solid cylinder rolls without slipping at constant velocity across a horizontal surface. In Exploration 11.3, we looked at how the net velocity of any point on such a rolling object can be determined. Is there any point on the cylinder with a net velocity directed in exactly the opposite direction as the cylinder's translational velocity? Briefly justify your answer.

6. You take a photograph of a bicycle race. Later, when you get home and look at the photo, you notice that some parts of each bicycle wheel in your photo are blurred, while others are not, or not as badly blurred. Compare the sharpness of the center of a wheel, the top of a wheel, and the bottom of a wheel. (a) Which point do you expect to be the most blurred? Why? (b) Which point do you expect to be in the sharpest focus? Why?

7. You have a race between two objects that have the same mass and radius by rolling them, without slipping, up a ramp. One object is a uniform solid sphere while the other is a ring. (a) Sketch a free-body diagram for one of the objects as it rolls without slipping up the ramp. (b) If the two objects have the same velocity at the bottom of the ramp, which object rolls farther up the ramp before turning around? Briefly justify your answer.

8. Repeat part (b) of the previous exercise, but this time the objects have the same total kinetic energy at the bottom of the ramp.

9. A figure skater is whirling around with her arms held out from her body. (a) What happens to her angular speed when she pulls her arms in close to her body? Why? (b) What happens to the skater's kinetic energy in this process? Explain your answer.

10. A solid cylinder is released from rest at the top of the ramp, and the cylinder rolls without slipping down the ramp. Defining the zero for gravitational potential energy to be at the level of the bottom of the ramp, the cylinder has a gravitational potential energy of 24 J when it is released. Draw a set of energy bar graphs to represent the cylinder's gravitational potential energy, translational kinetic energy, rotational kinetic energy, and total mechanical energy when the cylinder is (a) at the top of the ramp; (b) halfway down the ramp; and (c) at the bottom of the ramp.

11. Repeat Exercise 10, replacing the solid cylinder by a ring.

12. You have a race between a uniform solid sphere and a basketball, by releasing both objects from rest at the top of an incline. Which object reaches the bottom of the ramp first, assuming they both roll without slipping? Justify your answer.

Exercises 13 – 16 are designed to give you practice solving typical Newton's second law for rotation problems. For each exercise start with the following steps: (a) draw a diagram; (b) draw one or more free-body diagrams, as appropriate; (c) choose an appropriate rotational coordinate system and apply Newton's second law for rotation; (d) apply Newton's second law.

13. A block with a mass of 500 g is at rest on a frictionless table. A horizontal string tied to the block passes over a pulley mounted on the edge of the table, and the end of the string hangs down vertically below the pulley. The pulley is a uniform solid disk with a mass of 2.0 kg and a radius of 20 cm that rotates with no friction about an axis through its center. You then exert a constant force of 4.0 N down on the end of the string. The goal is to determine the acceleration of the block. Use $g = 10$ m/s². Parts (a) – (d) as described above. (e) Find the block's acceleration.

14. Repeat the previous exercise, but now there is some friction between the block and the table. The coefficient of static friction is $\mu_S = 0.40$, while the coefficient of kinetic friction is $\mu_K = 0.30$.

15. A 2.0-m long board is placed with one end on the floor and the other resting on a box that has a height of 30 cm. A uniform solid sphere is released from rest from the higher end of this board and rolls without slipping to the lower end. The goal is to determine how long the sphere takes to move from one end of the board to the other. Parts (a) – (d) as described above. (e) Combine your equations to determine the sphere's acceleration. (f) How long does it take the sphere to reach the lower end of the board?

16. A uniform solid cylinder with a mass of 2.0 kg rests on its side on a horizontal surface. A ribbon is wrapped around the outside of the cylinder with the end of the ribbon coming away from the cylinder horizontally from its highest point. When you exert a constant force of 4.0 N on the cylinder, the cylinder rolls without slipping in the direction of the force. The goal of this exercise is to determine the cylinder's acceleration. Parts (a) – (d) as described above. (e) Find the magnitude and direction of the force of friction exerted on the cylinder. (f) Find the cylinder's acceleration.

Exercises 17 – 21 deal with rolling situations.

17. A particular wheel has a radius of 50 cm. It rolls without slipping exactly half a rotation. (a) What is the translational distance moved by the wheel? (b) Considering motion due to the wheel's rotation only, what is the distance traveled by a point on the outer edge of the wheel? (c) Consider now the magnitude of the total displacement experienced by the point on the outer edge of the wheel that started at the top of the wheel. Is this equal to the sum of your two answers from (a) and (b)? Explain why or why not. (d) Work out the magnitude of the displacement of the point referred to in (c).

18. One end of a 2.0-m long board rests on a cylinder that has been placed on its side on a horizontal surface. You hold the other end of the board so the board is horizontal. When you walk forward, the cylinder rolls so the board does not slip on the cylinder, and the cylinder also rolls without slipping across the floor. When you move forward 1.0 m, how far does the cylinder move?

19. A particular point on a wheel is halfway between the center and the outer edge. When the point is at the same distance from the ground as the center of the wheel, the point's speed is 25.2 m/s. If the wheel has a radius of 35.0 cm and is rolling without slipping, find the translational speed of the wheel.

20. A solid sphere is released from rest and rolls without slipping down a ramp inclined at 12° to the horizontal. What is the sphere's speed when it is 1.0 m (measuring vertically) below the level it started?

21. Return to the situation described in Exercise 20, but this time the incline is changed to 6°. The sphere again rolls without slipping starting from rest. Comparing the sphere's speed in both cases when it is 1.0 m (measuring vertically) below its starting point, in which case is the sphere moving faster? Rather then doing another calculation, see if you can come up with a conceptual argument to justify your answer.

Exercises 22 – 31 are modeled after similar exercises in previous chapters. Note the similarities between how we analyze rotational situations and how we analyze straight-line motion situations.

22. Two uniform solid disks, A and B, are initially at rest. The mass of disk B is two times larger than that of disk A. Identical net torques are then applied to the two disks, giving them each an angular acceleration as they rotate about their centers. Each net torque is removed once the object it is applied to has rotated through two revolutions. After both net torques are removed, how do: (a) the kinetic energies compare? (b) the angular speeds compare? (c) the angular momenta compare? (Compare this to Exercise 9 in Chapter 6.)

23. Repeat Exercise 22, assuming that both net torques are removed after the same amount of time instead. (Compare this to Exercise 10 in Chapter 6.)

24. Two identical grinding wheels of mass m and radius r are spinning about their centers. Wheel A has an initial angular speed of ω, while wheel B has an initial angular speed of 2ω. Both wheels are being used to sharpen tools. As shown in Figure 11.24, in both cases the tool is being pressed against the wheel with a force F directed toward the center of the wheel, and the coefficient of kinetic friction between the wheel and the tool is μ_K.

The tool does not move from the position shown in the diagram. (a) If it takes wheel A a time T to come to a stop, how long does it take for wheel B to come to a stop? (b) Find an expression for T in terms of the variables specified in the exercise. (c) If wheel A rotates through an angle θ before coming to rest, through what angle does wheel B rotate before coming to rest? (d) Find an expression for θ in terms of the variables specified in the exercise. (Compare this to Exercise 52 in Chapter 6.)

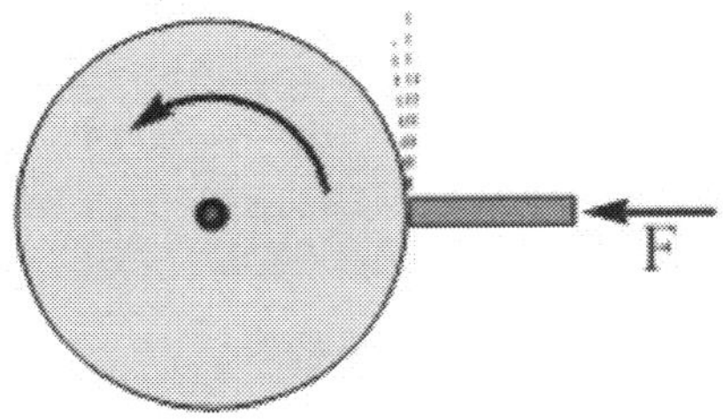

25. Return to the situation described in Exercise 24. How would T, the stopping time for wheel A, change if (a) m was doubled? (b) ω was doubled? (c) μ_K was doubled?

(Compare this to Exercise 53 in Chapter 6.)

Figure 11.24: Sharpening a tool by holding it against a grinding wheel, for Exercises 24 – 26.

26. Return to the situation described in Exercise 24. How would θ, the angle wheel A rotates through before stopping, change if (a) m was doubled? (b) ω was doubled? (c) μ_K was doubled? (Compare this to Exercise 54 in Chapter 6.)

27. You pick up a bicycle wheel, with a mass of 800 grams and a radius of 40 cm, and spin it so the wheel rotates about its center. Assume that the mass of the wheel is concentrated in the rim. The initial angular speed is 5.0 rad/s, but after 10 s the angular speed is 3.0 rad/s. The goal here is to determine the magnitude of the frictional torque acting to slow the wheel, assuming it to be constant. (a) Sketch a diagram of the situation. (b) Choose a positive direction, and show this on the diagram. (c) Draw a free-body diagram of the wheel, focusing on the torque(s) acting on the wheel. (d) Write an expression for the net torque acting on the wheel. (e) Write an expression representing the wheel's change in angular momentum over the 10-second period. (f) Use the equation $\vec{\tau}\,\Delta t = \Delta \vec{L}$ to relate the expressions you wrote down in parts (d) and (e). (g) Solve for the frictional torque acting on the wheel. (Compare this to Exercise 23 in Chapter 6.)

28. At a time $t = 0$, a bicycle wheel with a mass of 4.00 kg has an angular velocity of 5.00 rad/s directed clockwise. For the next 8.00 seconds it then experiences a net torque, as shown in the graph in Figure 11.25 (taking clockwise to be positive). The wheel rotates about its center, and we can treat the wheel as a ring with a radius of $\dfrac{1}{\sqrt{2}}$ m. (a) Sketch a graph of the wheel's angular momentum as a function of time. (b) What is the cart's maximum angular speed during the 8.00-second interval the varying torque is being applied? At what time does the cart reach this maximum speed? (c) What is the cart's minimum angular speed during the 8.00-second interval the varying torque is being applied? At what time does the cart reach this minimum speed? (Compare this to Exercise 24 in Chapter 6.)

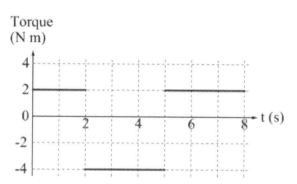

Figure 11.25: A plot of the torque applied to a bicycle wheel as a function of time, for Exercise 28.

29. Two blocks are connected by a string that passes over a frictionless pulley, as shown in Figure 11.26. Block A, with a mass $m_A = 2.0$ kg, rests on a ramp measuring 3.0 m vertically and 4.0 m horizontally. Block B hangs vertically below the pulley. The pulley has a mass of 1.0 kg, and can be treated as a uniform solid disk that rotates about its center. Note that you can solve this exercise entirely using forces and the constant-acceleration equations, but see if you can apply energy ideas instead. Use $g = 10$ m/s². When the system is released from rest, block A accelerates up the slope and block B accelerates straight down. When block B has fallen through a height $h = 2.0$ m, its speed is $v = 6.0$ m/s. (a) At any instant in time, how does the speed of block A compare to that of block B? (b) Assuming there is no friction acting on block A, what is the mass of block B? (Compare this to Exercise 44 in Chapter 7.)

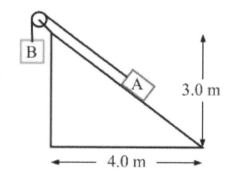

Figure 11.26: Two blocks connected by a string passing over a pulley, for Exercises 29 and 30.

30. Repeat Exercise 29, this time accounting for friction. If the coefficient of kinetic friction for the block A – ramp interaction is 0.625, what is the mass of block B? (Compare this to Exercise 45 in Chapter 7.)

31. A uniform solid sphere of mass m is released from rest at a height h above the base of a loop-the-loop track, as shown in Figure 11.27. The loop has a radius R. What is the minimum value of h necessary for the sphere to make it all the way around the loop without losing contact with the track? Express your answer in terms of R, and assume that the sphere's radius is much smaller than the loop's. (Compare this to Exercise 49 in Chapter 7.)

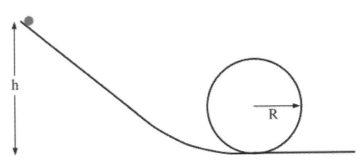

Figure 11.27: A solid sphere released from rest from a height h above the bottom of a loop-the-loop track, for Exercise 31.

Exercises 32 – 34 deal with angular momentum conservation.

32. A beetle with a mass of 20 g is initially at rest on the outer edge of a horizontal turntable that is also initially at rest. The turntable, which is free to rotate with no friction about an axis through its center, has a mass of 80 g and can be treated as a uniform disk. The beetle then starts to walk around the edge of the turntable, traveling at an angular velocity of 0.060 rad/s clockwise **with respect to the turntable**. (a) Qualitatively, what does the turntable do while the beetle is walking? Why? (b) With respect to you, motionless as you watch the beetle and turntable, what is the angular velocity of the beetle? What is the angular velocity of the turntable? (c) If a mark was placed on the turntable at the beetle's starting point, how long does it take the beetle to reach the mark? (d) Upon reaching the mark, the beetle stops. What does the turntable do? Why?

33. A bullet with a mass of 12 g is fired at a wooden rod that hangs vertically down from a pivot point that passes through the upper end of the rod. The bullet embeds itself in the lower end of the rod and the rod/bullet system swings up, reaching a maximum angular displacement of 60° from the vertical. The rod has a mass of 300 g, a length of 1.2 m, and we can assume the rod rotates without friction about the pivot point. What is the bullet's speed when it hits the rod? Assume the bullet is traveling horizontally when it hits the rod, and use $g = 10 \text{ m/s}^2$.

34. A particular horizontal turntable can be modeled as a uniform disk with a mass of 200 g and a radius of 20 cm that rotates without friction about a vertical axis passing through its center. The angular speed of the turntable is 2.0 rad/s. A ball of clay, with a mass of 40 g, is dropped from a height of 35 cm above the turntable. It hits the turntable at a distance of 15 cm from the middle, and sticks where it hits. Assuming the turntable is firmly supported by its axle so it remains horizontal at all times, find the final angular speed of the turntable-clay system.

Use conservation of energy to solve Exercises 35 – 38. For each exercise begin by (a) writing down the energy conservation equation and choosing a zero level for gravitational potential energy; (b) identifying the terms that are zero and eliminating them; (c) writing out expressions for the remaining terms, remembering to account for both translational kinetic energy and rotational kinetic energy.

35. A uniform solid disk is released from rest at the top of a ramp, and rolls without slipping down the ramp. The goal of the exercise is to determine the disk's speed when it reaches a level 50 cm below (measured vertically) its starting point. Parts (a) – (c) as described above. (d) What is that speed?

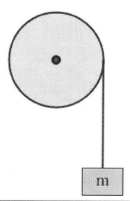

36. The pulley shown in Figure 11.28 has a mass $M = 2.0$ kg and radius $R = 50$ cm, and can be treated as a uniform solid disk that can rotate about its center. The block (which has a mass of 800 g) hanging from the string wrapped around the pulley is then released from rest. The goal of the exercise is to determine the speed of the block when it has dropped 1.0 m. Parts (a) – (c) as described above. (d) What is the block's speed after dropping through 1.0 m?

Figure 11.28: A block and a pulley, for Exercise 36.

37. A uniform solid sphere with a mass $M = 1.0$ kg and radius $R = 40$ cm is mounted on a frictionless vertical axle that passes through the center of the sphere. The sphere is initially at rest. You then pull on a string wrapped around the sphere's equator, exerting a constant force of 5.0 N. The string unwraps from the sphere when you have moved the end of the string through a distance of 2.0 m. The goal of the exercise is to determine the resulting angular speed of the sphere. Parts (a) – (c) as described above. (d) What is the resulting angular speed?

38. A uniform solid sphere with a mass of $M = 1.6$ kg and radius $R = 20$ cm is rolling without slipping on a horizontal surface at a constant speed of 2.1 m/s. It then encounters a ramp inclined at an angle of 10° with the horizontal, and proceeds to roll without slipping up the ramp. The goal of this exercise is to determine the distance the sphere rolls up the ramp (measured along the ramp) before it turns around. Parts (a) – (c) as described above. (d) How far does the sphere roll up the ramp? (e) Which of the values given in this exercise did you not need to find the solution?

General Problems and Conceptual Questions.

39. Figure 11.29 shows four different cases involving a uniform rod of length L and mass M is subjected to two forces of equal magnitude. The rod is free to rotate about an axis that either passes through one end of the rod, as in (a) and (b), or passes through the middle of the rod, as in (c) and (d). The axis is marked by the red and black circle, and is perpendicular to the page in each case. This is an overhead view, and we can neglect any effect of the force of gravity acting on the rod. If the rod has a length of 1.0 m, a mass of 3.0 kg, and each force has a magnitude of 5.0 N, determine the magnitude and direction of the angular acceleration of the rod in (a) Case (a); (b) Case (b); (c) Case (c); (d) Case (d).

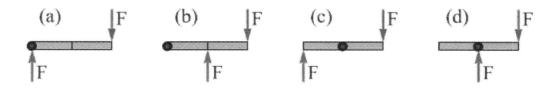

Figure 11.29: Four situations involving a uniform rod that can rotate about an axis being subjected to two forces of equal magnitude. For Exercise 39.

40. A pulley has a mass M, a radius R, and is in the form of a uniform solid disk. The pulley can rotate without friction about a horizontal axis through its center. As shown in Figure 11.30, the string wrapped around the outside edge of the pulley is subjected to an 8.0 N force in case 1, while in case 2 a block with a weight of 8.0 N hangs down from the string. If $M = 2.0$ kg and $R = 50$ cm, calculate the angular acceleration of the pulley in (a) case 1; (b) case 2.

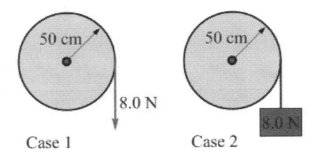

Case 1   Case 2

Figure 11.30: A frictionless pulley with a string wrapped around its outer edge. The string is subjected to an 8.0 N force in case 1, while in case 2 a block with a weight of 8.0 N hangs from the end of the string. For Exercise 40.

41. Atwood's machine is a system consisting of two blocks that have different masses connected by a string that passes over a frictionless pulley, as shown in Figure 11.31. The pulley has a mass m_p. Compare the tension in the part of the string just above the block with the larger mass, M, to that in the part of the string just above the block with the smaller mass, m, in the following cases: (a) You hold on to the smaller-mass block to keep the system at rest; (b) The system is released from rest; (c) You hold on to the smaller-mass block and pull down so the blocks move with constant velocity. Justify your answers in each case.

42. Atwood's machine is a system consisting of two objects connected by a string that passes over a frictionless pulley, as shown in Figure 11.31. In Chapter 5, we neglected the effect of the pulley, but now we know how to account for the pulley's impact on the system. (a) If the two objects have masses of M and m, with $M > m$, and the pulley is in the shape of a uniform solid disk and has a mass m_p, derive an expression for the acceleration of either block, in terms of the given masses and g. (b) What does the expression reduce to in the limit where the mass of the pulley approaches zero? (c) How does accounting for the fact that the pulley has a non-zero mass affect the magnitude of the acceleration of a block?

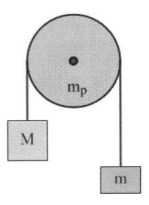

Figure 11.31: Atwood's machine – a device consisting of two objects connected by a string that passes over a pulley. For Exercises 41 – 43.

43. Consider again the Atwood's machine described in Exercise 42 and pictured in Figure 11.31. (a) If $M = 500$ g, $m = 300$ g, and the pulley mass is $m_p = 400$ g, what is the magnitude of the acceleration of one of the blocks? (b) If the system is released from rest, what is the angular velocity of the pulley 2.0 seconds after the motion begins? The pulley has a radius of 10 cm.

44. A particular double-pulley consists of a small pulley of radius 20 cm mounted on a large pulley of radius 50 cm, as shown in Figure 11.32. A block of mass 2.0 kg hangs from a string wrapped around the large pulley. To keep the system at rest, what mass block should be hung from the small pulley?

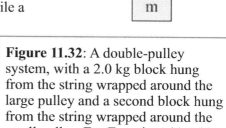

45. A particular double pulley consists of a small pulley of radius 20 cm mounted on a large pulley of radius 50 cm, as shown in Figure 11.32. The pulleys rotate together, rather than independently. A block of mass 2.0 kg hangs from a string wrapped around the large pulley, while a second block of mass 2.0 kg hangs from the small pulley. Each pulley has a mass of 1.0 kg and is in the form of a uniform solid disk. Use $g = 9.8$ m/s^2. (a) What is the acceleration of the block attached to the large pulley? (b) What is the acceleration of the block attached to the small pulley?

Figure 11.32: A double-pulley system, with a 2.0 kg block hung from the string wrapped around the large pulley and a second block hung from the string wrapped around the small pulley. For Exercises 44 – 46.

46. A pulley consists of a small uniform disk of radius 0.50 m mounted on a larger uniform disk of radius 1.0 m. Each disk has a mass of 1.0 kg. The pulley can rotate without friction about an axis through its center. As shown in Figure 11.32, a block with mass $m = 1.0$ kg hangs down from the larger disk while a block of mass M hangs down from the smaller disk. If the angular acceleration of the system has a magnitude of 1.0 rad/s^2 what is the value of M? Consider all possible answers, and use $g = 10$ m/s^2.

47. A yo-yo consists of two identical disks, each with a mass of 40 g and a radius of 4.0 cm, joined by a small cylindrical axle with negligible mass and a radius of 1.0 cm. When the yo-yo is released it essentially rolls without slipping down the string wrapped around the axle. If the end of the string (the one you would hold) remains fixed in place, determine the acceleration of the yo-yo. Use $g = 9.8$ m/s^2.

48. As shown in Figure 11.33, blocks A and B are connected by a massless string that passes over the outer edge of a pulley that is a uniform solid disk. The mass of block A is equal to that of block B; the mass of the pulley, coincidentally, is also the same as that of block A. When the system is released from rest it experiences a constant (and non-zero) acceleration. There is no friction between block A and the surface. Use $g = 10.0$ m/s^2. (a) What is the acceleration of the system? (b) The two parts of the string have different tensions. In which part of the string is the tension larger, between block A and the pulley or between the pulley and block B? Briefly justify your answer. (c) If the tension in one part of the string is 3.00 N larger than the tension in the other part, what are the values of the tensions in the two parts of the string?

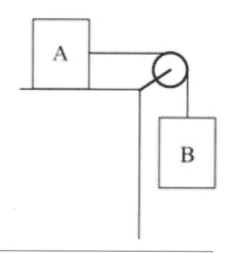

Figure 11.33: Two blocks connected by a string passing over a pulley, for Exercises 48 and 49.

49. Repeat Exercise 48, but this time there is some friction between block A and the surface, with a coefficient of kinetic friction of 0.500.

50. A spool has a string wrapped around its axle, with the string coming away from the underside of the spool. The spool is on a ramp inclined at 20° with the horizontal, as shown in Figure 11.34. There is no friction between the spool and the ramp. Assuming you can exert as much or as little force on the end of the string as you wish (always directed up the slope) which of the following situations are possible? If so, explain how the situation could be achieved; if not, explain why not. (a) The spool remains completely motionless. (b) The spool rotates about its center but does not move up or down the ramp. (c) The spool has no rotation but moves down the ramp.

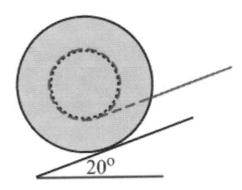

Figure 11.34: A spool on a ramp inclined at 20° with the horizontal. The string wrapped around the spool's axle comes away from the spool on its underside. Any force exerted on the end of the string is exerted up the slope. For Exercises 50 and 51.

51. Return to the situation described in Exercise 50 and shown in Figure 11.34, but now there is friction between the spool and the ramp. Is it possible for the spool to remain completely motionless now? If so, explain how. If not, explain why not.

52. Consider the spool shown in Figure 11.35. The spool has a radius R, while the spool's axle has a radius of $R/2$. There is some friction between the spool and the horizontal surface it is on, so that when a modest tension is exerted on the string the spool may roll one way or the other without slipping. It turns out that when the angle θ between the string and the vertical is larger than some critical value θ_C, the spool rolls without slipping one way; when $\theta < \theta_C$, the spool rolls the other way, and when $\theta = \theta_C$, the spool remains at rest. (a) Find θ_C. (b) Which way does the spool roll when $\theta < \theta_C$?

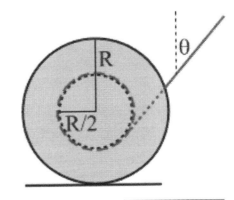

Figure 11.35: A spool on a horizontal surface, for Exercise 52.

53. A spool consists of two disks, each of radius R and mass M, connected by a cylindrical axle of radius $R/2$ and mass M. When an upward force of magnitude F is exerted on a string wrapped around the axle, the spool will roll without slipping as long as F is not too large. (a) What is the spool's rotational inertia, in terms of M and R, about an axis through its center? (b) If the coefficient of static friction between the spool and the horizontal surface it is on is 0.50, what is the maximum value F can be, in terms of M and g, for the spool to roll without slipping?

54. A solid sphere rolls without slipping when it is released from rest at the top of a ramp that is inclined at 30° with respect to the horizontal, but, if the angle exceeds 30°, the sphere slips as it rolls. Calculate the coefficient of static friction between the sphere and the incline.

55. While fixing your bicycle, you remove the front wheel from the frame. A bicycle wheel can be approximated as a ring, with all the mass of the wheel concentrated on the wheel's outer edge. The wheel has a mass M, a radius R, and it is initially spinning at a particular angular velocity. There is a constant frictional torque that is causing the wheel to slow down, however. You also have a uniform solid disk of the same mass and radius as the bicycle wheel. It also has the same initial angular velocity and the same frictional torque as the wheel. Which of these objects will spin for the longer time? Justify your answer.

56. As shown in Figure 11.36, a bowling ball of mass M and radius $R = 20.0$ cm is released with an initial translational velocity of $\vec{v}_0 = 14.0$ m/s to the right and an initial angular velocity of $\vec{\omega}_0 = 0$. The bowling ball can be treated as a uniform solid sphere. The coefficient of kinetic friction between the ball and the surface is $\mu_K = 0.200$. The force of kinetic friction causes a linear acceleration, as well as a torque that causes the ball to spin. The ball slides along the horizontal surface for some time, and then rolls without slipping at constant velocity after that. Use $g = 10$ m/s². (a) Draw the free-body diagram of the ball showing all the forces acting on it while it is sliding. (b) What is the acceleration of the ball while it is sliding? (c) What is the angular acceleration of the ball while it is sliding? (d) How far does the ball travel while it is sliding? (e) What is the constant speed of the ball when it rolls without slipping?

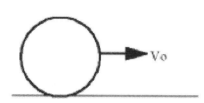

Figure 11.36: The initial state of a bowling ball, for Exercise 56.

57. Figure 11.37 shows the side view of a meter stick that can rotate without friction about an axis passing through one end. Pennies (of negligible mass in comparison to the mass of the meter stick) have been placed on the meter stick at regular intervals. When the meter stick is released from rest, it rotates about the axis. Some of the pennies remain in contact with the meter stick while some lose contact with it. (a) Which pennies do you expect to lose contact with the meter stick, the ones close to the axis or the ones farther from it? (b) Determine the initial acceleration of the right end of the meter stick, and of the center of the meter stick, to help justify your conclusion in (a).

Figure 11.37: A side view of a meter stick, with pennies resting on it at regular intervals, that is initially held horizontal. The meter stick can rotate without friction about an axis through the left end. For Exercises 57 and 58.

58. Return to the situation described in Exercise 57 and shown in Figure 11.37. Assuming the axis is at the 0-cm mark of the meter stick, determine the point on the meter stick beyond which the pennies will lose contact with the meter stick when the system is released from rest.

59. A particularly large playground merry-go-round is essentially a uniform solid disk of mass $4M$ and radius R that can rotate with no friction about a central axis. You, with a mass M, are a distance of $R/2$ from the center of the merry-go-round, rotating together with it at an angular velocity of 2.4 rad/s clockwise (when viewed from above). You then walk to the outside of the merry-go-round so you are a distance R from the center, still rotating with the merry-go-round. Consider you and the merry-go-round to be one system. (a) When you walk to the outside of the merry-go-round, does the angular momentum of the system increase, decrease, or stay the same? Why? (b) Does the kinetic energy of the system increase, decrease, or stay the same? Why? (c) If you started running around the outer edge of the merry-go-round, at what angular velocity would you have to run to make the merry-go-round alone come to a complete stop? Specify the magnitude and direction.

60. Consider the following situations. For each, state whether you would apply energy methods, torque/rotational kinematics methods, or either to solve the exercise. You don't need to solve the exercise. (a) Find the final speed of a uniform solid sphere that rolls without slipping down a ramp inclined at 8.0° with the horizontal, if the sphere is released from rest and the vertical component of its displacement is 1.0 m. (b) Find the time it takes the sphere in (a) to reach the bottom of the ramp. (c) Find the number of rotations the sphere in (a) makes as it rolls down the ramp, if the sphere's radius is 15 cm.

61. Return to Exercise 60, and this time solve each part.

62. The planet Earth orbits the Sun in an orbit that is roughly circular. Assuming the orbit is exactly circular, which of the following is conserved as the Earth travels along its orbit? (a) Its momentum? (b) Its angular momentum, relative to an axis passing through the center, and perpendicular to the plane, of the orbit? (c) Its translational kinetic energy? (d) Its gravitational potential energy? (e) Its total mechanical energy? For any that are not conserved, explain why.

63. A typical comet orbit ranges from relatively close to the Sun to many times farther than the Sun. Which of the following is conserved as the comet travels along its orbit? (a) Its angular momentum, relative to an axis passing through the Sun, and perpendicular to the plane, of the orbit? (b) Its translational kinetic energy? (c) Its gravitational potential energy? (d) Its total mechanical energy? For any that are not conserved, explain why.

64. Two of your classmates, Alex and Shaun, are carrying on a conversation about a physics problem. Comment on each of their statements.

> *Alex: In this situation, we have to draw the free-body diagram for a sphere that is rolling without slipping up an incline. OK, so, there's a force of gravity acting down, and a normal force perpendicular to the surface. Is there a friction force?*

> **Shaun: Don't we need another force directed up the incline, in the direction of motion?**

> *Alex: I don't think so. I think we just need to add a kinetic friction force down the incline, opposing the motion.*

> **Shaun: Wait a second – isn't it static friction? Isn't it always static friction when something rolls without slipping?**

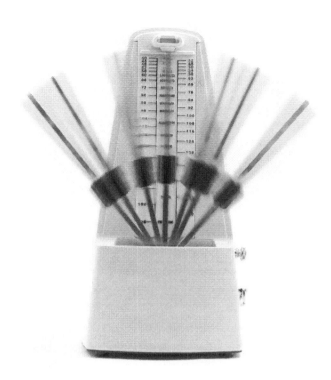

The photograph shows a time-sequence of a metronome. Understanding the motion of such oscillating systems is the main focus of this chapter.

Photo credit:
Libby Chapman / iStockPhoto.

Chapter 12 – Simple Harmonic Motion

CHAPTER CONTENTS

We now turn our attention to oscillating systems, such as an object bobbing up and down on the end of a spring, or a child swinging on a playground swing. We'll focus on a simple model, in which the total mechanical energy is constant. This is a reasonable starting point for most oscillating systems. Our own starting point, however, will be to consider how to incorporate springs into our force and energy perspectives. One of the key things to remember as you work through this chapter is that, while we are certainly covering some new material, you can go a long way towards understanding oscillating systems by applying ideas that you are already familiar with, including forces and, especially, energy.

There are many practical applications of oscillating systems, or systems that make use of springs and pendulums. The oscillations of a quartz crystal are at the heart of how a digital watch keeps time. Elevator shafts generally have a large spring at the bottom, just in case something happens to the elevator. People fasten bungee cords to their legs and feet and jump off bridges, relying on the springiness of the bungee cords to keep them from crashing into the ground. Springs and shock absorbers in cars are designed so that passengers do not oscillate very much when a car goes over a bump. While we will not get into all the details of all of these applications, you should come away from this chapter with a good conceptual understanding of the basic principles that underlie all such systems.

12-1 Hooke's Law

We probably all have some experience with springs. One observation we can make is that it doesn't take much force to stretch or compress a spring a small amount, but the more we try to compress or stretch it, the more force we need. We'll use a model of an ideal spring, in which the magnitude of the force associated with stretching or compressing the spring is proportional to the distance the spring is stretched or compressed.

The equation describing the proportionality of the spring force with the displacement of the end of the spring from its natural length is known as Hooke's law.

$$\vec{F}_{Spring} = -k\,\vec{x}\,.$$ (Equation 12.1: **Hooke's Law**)

The negative sign is associated with the restoring nature of the force. When you displace the end of the spring in one direction from its equilibrium position, the spring applies a force in the opposite direction, essentially in an attempt to return the system toward the equilibrium position (the position where the spring is at its natural length, neither stretched nor compressed). The force applied by the spring is proportional to the distance the spring is stretched or compressed relative to its natural length.

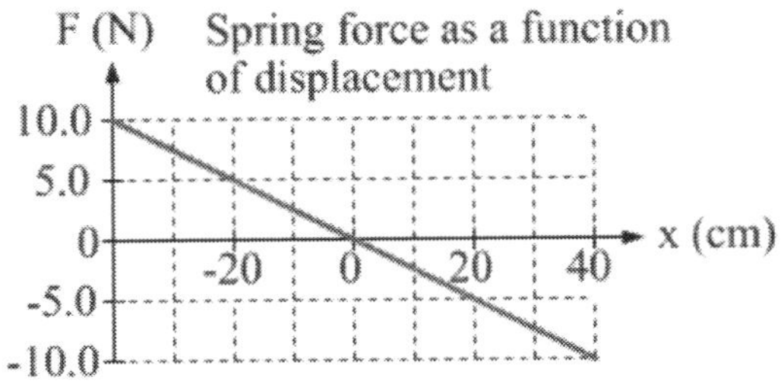

The k in the Hooke's law equation is known as the **spring constant**. This is a measure of the stiffness of the spring. Say you have two different springs and you stretch them the same amount from equilibrium. The one that requires more force to maintain that stretch has the larger spring constant. Figure 12.1 shows the Hooke's law relationship as a graph of force as a function of the amount of compression or stretch of a particular spring from its natural length.

Figure 12.1: A graph of the force applied by a particular spring as a function of the displacement of the end of the spring from its equilibrium position.

The Hooke's law relationship is illustrated in Figure 12.2, where $x = 0$ means the spring is neither stretched nor compressed from its natural length. A block attached to spring has been released and is oscillating on a frictionless surface. Free-body diagrams are shown in Figure 12.2, illustrating how the force exerted by the spring on the block depends on the displacement of the end of the spring from its equilibrium position.

Figure 12.2: A block attached to an ideal spring oscillates on a frictionless surface. By looking at the free-body diagrams of the block when the block is at various positions, we can see that the force applied by the spring on the block is proportional to the displacement of the end of the spring from its equilibrium position, and opposite in direction to that displacement.

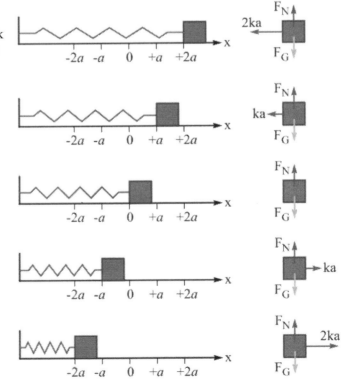

EXAMPLE 12.1 – Initial acceleration of a block

A block of mass 300 g is attached to a horizontal spring that has a spring constant of 6.0 N/m. The block is on a horizontal frictionless surface. You release the block from rest when the spring is stretched by 20 cm.

    (a) Sketch a diagram of the situation, and a free-body diagram of the block immediately after you release the block.
    (b) Determine the block's initial acceleration.
    (c) What happens to the block's free-body diagram as the block moves to the left?

SOLUTION

(a) The diagram and free-body diagram are shown in Figure 12.3. After you release the block, only three forces act on the block. The downward force of gravity is balanced by the upward normal force applied by the surface. The third force is the force applied by the spring. The spring force is directed to the left because the end of the spring has been displaced to the right from its equilibrium position.

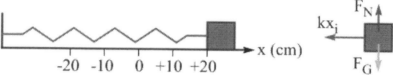

Figure 12.3: A diagram of the block and spring, and the free-body diagram of the block, immediately after the block is released from rest.

(b) Here we can apply Newton's Second Law horizontally, $\sum \vec{F}_x = m\vec{a}$, taking right to be the positive x-direction. This gives: $-F_{Spring} = m\vec{a}$.

Now we can bring in Equation 12.1, $\vec{F}_{Spring} = -k\vec{x}$, to get: $-kx_i = m\vec{a}_i$.

Note that we use only one minus sign in the equation because we're substituting for the magnitude of the spring force only. The one minus sign represents the direction of the spring force, which is to the left. Solving for the block's initial acceleration gives:

$$\vec{a}_i = +\frac{kx_i}{m} = -\frac{(6.0\,\text{N/m})(0.20\,\text{m})}{0.30\,\text{kg}} = -4.0\,\text{N/kg}.$$

The initial acceleration is 4.0 N/kg to the left.

(c) As the block moves to the left, nothing changes about the vertical forces, but the spring force steadily decreases in magnitude because the stretch of the spring steadily decreases. Once the block goes past the equilibrium position, the spring force points to the right, and increases in magnitude as the compression increases. The dependence of the spring force on the block's position is shown, for five different positions, in Figure 12.2.

Related End-of-Chapter Exercises: 16, 55.

Essential Question 12.1: Let's say we estimated the time it takes the block in Example 12.1 to reach equilibrium, by assuming the block's acceleration is constant at 4.0 N/kg to the left. Is our estimated time smaller than or larger than the time it actually takes the block to reach the equilibrium point?

Answer to Essential Question 12.1: This estimated time is less than the actual time. The closer the block gets to the equilibrium position, the smaller the force that is exerted on it by the spring, and the smaller the magnitude of the block's acceleration. Because the block generally has a smaller acceleration than the acceleration we used in the constant-acceleration analysis, it will take longer to reach equilibrium than the time we calculated with the constant-acceleration analysis. Thus, remember not to use constant-acceleration equations in harmonic motion situations! We'll learn how to calculate exact times in sections 12-4 to 12-6.

12-2 Springs and Energy Conservation

Now that we have seen how to incorporate springs into a force perspective, let's go on to consider how to fit springs into what we know about energy.

EXPLORATION 12.2 – Another kind of potential energy

Step 1 – *Attach a block to a spring, and position the block so that the spring is stretched. Let's neglect friction, so when you release the block from rest it oscillates back and forth about the equilibrium position. What is going on with the energy of the system as the block oscillates?* As the block oscillates, its speed increases from zero to some maximum value, then decreases to zero again, and keeps doing this over and over. The kinetic energy of the system does exactly the same thing, since it is proportional to the square of this speed. Where does the energy go when the kinetic energy decreases, and where does it comes from when the kinetic energy increases?

The energy is stored as potential energy in the spring. This is similar to what happens when we throw a ball up into the air. As the ball rises, the ball's loss of kinetic energy is offset by the gain in the gravitational potential energy of the Earth-ball system, and then that potential energy is transformed back into kinetic energy. Compressed or stretched springs also store potential energy. Such energy is known as **elastic potential energy**.

Step 2 - *Consider the graph of force, as a function of the displacement of the end of the spring, shown in Figure 12.4. As we did in Chapter 6, defining the change in gravitational potential energy to be the negative of the work done by gravity on an object, find an expression for the change in elastic potential energy as the end of the spring is displaced from its equilibrium position (x = 0) to some arbitrary final position x. Make use of the fact that work is the area under the force-versus-position graph in Figure 12.4.*

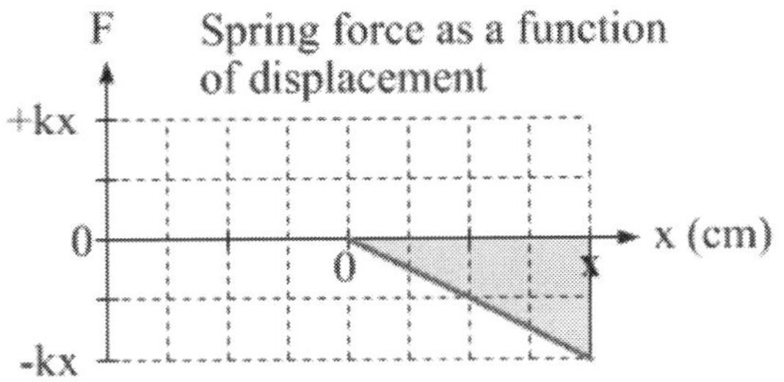

Figure 12.4: The work done by a spring when its end is displaced from the equilibrium position to a point x away from equilibrium is represented by the shaded area in the graph.

The area in question is that of the right-angled triangle shown in Figure 12.4. The area is negative because the force is negative the entire time. The area under the curve is given by:

$$\text{area} = -\frac{1}{2}\,\text{base} \times \text{height} = -\frac{1}{2}x(kx) = -\frac{1}{2}kx^2 .$$

This area represents the work done by the spring. This work is negative because the spring force is opposite in direction to the displacement. Because ΔU_e, the change in the elastic potential energy, is the negative of the work, we have $\Delta U_e = \frac{1}{2}kx^2$ in this case.

Step 3 – *How much elastic potential energy is stored in the spring when the spring is at its natural length?* None. If we attach a block to such a spring and release the block from rest, no motion occurs because the system is at equilibrium. There is no transformation of elastic potential energy into kinetic energy because the system has no elastic potential energy when the spring is at its natural length – the equilibrium position is the zero for elastic potential energy.

Step 4 – *Combine the results from parts 2 and 3 to determine the expression for the elastic potential energy stored in a spring when the end of the spring is displaced a distance x from its equilibrium position.* In step 3 we found the change in elastic potential energy in displacing the end of the spring from its equilibrium position to a point x away from equilibrium to be $\Delta U_e = \frac{1}{2}kx^2$. This change in elastic potential energy is equal to the final elastic potential energy minus the initial elastic potential energy. However, we found the initial elastic potential energy to be zero in step 3, which means the expression for elastic potential energy is simply:

$$U_e = \frac{1}{2}kx^2 .$$ (Equation 12.2: **Elastic potential energy**)

Key ideas: Compressed or stretched springs store energy – this is known as elastic potential energy. For an ideal spring, the elastic potential energy is $U_e = \frac{1}{2}kx^2$.
Related End-of-Chapter Exercises: 9, 48.

Now that we know the form of the elastic potential energy equation, we can incorporate springs into the conservation of energy equation we first used in chapter 7:

$$K_i + U_i + W_{nc} = K_f + U_f .$$ (Equation 7.1)

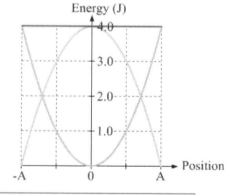

Graphs of the energies as a function of position are interesting. Consider a block attached to a spring. The block is oscillating back and forth on a frictionless surface, so the total mechanical energy stays constant. An easy way to graph the kinetic energy is to exploit energy conservation, $E = K + U$. Solving for the kinetic energy as a function of position gives:

$$K = E - U = E - \frac{1}{2}kx^2 .$$

Graphs of the energies as a function of position are shown in Figure 12.5, for a situation in which the total mechanical energy is 4.0 J. After tracing out the complete energy curves over half an oscillation, the system re-traces these energy-versus-position plots as the block oscillates.

Figure 12.5: Graphs of the system's kinetic energy (zero at -A and A), elastic potential energy (zero at x = 0), and total mechanical energy (constant), as a function of position. The system traces over each of the energy graphs every half oscillation.

Essential Question 12.2: Consider a system consisting of a block attached to an ideal spring. The block is oscillating on a horizontal frictionless surface. When the block is 20 cm away from the equilibrium position, the elastic potential energy stored in the spring is 24 J. What is the elastic potential energy when the block is 10 cm away from equilibrium?

Answer to Essential Question 12.2: To answer this question, we can use the fact that the elastic potential energy is proportional to x^2. Doubling x, the distance from equilibrium, increases the elastic potential energy by a factor of 4. Thus, the elastic potential energy is 6 J when $x = 10$ cm.

12-3 An Example Involving Springs and Energy

EXAMPLE 12.3 – A fast-moving block
        (a) A block of mass m, which rests on a horizontal frictionless surface, is attached to an ideal horizontal spring. The block is released from rest when the spring is stretched by a distance A from its natural length. What is the block's maximum speed during the ensuing oscillations?
        (b) If the block is released from rest when the spring is stretched by $2A$ instead, how does the block's maximum speed change?

SOLUTION
        (a) Let's begin, as usual, with a diagram of the situation (see Figure 12.6). When will the block achieve its maximum speed? Maximum speed corresponds to maximum kinetic energy, which corresponds to minimum potential energy. The gravitational potential energy is constant, since there is no up or down motion, so we can focus on the elastic potential energy. The elastic potential energy is a minimum (zero, in fact) when the block passes through equilibrium, where the spring is at its natural length. Energy bar graphs for the two points are shown in Figure 12.6.

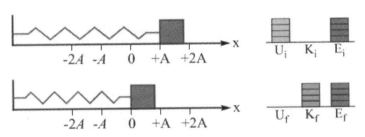

Figure 12.6: Diagrams of the block at the release point and at the equilibrium position, and the corresponding energy bar graphs.

        Let's continue with the energy analysis by writing out the conservation of energy equation: $K_i + U_i + W_{nc} = K_f + U_f$. The initial point is the point from which the block is released, while the final point is the equilibrium position.

        $K_i = 0$, because the block is released from rest from the initial point.

        $W_{nc} = 0$, because there is no work being done by non-conservative forces.

        We can neglect gravitational potential energy, because there is no vertical motion. This gives $U_f = 0$, because the elastic potential energy is also zero at the final point.

        We have thus reduced the energy equation to: $U_{ei} = K_f$. This gives:

$$\frac{1}{2}kA^2 = \frac{1}{2}mv_{max}^2.$$ Solving for the maximum speed gives: $v_{max} = A\sqrt{\dfrac{k}{m}}$.

        Is this answer reasonable? The maximum speed is larger if we start the block farther from equilibrium (where the spring exerts a larger force); if we increase the spring constant (also increasing the force); or if we decrease the mass (increasing acceleration). This all makes sense.

        (b) If we start the block from $2A$ away from equilibrium, we simply replace A in our equation above by $2A$, showing us that the maximum speed is twice as large:

$$v_{max}' = 2A\sqrt{\frac{k}{m}}.$$

Related End-of-Chapter Exercises: 4, 5.

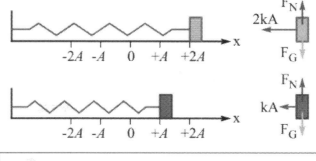

We can make an interesting generalization based on further analysis of the situation in Example 12.3. Take two blocks, one red and one blue but otherwise identical, and two identical springs. Attach each block to one of the springs, and place these two block-spring systems on frictionless horizontal surfaces. As shown in Figure 12.7, we will release one block from rest from a distance A from equilibrium and the other from a distance $2A$ from equilibrium. If the blocks are released simultaneously, which block reaches the equilibrium point first?

Figure 12.7: Identical blocks attached to identical springs. The blocks are released from rest simultaneously. Block 2, at the top, is released from a distance $2A$ from equilibrium. Block 1 is released from a distance A from equilibrium. The initial free-body diagrams are also shown.

Block 2 has an initial acceleration twice as large as that of block 1, because block 2 experiences a net force that is twice as large as that experienced by block 1. The accelerations steadily decrease, because the spring force decreases as the blocks get closer to equilibrium, but we can neglect this change if we choose a time interval that is sufficiently small.

At the end of this time interval, Δt, what is the speed of each block? We're choosing a small time interval so that we can apply a constant-acceleration analysis. Remembering that the blocks are released from rest, so $v_i = 0$, we have:

for block 1, $\qquad \vec{v}_1 = \vec{v}_{i1} + \vec{a}_1 \Delta t = \vec{a}_1 \Delta t$;

for block 2, $\qquad \vec{v}_2 = \vec{v}_{i2} + \vec{a}_2 \Delta t = \vec{a}_2 \Delta t = 2\vec{a}_1 \Delta t = 2\vec{v}_1$.

What about the distance each block travels? Here we can apply another constant acceleration equation:

for block 1 $\qquad \Delta \vec{x}_1 = \vec{v}_{i1} \Delta t + \frac{1}{2} \vec{a}_1 \left(\Delta t \right)^2 = \frac{1}{2} \vec{a}_1 \left(\Delta t \right)^2$;

for block 2 $\qquad \Delta \vec{x}_2 = \vec{v}_{02} \Delta t + \frac{1}{2} \vec{a}_2 \left(\Delta t \right)^2 = \frac{1}{2} \vec{a}_2 \left(\Delta t \right)^2 = \frac{1}{2} \left(2\vec{a}_1 \right) \left(\Delta t \right)^2 = 2\Delta \vec{x}_1$.

At the end of the time interval, block 1 is $A - \Delta x_1$ from equilibrium and block 2 is exactly twice as far from equilibrium as block 1, at $2A - 2\Delta x_1 = 2\left(A - \Delta x_1 \right)$ from equilibrium. Thus, after this small time interval has passed, block 2 is still twice as far from equilibrium as block 1, its velocity is twice as large, and its acceleration is twice as large. We could keep the process going, following the two blocks as time goes by, and we would find this always to be true, that block 2's velocity, acceleration, and displacement from equilibrium, is always double that of block 1. This is true at all times, even after the blocks pass through their equilibrium positions to the far side of equilibrium.

This leads to an amazing conclusion – *that the two blocks take exactly the same time to reach equilibrium* (and to complete one full cycle of an oscillation). This is because block 2 experiences twice the displacement of block 1, but its average velocity is also twice as large. Because the time is the distance divided by the average velocity, these factors of two cancel out.

Essential Question 12.3: Above we analyzed the situation of two identical (aside from color) blocks, oscillating on identical springs, and found the time to reach equilibrium (or to complete one full oscillation) to be the same. Was that just a coincidence that happened to work out because the starting displacements from equilibrium were in a 2:1 ratio, or can we generalize and say that the time is the same no matter where the block is released?

Answer to Essential Question 12.3: In fact, this result is generally true. As long as the spring is ideal, then **the time it takes a block to move through one complete oscillation is independent of the amplitude of the oscillation**. The **amplitude** is defined as the maximum distance an object gets from its equilibrium position during its oscillatory motion.

12-4 The Connection with Circular Motion

So far we have looked at how to apply force and energy ideas to springs. Let's now explore an interesting connection between what is called **simple harmonic motion** (oscillatory motion without any loss of mechanical energy), and uniform circular motion.

EXPLORATION 12.4 – Connecting circular motion to simple harmonic motion

Take the two spring-block systems we investigated at the end of the previous section and place them beside a large turntable that is rotating about a vertical axis. Set the constant angular speed of the turntable so that the turntable undergoes one complete revolution in the time it takes the blocks on the springs to move through one complete oscillation. As shown in Figure 12.8, there are two disks on the turntable, one a distance A from the center and the other a distance $2A$ from the center. The blocks are simultaneously released from rest at the instant the disks pass through the position shown in the figure.

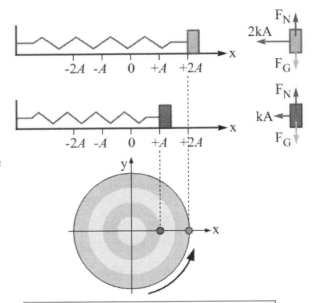

Another amazing thing happens. As the disks spin at constant angular velocity and the blocks oscillate back and forth, the motion of block 1 matches the motion of disk 1, while the motion of block 2 matches the motion of disk 2. The position of the left-hand side of each block *is at all times* equal to the x-coordinate of the position of the center of its corresponding disk, taking the origin to be at the center of the turntable and using the x-y coordinate system shown in Figure 12.8.

Figure 12.8: Comparing systems – two disks on a rotating turntable and two oscillating block-and-spring systems.

Step 1 – *Sketch two separate motion diagrams, one showing the successive positions of disk 1 and the other showing the successive positions of disk 2, as the turntable undergoes one complete revolution. Plot the positions at regular time intervals which, because the disk rotates at a constant rate, correspond to regular angular displacements.* Motion diagrams for the disks are shown in Figure 12.9, showing positions at 30° intervals.

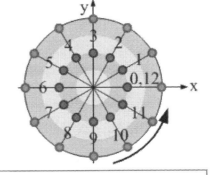

Step 2 – *Now add motion diagrams for the two blocks, sketching their positions so they agree with the statement above, that the left-hand side of each block is at all times equal to the x-coordinate of the position of the center of its corresponding disk.* These motion diagrams are shown in Figure 12.10.

Figure 12.9: Motion diagrams for the two disks, showing their positions at 30° intervals. Because the turntable (and each disk) rotates at a constant rate, these equal angular displacements correspond to the equal time intervals we're used to seeing on motion diagrams.

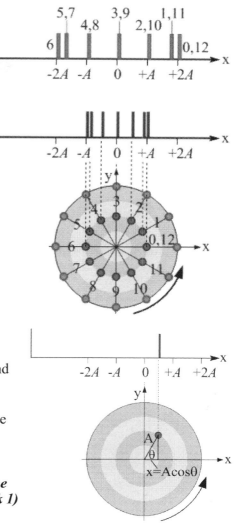

Figure 12.10: Motion diagrams for the two blocks, showing that the motion of a block experiencing simple harmonic motion exactly matches the motion of a well-chosen object experiencing uniform circular motion. The springs have been removed from the picture for clarity, and the motion diagrams for the blocks show the successive positions of the left-hand side of each block during its motion. The motion of a block matches the x-component of the motion of the corresponding disk on the turntable.

Step 3 – *Measuring angles counterclockwise from the positive x-axis, write an equation giving the x-coordinate of disk 1 as a function of time. Hint: first write out the x-coordinate in terms of an arbitrary angle the turntable has rotated through, and then express that angle in terms of time and the turntable's constant angular speed* ω. Figure 12.11 shows the position of disk 1 when the turntable has rotated through some arbitrary angle θ from its initial position. Its x-position at this angle can be found from the adjacent side of the right-angled triangle: $\vec{x} = A\cos(\theta)$. Because the angular velocity is constant, however, and the initial angle $\theta_i = 0$ we can express the angle as:

$\theta = \theta_i + \omega t = 0 + \omega t = \omega t$. Substituting this into our expression for the disk's x-position gives: $\vec{x} = A\cos(\omega t)$.

Step 4 – *Based on the results above, what is the equation giving the x-position of block 1 (actually, the position of the left edge of block 1) as a function of time? What is the equation giving the position of block 2 as a function of time?* Because the motion of block 1 matches exactly the x-component of the motion of disk 1, the equation that gives the disk's x-position must also gives the block's x-position. Thus, for block 1 we have:

Figure 12.11: The position of block 1 and disk 1 after the turntable has rotated through an arbitrary angle θ.

$$\vec{x} = A\cos(\omega t) \quad . \qquad \text{(Eq. 12.3: Position-versus-time for simple harmonic motion)}$$

Using the convention introduced earlier in this book, in which a + or – sign is used to represent the direction of a vector in one dimension, the right-hand side of equation 12.3 can be viewed as a vector quantity, with the sign hidden in the cosine. We get a positive sign for some values of time and a negative sign for others. The equation for block 2 is virtually identical to that of block 1, with the only change being the extra factor of 2. For block 2: $\vec{x} = 2A\cos(\omega t)$.

Key ideas: There is an interesting connection between simple harmonic motion and uniform circular motion. One-dimensional simple harmonic motion matches one component of a carefully chosen two-dimensional uniform circular motion. This allows us to write an equation of motion for an object experiencing simple harmonic motion: $\vec{x} = A\cos(\omega t)$. In this context, ω is known as the angular frequency. **Related End-of-Chapter Exercises: 42, 43.**

Essential Question 12.4: We showed above how the position of a block oscillating on a spring matches one component of the position of an object experiencing uniform circular motion. Can we make similar conclusions about the velocity and acceleration of the block on the spring?

Answer to Essential Question 12.4: Absolutely. All aspects of the motion of the oscillating block match one component of the motion of the object experiencing uniform circular motion. If the position of the block is given by $\vec{x} = A\cos(\omega t)$, then its velocity and acceleration are given by:

$$\vec{v}_{x,disk1} = \vec{v}_{block1} = -v\sin(\omega t) = -A\omega\sin(\omega t) \quad \text{and} \quad \vec{a}_{x,disk1} = \vec{a}_{block1} = -\frac{v^2}{A}\cos(\omega t) = -A\omega^2\cos(\omega t).$$

Here, v represents the constant speed of disk 1 as it moves in uniform circular motion.

12-5 Hallmarks of Simple Harmonic Motion

Simple harmonic motion (often referred to as SHM) is a special case of oscillatory motion. An object oscillating in one dimension on an ideal spring is a prime example of SHM. The characteristics of simple harmonic motion include:

- A force (and therefore an acceleration) that is opposite in direction, and proportional to, the displacement of the system from equilibrium. Such a force, that acts to restore the system to equilibrium, is known as a **restoring force**.
- No loss of mechanical energy.
- An angular frequency ω that depends on properties of the system.
- Position, velocity, and acceleration given by Equations 12.3 – 12.5:

$$\vec{x} = A\cos(\omega t) \quad . \qquad \text{(Equation 12.3: \textbf{Position in simple harmonic motion})}$$

$$\vec{v} = -v_{max}\sin(\omega t) = -A\omega\sin(\omega t). \qquad \text{(Equation 12.4: \textbf{Velocity in SHM})}$$

$$\vec{a} = -a_{max}\cos(\omega t) = -\frac{v^2}{A}\cos(\omega t) = -A\omega^2\cos(\omega t). \qquad \text{(Eq. 12.5: \textbf{Acceleration in SHM})}$$

The above equations apply if the object is released from rest from $\vec{x} = +A$ at $t = 0$. Starting the block with different initial conditions requires a modification of the equations.

Combining Equations 12.3 and 12.5, in any simple harmonic motion system we see that the acceleration is opposite in direction, and proportional to, the displacement:

$$\vec{a} = -\omega^2\,\vec{x}\,. \qquad \text{(Equation 12.6: \textbf{Connecting acceleration and displacement in SHM})}$$

In general, the angular frequency (ω), frequency (f), and period (T) are connected by:

$$\omega = 2\pi f = \frac{2\pi}{T}\,. \qquad \text{(Eq. 12.7: \textbf{Relating angular frequency, frequency, and period})}$$

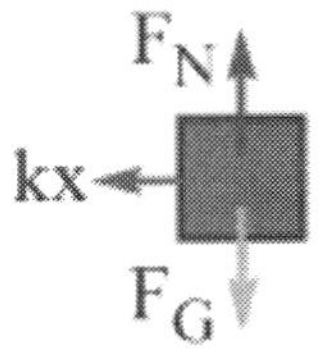

What determines the angular frequency ω in a particular situation? Let's return to the free-body diagram of a block on a spring, shown in Figure 12.12.

Applying Newton's second law horizontally, $\sum \vec{F}_x = m\vec{a}$, we get:

$$-k\,\vec{x} = m\vec{a}\,.$$

Re-arranging gives $\vec{a} = -(k/m)\vec{x}$. Comparing this result to the general

Figure 12.12: The free-body diagram of a block connected to a spring of spring constant k. The block is displaced to the right of the equilibrium point by a distance x.

SHM Equation 12.6 tells us that, for a mass on an ideal spring, $\omega^2 = k/m$, or:

$$\omega = \sqrt{\frac{k}{m}}\,. \qquad \text{(Equation 12.8: \textbf{Angular frequency for a mass on a spring}).}$$

This is a typical result, that the angular frequency is given by the square root of a parameter related to the restoring force (or torque, in rotational motion) divided by the inertia.

EXAMPLE 12.5 – Plotting graphs of position, velocity, and acceleration versus time

Once again, let's attach a block to a spring and release the block from rest from a position $\bar{x} = +A$ (relative to $\bar{x} = 0$, which is the equilibrium position). The block oscillates back and forth with a period of $T = 4.00$ s.

(a) Plot graphs of the block's position, velocity, and acceleration as a function of time over two complete oscillations.

(b) Compare the position graph to the velocity graph.

(c) How does the acceleration graph compare to the position graph?

SOLUTION

(a) We can make use of Equations 12.3 – 12.5 to plot the graphs. Before doing so, we can solve for the angular velocity ω, using:

$$\omega = \frac{2\pi}{T} = \frac{2\pi \text{ rad}}{4.00 \text{ s}} = 1.57 \text{ rad/s}.$$

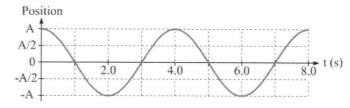

Also, it makes it easier to plot the graphs if we remember that, if the block is released from rest, it returns to its starting point after one period; after half a period it comes instantaneously to rest on the far side of equilibrium; and at times of T/4 and 3T/4 it is passing through equilibrium at its maximum speed. Determining when each graph passes through zero, when it reaches its largest positive and negative values, and then connecting these points with sinusoidally oscillating graphs, gives the results shown in Figure 12.13.

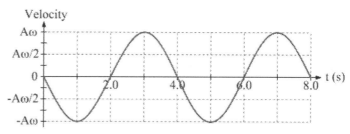

(b) Comparing the position and velocity graphs in Figure 12.13, we can see that the block's speed is maximum when the block's displacement from equilibrium is zero. Conversely, the block's speed is zero when the magnitude of the block's displacement from equilibrium is maximized. These observations are consistent with what is taking place with the energy. The kinetic energy is proportional to the speed squared and the elastic potential energy is proportional to the square of the magnitude of the displacement from equilibrium. Kinetic energy is maximum when the elastic potential energy is zero, and vice versa.

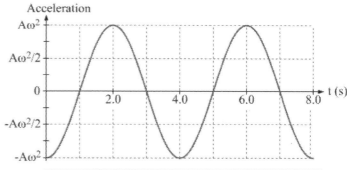

Figure 12.13: Graphs of the position, velocity, and acceleration, as a function of time, of the block in Example 12.5 for two complete oscillations of the block.

(c) Comparing the position and acceleration graphs, we see that one is the opposite of the other, in the sense that when the position is positive the acceleration is negative, and vice versa. This is expected because one of the hallmarks of simple harmonic motion is that $\bar{a} = -\omega^2 \bar{x}$.

Related End-of-Chapter Exercises: 32, 41, 46.

Essential Question 12.5: Return to the situation described in Example 12.5, but now increase the angular frequency by a factor of 2. We can accomplish this by either changing only the spring constant or by changing only the mass. Can we tell which one was changed by looking at the resulting graphs of position, velocity, and/or acceleration as a function of time? Assume that the block is released from rest from the same point it was in Example 12.5, and that the equilibrium position remains the same.

Answer to Essential Question 12.5: We cannot tell. Any one of the three graphs can be used to determine that the angular frequency has changed, because they all involve ω, but none of the graphs can tell us whether we adjusted the spring constant or the mass.

12-6 Examples Involving Simple Harmonic Motion

EXAMPLE 12.6A – Energy graphs
    Take a 0.500-kg block and attach it to a spring. We would like the block to undergo oscillations that have a period (the time for one complete oscillation) of 4.00 seconds.
    (a) What should the spring constant be?
    (b) We'll release the block from rest from a distance A from the equilibrium point so that the block has a speed of 4.00 m/s when it passes through equilibrium. Over two complete oscillations, plot the system's elastic potential energy, kinetic energy, and total mechanical energy as a function of time.

SOLUTION
(a) Let's first apply Equation 12.7, $\omega = 2\pi / T$, to find the angular frequency. This gives:

$$\omega = \frac{2\pi \ \text{rad}}{4.00\,\text{s}} = \frac{\pi}{2.00}\,\text{rad/s}.$$

    Using Equation 12.7, $\omega = \sqrt{\dfrac{k}{m}}$, we get:

$$k = \omega^2 m = \frac{\pi^2 (0.500)}{4.00}\,\text{rad}^2\,\text{kg/s}^2 = 1.23\,\text{N/m},$$

    where we treated the factor of radians as being dimensionless.

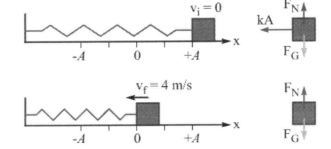

    (b) A diagram of the situation is shown in Figure 12.14. Let's solve for the maximum kinetic energy, which equals the mechanical energy:

Figure 12.14: A diagram of the block and spring, showing the initial situation and the situation as the block passes through equilibrium.

$$K_{max} = \frac{1}{2} m v_{max}^2 = \frac{1}{2}(0.500\,\text{kg})(4.00\,\text{m/s})^2 = 4.00\,\text{J}.$$

    The maximum potential energy is also 4.00 J, because the energy oscillates between potential and kinetic, and the total mechanical energy is conserved.

    Using Equation 12.4, $\bar{v} = -v_{max} \sin(\omega t)$, we can write the kinetic energy as a function of

time as $K = \dfrac{1}{2} m v^2 = \dfrac{1}{2} m v_{max}^2 \sin^2(\omega t) = (4.00\,\text{J})\sin^2\left(\dfrac{\pi}{2.00\,\text{s}} t\right)$.

    Because the block takes 4.00 s to complete one oscillation, at $t = 0$ and $t = 4.00$ s it is instantaneously at rest at the starting point. At $t = 2.00$ s (halfway through the cycle) the block is instantaneously at rest on the far side of equilibrium. At each of these times the kinetic energy is 0 and the elastic potential energy is 4.00 J. Conversely, at $t = 1.00$ s and 3.00 s it passes through equilibrium, where the elastic potential energy is zero and the kinetic energy is its maximum value of 4.00 J. Graphs of the various energies as a function of time are shown in Figure 12.15. Note that, at all times, the sum of the kinetic and potential energies is 4.00 J.

Related End-of-Chapter Exercises: 27, 40.

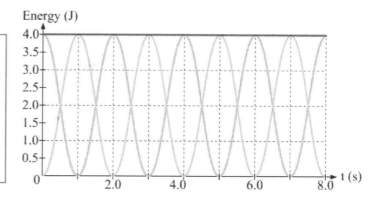

Figure 12.15: Graphs of the block's kinetic energy (zero at $t = 0$ s), elastic potential energy (zero at $t = 1.0$ s), and total mechanical energy (constant), as a function of time over 8.00 s, the time for two complete oscillations. The kinetic and potential energies go through two cycles for every one complete oscillation of the block. Compare this graph to Figure 12.5, which shows the energies as a function of position.

EXAMPLE 12.6B – An eighth of the motion

Attach an object to an ideal spring and set it oscillating. The object is released from rest from a distance A from equilibrium. The object travels a distance A to the equilibrium position, ¼ of the entire distance for one complete oscillation, in ¼ of the period. How long does it take the object to travel from A away from equilibrium to $A/2$ from equilibrium, 1/8 of the entire distance covered in one complete oscillation?

SOLUTION

A diagram of the situation is shown in Figure 12.16. Position 1 is where the block is released from rest. Position 2 is halfway between the release point and the equilibrium position, which is position 3.

Because the block's speed increases as the block approaches equilibrium, the block's average speed as it moves from position 1 to position 2 is less than its average speed as it moves from position 2 to position 3. Thus, because of this low average speed, the time it takes to move from position 1 to position 2 is larger than $T/8$, 1/8 of the total time.

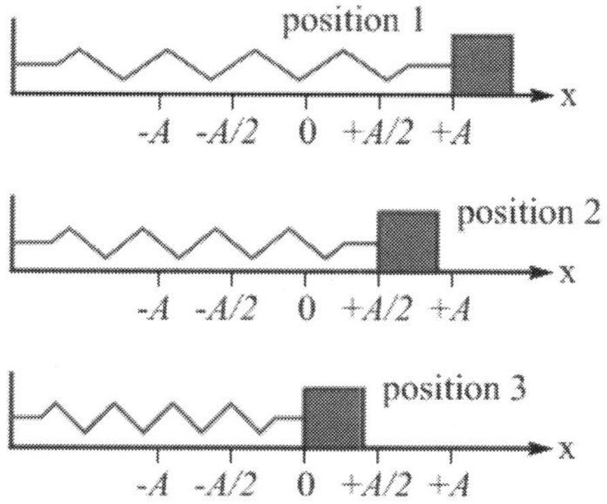

Figure 12.16: A diagram of the position of the block as it moves from its release point (position 1) to the equilibrium position (position 3). Position 2 is exactly halfway between the release point and equilibrium.

Using Equation 12.3, $\bar{x} = A\cos(\omega t)$, let's find the time it takes. At position 2, the block's position is $+A/2$. Using this in Equation 12.3 gives:

$$+\frac{A}{2} = A\cos(\omega t).$$ Dividing by A gives: $+\frac{1}{2} = \cos(\omega t).$

We can use the relation $\omega = 2\pi / T$ to re-write the equation: $+\frac{1}{2} = \cos\left(\frac{2\pi}{T}t\right).$

The logical next step is to take the inverse cosine of both sides. Here it is critical to remember that ωt has units of radians. ***Any time we use the three equations involving time (Equations 12.3 – 12.5), we need to work in radians.*** Thus, when we determine $\cos^{-1}(+1/2)$ we will write the result as $\pi/3$ radians instead of 60°.

Taking the inverse cosine of both sides, then, gives: $\dfrac{\pi}{3} = \dfrac{2\pi}{T}t$.

This gives us a time of t = $T/6$. As we concluded above, the time is larger than $T/8$.
Related End-of-Chapter Exercises: 34, 50.

Essential Question 12.6: Return to the situation described in Example 12.6A, but now increase the angular frequency by a factor of 2. Let's say we achieve the change in angular frequency by changing either the spring constant or the mass, but not both. Can we tell which one was changed, if the graphs of energy as a function of time still reach a maximum of 4.0 J when the block is released from rest from a distance A from equilibrium?

Answer to Essential Question 12.6: Consider Equation 12.8, $\omega = \sqrt{k/m}$. We can double the angular frequency by increasing the spring constant by a factor of 4, or by decreasing the mass by a factor of 4. Because the object is released from rest, the initial energy is all elastic potential energy, given by $U_i = 0.5 k A^2$. We have not changed A, so if the total energy stayed the same we must not have changed the spring constant k. Thus we must have changed the mass.

12-7 The Simple Pendulum
Another classic simple harmonic motion system is the simple pendulum, which is an object with mass that swings back and forth on a string of negligible mass.

EXAMPLE 12.7 – Pendulum speed limit
A ball of mass m is fastened to a string with a length L. Initially the ball hangs vertically down from the string in its equilibrium position. The ball is then displaced so the string makes an angle of θ with the vertical, and then released from rest.
    (a) What is the height of the ball above the equilibrium position when it is released?
    (b) What is the speed of the ball when it passes through equilibrium?

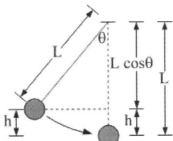

Figure 12.17: A diagram showing the point from which the ball is released and the equilibrium position.

SOLUTION
    (a) Consider the geometry of the situation, shown in Figure 12.17. The key to finding the height of the ball is to consider the right-angled triangle in Figure 12.17. The vertical side of the triangle measures $L\cos\theta$. Because the string measures L, the height of the ball above equilibrium is $h = L - L\cos\theta = L(1-\cos\theta)$.

    (b) Let's apply energy conservation, starting as usual with: $K_i + U_i + W_{nc} = K_f + U_f$.
The initial point is the release point, while the final point is the equilibrium position.
    $K_i = 0$, because the ball is released from rest from the initial point.

    $W_{nc} = 0$, because there is no work being done by non-conservative forces.

    We can define the ball's gravitational potential energy to be zero at the equilibrium point, giving $U_f = 0$.

    The equation thus reduces to $U_i = K_f$, which gives: $mgh = \dfrac{1}{2}mv_f^2$.

    The mass cancels out, so the speed does not depend on the mass. Solving for the speed:
$$v_f = \sqrt{2gh} = \sqrt{2gL(1-\cos\theta)} \, .$$

    Note that we have seen this $v_f = \sqrt{2gh}$ result before, such as in cases in which an object falls straight down from rest, or when water leaks out a hole in a container.

Related End-of-Chapter Exercises: 11, 29.

EXPLORATION 12.7 – Torques on the pendulum
    A simple pendulum consists of a ball of mass m that hangs down vertically from a string. The ball is displaced by an angle θ from equilibrium and released from rest.

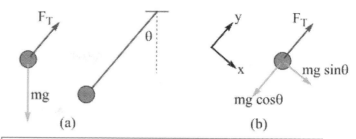

(a)   (b)

Step 1 – *Draw a free-body diagram for the ball immediately after it is released.* The free-body diagram is drawn in Figure 12.18. There is a downward force of gravity, and a force of tension directed away from the ball along the string. Using a coordinate system aligned with the string, we can split the force of gravity into components, one component opposite to the tension and the other component giving an acceleration toward the equilibrium position (see Figure 12.18(b)).

Figure 12.18: Figure (a) shows the free-body diagram, with a force of tension and a force of gravity acting on the ball. Figure (b) shows the force of gravity in components, using a coordinate system in which one axis is parallel to the string.

Step 2 – *Apply Newton's second law for rotation to find a relationship between the angular acceleration of the ball and the ball's angular displacement (measured from the vertical).* Take torques about the axis perpendicular to the page passing through the upper end of the string. There is no torque about this axis from the tension or from the component of the force of gravity parallel to the string. The only torque comes from the component of the force of gravity that acts perpendicular to the string. Applying $\tau = r\,F\sin\phi$, where $r = L$, $F = mg\sin\theta$, and $\phi = 90°$, the torque has a magnitude of $\tau = Lmg\sin\theta$. Taking counterclockwise to be positive, the torque is negative. Applying Newton's Second Law for Rotation, $\sum \vec{\tau} = I\vec{\alpha}$, we get: $-Lmg\sin\theta = I\vec{\alpha}$.

The rotational inertia of the ball is $I = mL^2$, giving: $-Lmg\sin\theta = mL^2\vec{\alpha}$.

Canceling the mass (thus, the mass does not matter) and a factor of L gives:

$$\vec{\alpha} = -\frac{g}{L}\sin\theta \ .$$ (Equation 12.9: **Angular acceleration of a simple pendulum**)

Step 3 – *Use the small-angle approximation,* $\sin\theta \approx \theta$, ***to find an expression for the angular frequency of the pendulum.*** We can say that $\sin\theta \approx \theta$ if θ is given in radians, and the angle is less than about 10° (about 1/6 radians). Using the small-angle approximation in Equation 12.9:

$$\vec{\alpha} = -\frac{g}{L}\vec{\theta} \ .$$ (Equation 12.10: **For a simple pendulum at small angles**)

The general simple harmonic relationship, $\vec{a} = -\omega^2 \vec{x}$, can be transformed to an analogous general equation for rotational motion, $\vec{\alpha} = -\omega^2\vec{\theta}$. Equation 12.10 fits this form, so:

$$\omega = \sqrt{\frac{g}{L}} \ .$$ (Eq. 12.11: **Angular frequency for a simple pendulum at small angles**)

For a pendulum, gravity provides the restoring force, so it makes sense that the angular frequency is larger if g is larger. Conversely, increasing L means the pendulum has farther to travel to reach equilibrium, reducing the angular frequency.

Key ideas: For small-angle oscillations, the motion of a simple pendulum is simple harmonic. Large-angle oscillations are not simple harmonic because the restoring torque is not proportional to the angular displacement.   **Related End-of-Chapter Exercises: 57, 58.**

Essential Question 12.7: Compare the free-body diagram of a ball of mass *m*, hanging at rest from a string of length *L*, to that of the same system oscillating as a pendulum, when the ball passes through equilibrium. Make note of any differences between the two free-body diagrams.

Answer to Essential Question 12.7: The two free-body diagrams are shown in Figure 12.19. When the ball is at rest, its acceleration is zero. Applying Newton's second law tells us that, in this case, the force of tension exactly balances the force of gravity, so $F_T = mg$. When the ball is

oscillating, it is moving along a circular arc as it passes through the equilibrium position. In this case there is a non-zero acceleration, the centripetal acceleration directed toward the center of the circular arc. To produce the upward acceleration the upward force of tension must be larger than the downward force of gravity. Applying Newton's second law shows that the force of tension increases to:

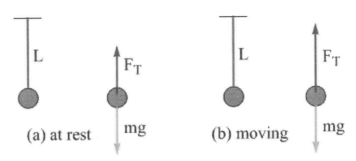

$$F_T = mg + m\frac{v^2}{L}.$$

Figure 12.19: Free-body diagrams for the ball on the string when it is (a) at rest, and (b) passing through equilibrium with a speed v.

Chapter Summary

Essential Idea

Harmonic oscillations are important in many applications, from musical instruments to clocks and, in the human body, from walking to the creation of sounds with our vocal cords. Even though we have focused on two basic models in this chapter, the block on the spring and the simple pendulum, the same principles apply in many real-life situations.

Springs

An ideal spring obeys Hooke's Law, $\vec{F}_{Spring} = -k\vec{x}$.   (Equation 12.1: **Hooke's Law**)

k is the spring constant, a measure of the stiffness of the spring.

Springs that are stretched or compressed store energy. This energy is known as elastic potential energy.

$$U_e = \frac{1}{2}kx^2.$$ (Equation 12.2: **Elastic potential energy for an ideal spring**)

When an object oscillates on a spring, the angular frequency of the oscillations depends on the mass of the object and the spring constant of the spring.

$$\omega = \sqrt{\frac{k}{m}}.$$ (Equation 12.8: **The angular frequency of a mass on a spring**)

Simple Harmonic Motion and Energy Conservation

Energy conservation is a useful tool for analyzing oscillating systems. When springs are involved we use elastic potential energy, an idea introduced in this chapter. To analyze a pendulum in terms of energy conservation nothing new whatsoever is needed.

Hallmarks of Simple Harmonic Motion

The main features of a system that undergoes simple harmonic motion include:
- No loss of mechanical energy.
- A restoring force or torque that is proportional to, and opposite in direction to, the displacement of the system from equilibrium.

In this situation the acceleration of the system is related to its position by:

$$\vec{a} = -\omega^2 \vec{x} \,, \qquad \text{(Eq. 12.6: The connection between acceleration and displacement)}$$

where the angular frequency ω is generally given by the square root of some elastic property of the system (such as the spring constant) divided by an inertial property (such as the mass).

Time and Simple Harmonic Motion

When we are interested in how a simple harmonic oscillator evolves over time the following equations are extremely useful. These were derived by looking at the connection between simple harmonic motion and one component of the motion of an object experiencing uniform circular motion.

$$\vec{x} = A\cos(\omega t) \,. \qquad \text{(Equation 12.3: Position in simple harmonic motion)}$$

$$\vec{v} = -A\omega\sin(\omega t). \qquad \text{(Equation 12.4: Velocity in simple harmonic motion)}$$

$$\vec{a} = -A\omega^2\cos(\omega t). \qquad \text{(Eq. 12.5: Acceleration in simple harmonic motion)}$$

Equations 12.3 – 12.5 apply when the object is released from rest at $t = 0$ from a distance A from equilibrium.

In general, the angular frequency (ω), frequency (f), and period (T) are connected by:

$$\omega = 2\pi f = \frac{2\pi}{T}. \qquad \text{(Eq. 12.7: Relating angular frequency, frequency, and period)}$$

The Simple Pendulum

A simple pendulum, consisting of an object on the end of a string, is another good example of an oscillating system. As long as the amplitude of the oscillations is small (less than about 10°) and mechanical energy is conserved then the motion is simple harmonic. For larger angles the motion diverges from simple harmonic because the restoring torque is not directly proportional to the angular displacement.

$$\omega = \sqrt{\frac{g}{L}}. \qquad \text{(Equation 12.11: Angular frequency of a simple pendulum)}$$

End-of-Chapter Exercises

Exercises 1 – 12 are conceptual questions that are designed to see if you have understood the main concepts of the chapter.

1. When a spring is compressed 10 cm, compared to its natural length, the spring exerts a force of 5 N. What is the spring force when the spring is stretched by 10 cm compared to its natural length, instead?

2. A block on a horizontal surface is attached to an ideal horizontal spring, as shown in Figure 12.20. When the block compresses the spring by 10 cm, the spring exerts a force of 10 N on the block. The block is then moved either left or right to a new position, where the force the spring exerts on the block has a magnitude of 20 N. How far has the block been moved? State all possible answers.

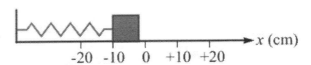

Figure 12.20: A block connected to an ideal horizontal spring. The block initially compresses the spring by 10 cm. For Exercises 2 and 3.

3. A block on a frictionless horizontal surface is attached to an ideal horizontal spring, as shown in Figure 12.20. When the block compresses the spring by 10 cm, the elastic potential energy stored in the spring is 10 J. The block is then moved either left or right to a new position, where the elastic potential energy stored in the spring is 40 J. How far has the block been moved? State all possible answers.

4. A small ball is loaded into a spring gun, compressing the spring by a distance A. When the trigger is pressed, the ball emerges from the gun with a speed v. The ball is loaded into the gun again, this time compressing the spring by a distance $2A$. With what speed will the ball emerge from the gun this time? Justify your answer.

5. A block is attached to a spring and the system is placed on a horizontal frictionless surface with the other end of the spring anchored firmly to a wall. The block is then displaced from equilibrium until 8.0 J of elastic potential energy has been stored in the spring. The block is then released from rest and the block oscillates back and forth about the equilibrium position. Sketch energy bar graphs for this system, showing the elastic potential energy, kinetic energy, and total mechanical energy when the system is (a) at the point where it is released from rest; (b) halfway between that point and the equilibrium position; (c) at the equilibrium position.

6. A block on a spring experiences simple harmonic motion with amplitude A and period T. For one complete oscillation, determine (a) the block's displacement; (b) the total distance traveled by the block; (c) the block's average velocity; (d) the block's average speed; (e) the block's average acceleration.

7. Consider the following four cases. In each case, the block experiences simple harmonic motion of amplitude A.
   Case 1: a block of mass m connected to a spring of spring constant k.
   Case 2: a block of mass m connected to a spring of spring constant $2k$.
   Case 3: a block of mass $2m$ connected to a spring of spring constant k.
   Case 4: a block of mass $2m$ connected to a spring of spring constant $2k$.
   Rank these cases, from largest to smallest, based on (a) their angular frequency; (b) their total mechanical energy; (c) the maximum speed reached by the block during its motion. Your answers should have a form similar to 3>1>2=4.

8. You have three blocks, of mass m, $2m$, and $3m$, and three springs of spring constant k, $2k$, and $3k$. You can attach any one of the blocks to any one of the springs, displace the block from equilibrium by a distance of A, $2A$, or $3A$, and release the block from rest so it experiences simple harmonic motion. Which combination of these three parameters (mass, spring constant, and amplitude) results in oscillations with the largest (a) angular frequency? (b) period? (c) total mechanical energy? (d) speed when the block passes through equilibrium?

9. A block of mass m is connected to a spring with a spring constant k, and displaced a distance A from equilibrium. Upon being released from rest, the block experiences simple harmonic motion. Let's say you wanted to double the mechanical energy of the system. (a) Could you accomplish this by changing the mass, but keeping everything else the same? If so, what would the new mass be? (b) Could you accomplish this by changing the spring constant, but keeping everything else the same? If so, what would the new spring constant be? (c) Could you accomplish this by changing the amplitude of the oscillation, but keeping everything else the same? If so, what would the new amplitude be?

10. Repeat Exercise 9, except this time you want to double the angular frequency instead of the energy.

11. Consider the following four simple pendula. In each case, the pendulum experiences simple harmonic motion with a maximum angular displacement θ_{max}, where θ_{max} is small enough that the small-angle approximation can be used.
    Case 1: a pendulum consisting of a ball of mass m on a string of length L.
    Case 2: a pendulum consisting of a ball of mass m on a string of length $2L$.
    Case 3: a pendulum consisting of a ball of mass $2m$ on a string of length L.
    Case 4: a pendulum consisting of a ball of mass $2m$ on a string of length $2L$.
    Rank these cases, from largest to smallest, based on (a) their angular frequency; (b) their total mechanical energy; (c) the maximum speed reached by the block during its motion. Your answers should have the form 3>1>2=4.

12. Return to Exercise 11. Now rank the cases, from largest to smallest, based on the tension in the string when the ball passes through the equilibrium position.

Exercises 13 – 17 deal with various situations involving ideal springs.

13. A spring hangs vertically down from a support, with a ball with a weight of 6.00 N hanging from the spring's lower end. If the ball remains at rest and the spring is stretched by 20.0 cm with respect to its natural length, what is the spring constant of the spring?

14. Consider again the situation described in Exercise 13. You now take the spring, cut it in half, and hang the same ball from one half of the spring (the other half you don't use at all) so the ball again remains at rest as it hangs vertically from the spring. (a) How much is the spring stretched from its natural length? Briefly justify your answer. (b) How does the spring constant of this new spring compare to the spring constant of the original spring?

15. A small ball with a mass of 50 g is loaded into a spring gun, compressing the spring by 12 cm. When the trigger is pressed, the ball emerges horizontally from the barrel at a height of 1.4 m above the floor. It then strikes the floor after traveling a horizontal distance of 2.5 m. Use $g = 9.8$ m/s². Assuming all the energy stored in the spring is transferred to the ball, determine the spring constant of the spring.

16. A block of mass M is connected to a spring of spring constant k. The system is placed on a frictionless horizontal surface and the other end of the spring is firmly fixed to a wall so the block, when displaced from equilibrium a distance A and then released from rest, will experience simple harmonic motion. A second block of mass m is then placed on top of the first block. The coefficient of static friction associated with the interaction between the two blocks is μ_s . What is the maximum value A can be so the blocks oscillate together without the top block slipping on the bottom block?

17. A block of mass $M = 0.800$ kg is connected to a spring of spring constant $k = 2.00$ N/m. The system is placed on a frictionless horizontal surface and the other end of the spring is firmly fixed to a wall so the block, when displaced from equilibrium a distance A and then released from rest, will experience simple harmonic motion. A second block of mass $m = 0.600$ kg is then placed on top of the first block. The coefficient of static friction associated with the interaction between the two blocks is $\mu_s = 0.500$. Use $g = 9.80$ m/s².

(a) What is the maximum value A can be so the blocks oscillate together without the top block slipping on the bottom block? (b) What is the angular frequency in this situation?

Exercises 18 – 23 deal with various aspects of the situation shown in Figure 12.21.

18. A block with a mass of 0.500 kg is released from rest from the top of a ramp that has the form of a 3-4-5 triangle, measuring 3.00 m high and having a base of 4.00 m, as shown in Figure 12.21. The block then slides down the incline, encountering a spring with a spring constant of 5.00 N/m after sliding for 2.50 m. Neglect friction and use $g = 9.80$ m/s². Where does the block reach its maximum speed, at the point it first makes contact with the spring, or at a point higher up the ramp or lower down the ramp than where it first makes contact with the spring? Briefly justify your answer.

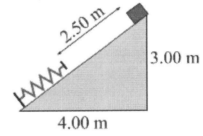

19. Return to the situation described in Exercise 18. Determine (a) how far the block has slid down the ramp when the block reaches its highest speed; (b) the value of this maximum speed.

Figure 12.21: A block released from rest from the top of a ramp slides for a distance of 2.50 m before it encounters a spring. For Exercises 18 – 23.

20. Return to the situation described in Exercise 18. Find the maximum compression of the spring in this situation.

21. Return to the situation described in Exercise 18, but now we'll add friction between the block and the ramp. The coefficient of kinetic friction between the block and the ramp is 0.400. Where does the block reach its maximum speed now, at exactly the same point on the ramp it did in Exercise 18, or at some place higher up or lower down than this point? Briefly justify your answer.

22. Return to the situation described in Exercise 18, but now we'll add friction between the block and the ramp. The coefficient of kinetic friction between the block and the ramp is 0.400. Determine (a) how far the block has slid down the ramp when the block reaches its highest speed; (b) the value of this maximum speed.

23. Return to the situation described in Exercise 18, but now we'll add friction between the block and the ramp. The coefficient of kinetic friction between the block and the ramp is 0.400. Find the maximum compression of the spring in this situation.

Exercises 24 – 30 deal with energy and energy conservation in oscillating systems.

24. A block with a mass of 0.500 kg that is attached to a spring is oscillating back and forth on a frictionless horizontal surface. The period of the oscillations is 2.00 s. When the block is 30.0 cm from its equilibrium position, its speed is 1.20 m/s. What is the amplitude of the oscillations?

25. Consider again the system described in Exercise 24. At a time of $T/4$ after being released from rest, the block is passing through the equilibrium position. At a time of $T/8$ after being released, determine the system's (a) elastic potential energy; (b) kinetic energy.

26. Consider again the system described in Exercise 24, but now we'll make it more realistic. There is a small coefficient of friction associated with the interaction between the block and the surface. This means that, over time, the amplitude of the oscillations decrease until eventually the block comes to rest and remains at rest. Approximately how much work is done by friction on the block during this process?

27. A block is attached to a spring, displaced from equilibrium a distance of 0.800 m, and released from rest. It then oscillates on a frictionless horizontal surface with a period of 4.00 s. At the instant the block is released from rest, the energy in the system is all elastic potential energy, as shown in the set of energy bar graphs in Figure 12.22(a). (a) At how many locations in the subsequent oscillations do we get the set of energy bar graphs shown in Figure 12.22(b)? (b) Determine the distance of each of these locations from the equilibrium position.

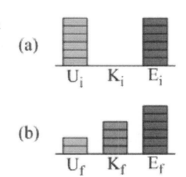

28. Return to the situation described in Exercise 27. (a) During one complete oscillation, at how many different times does the energy correspond to the set of energy bar graphs shown in Figure 12.22(b)? (b) Assuming the block is released from rest at $t = 0$, determine all the times during the first complete oscillation when the system's energy corresponds to the energy bar graphs shown in Figure 12.22(b).

Figure 12.22: Energy bar graphs for a block on a spring at (a) its release point, and (b) some other point. For Exercises 27 and 28.

29. A ball is tied to a string to form a simple pendulum with a length of 1.20 m. The ball is displaced from equilibrium by some angle θ_i and released from rest. In the subsequent oscillations, the ball's maximum speed is 2.50 m/s. (a) From what height above equilibrium was the ball released? (b) What is θ_i? Use $g = 9.80$ m/s².

30. As shown in Figure 12.23 a simple pendulum with a length of 1.00 m is released from rest from an angle of 20° measured from the vertical. When the ball passes through its equilibrium position the string hits a peg, effectively shortening the length of the pendulum to 50.0 cm. (a) How does the maximum height reached by the pendulum on the right compare to the height of the ball at its release point on the left? Justify your answer. (b) What is the maximum angle, measured from the vertical, of the string when the ball is on the right?

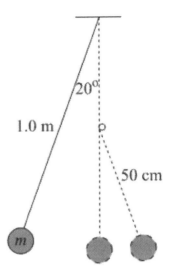

Figure 12.23: A simple pendulum swings down through its equilibrium position, and then hits a peg that effectively shortens its length to 50 cm. For Exercises 30 and 31.

Exercises 31 – 35 deal with time and simple harmonic motion.

31. Return to the situation described in Exercise 30 and shown in Figure 12.23. What is the period of one complete oscillation for this pendulum? Use $g = 9.80$ m/s^2.

32. A block with a mass of 0.600 kg is connected to a spring, displaced in the positive direction a distance of 50.0 cm from equilibrium, and released from rest at $t = 0$. The block then oscillates without friction on a horizontal surface. The first time the block is a distance of 15.0 cm from equilibrium is at $t = 0.200$ s. Determine (a) the period of oscillation; (b) the value of the spring constant; (c) the block's velocity at $t = 0.200$ s; and (d) the block's acceleration at $t = 0.200$ s.

33. Repeat Exercise 32, but now $t = 0.200$ s represents the second time the block is a distance of 15.0 cm from equilibrium.

34. A block on a spring is released from rest from a distance A from equilibrium at $t = 0$. The block then experiences simple harmonic motion with a period T. Determine all the times during the first complete oscillation when the block is a distance $A/4$ from equilibrium.

35. A block with a mass of 0.800 kg is connected to a spring, displaced in the positive direction a distance of 40.0 cm from equilibrium, and released from rest at $t = 0$. The block then oscillates without friction on a horizontal surface. At a time of $t = 0.500$ s, the block is 30.0 cm from equilibrium. If the block's period of oscillation is longer than 0.500 s, determine the spring constant of the spring. Find all possible answers.

Exercises 36 – 39 combine collisions with simple harmonic motion situations.

36. A wheeled cart with a mass of 0.50 kg is rolling along a horizontal track at a constant velocity of 2.0 m/s when it experiences an elastic collision with a second identical cart that is initially at rest, and attached to a spring with a spring constant of 4.0 N/m. This situation is illustrated in Figure 12.24. After the collision, the second cart moves to the right. (a) What is the first cart doing after the collision? Briefly justify your answer. (b) What is the maximum compression of the spring? (c) There is a second collision between the carts. What is each cart doing after the second collision?

Figure 12.24: A cart collides with a second cart that is initially at rest and attached to a spring. For Exercises 36 and 37.

37. A wheeled cart with a mass of 0.50 kg is rolling along a horizontal track at a constant velocity of 2.0 m/s when it experiences a collision with a second identical cart that is initially at rest, and attached to a spring with a spring constant of 4.0 N/m. This situation is illustrated in Figure 12.24. After the collision, the carts stick together and move as one unit. What is the maximum compression of the spring?

38. In a spring version of the ballistic pendulum situation we looked at in Chapter 7, a wooden block with a mass of 0.500 kg is attached to a spring with a spring constant of $k = 600$ N/m. As shown in Figure 12.25, the system is placed on a frictionless horizontal surface with the block at rest at the equilibrium position. A bullet with a mass of 30.0 g is fired at the block. The bullet gets embedded in the block and, after the collision, the block experiences simple harmonic motion with an amplitude of 15.0 cm. Assuming the bullet's velocity is horizontal at the instant the collision takes place, what is the speed of the bullet just before it hits the block?

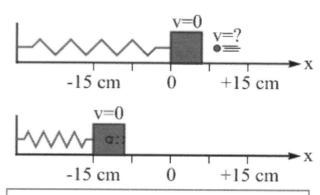

Figure 12.25: A bullet embeds itself in a block. The subsequent oscillations take place with an amplitude of 15.0 cm. For Exercise 38.

39. As shown in Figure 12.26, a ball of mass m is tied to a string to form a simple pendulum. The ball is displaced from equilibrium so the angle between the string and the vertical is 60°, and is then released from rest. It swings down, and at its lowest point it collides with a second ball of mass $4m$ that is initially at rest on the edge of a table. If the collision is elastic and the second ball strikes the floor at a point 1.50 m vertically lower and 1.20 m horizontally from where it started, find the length of the string the first ball is attached to.

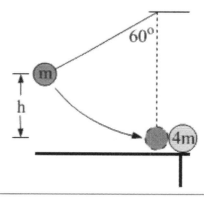

Figure 12.26: A ball of mass m, tied to a string to form a simple pendulum, swings down and collides elastically with a ball of mass $4m$. For Exercise 39.

General Problems and Conceptual Questions

40. A block of mass m is connected to a spring of spring constant k, displaced a distance A from equilibrium and released from rest. An identical block is connected to a spring of spring constant $4k$ and released from rest so its total mechanical energy is equal to that of the first block-spring system. (a) Assuming the blocks are simultaneously released from rest and that they each experience simple harmonic motion horizontally, sketch graphs of the total mechanical energy, elastic potential energy, and kinetic energy as a function of displacement from equilibrium. (b) Repeat part (a), but this time sketch graphs of the three types of energy as a function of time instead.

41. Return to the situation described in Exercise 40. Now plot graphs, as a function of time, of the (a) position; (b) velocity; and (c) acceleration for both of the blocks.

42. You are trying to demonstrate to your friend the connection between uniform circular motion and simple harmonic motion. You have a motorized turntable that spins at a rate of exactly 1 revolution per second, and you glue a ball to the turntable at a distance of 50 cm from the center. You then take a small bucket with a mass of 100 g and connect it to a spring that has a spring constant of 4.00 N/m. The bucket will then oscillate back and forth on a frictionless surface. (a) What mass of sand should you place in the bucket so the period of oscillation of the bucket matches the period of revolution of the ball on the turntable? (b) Assuming you have lined up the equilibrium position of the bucket on the spring with the center of the turntable, what amplitude should you give the bucket so its motion exactly matches one component of the motion of the ball on the turntable?

43. You match the motion of two objects, a ball glued to a turntable that is rotating at a constant angular velocity and a block oscillating on a frictionless surface because it is connected to a spring. (a) If the period of the block's oscillations is 2.5 s, what is the angular frequency of the turntable? (b) If you replace the spring by a spring with double the original spring constant, what should the angular frequency of the turntable be?

44. Consider the two graphs shown in Figure 12.27 for a block that oscillates back and forth on a frictionless surface because it is connected to a spring. The graph on the left shows the block's displacement from equilibrium, as a function of time, for one complete oscillation of the block. The graph on the right shows the elastic potential energy stored in the spring, as a function of time, over the same time period. (a) What is the spring constant of the spring? (b) What is the mass of the block? (c) What is the maximum speed reached by the block as it oscillates?

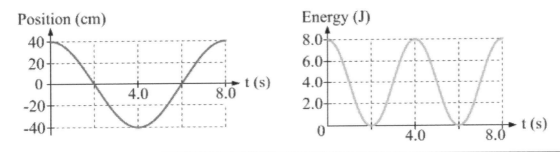

Figure 12.27: Graphs of position as a function of time and elastic potential energy as a function of time for a block oscillating horizontally on a spring. For Exercises 44 and 45.

45. Using only the information available to you in the position vs. time graph shown in Figure 12.27, determine (a) the maximum speed of the oscillating block; (b) the magnitude of the maximum acceleration of the oscillating block.

46. A block with a mass of 500 g is connected to a spring with a spring constant of 2.00 N/m. You start the motion by hitting the block with a stick so that, at $t = 0$, the block is at the equilibrium position but has an initial velocity of 2.00 m/s in the positive direction. The block then oscillates back and forth without friction. Over two complete cycles of the resulting oscillation, plot, as a function of time, the block's (a) position; (b) velocity; and (c) acceleration.

47. Equations 12.3 – 12.5 are ideal for describing the motion of an object experiencing simple harmonic motion after having been displaced from equilibrium and released from rest at $t = 0$. Modify the three equations so they match the motion described in Exercise 46.

48. As shown in Figure 12.28, two springs are separated by a distance of 1.60 m when the springs are at their equilibrium lengths. The spring on the left has a spring constant of 15.0 N/m, while the spring on the right has a spring constant of 7.50 N/m. A block, with a mass of 400 g, is then placed against the spring on the right, compressing it by a distance of 20.0 cm, which also happens to be the width of the block. The block is then released from rest. (a) Assuming all the energy initially stored in the spring is transferred to the block, and that the horizontal surface is frictionless, how long will it take until the block returns to its release point? (b) Repeat the question, but now assume that the block is held against the spring on the left, and released from rest after compressing that spring by 20 cm.

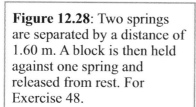 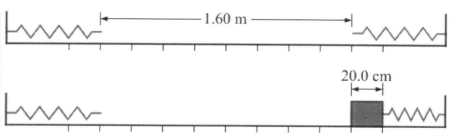

Figure 12.28: Two springs are separated by a distance of 1.60 m. A block is then held against one spring and released from rest. For Exercise 48.

49. A ballistic cart is a cart containing a ball on a compressed spring. In a popular demonstration, the cart is rolled with a constant horizontal velocity past a trigger, which causes the spring to be released, firing the ball vertically (with respect to the cart) into the air. The ball then lands in the cart again 0.60 s later. If the ball has a mass of 22 grams and the spring is initially compressed by 7.5 cm, determine the spring constant of the spring. Assume all the energy stored in the spring is transferred to the ball, and use $g = 9.8 \text{ m/s}^2$.

50. A block with a mass of 0.600 kg is connected to a spring with a spring constant of 4.50 N/m. The block is displaced a distance A from equilibrium and released from rest. How long after being released is the block first (a) at the equilibrium position? (b) at a point $A/4$ from equilibrium?

51. A block on a spring of spring constant $k = 12.0$ N/m experiences simple harmonic motion with a period of 1.50 s. What is the block's mass?

52. Consider again the situation described in Exercise 51. Such a system is used by astronauts in orbit to measure their own masses. Do a web search for "body mass measurement device" (the name of this system) and write a paragraph or two describing how it works.

53. Among the many things that Galileo Galilei is known for are his observations about pendula. Do some research about Galileo and write a paragraph or two describing his contributions to our understanding of the simple pendulum.

54. A particular wooden block floats in water with 30% of its volume submerged. You then push the block farther under the water so that 40% of its volume is submerged. When you let go the block bobs up and down. (a) For simple harmonic motion, there must be a restoring force proportional, and opposite in direction, to the displacement from equilibrium. Considering the net force on the block from combining the buoyant force and the force of gravity, does that net force fit the requirement necessary for simple harmonic motion? (b) Write an expression for the angular frequency of the block's oscillations.

55. Return again to the situation described in Exercise 54. The block has a mass of 0.30 kg. (a) Draw a free-body diagram showing the forces acting on the block immediately after it is released from rest. (b) Using $g = 10$ m/s², determine the block's initial acceleration. (c) Describe what happens to the block's free-body diagram as the block moves.

56. You are at the playground with a young boy who has a mass of 20 kg. When the boy is on a swing, you observe that you push him exactly once every 2.0 s. How long are the ropes attaching the swing to its support? Use $g = 9.8$ m/s².

57. A simple pendulum consists of a ball with a mass of 0.500 kg attached to a string of length L. The ball is displaced from equilibrium so that, when the ball is released from rest, it is at a level 1.00 m above its equilibrium position, and the string makes a 60° angle with the vertical. Use $g = 9.80$ m/s². (a) What is the length of the string? (b) Apply energy conservation to find the speed of the ball as it passes through the equilibrium position. (c) Using the small-angle approximation, it can be shown that the maximum speed of the pendulum ball is given by $v_{max} = L\theta_{max}\omega$. Making sure that your units are correct, use this equation to check your answer to part (b). (d) Your results in parts (b) and (c) should be close but should not agree exactly. Comment on which answer is better and why there is any disagreement at all.

58. Return to the situation described in Exercise 57. What is the tension in the string when the ball passes through the equilibrium position?

59. Return to the situation described in Exercise 57. After many oscillations, air resistance and friction eventually bring the pendulum to a stop. What is the total work done by resistive forces in this situation?

60. As shown in Figure 12.29, two simple pendula are identical except that the mass of the ball on one pendulum is 3 times the mass of the ball on the other. Each pendulum has a length of 1.5 m. The pendula are displaced by angles of 20°, but in opposite directions, and simultaneously released from rest. The balls then experience an elastic collision with one another. (a) What is the velocity of each ball immediately after the first collision? (b) The balls experience a second elastic collision. What is the maximum angular displacement reached by each ball as a direct result of this second collision? (c) Describe, in general, how the motion proceeds after that.

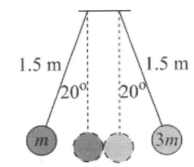

Figure 12.29: Two simple pendula, one with a ball of mass m and the other with a ball of mass $3m$, are displaced by 20° and released from rest. They swing down and collide with one another. What happens next? For Exercises 60 and 61.

61. Consider again the situation described in Exercise 60. Determine the time taken by each phase of the motion.

62. A grandfather clock uses a pendulum to keep time. For this exercise, treat the pendulum as a simple pendulum and use $g = 9.80$ m/s². (a) How long should the pendulum be if its period needs to be exactly 4 seconds for the clock to keep accurate time? (b) You now have two identical grandfather clocks, both set to keep accurate time on Earth. You take one to the Moon, where the magnitude of the gravitational field is 1/6 what it is on Earth. The clocks are started simultaneously when they both read 12 o'clock. One hour later, the clock on the Earth reads 1 o'clock. What is the time shown on the clock on the Moon at that instant?

The photo shows molten lava meeting the ocean on a beach on the Big Island in Hawaii. We have at least two indications here that the lava is very hot. First, some of it is glowing orange, the color indicative of the high temperature of the lava. Second, when the lava meets the water, enough energy is transferred to the water to vaporize some of it, producing the steam we see in the photograph. Photo credit: Ashok Rodrigues/ iStockPhoto.

Chapter 13 – Thermal Physics: A Macroscopic View

CHAPTER CONTENTS

In the next three chapters, we will focus on thermal physics, investigating concepts of temperature and heat (in this chapter); examining how the basic principles from earlier chapters relate to thermal physics (in Chapter 14); and looking at how to connect the macroscopic and microscopic views, and energy conservation, in our studies of thermodynamics (in Chapter 15).

In this chapter, our main focus will be on calorimetry, which involves bringing two or more objects, or materials, of different temperatures together. Understanding calorimetry allows us to predict the final temperature, when equilibrium is reached, and it helps us to answer questions such as why we need a relatively small piece of ice to cool a drink. In addition, we will examine the phenomena of thermal expansion, and discuss issues related to how energy is transferred between objects, or within one object, when there is a temperature difference.

13-1 Temperature Scales

In the next chapter, we'll come to a fundamental understanding of what temperature is. Let's begin our investigation of thermal physics, however, by discussing various temperature scales and how to measure temperature.

Temperature is something that has an important impact on our daily lives. In the United States, temperatures in a weather forecast are generally specified in Fahrenheit. In most of the rest of the world, however, such temperatures are given in Celsius. It is useful to know how to convert a temperature from one unit to another. The picture of the thermometer in Figure 13.1 helps us to understand the conversion process to go from Fahrenheit to Celsius, or vice versa, as well as the conversion between Celsius and Kelvin. Note that a temperature of −40°F is the same as a temperature of −40°C. Starting there, every change by 5°C is equivalent to a change of 9°F. This is where the factor of 9/5, or 5/9, in the conversion equations comes from.

$$T_C = \left(\frac{5°C}{9°F}\right)\left(T_F - 32°F\right). \quad \text{(Equation 13.1: \textbf{Converting from Fahrenheit to Celsius})}$$

$$T_F = 32°F + \left(\frac{9°F}{5°C}\right)T_C. \quad \text{(Equation 13.2: \textbf{Converting from Celsius to Fahrenheit})}$$

The Celsius scale is convenient for measuring everyday temperatures. The scale is based on the properties of water, with 0°C corresponding to the freezing point of water, and 100°C corresponding to its boiling point (at standard atmospheric pressure, at least). In scientific work temperatures are usually measured in Celsius, or in Kelvin. The Kelvin scale is an absolute temperature scale, because its zero corresponds to absolute zero. Because an increase by 1°C corresponds to an increase of 1K, it is easy to convert between the two scales.

$$T_C = T_K - 273.16°. \quad \text{(Equation 13.3: \textbf{Converting from Kelvin to Celsius})}$$

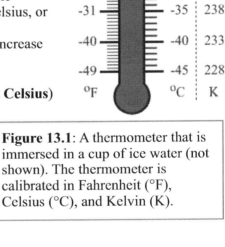

In this three-chapter sequence on thermal physics, we will use several equations that involve either the temperature T or the change in temperature ΔT. When an equation involves T the temperature is an absolute temperature – use a temperature in Kelvin. When an equation has ΔT, we can use Celsius or Kelvin, because the change in temperature is the same on the two scales.

Figure 13.1: A thermometer that is immersed in a cup of ice water (not shown). The thermometer is calibrated in Fahrenheit (°F), Celsius (°C), and Kelvin (K).

Measuring Temperature

There are many devices that can be used to measure temperature, such as the glass thermometer illustrated in Figure 13.1. These used to be filled with mercury but, with mercury now known to have negative health and environmental effects, such thermometers are now usually filled with alcohol. Almost any property of a material that changes with temperature can be exploited to make a temperature-measuring device. Examples include:

- A thermocouple – Two different kinds of metal are bonded together at two junctions. A temperature difference between the two junctions gives rise to a voltage (we'll explore that concept later in the book) that corresponds to that temperature difference.
- A gas-law thermometer – The pressure of a gas in a bulb of fixed volume can be directly converted to temperature via the ideal gas law (see Chapter 14).
- A thermopile – The higher its temperature, the more energy an object gives off. A thermopile picks up such radiated energy, which can be converted to temperature.

How does an alcohol-filled thermometer measure temperature? The thermometer, initially at room temperature (about 20°C), is placed in some hot water. To measure the water temperature, several things happen, which we will explore in the next three chapters:

- Energy is transferred from the water to the thermometer, raising the temperature of the thermometer (and lowering the temperature of the water by a small amount). This transferred energy is known as **heat**. Energy is always spontaneously transferred from the higher-temperature object to the lower-temperature object. Energy can only be transferred in the reverse direction with the aid of something else, such as a heat pump.

- The energy is transferred through the glass wall of the thermometer into the alcohol, through a process known as **thermal conduction**.

- The transfer of energy continues until **thermal equilibrium** is reached (when the thermometer reaches the same temperature as the water).

- As the temperature of the alcohol increases, the alcohol expands, causing the level of alcohol to rise in the thermometer. We exploit this expansion to measure temperature.

We will explore this last process, known as **thermal expansion**, in the next section.

Related End-of-Chapter Exercises: 18 and 20

A note about "heat" and "temperature"
We will discuss the ideas of heat and temperature in more detail in this chapter and in Chapters 14 and 15, but it is not too early to begin distinguishing between these two concepts. The word heat, as a noun, is reserved for a transfer of energy because of a temperature difference. Thus, it is incorrect to say that a hot object contains heat. A hot object has a high temperature, and thus has more **internal energy** (energy associated with the motion of the object's atoms and molecules) when it is warmer than when it is cooler. Temperature, as we will see, is a direct measure of this internal energy.

To add to the confusion, we also use "heat" as a verb, such as "I put the kettle on the stove to heat water so I could make tea." However, note that the meaning of heat as a transfer of energy is preserved here. In this situation, the water's internal energy, and therefore its temperature, increases because of heat (energy transferred to it) from, for instance, a high-temperature element on the stove.

Essential Question 13.1: Consider an alcohol thermometer of the kind drawn in Figure 13.1. Increasing the temperature causes both the alcohol and the glass shell of the thermometer to expand. Predict whether you think the cavity inside the glass (the volume occupied by the alcohol, and the space above the alcohol) increases or decreases because of the expansion of the glass. How does this affect the level of the alcohol? Is the level higher or lower because of the glass expansion than would be achieved by the expansion of the alcohol alone?

Answer to Essential Question 13.1: We'll address this issue further in the next section, but a prediction that the cavity volume decreases because of the expansion of the glass is consistent with an alcohol level higher than it would be if only the alcohol expanded. Increasing the cavity volume corresponds to an alcohol level lower than it would be if only the alcohol expanded.

13-2 Thermal Expansion

When an object's temperature changes, we assume the change in length ΔL experienced by each dimension of the object is proportional to its change in temperature ΔT. This model is valid as long as ΔT is not too large. We can express this linear relationship as an equation:

$$\Delta L = L_i \alpha \Delta T \, , \qquad \text{(Equation 13.4: Length change from thermal expansion)}$$

where L_i is the original length and α is known as the linear thermal expansion coefficient, which depends on the material (see Table 13.1).

Another way to express the linear nature of the model is to relate the initial length L_i to the new length L. Because $L = L_i + \Delta L$ we can express the new length as:

$$L = L_i + L_i \alpha \Delta T = L_i \left(1 + \alpha \Delta T\right). \text{ (Equation 13.5)}$$

Consider now a rectangular object with an area A_i that is the product of its height H_i and its width W_i. Both the height and width change with temperature, so the new area is:

Material	α ($\times 10^{-6}$ /°C)
Aluminum	23
Brass	19
Copper	17
Diamond	1
Glass	8.5
Gold	14
Iron or steel	12
Lead	29
Stainless steel	17

Table 13.1: Values of α, the linear thermal expansion coefficient, for various materials at 20°C.

$$A = HW = H_i \left(1 + \alpha \Delta T\right) W_i \left(1 + \alpha \Delta T\right) = H_i W_i \left(1 + \alpha \Delta T\right)\left(1 + \alpha \Delta T\right) = A_i \left[1 + 2\alpha \Delta T + \left(\alpha \Delta T\right)^2\right]$$

Because values of α are small, for small temperature changes $\alpha \Delta T$ is significantly less than 1. Squaring a number less than 1 makes it even smaller, so the $\left(\alpha \Delta T\right)^2$ term in the previous equation is negligible compared to the $2\alpha \Delta T$ term. Thus we can write the area equation as:

$$A = A_i \left(1 + 2\alpha \Delta T\right). \qquad \text{(Equation 13.6: Area thermal expansion)}$$

A similar process can be followed to show that the equation for the volume V resulting from imposing a temperature change ΔT on an object of original volume V_i is:

$$V = V_i \left(1 + 3\alpha \Delta T\right). \qquad \text{(Equation 13.7: Volume thermal expansion)}$$

EXAMPLE 13.2 – Shrink to fit

You are building a plane, but the stainless steel rivets you are using will not fit through the holes in the skin of the plane. At 20°C, each rivet has a diameter of 12.050 mm, while the diameter of a hole is 12.000 mm. To what temperature should you cool a rivet so it fits in a hole?

SOLUTION

We need to find the temperature change required so the diameter of the rivet equals the diameter of the hole. To find this we can use Equation 13.5, $L = L_0\left(1 + \alpha\,\Delta T\right)$, with the thermal expansion coefficient from Table 13.1 for stainless steel, $\alpha = 17 \times 10^{-6}/°C$. Solving for ΔT gives:

$$\Delta T = \frac{L - L_0}{L_0\,\alpha} = \frac{12.000\,\text{mm} - 12.020\,\text{mm}}{(12.020\,\text{mm})\left(17 \times 10^{-6}/°C\right)} = \frac{-0.020\ \text{mm}}{204.34 \times 10^{-6}\ \text{mm}/°C} = -98°C .$$

Because the initial temperature is 20°C, we get a final temperature of –78°C. Note that the equation involves ΔT, so we can work entirely in Celsius.

Also, there is no reason to convert the lengths to meters because the length units cancel out. One way to cool the rivets this much would be to immerse them in liquid nitrogen, which has a temperature of about –196°C, or 77K.

EXPLORATION 13.2 – What do holes do?

Imagine trying to fit a brass ball through a brass ring that has an inner radius just a little smaller than the ring. Should we heat the ring or cool it so that the ball can pass through?

Step 1 – *Take a solid disk with a radius equal to the outer radius of the ring. Draw a circle on this disk so the radius of the circle matches the inner radius of the ring. What happens when the temperature of the disk increases?* A picture of this situation is shown in Figure 13.2. The disk expands when its temperature increases, and so does the circle drawn on the disk. Removing the material inside the circle leaves a ring.

Step 2 – *Reverse the order of the actions. First remove the inner part of the circle and then increase the temperature. What is the result?* Figure 13.2 shows that changing the order of the actions makes no difference. The ring is the same size in both cases – a hole expands or contracts as if it were solid.

> **Key idea for holes**: Holes and cavities in materials expand and contract as if they were made from the surrounding material.
> **Related End-of-Chapter Exercises: 17 – 19.**

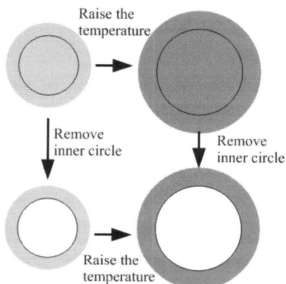

Figure 13.2: When the temperature of a disk increases, the entire disk expands, including the circle that was drawn on the disk. Removing the area within the circle leaves a ring with an inside diameter larger than the original diameter of the circle. Doing this in reverse order, removing the inside of the circle and then raising the temperature, has the same result. Holes expand as if they are filled with the surrounding material.

Increasing the temperature of a solid makes its atoms vibrate more energetically, so each atom effectively needs more space. If we imagine a circle of atoms around the inside diameter of the ring, as in Figure 13.3, the atoms spread out when heated, making the hole in the ring larger.

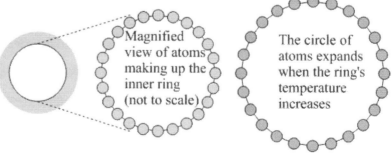

Essential Question 13.2: An iron ring will not quite fit over an aluminum cylinder. Can you fit the ring over the cylinder by increasing or decreasing the temperature of both objects at the same time? Explain.

Figure 13.3: An object expands when its temperature increases because its atoms vibrate more energetically. This causes the circle of atoms along the inside diameter of a ring to expand when the temperature increases.

Answer to Essential Question 13.2: The key here is that aluminum has a larger expansion coefficient than iron (see Table 13.1), so the aluminum expands or contracts more than does the iron ring for the same change in temperature. Increasing the temperature would make the mismatch worse but, if we decrease the temperature enough, both objects will contract, with the aluminum contracting more so the iron ring will slide onto the cylinder.

13-3 Specific Heat

Heat is energy transferred between objects that are at different temperatures, transferred from the higher-temperature object to the lower-temperature object. Although this is a form of energy we have not discussed in previous chapters, we will deal with it in much the same way we dealt with other forms of energy. Conservation of energy, for instance, will be a very useful tool, as it was previously.

The type of question we'll deal with next is the following: what is the equilibrium temperature when a 500-gram lead ball at 100°C is added to 400 g of water that is at room temperature, 20°C? To answer this kind of question we will assume that no energy is transferred into or out of the system, and we will also use a simple model that says that when energy in the form of heat is added to or removed from a substance, the substance's change in temperature is proportional to the amount of heat added or removed. The equation that goes with this statement is the following, where the symbol Q is used to represent heat, with units of energy:

$$Q = mc\Delta T, \qquad \text{(Eq. 13.8: The heat required to change an object's temperature)}$$

where m is the mass of the substance, c is known as the specific heat capacity (a constant that depends on the material), and ΔT is the temperature change. Note that, if heat is added to a substance, Q and ΔT are both positive; if heat is removed, Q and ΔT are both negative. Table 13.2 shows values of specific heat capacity for various materials. Equation 13.8 is generally valid as long as the substance does not change phase (such as from solid to liquid). We will learn how to deal with phase changes in the next section.

Material	c, specific heat capacity (J / kg °C)
Aluminum	900
Brass	377
Copper	385
Gold	128
Iron	449
Lead	129
Water (gas)	1850
Water (liquid)	4186
Water (solid)	2060

Table 13.2: Specific heat capacities for several solid metals, and for three common forms of water. The value for gaseous water is valid at standard atmospheric pressure.

Specific heat capacities are sometimes given in calories / (g °C). One calorie is the amount of heat needed to change the temperature of 1 g of water by 1°C. Thus 1000 cal is needed to change the temperature of 1 kg (1000 g) of water by 1°C. Looking at the specific heat capacity of liquid water in Table 13.2, we can see that 4186 J is equivalent to 1000 cal, giving us a conversion factor of 4.186 J/cal.

EXPLORATION 13.3 – The large heat capacity of water

A 500-gram lead ball that is initially at 100°C is added to an unknown mass of water that is initially at room temperature, 20°C, in a Styrofoam cup. After allowing the system to come to thermal equilibrium (i.e., waiting until the lead ball and the water have the same temperature), the final temperature of the system is measured to be 60°C, exactly halfway between the initial temperatures of the lead ball and the water. Based on this, determine the mass of water in the system.

Step 1 – *Using Table 13.2, use a qualitative argument to predict how the mass of water compares to the mass of the lead ball.* Both the lead and the water experience a temperature change of the same magnitude, with the temperature of the lead falling 40°C and the water temperature rising 40°C. If their heat capacities were equal, therefore, their masses would also be equal. From Table 13.2, however, we see that their heat capacities differ by a large factor. The heat capacity of lead (chemical symbol Pb) is $c_{Pb} = 129\,J/kg°C$ while the heat capacity of water (H_2O), is much larger, at $c_{H_2O} = 4186\,J/kg°C$. Refer to equation 13.8, $Q = mc\Delta T$. The heat Q has the same magnitude for both, because it is the energy transferred from the lead to the water. Therefore, the larger the specific heat capacity, the smaller the mass. In this case, then, the mass of water must be considerably smaller than the mass of the lead ball.

Step 2 – *Use equation 13.8, and energy conservation, to find the mass of water in the system.* In the static equilibrium problems we analyzed in Chapter 10 we used the equations $\sum \vec{F} = 0$ and $\sum \vec{\tau} = 0$. We can apply a similar equation in this thermal equilibrium situation:

$$\sum Q = 0.$$ (Equation 13.9: **An equation for thermal equilibrium**)

This is really a simple statement of conservation of energy. In the case of the lead ball and the water, it means that all the heat transferred out of the lead ball (this Q is negative) is transferred to the water (that Q is positive). Combining equations 13.8 and 13.9 gives:

$$Q_{Pb} + Q_{H_2O} = 0;$$

$$m_{Pb}\,c_{Pb}\,\Delta T_{Pb} + m_{H_2O}\,c_{H_2O}\,\Delta T_{H_2O} = 0.$$

Solving for the unknown mass of the water gives:

$$m_{H_2O} = \frac{-m_{Pb}\,c_{Pb}\,\Delta T_{Pb}}{c_{H_2O}\,\Delta T_{H_2O}} = \frac{-m_{Pb}\,c_{Pb}\left(T_{f,Pb} - T_{i,Pb}\right)}{c_{H_2O}\left(T_{f,H_2O} - T_{i,H_2O}\right)},$$

where the subscripts i and f on the temperatures correspond to the initial and final values, respectively. Plugging in the numerical values gives:

$$m_{H_2O} = \frac{-(500\,g)(129\,J/kg°C)(60°C - 100°C)}{(4186\,J/kg°C)(60°C - 20°C)} = 15.4\,g.$$

As we predicted, the mass of water is much smaller than the mass of the lead ball. This is because of the unusually large specific heat capacity of liquid water.

> **Key ideas regarding thermal equilibrium**: In a thermal equilibrium situation we can use a simple energy conservation relationship, $\sum Q = 0$. We also observed that liquid water has a large specific heat capacity.   **Related End-of-Chapter Exercises: 22 – 25.**

Note that, in Exploration 13.3, there is no need to convert the temperatures from Celsius to Kelvin, because the equation for Q involves ΔT and not T. In addition, there is no need to convert the given mass of the lead ball from grams to kilograms, even though the specific heat capacities are specified in kg. In the end, the units in the two specific heats cancel out, so the mass of the water comes out in the units the mass of the lead ball is given in.

Essential Question 13.3: Many people learn to do thermal equilibrium problems by setting the heat lost by one or more parts of a system equal to the heat gained by other parts. Compare and contrast that method to the $\sum Q = 0$ used in Exploration 13.3.

Answer to Essential Question 13.3: First, the two methods are completely equivalent, and they should give the same answer. There is a difference, however, in how signs are handled for the ΔT 's. In the $\sum Q = 0$ method, each ΔT is specified as a final temperature minus an initial temperature, so a ΔT can be positive (if the temperature increases) or negative (if the temperature decreases). In the heat lost vs. heat gained method, all the ΔT 's are taken to be positive. In the case of the lead ball and water, the analysis would proceed like this:

$$Q_{gained} = Q_{lost} \qquad \text{so} \qquad m_{H_2O}\, c_{H_2O} \left|\Delta T_{H_2O}\right| = m_{Pb}\, c_{Pb} \left|\Delta T_{Pb}\right|.$$

Using the absolute values of the ΔT 's ensures that every term in the equation is positive.

13-4 Latent Heat

Let's expand our knowledge of thermal equilibrium problems by learning to handle phase changes. In general, a change of phase is associated with a relatively large amount of energy, and occurs at a particular temperature. To melt ice, for instance, first heat is added to raise the ice's temperature to 0°C, the melting point. Adding more heat breaks bonds and gradually transforms the solid water into liquid water. The transformation takes place at a constant temperature of 0°C.

Because there is no change in temperature associated with a phase change, the equation we use has a different form than the $Q = mc\Delta T$ equation we use for the heat required to change a substance's temperature. The amount of heat associated with a phase change is given by:

$$Q = mL_f, \qquad \text{(Equation 13.10: \textbf{Heat for a liquid-solid phase change})}$$

where L_f is known as the **latent heat of fusion,** or,

$$Q = mL_v, \qquad \text{(Equation 13.11: \textbf{Heat for a gas-liquid phase change})}$$

where L_v is known as the **latent heat of vaporization**.

As shown in Table 13.3, latent heat values depend on the material. To melt or vaporize a substance requires that heat is added, while heat must be removed from a substance to solidify or condense it. Table 13.4 shows melting points and latent heats of fusion of some common metals.

Material	Melting point	Latent heat of fusion (kJ/kg)	Boiling point	Latent heat of vaporization (kJ/kg)
Water	0°C	335	100°C	2272
Nitrogen	-210°C	25.7	-196°C	200
Oxygen	-219°C	13.9	-183°C	213
Ethyl alcohol	-114°C	108	-78.3°C	855

Table 13.3: A table of melting points, boiling points, and latent heats for various materials.

Material	Melting point	Latent heat of fusion (kJ/kg)
Aluminum	933.5K	396
Copper	1356.6K	209
Gold	1337.58K	64
Iron	1808K	250
Lead	600.65K	23

Table 13.4: A table of melting points and latent heats of fusion for various metals.

EXPLORATION 13.4 – Temperature vs. time

  100 grams of ice at –20°C is put in a pot on a burner on the stove. The burner transfers energy to the water at a rate of 400 W. The ice melts, and eventually all the water boils away.

Step 1 – *How many different heat terms do we need to keep track of in this process? Describe them in words.* There are four heat terms we need to keep track of. These include:

1. The heat needed to increase the temperature of the ice to the melting point.
2. The heat needed to melt the ice at 0°C.
3. The heat needed to raise the liquid water to the boiling point.
4. The heat needed to boil the water at 100°C.

Step 2 – *Plot a graph of the temperature of the water as a function of time, starting at t = 0 when the temperature is at –20°C.* To draw the graph it's helpful to find the time each of the four steps takes in the process. Because energy is power multiplied by time, the times can be found by dividing the heat in each case by the power.

Raising the temperature of the ice takes $t_1 = \dfrac{mc\Delta T}{P} = \dfrac{(0.100\,\text{kg})(2060\,\text{J}/\text{kg}°\text{C})(+20°\text{C})}{400\,\text{J}/\text{s}} = 10.3\,\text{s}$.

Melting the ice takes $t_2 = \dfrac{mL_f}{P} = \dfrac{(0.100\,\text{kg})(3.33\times10^5\,\text{J}/\text{kg})}{400\,\text{J}/\text{s}} = 83.25\,\text{s}$.

Raising the water by 100°C takes $t_3 = \dfrac{mc\Delta T}{P} = \dfrac{(0.100\,\text{kg})(4186\,\text{J}/\text{kg}°\text{C})(+100°\text{C})}{400\,\text{J}/\text{s}} = 105\,\text{s}$.

Boiling the water takes $t_4 = \dfrac{mL_V}{P} = \dfrac{(0.100\,\text{kg})(2.256\times10^6\,\text{J}/\text{kg})}{400\,\text{J}/\text{s}} = 564\,\text{s}$.

  The total time required to vaporize all the water is $t = 762$ s, which is equivalent to 12.7 minutes. The water spends most of the time changing phase, particularly while it is vaporizing. This tells us that it takes a lot of energy to change phase. We can also see the large fraction of time spent changing phase on the graph of the temperature as a function of time in Figure 13.4.

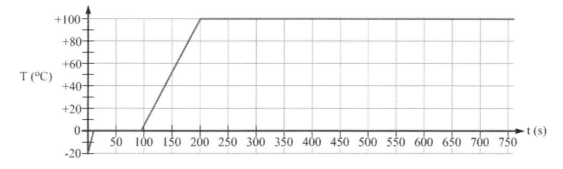

Figure 13.4: A plot of the temperature as a function of time (in seconds) for a sample of water that starts as a solid at –20°C and ends up boiling away completely. Note the large fraction of the time the water spends changing phase, in the two constant-temperature sections of the graph.

Key idea regarding phase changes: Changes of phase are generally associated with a relatively large amount of energy.   **Related End-of-Chapter Exercises: 4, 31, 32.**

Essential Question 13.4: Return to the graph in Figure 13.4. Why is the slope of the graph constant when the sample's temperature is changing? What is the slope of the graph equal to?

Answer to Essential Question 13.4: Let's begin by solving for the slope. For a particular time interval Δt, we have $mc\Delta T = P\Delta t$. The slope is equal to the rise (ΔT) over the run (Δt):

$$\frac{\Delta T}{\Delta t} = \frac{P}{mc}.$$

Until the phase changes, the power, mass, and specific heat capacity, c, are constant, so the slope is constant. Also, the slope is inversely proportional to c. A lower specific heat capacity (which water has in the solid phase) corresponds to a larger slope, so the temperature rises faster. It does not take as much energy to change temperature when the specific heat is lower.

13-5 Solving Thermal Equilibrium Problems

Let's summarize the various steps involved in solving a thermal equilibrium problem.

A General Method for Solving a Thermal Equilibrium Problem
1. Write out in words a brief description of the various heats involved.
2. Use the equation $\sum Q = 0$ to write out an equation of the form $Q_1 + Q_2 + \ldots = 0$, where each Q corresponds to one of the brief descriptions from step 1.
3. Use $Q = mc\Delta T$ to write expressions for each Q associated with a change in temperature, ΔT. Express each ΔT as $T_{final} - T_{initial}$ and the sign of that term will be built into the ΔT.
4. Use the equation $Q = mL_f$ or $Q = mL_v$ to write an expression for the heat associated with a phase change. Use a plus sign if heat is added to produce the phase change (melting or vaporizing) and a negative sign if heat is removed (solidifying or condensing).
5. Solve the resulting equation for the unknown variable.

This method applies when no heat is transferred between a system and its surroundings. If a heat transfer is involved, we can generalize the method to account for that. In that case, $\sum Q = q$, where q is the heat added to (positive q) or removed from (negative q) the system.

Let's do an example in which we combine changes of temperature and changes of phase.

EXPLORATION 13.5 – Freezing the punch

You have 2.00 liters of fruit punch at 20.0°C that you are trying to cool to get ready for a party. Because fruit punch is mostly water let's assume that the relevant specific heat capacities and latent heats for water can also be used for the fruit punch, and that the freezing point is 0°C.

Step 1 - *To cool the fruit punch quickly, you pour it into a bowl of ice that is initially –15.0°C. There is so much ice, however, that all the punch freezes and the whole mixture comes to a final temperature of –5.0°C. Find the mass of ice in the bowl originally, assuming no energy is transferred between the ice-fruit punch system and the bowl or the surrounding environment.*

This is a thermal equilibrium problem, so we apply Equation 13.9, $\sum Q = 0$, to solve it.

Let's first just write out in words the different Q's we need to include in the analysis:
1. The heat associated with increasing the temperature of the ice to –5.0 °C.
2. The heat associated with decreasing the temperature of the fruit punch to the freezing point, 0 °C.
3. The heat associated with freezing the fruit punch.
4. The heat associated with decreasing the solid fruit punch from 0 °C to the final temperature of –5.0 °C.

Let's use m_i to represent the mass of the ice, and m_{fp} to represent the mass of the fruit punch. Use the information given in the problem to determine the mass of the fruit punch, assuming again that fruit punch is basically water. Water's density is such that 1 ml of water has a mass of 1 g. With 2.00 liters of punch we have 2000 ml, which is $m_{fp} = 2000$ g = 2.00 kg.

Setting up the $\sum Q = 0$ equation gives:

$$Q_1 + Q_2 + Q_3 + Q_4 = 0;$$

$$m_i c_{solid} \left(-5°C - -15°C \right) + m_{fp} c_{liquid} \left(0°C - 20°C \right) - m_{fp} L_f + m_{fp} c_{solid} \left(-5°C - 0°C \right) = 0.$$

In the last term, we use the specific heat capacity for solid water, because that term is the heat required to bring the solid punch from the freezing point to the final temperature of $-5°C$.

When using $\sum Q = 0$ the signs on the ΔT terms come naturally from the relationship $\Delta T = T_{final} - T_{initial}$. The sign on a heat associated with a phase change, however, needs to be put in explicitly. In this case the fruit punch is solidifying, so heat is removed from the punch. This is why the $m_{fp} L_f$ term has a negative sign. Solving the equation for the mass of the ice gives:

$$m_i = \frac{m_{fp} \left[c_{liquid} \left(20°C \right) + L_f + c_{solid} \left(5°C \right) \right]}{c_{solid} \left(10°C \right)}.$$

$$m_i = \frac{2.00 \text{ kg} \left[(4186 \text{ J / kg °C})(20°C) + (3.33 \times 10^5 \text{ J/kg}) + (2060 \text{ J / kg °C})(5°C) \right]}{(2060 \text{ J / kg °C})(10°C)} = 41.5 \text{ kg.}$$

The fact that the mass required is so large reflects the fact that it takes a great deal of energy to freeze the punch.

Step 2 - *Because nobody can drink the solid fruit punch, you start again with a new 2.00-liter batch of punch at 20.0°C. Calculate the mass of ice, at $-15.0°C$, you should add to the punch so that the final temperature is 0°C and all the ice melts.*
This time we need three terms in the heat equation. They are:
1.   The heat to increase the temperature of the ice to the melting point, 0°C.
2.   The heat needed to melt all the ice at 0°C.
3.   The heat associated with decreasing the temperature of the fruit punch to 0°C.

Write out these terms in the $\sum Q = 0$ analysis to solve for m_i', the new mass of the ice:

$$Q_1 + Q_2 + Q_3 = 0;$$

$$m_i' c_{solid} \left(0°C - -15°C \right) + m_i' L_f + m_{fp} c_{liquid} \left(0°C - 20°C \right) = 0.$$

Note that the term involving the phase change is positive, because heat is added to the ice to cause it to melt. Solving for the mass of ice this time gives:

$$m_i' = \frac{m_{fp} c_{liquid} \left(20°C \right)}{c_{solid} \left(15°C \right) + L_f} = \frac{(2.00 \text{kg})(4186 \text{J/kg°C})(20°C)}{(2060 \text{J/kg°C})(20°C) + 3.33 \times 10^5 \text{J/kg}} = 0.447 \text{ kg.}$$

This is a more reasonable amount of ice than the more than 20 kg we found in step 1!

Key idea for solving thermal equilibrium problems: The $\sum Q = 0$ method can be used in cases involving changes of phase. **Related End-of-Chapter Exercises**: Exercises 43 – 45.

Essential Question 13.5: Return to step 2 of Exploration 13.5. Is 0.447 kg of ice the only mass of ice that gives a final temperature of 0°C? Explain your answer qualitatively.

Answer to Essential Question 13.5: A whole range of masses of ice can produce a mixture with a final temperature of 0°C. The mixture could be all liquid, as in Exploration 13.5, but by increasing the mass of the ice we could have some ice left, all the ice left, some liquid solidifying, or even all the liquid solidifying. We get such a range because the final temperature is 0°C, the temperature of the phase change. This issue is explored further in end-of-chapter exercise 45.

13-6 Energy-Transfer Mechanisms

Let's investigate in more detail the mechanisms by which energy is transferred when there is a temperature difference. In Exploration 13.5, we did not concern ourselves with exactly how energy is transferred from the burner through the pot to the water. That process is known as thermal conduction. Two other important energy-transfer processes are convection and radiation. Let's first discuss those two processes qualitatively, and then discuss conduction in more detail.

Convection

Energy transfer in fluids generally takes place via convection, in which flowing fluid carries energy from one place to another. Convection currents are produced by temperature differences. Hotter (less dense) parts of the fluid rise, while cooler (more dense) areas sink. Birds and gliders make use of upward convection currents to rise, and we also rely on convection to remove ground-level pollution. Forced convection, where the fluid is forced to flow, is often used for heating (e.g., forced-air furnaces) or cooling (e.g., fans, automobile cooling systems).

Thermal Radiation

Thermal radiation involves energy transferred via electromagnetic waves. Often this is infrared radiation (which we can detect with the backs of our hands), but it can also be visible light or radiation of higher energy.

All objects continually absorb energy and radiate it away again. When everything is at the same temperature, the amount of energy received is equal to the amount given off and no changes in temperature occur. If an object emits more than it absorbs, though, it tends to cool down unless some other process replenishes its energy. For an object with a temperature T (in Kelvin) and a surface area A, the net rate of radiated energy P_{net} depends strongly on temperature:

$$P_{net} = P_{radiated} - P_{absorbed} = A\varepsilon\sigma\left(T^4 - T_{env}^4\right),$$ (Equation 13.12: **Net radiated power**)

where T_{env} is the temperature of the surrounding environment; A is the object's surface area; the Stefan-Boltzmann constant has a value $\sigma = 5.67\times10^{-8}\,\text{W}/\text{m}^2$; and ε is the emissivity, which depends on the object. If an object readily absorbs or emits radiation, its emissivity is close to 1 (if it is exactly 1 the object is a **perfect blackbody**); if an object is highly reflective and does not absorb (or emit) radiation readily, its emissivity is close to 0. The best absorbers are also the best emitters. Black objects heat up faster than shiny ones, but they cool down faster too.

Thermal Conduction

Thermal conduction involves energy being transferred from a hot region to a cooler region through a material, without a net flow of the material. At the hotter end, the atoms, molecules, and electrons vibrate with more energy than they do at the cooler end. The energy flows through the material, passed along by these vibrations. One thing the rate at which energy is conducted through a uniform slab of material depends on is the temperature difference between the two faces of the slab. The rate of energy transfer is directly proportional to the temperature difference. Let's examine other contributing factors.

EXPLORATION 13.6 – The house beside the ice factory

You're planning to build a house, so you buy a small plot of land that is right next to a factory that makes ice. To get maximum use out of your land, one wall of your house will be built against the wall of the ice factory, where the ice is stored at –20°C. This might help keep your house cool in the summer, but you're more worried about staying warm in the winter.

Step 1 - *You plan to build a rectangular house. To minimize the rate at which energy is transferred from inside your house, at +20°C, to the ice factory in the winter, should you place a large-area wall or a small-area wall of your house against the wall of the ice factory? Why?* Imagine drawing a grid of equal-area squares on the wall. The more squares you have (that is, the larger the area of the wall), the more heat is transferred. To minimize your heat transfer, you need to minimize the wall area, because the rate of heat transfer is proportional to the area, A.

Step 2 – *You put a layer of insulating material in the walls of your house. To be most effective, should the material be thin or thick?* The thicker the insulation, the lower the rate of heat transfer. The rate of heat transfer is inversely proportional to L, the thickness of the insulation.

Step 3 – *While shopping at Insulation World you notice that they have slabs of copper on sale for 10% less than the slabs of Styrofoam insulation you intended to buy. Check the thermal conductivity of these materials in Table 13.5 below. Should you save 10% and buy the copper?* Absolutely not! The larger the thermal conductivity, the more effectively the material transfers heat. This is why copper is used in the base of pots and pans, efficiently transferring heat through the pot to the food. In insulating your house, however, you want to minimize the rate of heat transfer, so choose the material with the lowest thermal conductivity you can find.

Key idea: The rate at which heat is transferred (i.e., the power, P) through a slab of area A, thermal conductivity k, and thickness L is $P = \dfrac{kA}{L}(T_H - T_L)$.   (Equation 13.13)

T_H is the temperature at the higher-temperature face of the slab, while T_L is the temperature at the lower-temperature face.   **Related End-of-Chapter Exercises: 61, 62.**

R-values of insulation

Insulating materials are rated in terms of their resistance to conduction, or their R-value, also known as their thermal resistance. The larger the R-value, the better the insulating properties. R-values also make some calculations easier, because the total R-value of two materials placed back to back is the sum of their individual R-values. The R-value for a layer of material is found by dividing the material's thickness, L, by its thermal conductivity, k:

$$R = \frac{L}{k}. \quad \text{(Eq. 13.14: \textbf{R-values for insulation})}$$

Material	Thermal Conductivity, k [W/(m K)]
Air	0.0262
Aluminum	237
Brass	120
Copper	401
Diamond	895-2300
Ice	2.2
Glass	1.35
Styrofoam	0.033

Table 13.5: Thermal conductivities for various materials. These values are valid at 25°C, except for the value for ice which applies to ice at 0°C.

Essential Question 13.6: In the United States, R-values for insulation are specified in units of °F h ft^2 / BTU. BTU stands for British Thermal Unit, where 1 BTU = 1055 J. You have some insulation rated at R-15, so its R-value is 15 °F h ft^2 / BTU. What is this in units of K m^2 / W?

Answer to Essential Question 13.6: The conversion factors we need here include 1 BTU = 1055 J; 1 h = 3600 s; 1 m = 3.281 ft; and a change of 1 K is equivalent to a change of 1.8°F. Putting all this together gives us an overall conversion factor of:

$$1\frac{°F\,h\,ft^2}{BTU} \times \frac{1\,K}{1.8°F} \times \frac{3600\,s}{1\,h} \times \left(\frac{1\,m}{3.281\,ft}\right)^2 \times \frac{1\,BTU}{1055\,J} = 0.1761\frac{K\,m^2}{W}.$$

Converting 15 °F h ft² / BTU requires multiplying the 15 by the conversion factor above, giving an R-value of 2.64 K m² / W.

Chapter Summary

Essential Idea: Understanding thermal expansion and energy transfer

In this chapter, we applied simple models to help us understand thermal expansion (the change in size an object experiences when its temperature changes) and energy transfer, especially the process of thermal conduction (energy transferred through a solid object because of a temperature difference).

Temperature scales

Temperature can be measured on various scales. Among the more common temperature scales are the Fahrenheit scale, used in daily life in the United States; the Celsius scale used in daily life in the rest of the world; and the Kelvin scale, which is used in scientific work.

$$T_C = \left(\frac{5°C}{9°F}\right)(T_F - 32°F). \quad \text{(Equation 13.1: \textbf{Converting from Fahrenheit to Celsius})}$$

$$T_F = 32°F + \left(\frac{9°F}{5°C}\right)T_C. \quad \text{(Equation 13.2: \textbf{Converting from Celsius to Fahrenheit})}$$

$$T_C = T_K - 273.16°. \quad \text{(Equation 13.3: \textbf{Converting from Kelvin to Celsius})}$$

Thermal Expansion

When an object's temperature changes, its dimensions also generally change, with most materials expanding as the temperature increases. The model we applied to understand this assumes that each dimension of the object experiences a change in length that depends on the change in temperature, the initial length L_i, and the material the object is made from.

$$\Delta L = L_i \alpha \Delta T \quad , \quad \text{(Equation 13.4: \textbf{Length change from thermal expansion})}$$

where α is the thermal expansion coefficient, which depends on the material.

Equations giving the new length, area, or volume are:

$$L = L_i + L_i \alpha \Delta T = L_i(1 + \alpha \Delta T). \quad \text{(Equation 13.5: \textbf{Length thermal expansion})}$$

$$A = A_0(1 + 2\alpha \Delta T). \quad \text{(Equation 13.6: \textbf{Area thermal expansion})}$$

$$V = V_0\left(1 + 3\alpha\,\Delta T\right).$$ (Equation 13.7: **Volume thermal expansion**)

A General Method for Solving a Thermal Equilibrium Problem
1. Write out in words a brief description of the various heats involved. This helps to keep track of all the terms.
2. Use the equation $\sum Q = 0$ to write out an equation of the form $Q_1 + Q_2 + \ldots = 0$, where each Q corresponds to one of the brief descriptions you wrote in step 1.
3. Use the equation $Q = mc\,\Delta T$ to write expressions for each Q associated with a change in temperature. Express each ΔT as $T_{final} - T_{initial}$ and the sign of that heat term will be built into that temperature change.
4. Use the equation $Q = mL_f$ or $Q = mL_v$ to write an expression for the heat associated with a change of phase. Use a plus sign if heat must be added to produce the phase change (melting or vaporizing) and a negative sign if heat must be removed (solidifying or condensing).
5. Solve the resulting equation for the unknown.

The method above applies when no heat is transferred between a system and its surroundings. If heat is transferred into or out of a system, we can say that $\sum Q = q$, where q represents heat added to (q is positive) or removed from (q is negative) the system.

Energy-Transfer Mechanisms
Three mechanisms of energy transfer, driven by temperature differences, include:

1. Convection - energy is carried by a flowing fluid.

2. Thermal radiation – energy is given off in the form of electromagnetic radiation. Power radiated by an object is strongly dependent on how its temperature compares to the temperature of its surroundings.

$$P_{net} = P_{radiated} - P_{absorbed} = A\varepsilon\sigma\left(T^4 - T_{env}^4\right),$$ (Equation 13.12: **Net radiated power**)

Where T_{env} is the temperature of the surrounding environment; A is the surface area of the object; the Stefan-Boltzmann constant σ has a value $\sigma = 5.67\times10^{-8}\ \mathrm{W/m^2}$; and ε is known as the emissivity, which depends on the object. If an object readily absorbs or emits radiation then its emissivity is close to 1.

3. Thermal conduction – energy is transferred through a material by the vibrations of atoms and molecules. The rate at which energy is transferred (i.e., the power, P) through a slab of area A, thermal conductivity k, and thickness L is:

$$P = \frac{kA}{L}\left(T_H - T_L\right).$$ (Equation 13.13: **Thermal conduction**)

T_H is the temperature at the higher-temperature face of the slab, while T_L is the temperature at the lower-temperature face.

End-of-Chapter Exercises

Exercises 1 – 12 are conceptual questions that are designed to see if you have understood the main concepts of the chapter.

1. An American named Bill is visiting London, England. Listening to the weather forecast, Bill hears that the temperature will be 25°. Back home in New York, this would be a relatively cold day, so Bill puts on warm clothing. When Bill goes out of his hotel on his way to Trafalgar Square, however, he realizes that the temperature is much warmer than he thought. (a) What happened? (b) On the temperature scale that Bill is used to, what is the temperature?

2. A density ball is a ball that is weighted so its density is very similar to that of water. The ball floats in cold water but sinks in warm water. Explain why.

3. A solid iron disk is rotating without friction about an axle through its center. The Sun then comes out from behind a cloud and increases the temperature of the disk. You notice that the disk's angular velocity changes a little. What is responsible for the change, and does the disk speed up or slow down?

4. The graph in Figure 13.5 shows the temperature of a sample of an unknown material as a function of time. Heat is either being transferred into or out of this sample at a steady rate. (a) Is heat being transferred into or out of the sample? How do you know? (b) Does it look like the material undergoes a phase change? How can you tell? (c) If, over the period shown in the graph, the sample is in the solid phase for part of the time and is in the liquid phase for another part of the time, how do its two specific heat capacities compare? If possible, find the ratio of the specific heat capacity in the solid phase to the specific heat capacity in the liquid phase.

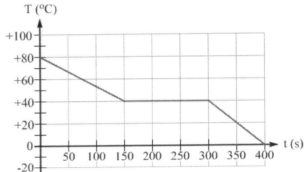

Figure 13.5: A graph of temperature as a function of time for a sample to which heat is being transferred at a constant rate, for Exercise 4.

5. A bimetallic strip is actually made from strips of two different metals that are bonded together back-to-back, as shown in Figure 13.6. The two strips have the same length at room temperature, 20°C. If the temperature changes, the two strips expand or contract different amounts, causing the bimetallic strip to bend into a circular arc, as shown in Figure 13.6, with the longer strip on the outside of the arc. (a) If the bimetallic strip in Figure 13.6 is composed of a strip of brass bonded to a strip of iron, which side is which? Why? (b) Bimetallic strips are often used as switches in thermostats, turning a furnace on if the temperature falls below a certain pre-set minimum value and turning the furnace off again when the temperature has risen sufficiently. Briefly explain how this process works.

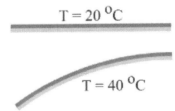

Figure 13.6: A bimetallic strip, for Exercise 5.

6. You have three blocks of equal mass. Block A is made of aluminum; block B is made of gold; and block C is made of copper. Each block starts at room temperature, 20°C. 100 J of energy is then added to each of the blocks. Rank the blocks based on their final temperatures, from largest to smallest.

7. You have three blocks. Block A is made of aluminum, block B is made of gold, and block C is made of copper. The blocks start at the same temperature. When 50 J of heat is added to each block, the blocks also have the same final temperature. Rank the blocks based on their masses, from largest to smallest.

8. You have three blocks of equal mass. Block A is made of aluminum; block B is made of gold; and block C is made of copper. Each block is initially at 80°C. You also have three identical Styrofoam cups, each containing the same amount of water at 10°C. You add one block to each of the cups and measure the final temperature. Assuming no heat is transferred to the cup or the surroundings, rank the final temperatures from highest to lowest.

9. You have three blocks. Block A is made of aluminum; block B is made of gold; and block C is made of copper. Each block is initially at 80°C. You also have three identical Styrofoam cups, each containing the same amount of water at 10°C. You add one block to each of the cups and measure the final temperature, which is exactly the same in each case. Assuming no heat is transferred to the cup or the surroundings, rank the blocks from largest to smallest based on their masses.

10. You decide to have spaghetti for dinner, so you fill a large pot with water and place it on the stove. While you are waiting for the water to boil, you decide to do an experiment, so you place a couple of drops of food coloring in the water to observe what the water is doing. You observe that the water is moving, with some water rising from the bottom of the pot to the top, moving sideways at the top, and then gradually falling back down to the bottom of the pot again. Which of the three heat transfer mechanisms are you observing here? Explain how it works.

11. You take two identical shiny metal cans and paint one black. You then place them out in the sun on a hot summer's day, and measure their temperatures as a function of time. (a) If the cans are initially at 20°C, which can reaches 30°C first? Why? (b) Later on, you take the cans inside and fill them both with hot water so both cans are initially at 95°C. Again you measure their temperatures as a function of time. Which can reaches 85°C first? Why?

12. Thin films of diamond are used on computer chips to ensure that the chips do not get too hot. Why is diamond such a good material for this application?

Exercises 13 – 16 deal with temperature scales and conversions between temperature scales.

13. What is the conversion equation to transform from (a) Fahrenheit to Kelvin? (b) Kelvin to Fahrenheit?

14. (a) What is the Rankine temperature scale? What is the equation for converting from the Rankine scale to (b) the Fahrenheit scale? (c) the Kelvin scale?

15. While visiting Toronto, Canada, James buys a cake mix at the grocery store. When James gets home to Los Angeles he tries baking the cake. He carefully follows the instructions on the package, including baking the cake at 250° for 45 minutes, but the cake is a disaster. (a) Explain what happened. (b) Did James burn the cake or was it underdone? (c) To what temperature should James have set his oven?

16. Liquid nitrogen boils at a temperature of 77K. What is this in (a) Celsius? (b) Fahrenheit?

Questions 17 – 21 deal with thermal expansion.

17. An iron ring has an inner radius of 2.5000 cm and an outer radius of 3.5000 cm, giving the ring a thickness of 1.0000 cm. If the temperature of the ring is increased from 20°C to 80°C, what is the thickness of the ring?

18. Do you think it is important for bridge designers to worry about thermal expansion when they design bridges? Why? Carry out the following two calculations to support your answer. At 10°C, a particular steel bridge has a length of precisely 500 m. Find the length of the bridge on a hot summer day when the temperature reaches 40°C. Then find the length in the middle of winter when the temperature drops to –30°C.

19. Liquid water has a linear thermal expansion coefficient of 70 x 10⁻⁶ /°C. You absent-mindedly fill an aluminum pot to the brim with water. Both the pot and the water are at 20°C. You then want to bring the water to a boil, so you put the pot on the stove to heat the water. (a) Assuming the temperature of the water and the pot are always equal, what happens as the temperature increases? Does the water spill out of the pot or does the water level fall relative to the top of the pot? Explain. Neglect any loss of water because of evaporation. (b) When the temperature reaches 80°C, determine either what fraction of the original water has overflowed the pot, or what fraction of the volume inside the pot is no longer occupied by water.

20. In section 13.2, we derived the following expression for the area A resulting from imposing a temperature change ΔT on an object of original area A_0 :

$A = A_0 \left[1 + 2\alpha \, \Delta T + (\alpha \, \Delta T)^2 \right]$. We then argued that the $(\alpha \, \Delta T)^2$ inside the bracket is negligible in comparison to the $2\alpha \, \Delta T$ term. Let's check that to see the effect of neglecting the last term. Start with an aluminum cylinder with a cross-sectional area of 5.000000 cm². Compute the new cross-sectional area when the temperature is increased by 100°C by (a) using the complete area equation above; (b) using the approximation $A = A_0 \left[1 + 2\alpha \, \Delta T \right]$. (c) What is the percentage difference between your two answers? (d) Based on this, do you think it is reasonable to neglect the $(\alpha \, \Delta T)^2$ term?

21. A solid iron disk is rotating without friction about an axle through its center. The Sun then comes out from behind a cloud and increases the temperature of the disk. You notice that the disk's angular velocity changes a little. When the temperature is 20°C, the disk's radius is 15.00 cm, and the angular speed of the disk is 20.00 rad/s. What is the disk's angular speed when its temperature is 60°C?

Exercises 22 – 27 are designed to give you practice applying the general method for solving a thermal equilibrium problem. For each of these exercises, begin with the following steps:
   (a) Write out in words a brief description of the various heats involved.
   (b) Apply $\sum Q = 0$ to obtain an equation of the form $Q_1 + Q_2 + ... = 0$. Each Q corresponds to one of the brief descriptions you wrote in part (a).
   (c) For any temperature changes, apply the equation $Q = mc\Delta T$. Express each ΔT as $T_{final} - T_{initial}$. For any changes of phase, apply the equation $Q = mL_f$ or $Q = mL_v$. Use a positive sign if heat must be added to produce the phase change, and a negative sign if heat must be removed.

22. An aluminum block with a mass of 300 g and a temperature of 80°C is placed in a Styrofoam cup that contains 500 g of water at 10°C. Ignore any temperature change associated with the Styrofoam cup. Start by doing parts (a) – (c) as described above. (d) Find the equilibrium temperature.

23. Repeat Exercise 22, but now the water is in an aluminum container that has a mass of 200 g and is initially at the temperature of the water. The block, container, and water all come to the same final temperature.

24. A copper block with a mass of 500 g is cooled to 77K by being immersed in liquid nitrogen. The block is then placed in a Styrofoam cup containing some water that is initially at +50°C. Assume no heat is transferred to the cup or the surroundings. The goal of the exercise is to determine the mass of water in the cup, if the final temperature is +20°C. Start by doing parts (a) – (c) as described above. (d) Find the mass of the water.

25. Repeat Exercise 24, but this time the final temperature is –20°C.

26. Repeat Exercise 24, but this time the final temperature is 0°C. Start by doing parts (a) – (c) as described above. (d) Find the maximum mass of water in the cup. (e) Find the minimum mass of water in the cup.

27. Repeat Exercise 24, but this time the water is in an aluminum cup that has a mass of 300 g. Assume the temperature of the cup is equal to the temperature of the water at all times.

Exercises 28 – 32 are designed to give you practice solving problems in which heat (q) is added or removed from a system. Applying $\sum Q = q$ should help you answer these exercises.

28. A Styrofoam cup contains 200 g of water at 20°C. The cup is then placed in the freezer. The freezer can remove heat from the water at a steady rate of 50 W. (a) If we neglect any heat transfer involving the Styrofoam cup, how long does it take until the cup contains ice at –5°C? (b) Plot a graph of the temperature of the water as a function of time as the water cools from 20°C to –5°C.

29. Return to the situation described in Exercise 28, except now the water is placed in a 300 g aluminum container. The temperature of the container matches the water at all times and, because aluminum has a larger thermal conductivity than Styrofoam, the freezer removes heat from the aluminum-water system at a rate of 90 W. How long does it take the temperature of the system to drop from 20°C to –5°C now?

30. 500 g of water at 20°C is in a pot on the stove. An unknown mass of ice that is originally at –10°C is placed in an identical pot on the stove. Heat is then added to the two samples of water at precisely the same constant rate. You observe that both samples of water reach 80°C at the same time. (a) How does the mass of the ice in the second pot compare to the mass of the water in the first pot? (b) Which system reaches 90°C first? (c) Solve for the mass of the ice.

31. You have a 100 g block of lead that you intend to melt and then pour into a mold to form a bell. The lead is initially at room temperature, 20°C. You then add heat to the lead at a steady rate of 200 W. (a) How long does it take for the lead to reach its melting point? (b) How much additional time is required to completely melt the lead block? (c) Graph the temperature of the sample versus time, ending the graph when the lead is completely melted.

32. The graph in Figure 13.7 shows the temperature of a sample of an unknown material as a function of time. Heat is either being transferred into or out of this sample at a steady rate. Over the period shown in the graph, the sample is in the solid phase for part of the time and is in the liquid phase for another part of the time. (a) Is heat being transferred into or out of this sample? (b) What is the material's melting point? (c) What is the ratio of the material's specific heat when liquid to the specific heat when solid?

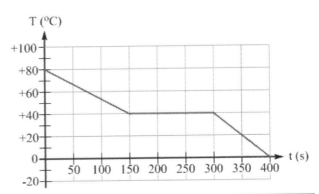

Figure 13.7: A graph of temperature as a function of time for a sample to which heat is being transferred at a constant rate, for Exercise 32.

Exercises 33 – 35 involve the energy-transfer mechanisms of convection, conduction, or radiation.

33. To keep yourself warm in the winter, you heat a solid metal ball and place it on a stand in the center of your room. The ball has a radius of 10.0 cm and an emissivity of 0.82. If your room has a temperature of 15°C, what is the net power radiated by the ball initially, when the ball's temperature is 200°C?

34. You have two rods that have the same dimensions but which are made from different materials. One rod is made of brass while the other is made of copper, and each rod is 1.00 m long. The rods are joined end-to-end, as shown in Figure 13.8. The far end of the brass rod is maintained at a temperature of 0°C, while the far end of the copper rod is maintained at a temperature of 90°C. The system is allowed to come to equilibrium (defined as each point on the rods reaching a constant temperature). What is the temperature at the point where the rods meet? Neglect thermal expansion.

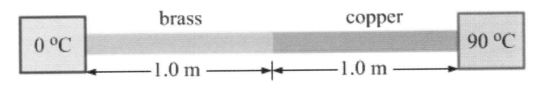

Figure 13.8: A brass rod and a copper rod are joined together end-to-end. The other end of the brass rod is kept at 0°C, while the far end of the copper rod is maintained at 90°C. For Exercises 34 and 35.

35. Return to the situation described in Exercise 34. Plot a graph of the temperature as a function of position along the rods, taking the cooler end of the brass rod to be the origin and the hotter end of the copper rod to be $x = +2.00$ m.

General Exercises and Conceptual Questions

36. The Fahrenheit and Celsius temperature scales are named after particular individuals. Do a little research about these people and write a paragraph or two describing each one.

37. A photograph of a Galileo thermometer is shown in Figure 13.9. Explain how such a thermometer works. What property of the liquid inside the thermometer is being exploited in this thermometer?

Figure 13.9: A photograph of a Galileo thermometer, for Exercise 37. Photo credit: Johanna Goodyear / iStockPhoto.

38. The "Mpemba effect" is the name for an interesting phenomenon, namely that in some circumstances warmer water can end up freezing before cooler water. Do some research on the Mpemba effect and write two or three paragraphs describing how this might be possible.

39. To make a cup of tea, you put 1000 g of water at 15°C into a kettle and bring the water to a boil. However, you only need 300 g of hot water to make your cup of tea. How much energy did you waste bringing the extra water to the boiling point?

40. In the second paragraph of Section 13.3, we ask the following question: what is the equilibrium temperature when a 500-gram lead ball at 100°C is added to 400 g of water that is at room temperature, 20°C? What is the answer?

41. A 500-gram lead ball that is initially at 100°C is added to 500 g of water that is initially at room temperature, 20°C, in a Styrofoam cup. The system is allowed to come to thermal equilibrium. Note that 60°C is exactly halfway between the initial temperatures of the lead ball and the water. (a) Come up with a qualitative argument for whether the final temperature is more than 60°C, less than 60°C, or equal to 60°C. (b) Find the final temperature.

42. A lead bullet with a mass of 25 g is fired into a target. The bullet completely melts upon impact. Assuming all the kinetic energy of the bullet goes into raising the bullet's temperature and then melting it, what is the speed of the bullet when it strikes the target?

43. You have 2.00 liters of fruit punch at 20.0°C that you are trying to cool to get ready for a party. Assume that the relevant specific heat capacities and latent heats for water can also be used for the fruit punch, and that the freezing point is 0°C. To cool the fruit punch quickly, you pour it over a large bowl of ice that is initially at –15.0°C. The mixture comes to a final temperature of +5.0°C. Find the mass of ice that was originally in the bowl, assuming no energy is transferred between the ice-fruit punch system and the bowl or the surrounding environment.

44. Repeat Exercise 43, but now account for the bowl. Assume the bowl is made from 300 g of aluminum and that the bowl is also initially at –15.0°C.

45. In part (b) of Exploration 13.3, we found that when 0.447 kg of ice at –15.0°C is added to 2.00 liters of fruit punch at 20.0°C and allowed to come to thermal equilibrium, the result is that all the ice melts and the final temperature of the mixture is 0°C. There is a whole range of masses of ice, however, that could have been added to the punch to achieve that same final temperature. (a) Does 0.447 kg represent the minimum or the maximum amount of ice at –15.0°C we can add to the punch to produce a final temperature of 0°C? Explain. (b) Determine the other end of the range, the other extreme in the amount of ice we can add to the punch and yet still achieve a final temperature of 0°C.

46. A block of ice has an unknown initial temperature. Heat is transferred to the ice, first bringing the ice to 0°C, then melting it, and then bringing the resulting water to 50°C. The total heat required for the two changes in temperature is equal to the heat associated with the melting. What was the block's initial temperature?

47. A copper block, with a mass of 1500 g, is cooled to 77K by being immersed in liquid nitrogen. The block is then transferred to a Styrofoam cup containing 1.20 liters of water at 50°C. Assuming no energy is transferred to the cup, determine the final temperature of the system.

48. Repeat Exercise 47, but this time the block is aluminum instead of copper.

49. Repeat Exercise 47, but this time assume the water is in an aluminum container that has a mass of 400 g, and that the temperature of the container is equal to the temperature of the water at all times.

50. You have three blocks of equal mass. Block A is made of aluminum; block B is made of gold; and block C is made of copper. Each block is initially at 80°C. The blocks are added, one at a time, to a Styrofoam cup containing 500 g of water at 10°C. The final temperature is 40°C. Assuming no heat is transferred to the cup or the environment, what is the mass of each block?

51. You have three balls of equal mass. Ball A is made of aluminum; ball B is made of gold; and ball C is made of copper. Each ball is initially at –50°C. You also have three identical Styrofoam cups, each containing equal amounts of water at 10°C. You add one ball to each of the cups and measure the final temperature. Assuming no heat is transferred to the cup or the surroundings, is there enough information provided to rank the final temperatures from highest to lowest? If so, provide the ranking. If not, explain why not.

52. Return to the situation described in Exercise 51. Is it possible for the final temperature in one of the cups to be below 0°C, the final temperature in another to be 0°C, and the final temperature in the remaining cup to be above 0°C? If so, come up with an example specifying the mass of the balls and the mass of the water in the cup. If not, explain why not.

53. James Prescott Joule carried out an experiment known as the mechanical equivalent of heat. Write a few paragraphs about Joule and the experiment.

54. The water at Niagara Falls drops through a height of 52 m. If the water's loss of gravitational potential energy shows up as an increase in temperature of the water, what is the temperature difference between the water at the top of the falls and the water at the bottom?

Figure 13.10: A photograph of Niagara Falls, for Exercise 54. Photo credit: photo from http://www.publicdomainpictures.wordpress.com

55. As part of an experiment, you fill a cardboard tube that has a length of 1.2 m with 200 g of lead shot (small lead balls) and seal the ends of the tube. Aligning the axis of the tube vertically, you then invert the tube 100 times. Predict what you observe for the temperature difference of the lead balls at the end of the experiment compared to what it was at the beginning.

56. Return to Exercise 55. Doing the experiment with lead balls can be something of a health risk, because you can breathe in lead dust if you open the tube to measure the temperature. You try the experiment with small copper balls instead of lead. (a) In which case would you observe a larger temperature change, when you used the copper balls or when you used the lead balls? Explain your answer. (b) If you take the ratio of the two temperature changes, what would you expect to find?

57. The Sun has a radius of 6.96×10^5 km, and an average temperature at its surface of 5780 K. (a) Calculate the power radiated by the Sun. (b) The distance from the Sun to the Earth is about 150 million km. Estimate the power from the incident sunlight on a 1.0 m² solar cell that is part of an array being used to provide energy for a satellite in orbit around the Earth.

58. The base of a copper-bottomed pot has a radius of 15 cm and a thickness of 3.0 mm. Its bottom surface is maintained at a temperature of 250°C by being placed on a hot burner on the stove. (a) If the pot contains 2.00 liters of water that is initially at 20°C determine the initial rate at which energy is transferred through the base of the pot to the water. (b) As the water temperature increases, does the rate at which energy is transferred through the base of the pot increase, decrease, or stay the same? Explain. (c) Calculate the rate of energy transfer when the water temperature is 95°C.

59. Return to the situation described in Exercise 58. Let's say it takes a time T to raise the water temperature to the boiling point. The process is repeated with a second pot in which everything is the same except for the fact that the base of the second pot is aluminum instead of copper. (a) Is the time it takes to bring the water to the boiling point in the second pot greater than, less than, or the same as T? Justify your answer. (b) How much time, in terms of T, does the process take in the second pot?

60. On a chilly November day, you go out for a hike to the top of a local mountain with your friend. The two of you dress in layers, but for your inner-most layer you are wearing a high-tech fabric that wicks moisture away from your skin while your friend is wearing a long-sleeved cotton shirt that is damp with perspiration by the time you reach the top of the mountain. Coming back down, you are quite comfortable, while your friend is feeling colder with each passing minute. Fortunately, you reach the lodge at the base of the mountain before your friend becomes hypothermic, and your friend is able to warm up again in front of a roaring fire. Explain what happened, given that the fabric you were wearing has a thermal conductivity of 0.06 W/(m K), dry cotton has a thermal conductivity of 0.04 W/(m K), and water has a thermal conductivity of 0.6 W/(m K).

61. The outer walls of your house have an R-value of 5.0 K m² / W, and a total area of 2000 m². Let's assume that, from the beginning of December to the end of February, the temperature outside the house is 0°C while you set your thermostat to maintain a constant temperature of 22°C inside the house. (a) How much energy is conducted through the walls of your house in this three-month period? Assume it is not a leap year. (b) How much energy would be conducted through the walls if you lowered the thermostat so as to maintain a constant temperature of 20°C? (c) Every kW-h of energy costs about 20 cents. First, convert a kW-h to joules, and then determine how much money you would save by keeping your thermostat at 20°C for the three months.

62. You have four square aluminum sheets, each with an area of 0.25 m² and a thickness of 5.0 mm, and four copper sheets, having exactly the same dimensions as the aluminum sheets. You plan to create a square piece of insulation, with an area of 1.00 m², by placing four of your sheets together in one layer, and layering the remaining four sheets on top of the first four. (a) To minimize the rate of energy transfer through your arrangement, should you place the four sheets of copper over the four sheets of aluminum, or should you stack sheets of aluminum together and stack the copper together? (b) Assuming a temperature difference of 20°C between the two faces of the arrangement, support your answer to part (a) by calculating the rate of energy transfer in the two cases.

63. What thickness of aluminum has the same R-value as 5.0 cm of Styrofoam?

64. You overhear two of your classmates discussing Essential Question 13.1. Comment on each of their statements.

Liam: *Did you notice that we never got the answer to Essential Question 13.1? Does the space inside the glass thermometer increase or decrease when the temperature goes up? Well, the glass expands, so it must fill in some of that space – that would cause the alcohol level to go up even more than it would if the glass did not change size.*

Sherry: Except, in Section 13.2 we looked at how holes expand when they're heated. Can't we apply that to the cavity inside the glass? That would mean the cavity volume increases when the temperature goes up, so the alcohol level is less than if the glass doesn't change size. Except, how do we know the level goes up at all when the temperature increases? Couldn't the level even go down, or stay the same? That doesn't sound like a very good thermometer!

Consider the balloons in the photograph. What is it that keeps the balloons inflated? What would happen to the balloons if we changed the temperature? These questions, and others, will concern us in this chapter. What these questions depend on is the fact that the balloons are full of gas (it could be air, or helium) and it is the behavior of the individual gas molecules that determine how inflated the balloon is.

Photo credit: Oleg Prikhodko /iStockPhoto.

Chapter 14 – Thermal Physics: A Microscopic View

CHAPTER CONTENTS

The main focus of this chapter is the application of some basic principles we learned earlier to thermal physics. This will give us some important insights into what temperature means, and into how systems behave on a microscopic level. The overall theme of the chapter is the connection between the microscopic and macroscopic properties of a system of gas. At the microscopic level, we're talking about the behavior of individual atoms and molecules.

Even a relatively small container that is full of air at atmospheric pressure may have at least 10^{20} gas molecules, with each of these molecules having a unique velocity. Fortunately for us, we don't have to keep track of what each of these molecules is doing. Instead, we can usually get away with characterizing this system of gas with a very small number of macroscopic parameters, such as the temperature and pressure. As we will see, both the temperature and the pressure of the system relate back to what is happening at the microscopic level, but we do not have to know all the details of what is going on microscopically to be able to determine the values of a system's temperature and pressure. Understanding these connections, however, can give us some insight into the behavior of such a gas system.

14-1 The Ideal Gas Law

Let's say you have a certain number of moles of ideal gas that fills a container that has a known volume. Such a system is shown in Figure 14.1.

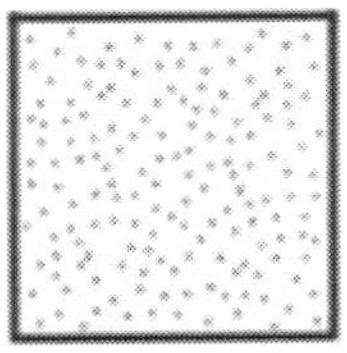

If you know the absolute temperature of the gas, what is the pressure? The answer can be found from the ideal gas law, which you may well have encountered before.

Figure 14.1: A container of ideal gas.

The ideal gas law connects the pressure P, the volume V, and the absolute temperature T, for an ideal gas of n moles:

$$PV = nRT,$$ (Equation 14.1: **The ideal gas law**)

where $R = 8.31$ J/(mol K) is the universal gas constant.

First of all, what is a mole? It is a not a cute, furry creature that you might find digging holes in your backyard. In this context, a mole represents an amount, and we use the term mole in the same way we use the word dozen. A dozen represents a particular number, 12. A mole also represents a particular number, 6.02×10^{23}, which we also refer to as Avogadro's number, N_A. Thus, a mole of something is Avogadro's number of those things. In this chapter, we generally want to know about the number of moles of a particular ideal gas. A toy balloon, for instance, has about 0.1 moles of air molecules inside it. Strangely enough, the number of stars in the observable universe can also be estimated at about 0.1 moles of stars.

In physics, we often find it convenient to state the ideal gas law not in terms of the number of moles but in terms of N, the number of atoms or molecules, where $N = nN_A$. Taking the ideal gas law and multiplying the right-hand side by N_A / N_A gives:

$$PV = nN_A \frac{R}{N_A} T = N \frac{R}{N_A} T.$$

The constant R / N_A has the value $k = 1.38 \times 10^{-23}$ J/K and is known as Boltzmann's constant. Using this in the equation above gives:

$$PV = NkT.$$ (Eq. 14.2: **Ideal gas law in terms of the number of molecules**)

Under what conditions is the ideal gas law valid? What is an ideal gas, anyway? For a system to represent an ideal gas it must satisfy the following conditions:

1. The system has a large number of atoms or molecules.
2. The total volume of the atoms or molecules should represent a very small fraction of the volume of the container.
3. The atoms or molecules obey Newton's Laws of motion; and they move about in random motion.
4. All collisions are elastic. The atoms or molecules experience forces only when they collide, and the collisions take a negligible amount of time.

The ideal gas law has a number of interesting implications, including –

Boyle's Law: at constant temperature, pressure and volume are inversely related;

Charles' Law: at constant pressure, volume and temperature are directly related;

Gay-Lussac's Law: at constant volume, pressure and temperature are directly related.

EXAMPLE 14.1 – Two containers of gas

The two sealed containers in Figure 14.2 contain the same type of ideal gas. Container 2 has twice the volume of container 1. Aside from that difference, the containers differ in only one of the following three parameters, pressure, number of moles, and temperature. (a) Could the containers differ only in pressure and volume? If so, explain how. (b) Could the containers differ only in the number of moles and volume? If so, explain how.

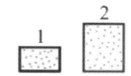

Figure 14.2: Two containers of ideal gas, one with twice the volume as the other.

SOLUTION

(a) Yes, if the number of moles of gas and the temperature are the same in each container, we must have the product of PV equal in the two containers, according to the ideal gas law. Thus, if container 2 has twice the volume as container 1, it must have half the pressure as container 1.

(b) Yes, if the pressure and the temperature are the same in each container, the number of moles of gas must be twice that in container 2 as it is in container 1. If we double the value of the volume, on the left side of Equation 14.1, we must double the value of *n*, the number of moles, on the right side of the equation, if everything else remains constant.

Prove to yourself that the containers could also differ only in volume and temperature.

An aside – Thinking about the rms average.

In Section 14.2, we will use the rms (root-mean square) average speed of a set of gas molecules. To gain some insight into the root-mean-square averaging process, let's work out the rms average of the set of numbers –1, 1, 3, and 5. The average of these numbers is 2. To work out the rms average, square the numbers to give 1, 1, 9, and 25. Then, find the average of these squared values, which is 9. Finally, take the square root of that average to find the rms average, 3.

Clearly, this is a funny way to do an average, because the average is 2 while the rms average is 3. There are two reasons why the rms average is larger than the average in this case. The first reason is that squaring the numbers makes everything positive – without this negative values cancel positive values when we add the numbers up. The second reason is that squaring the values weights the larger numbers more heavily (the 5 counts five times more than the 1 when doing the average, but 5^2 counts 25 times more than 1^2 when doing the rms average.) Note that we will discuss rms average values again later in the book when we talk about alternating current.

Related End-of-Chapter Exercises: 1, 2, 13, 17, 18.

Essential Question 14.1: A container of ideal gas is sealed so that it contains a particular number of moles of gas at a constant volume and an initial pressure of P_i. If the temperature of the system is then raised from 10°C to 30°C, by what factor does the pressure increase?

Answer to Essential Question 14.1: It is tempting to say that the pressure increases by a factor of 3, but that is incorrect. Because the ideal gas law involves T, not ΔT, we must use temperatures in Kelvin rather than Celsius. In Kelvin, the temperature rises from 283K to 303K. Finding the ratio of the final pressure to the initial pressure shows that pressure increases by a factor of 1.07:

$$\frac{P_f}{P_i} = \frac{nRT_f/V}{nRT_i/V} = \frac{T_f}{T_i} = \frac{303K}{283K} = 1.07 .$$

14-2 Kinetic Theory

We will now apply some principles of physics we learned earlier in the book to help us to come to a fundamental understanding of temperature. Consider a cubical box, measuring L on each side. The box contains N identical atoms of a monatomic ideal gas, each of mass m.

We will assume that all collisions are elastic. This applies to collisions of atoms with one another, and to collisions involving the atoms and the walls of the box. The collisions between the atoms and the walls of the box give rise to the pressure the walls of the box experience because the gas is enclosed within the box, so let's focus on those collisions.

Let's find the pressure associated with one atom because of its collisions with one wall of the box. As shown in Figure 14.3 we will focus on the right-hand wall of the box. Because the atom collides elastically, it has the same speed after hitting the wall that it had before hitting the wall. The direction of its velocity is different, however. The plane of the wall we're interested in is perpendicular to the x-axis, so collisions with that wall reverse the ball's x-component of velocity, while having no effect on the ball's y or z components of velocity. This is like the situation of the hockey puck bouncing off the boards that we looked at in Chapter 6.

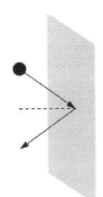

The collision with the wall changes the x-component of the ball's velocity from $+v_x$ to $-v_x$, so the ball's change in velocity is $-2v_x$ and its change in momentum is $\Delta \vec{p} = -2mv_x$, where the negative sign tells us that the change in the atom's momentum is in the negative x-direction.

Figure 14.3: An atom inside the box bouncing off the right-hand wall of the box.

In Chapter 6, we learned that the change in momentum is equal to the impulse (the product of the force $\vec{F}$ and the time interval Δt over which the force is applied). Thus:

$$\vec{F}_{\text{wall on molecule}} = \frac{-2mv_x}{\Delta t} .$$ (Equation 14.3: **The force the wall exerts on an atom**)

The atom feels an equal-magnitude force in the opposite direction (Newton's third law):

$$\vec{F}_{\text{molecule on wall}} = \frac{+2mv_x}{\Delta t} .$$ (Equation 14.4: **The force the atom exerts on the wall**)

What is this time interval, Δt? The atom exerts a force on the wall only during the small intervals it is in contact with the wall while it is changing direction. It spends most of the time not in contact with the wall, not exerting any force on it. We can find the time-averaged force the atom exerts on the wall by setting Δt equal to the time between collisions of the atom with that wall. Because the atom travels a distance L across the box in the x-direction at a speed of v_x, it takes a time of L/v_x to travel from the right wall of the box to the left wall, and the same amount of time to come back again. Thus:

$$\Delta t = \frac{2L}{v_x}. \qquad \text{(Equation 14.5: Time between collisions with the right wall)}$$

Substituting this into the force equation, Equation 14.4, tells us that the magnitude of the average force this one atom exerts on the right-hand wall of the box is:

$$\overline{F}_{\text{molecule on wall}} = \frac{2mv_x}{\Delta t} = \frac{2mv_x}{2L/v_x} = \frac{mv_x^2}{L}. \qquad \text{(Eq. 14.6: Average force exerted by one atom)}$$

To find the total force exerted on the wall we sum the contributions from all the atoms:

$$\overline{F}_{\text{on wall}} = \sum \frac{mv_x^2}{L} = \frac{m}{L} \sum v_x^2. \qquad \text{(Equation 14.7: Average force from all atoms)}$$

The Greek letter $\sum$ (sigma) indicates a sum. Here the sum is over all the atoms in the box.

If we have N atoms in the box then we can write this as:

$$\overline{F}_{\text{on wall}} = \frac{Nm}{L}\left(\frac{\sum v_x^2}{N}\right). \qquad \text{(Equation 14.8: Average force from all atoms)}$$

The term in brackets represents the average of the square of the magnitude of the x-component of the velocity of each atom. For a given atom if we apply the Pythagorean theorem in three dimensions we have $v_x^2 + v_y^2 + v_z^2 = v^2$. Doing this for all the atoms gives:

$$\sum v_x^2 + \sum v_y^2 + \sum v_z^2 = \sum v^2,$$

and there is no reason why the sum over the x-components would be any different from the sum over the y or z-components – there is no preferred direction in the box. We can thus say that $3\sum v_x^2 = \sum v^2$ or, equivalently, $\sum v_x^2 = \frac{1}{3}\sum v^2$.

Substituting this into the force equation, Equation 14.8, above gives:

$$\overline{F}_{\text{on wall}} = \frac{Nm}{3L}\left(\frac{\sum v^2}{N}\right). \qquad \text{(Equation 14.9: Average force on a wall)}$$

The term in brackets represents the square of the rms average speed. Thus:

$$\overline{F}_{\text{on wall}} = \frac{Nm}{3L}v_{rms}^2. \qquad \text{(Equation 14.10: Average force on a wall)}$$

By multiplying by 2 and dividing by 2, we can transform Equation 14.10 to:

$$\overline{F}_{\text{on wall}} = \frac{2N}{3L}\left(\frac{1}{2}mv_{rms}^2\right) = \frac{2N}{3L}K_{av}, \qquad \text{(Eq. 14.11: Force connected to kinetic energy)}$$

The term in brackets is a measure of the average kinetic energy, K_{av}, of the atoms.

Related End-of-Chapter Exercise: 36.

Essential Question 14.2: Why is the rms average speed, and not the average velocity, involved in the equations above? What is the average velocity of the atoms of ideal gas in the box?

Answer to Essential Question 14.2: The average velocity of the atoms is zero. This is because the motion of the atoms is random, and with a large number of atoms in the box there are as many, on average, going one way as the opposite way. Because velocity is a vector the individual vectors tend to cancel one another out. The average speed, however, is non-zero, and it makes sense that, the faster the atoms move, the more force they exert on the wall.

14-3 Temperature

Let's pick up where we left off at the end of the previous section. Because pressure is force divided by area, we can find the average pressure the atoms exert on the wall by dividing the average force by the wall area, L^2. This gives:

$$P = \frac{\overline{F}_{on\ wall}}{A} = \frac{2N}{3L^3}K_{av} .$$ (Equation 14.12: **Pressure in the gas**)

Now we have a factor of L^3, which is V, the volume of the cube. We can thus write Equation 14.12 as:

$$PV = N\left(\frac{2}{3}K_{av}\right).$$ (Equation 14.13: **The product PV**)

Compare Equation 14.13 to Equation 14.2, the ideal gas law in the form $PV = NkT$.

These equations must agree with one another, so we must conclude that:

$$\frac{2}{3}K_{av} = kT ,$$

or, equivalently,

$$K_{av} = \frac{3}{2}kT .$$ (Equation 14.14: **Average kinetic energy is directly related to temperature**)

This is an amazing result – it tells us what temperature is all about. Temperature is a direct measure of the average kinetic energy of the atoms in a material. It is further amazing that we obtained such a fundamental result by applying basic principles of physics (such as impulse, kinetic energy, and pressure) to an ideal gas. Consider now the following example.

EXAMPLE 14.3 – Two containers of ideal gas

Container A holds N atoms of ideal gas, while container B holds $5N$ atoms of the same ideal gas. The two containers are at the same temperature, T.
    (a) In which container is the pressure highest?
    (b) In which container do the atoms have the largest average kinetic energy? What is that average kinetic energy in terms of the variables specified above?
    (c) In which container do the atoms have the largest total kinetic energy? What is that total kinetic energy in terms of the variables specified above?

SOLUTION

(a) We don't know anything about the volumes of the two containers, so there is not enough information to say how the pressures compare. All we can say is that the product of the pressure multiplied by the volume is fives times larger in container B than in container A, because PV is proportional to the product of the number of atoms multiplied by the absolute temperature.

(b) The fact that the temperatures are equal tells us that the average kinetic energy of the atoms is the same in the two containers. Applying Equation 14.14, we get in each case:

$$K_{av} = \frac{3}{2}kT \, .$$

(c) The total kinetic energy is the average energy multiplied by the number of atoms, so container B has the larger total kinetic energy. Container B has a total kinetic energy of:

$$K_B = 5NK_{av} = 5N\frac{3}{2}kT = \frac{15}{2}NkT \, .$$

Related End-of-Chapter Exercises: 8, 10, 37.

Absolute zero

Another interesting concept contained in the ideal gas law is the idea of absolute zero. Let's say we seal a sample of ideal gas in a container that has a constant volume. The container has a pressure gauge connected to it that allows us to read the pressure inside. We then measure the pressure as a function of temperature, placing the container into boiling water (100 °C), ice water (0 °C), and liquid nitrogen (–196 °C). The pressures at these temperatures are 129 kPa, 93.9 kPa, and 26.6 kPa, respectively. Plotting pressure as a function of temperature results in the graph shown in Figure 14.4. We find that our three points, and other points we care to measure, fall on a straight line. Extrapolating this line to zero pressure tells us that the pressure equals zero at a temperature of –273 °C (also known as 0 K).

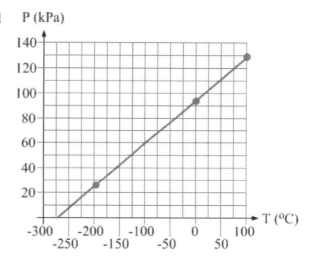

Figure 14.4: A graph of pressure as a function of temperature for a constant-volume situation. Extrapolating the graph to zero pressure shows that absolute zero corresponds to a temperature of –273 °C.

Based on the previous section, we would conclude that the pressure drops to zero at absolute zero because the atoms or molecules have no kinetic energy. This is not quite true, although applying ideas of quantum mechanics is necessary to understand why not. If the atoms and molecules stopped completely, we would be able to determine precisely where they are. Heisenberg's uncertainty principle, an idea from quantum mechanics, tells us that this is not possible, that the more accurately we know an object's position the more uncertainty there is in its momentum. The bottom line is that even at absolute zero there is motion, known as zero-point motion. Absolute zero can thus be defined as the temperature that results in the smallest possible average kinetic energy.

Essential Question 14.3: At a particular instant compare the kinetic energy of one particular atom in container A to that of one particular atom in container B. Which atom has the larger kinetic energy? The two containers are at the same temperature, and there are five times more atoms in container B than in container A.

Answer to Essential Question 14.3: There is no way we can answer this question. The ideal gas law, and kinetic theory, tells us about what the atoms are doing on average, but they tell us nothing about what a particular atom is doing at a particular instant in time. Atoms are continually colliding with one another and these collisions generally change both the magnitude and direction of the atom's velocity, and thus change the atom's kinetic energy. We can find the probability that an atom has a speed larger or smaller than some value, but that's about it.

14-4 Example Problems

EXPLORATION 14.4 – Finding pressure in a cylinder that has a movable piston

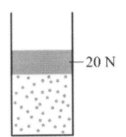

A cylinder filled with ideal gas is sealed by means of a piston. The piston is a disk, with a weight of 20.0 N, that can slide up or down in the cylinder without friction but which is currently at its equilibrium position. The inner radius of the cylinder, and the radius of the piston, is 10.0 cm. The top of the piston is exposed to the atmosphere, and the atmospheric pressure is 101.3 kPa. Our goals for this problem are to determine the pressure inside the cylinder, and then to determine what changes if the temperature is raised from 20°C to 80°C.

Figure 14.5: A diagram of the ideal gas sealed inside a cylinder by a piston that is free to move up and down without friction.

Step 1: *Picture the scene.* A diagram of the situation is shown in Figure 14.5.

Step 2: *Organize the data.* The best way to organize what we know in this case is to draw a free-body diagram of the piston, as in Figure 14.6. Three forces act on the piston: the force of gravity; a downward force associated with the top of the piston being exposed to atmospheric pressure; and an upward force from the bottom of the piston being exposed to the pressure in the cylinder.

Step 3: *Solve the problem.* The piston is in equilibrium, so let's apply Newton's second law, $\sum \vec{F} = m\vec{a} = 0$, to the piston. Choosing up to be positive gives:

$+PA - mg - P_{atm}A = 0$, where A is the cross-sectional area of the piston.

Solving for P, the pressure inside the cylinder, gives:

$$P = \frac{mg + P_{atm}A}{A} = \frac{mg}{\pi r^2} + P_{atm} = \frac{20.0 \text{ N}}{\pi (0.100 \text{ m})^2} + 101300 \text{ Pa} = 101900 \text{ Pa}.$$

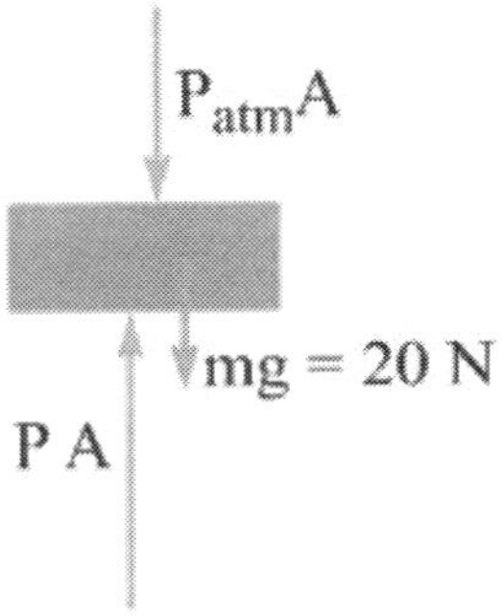

Figure 14.6: The free-body diagram of the piston, showing the forces acting on it.

The pressure inside the cylinder is not much larger than atmospheric pressure.

Step 4: *The temperature of the gas inside the piston is gradually raised from 20°C to 80°C, bringing the piston to a new equilibrium position. What happens to the pressure of the gas, and what happens to the volume occupied by the gas? Be as quantitative as possible.*

To answer the question about pressure we can once again draw a free-body diagram of the piston. However, the fact that the piston has changed position to a new equilibrium position in the cylinder changes nothing on the free-body diagram. Thus, the pressure in the cylinder is the same as it was before. The fact that the temperature increases, however, means the volume increases by the same factor. Because the pressure is constant, we can re-arrange the ideal gas law to:

$$\frac{P}{nR} = \frac{T}{V} = \text{constant}.$$

This tells us that $\dfrac{T_i}{V_i} = \dfrac{T_f}{V_f}$.

Re-arranging to find the ratio of the volumes, and using absolute temperatures, gives:

$$\frac{V_f}{V_i} = \frac{T_f}{T_i} = \frac{(273+80)\text{K}}{(273+20)\text{K}} = 1.20.$$

The volume expands by 20%, increasing by the same factor as the absolute temperature.

Key Idea for a cylinder sealed by a movable piston: When ideal gas is sealed inside a cylinder by a piston that is free to move without friction, the pressure of the gas is generally determined by balancing the forces on the piston's free-body diagram rather than from the volume or temperature of the gas.   **Related End-of-Chapter Exercises: 5, 26, 27, 30, 35.**

EXAMPLE 14.4 – Comparing two pistons

The two cylinders in Figure 14.7 contain an identical number of moles of the same type of ideal gas, and they are sealed at the top by identical pistons that are free to slide up and down without friction. The top of each piston is exposed to the atmosphere. One piston is higher than the other. (a) In which cylinder is the volume of the gas larger? (b) In which piston is the pressure higher? (c) In which piston is the temperature higher?

SOLUTION

(a) Cylinder 2 has a larger volume. Note that the volume in question is not the volume of the molecules themselves, but the volume of the space the molecules are confined to. In other words, it is the volume inside the cylinder itself, below the piston.

Figure 14.7: The cylinders contain the same number of moles of ideal gas, but the piston in cylinder 2 is at a higher level. The pistons are identical, are free to slide up and down without friction, and the top of each piston is exposed to the atmosphere.

(b) Despite the fact that the piston in cylinder 2 is at a higher level than the piston in cylinder 1, the pressure is the same in both cylinders is the same. This is because the free-body diagrams in Figure 14.8 applies to both pistons. The pressure in both cylinders exceeds atmospheric pressure by an amount that is just enough to balance the pressure associated with the downward force of gravity acting on the piston. The pressure is equal in both cases because the pistons are identical.

(c) Applying the ideal gas law tells us that the temperature is larger in cylinder 2, because $T = PV/nR$ and the only factor that is different on the right-hand side of that equation is the volume. In this case the absolute temperature is proportional to the volume.

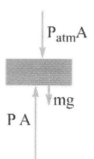

Related End-of-Chapter Exercises: 6, 7, 20 – 25, 28, 29.

Figure 14.8: The free-body diagram applies equally well to both pistons.

Essential Question 14.4: Piston 2, in Figure 14.7, could be the same piston as piston 1, but just at a later time. What could you do to move the system from the piston 1 state to the piston 2 state?

Answer to Essential Question 14.4: All we need to do is to increase the temperature of the piston. Based on our analysis in Exploration 14.4, raising the absolute temperature by 20% moves the piston from the state labeled Piston 1 to that labeled Piston 2.

14-5 The Maxwell-Boltzmann Distribution; Equipartition

We come now to James Clerk Maxwell, the Scottish physicist who determined that the probability a molecule in a container of ideal gas has a particular speed v is given by:

$$P(v) = 4\pi \left(\frac{M}{2\pi RT} \right)^{3/2} v^2 \, e^{-Mv^2/(2RT)} ,$$
(Equation 14.15: **Maxwell-Boltzmann distribution**)

where M is the molar mass (mass of 1 mole) of the gas.

This distribution of speeds is known as the Maxwell-Boltzmann distribution, and it is characterized by three speeds. These are, in decreasing order:

$$v_{rms} = \sqrt{\frac{3RT}{M}} ;$$
(Equation 14.16: **the rms speed**)

$$v_{av} = \sqrt{\frac{8RT}{\pi M}} ;$$
(Equation 14.17: **the average speed**)

$$v_{prob} = \sqrt{\frac{2RT}{M}} .$$
(Equation 14.18: **the most probable speed**)

Plots of the Maxwell-Boltzmann distribution are shown in Figure 14.9 for two different temperatures and two different monatomic gases, argon and helium. Table 14.1 shows the speeds characterizing the distributions. At low temperatures the molecules do not have much energy, on average, so the distribution clusters around the most probable speed. As temperature increases the distribution stretches out toward higher speeds. The area under the curve stays the same (it is the probability an atom has some velocity, which is 1) so the probability at the peak decreases.

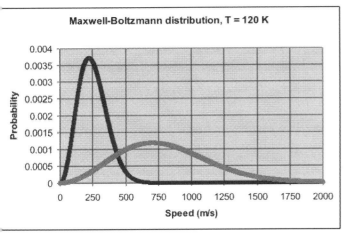
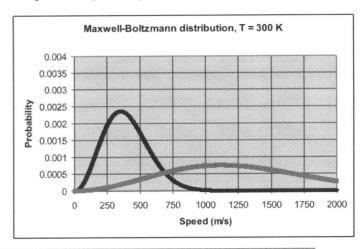

Figure 14.9: Maxwell-Boltzmann distributions at two different temperatures, 120 K and 300 K, for monatomic argon gas (the darker and taller curves, for argon with a molar mass of 40 g) and monatomic helium gas (lighter and shorter curves, with a molar mass of 4 g).

Table 14.1: The various speeds characterizing the Maxwell-Boltzmann distribution of speeds for monatomic argon gas, and for monatomic helium gas, at temperatures of 120 K and 300 K.

	v_{rms} (m/s)	v_{av} (m/s)	v_{prob} (m/s)
Argon, T = 120 K	273	252	223
Argon, T = 300 K	432	398	353
Helium, T = 120 K	865	797	706
Helium, T = 300 K	1367	1260	1116

The Equipartition Theorem

Earlier, we applied basic principles of mechanics to find that $K_{av} = (3/2)kT$. If we multiply by a factor of N, the number of atoms in the ideal gas, the equation becomes:

$$E_{int} = NK_{av} = \frac{3}{2}NkT = \frac{3}{2}nRT .$$ (Eq. 14.19: **Internal energy of a monatomic ideal gas**)

Equation 14.19 gives the total energy associated with the motion of the atoms in the ideal gas. This is known as the **internal energy**. The equipartition theorem states that all contributions to the internal energy contribute equally. For a monatomic ideal gas there are three contributions, coming from motion in the x, y, and z directions. Each direction thus contributes $(1/2)NkT$ to the internal energy. Each motion contributing to internal energy is called a **degree of freedom**. Thus:

the energy from each degree of freedom $= \frac{1}{2}Nkt = \frac{1}{2}nRt$. (Eq. 14.20)

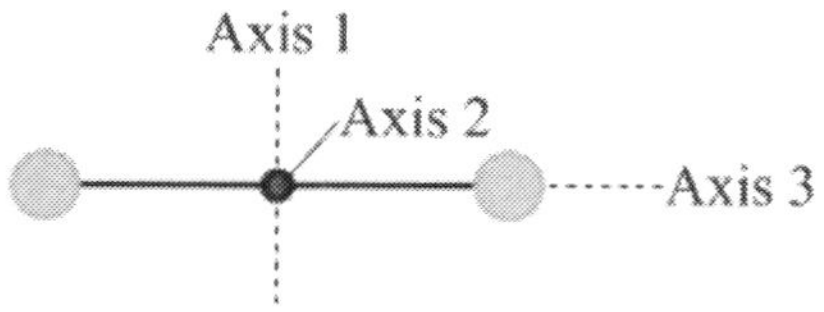

Consider a diatomic ideal gas, in which each molecule consists of two atoms. At low temperatures, only translational kinetic energy is important, but at intermediate temperatures (the range we will generally be interested in) rotation becomes important. As shown in Figure 14.10, rotational kinetic energy is important for rotation about two axes but can be neglected for the third axis because the rotational inertia is negligible for rotation about that axis. With five degrees of freedom, each counting for $(1/2)NkT$, the internal energy of a diatomic ideal gas is:

Figure 14.10: A diatomic molecule is modeled as two balls connected by a light rod. In addition to translating in three dimensions the molecule can rotate about axes 1 or 2, for a total of five degrees of freedom. There is no contribution to the internal energy from rotation about axis 3 because the molecule has negligible rotational inertia about that axis.

$$E_{int} = \frac{5}{2}NkT = \frac{5}{2}nRT .$$ (Eq. 14.21: **Internal energy of a diatomic ideal gas**)

At high temperatures, energy associated with the vibration of the atoms becomes important and there are two additional degrees of freedom (one associated with kinetic energy, one with elastic potential energy) to bring the coefficient in front of the NkT to 7/2.

Polyatomic molecules, at intermediate temperatures, have six degrees of freedom, translational kinetic energy in three dimensions, and rotational kinetic energy about three axes.

$$E_{int} = \frac{6}{2}NkT = 3NkT = 3nRT .$$ (Eq. 14.22: **Internal energy of a polyatomic ideal gas**)

Related End-of-Chapter Exercises: 38, 47, 48, 53.

Essential Question 14.5: Two containers have identical volumes, temperatures, and the same number of moles of gas. One contains monatomic ideal gas while the other has diatomic ideal gas. Which container has a higher pressure? In which does the gas have more internal energy?

Answer to Essential Question 14.5: To find the pressure we can apply the ideal gas law, in the form $P = nRT/V$. Because all the factors on the right-hand side are the same for the two containers the pressures must be equal. When applying the ideal gas law we do not have to worry about what the molecules consist of. We do have to account for this in determining which container has the larger internal energy, however. The internal energy for the monatomic gas is $E_{int} = (3/2)nRT$, while for the diatomic gas at room temperature it is $E_{int} = (5/2)nRT$. The monatomic ideal gas has 3/5 of the internal energy of the diatomic ideal gas.

14-6 The P-V Diagram

In Chapter 15, one of the tools we will use to analyze thermodynamic systems (systems involving energy in the form of heat and work) is the *P-V* diagram, which is a graph showing pressure on the *y*-axis and volume on the *x*-axis.

EXPLORATION 14.6 – Working with the *P-V* diagram

A cylinder of ideal gas is sealed by means of a cylindrical piston that can slide up and down in the cylinder without friction. The piston is above the gas. The entire cylinder is placed in a vacuum chamber, and air is removed from the vacuum chamber very slowly, slowly enough that the gas in the cylinder, and the air in the vacuum chamber, maintains a constant temperature (the temperature of the surroundings).

Step 1: *If you multiply pressure in units of kPa by volume in units of liters, what units do you get?*

$$1 \text{ kPa} \times 1 \text{ liter} = \left(1 \times 10^3 \text{ Pa}\right) \times \left(1 \times 10^{-3} \text{ m}^3\right) = 1 \text{ Pa m}^3 = 1 \text{ N m} = 1 \text{ J}.$$

Thus, the unit is the MKS unit the joule. This will be particularly relevant in the next chapter, when we deal with the area under the curve of the *P-V* diagram.

Step 2: *Complete Table 14.2, giving the pressure and volume of the ideal gas in the cylinder at various instants as the air is gradually removed from the vacuum chamber in which the cylinder is placed.*

State	Pressure (kPa)	Volume (liters)
1	120	1.0
2	80	
3		2.0
4		3.0
5	30	
6		6.0

Table 14.2: A table giving the pressure and volume for a system of ideal gas with a constant temperature and a constant number of moles of gas.

Using the ideal gas law, we can say that $PV = nRT =$ constant. In state 1, Table 14.2 tells us that the product of pressure and volume is 120 J. Thus the missing values in the table can be found from the equation $PV = 120$ J. In states 2 and 5, therefore, the gas occupies a volume of 1.5 liters and 4.0 liters, respectively. In states 3, 4, and 6, the pressure is 60 kPa, 40 kPa, and 20 kPa, respectively.

Step 3: *Plot these points on a P-V diagram similar to that in Figure 14.11, and connect the points with a smooth line. Note that such a line on a P-V diagram is known as an isotherm, which is a line of constant temperature.*

Figure 14.11: A blank *P-V* diagram.

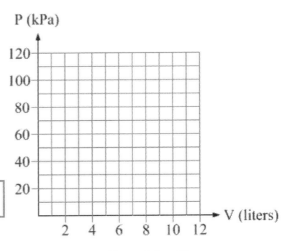

The *P-V* diagram with the points plotted, and the smooth line drawn through the points representing the isotherm, is shown in Figure 14.12.

Step 4: ***Repeat the process, but this time the absolute temperature of the gas is maintained at a value twice as large as that in the original process. Sketch that isotherm on the same P-V diagram.***

If the absolute temperature is doubled, the constant $P \times V$ must also double, from 120 J to 240 J. Starting with the original points we plotted, we can find points on the new isotherm by either doubling the pressure or doubling the volume. Several such points are shown on the modified *P-V* diagram in Figure 14.13, and we can see that this isotherm, at the higher temperature, is farther from the origin than the original isotherm. This is generally true, that the higher the temperature, the farther from the origin is the isotherm corresponding to that temperature.

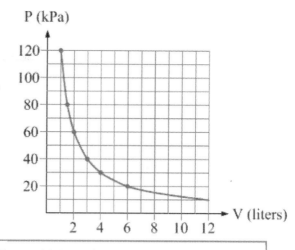

Figure 14.12: The *P-V* diagram corresponding to the points from Table 14.2. The smooth curve through the points is an isotherm, a line of constant temperature.

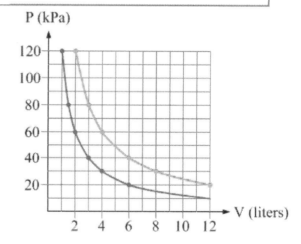

Key Ideas about P-V diagrams: The *P-V* diagram (the graph of pressure as a function of volume) for a system can convey significant information about the state of the system, including the pressure, volume, and temperature of the system when it is in a particular state. It can be helpful to sketch isotherms on the *P-V* diagram to convey temperature information – an isotherm is a line of constant temperature.

Related End-of-Chapter Exercises: 11, 12, 49 – 52.

Figure 14.13: A *P-V* diagram showing two different isotherms. The isotherm that is farther from the origin has twice the absolute temperature as the isotherm closer to the origin.

Essential Question 14.6: An isotherm on the *P-V* diagram has the shape it does because, from the ideal gas law, we are plotting pressure versus volume and the pressure is given by:

$$P = \frac{nRT}{V}.$$

For a particular isotherm, the value of *nRT* is constant, so an isotherm is a line with a shape similar to the plot of $1/V$ as a function of *V*. Let's say we now have two cylinders of ideal gas, sealed by pistons as in the previous Exploration. Cylinder *A*, however, has twice the number of moles of gas as cylinder *B*. We plot a *P-V* diagram for cylinder *A*, and plot the isotherm corresponding to a temperature of 300 K. We also draw a separate *P-V* diagram for cylinder *B*, and we find that the same points we connected to draw the 300 K isotherm on cylinder *A*'s *P-V* diagram are connected to form an isotherm on cylinder *B*'s *P-V* diagram. What is the temperature of that isotherm on cylinder *B*'s *P-V* diagram?

Answer to Essential Question 14.6: 600 K. An isotherm is a line connecting all the points satisfying the equation $PV = nRT =$ a particular constant that depends on n and T. Because we're talking about the same line on both P-V diagrams, we have $PV = n_A R T_A = n_B R T_B$. Solving for the temperature in cylinder B gives:

$$T_B = \frac{n_A R T_A}{n_B R} = \frac{n_A}{n_B} T_A = \frac{2n_B}{n_B} T_A = 2T_A = 2(300\text{K}) = 600\text{K} .$$

In this sense, then, the *P-V* diagrams for different ideal gas systems are unique, because the temperature of a particular isotherm depends on the number of moles of gas in the system.

14-7 Diffusion and Osmosis

In Chapter 9, we learned a little bit about how surface tension is important in the alveoli of the lungs. Another key process involved is diffusion. Each time we breathe in, oxygen-rich air fills the alveoli of the lungs. Some of this oxygen will then diffuse through the membrane between the alveoli and blood inside the capillaries, infusing the blood with oxygen. Carbon dioxide diffuses in the other direction, from the blood into the lungs, and we then breathe the carbon dioxide out. This can be a highly efficient process in the human body, because the total surface area inside the alveoli can approach 100 m², and the membrane thickness is extremely thin, generally several hundred nanometers.

Diffusion is a flow of molecules without requiring a net flow of a medium. For instance, in the case of the lungs described above, oxygen molecules diffuse from a region of high concentration of oxygen (in the lungs) to a region of lower concentration (in the blood). The carbon dioxide goes the other way because the high concentration of carbon dioxide is in the blood, and the lower concentration is in the lungs. From a physics perspective, diffusion simply comes from the random motion of molecules, as in an ideal gas.

The process of a molecule randomly moving is known as a **random walk**. This was studied by Robert Brown in 1827, hence the term **Brownian motion** for the motion of a small particle immersed in a fluid. Albert Einstein was also a key figure in our understanding of diffusion, as it was he who developed the theory of Brownian motion.

Another example of diffusion is a mylar balloon that is filled with helium. As time passes, helium atoms diffuse through the wall of the balloon, and the balloon gradually deflates.

Osmosis

The process of osmosis involves the diffusion of molecules of a solute through a membrane that is selectively permeable. Take a container that is divided by a semi-permeable membrane. The membrane allows molecules of the solvent (which might be water, for example) to pass through because these molecules are small. On the other hand, the solute molecules may not pass through because they are too large. If two different concentrations of the solution are placed in the two parts of the container, the solvent molecules will diffuse through the membrane from the low-concentration side to the high-concentration side (thereby diluting the high-concentration side, and increasing the concentration on the low-concentration side). We can refer to an osmotic pressure across the membrane that drives this flow - as shown in Figure 14.14, this osmotic pressure can balance a hydrostatic pressure difference between the two sides, coming from the difference in the height of the fluid columns.

Note that osmotic pressures can be rather large, up to many atmospheres of pressure, even. Because of this, osmosis is a key part of many biological systems.

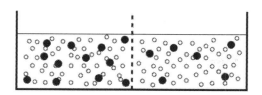

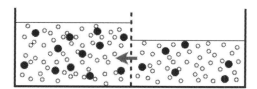

Figure 14.14: In the top diagram, the fluid levels are equal on both sides of a semi-permeable membrane, but the concentration is higher on the left. As shown in the bottom diagram, solvent molecules (small open circles) will then diffuse from right to left until the osmotic pressure balances the hydrostatic pressure. The membrane allows solvent molecules to pass through, but does not allow the larger (dark circles) solute molecules through.

A related phenomenon, which is important in many desalination plants (removing salt from water so that humans can drink it), is reverse osmosis, described below.

Reverse osmosis and desalination

Let's calculate the osmotic pressure of seawater. This is done by multiplying the molarity (M) of the solution, (the concentration in moles / liter), by the universal gas constant (R), in units of liter atm / (K moles), and multiplying that by the temperature (T) in Kelvin. A typical molarity of seawater is 1.1 moles / L. If we use a temperature of 300 K, the osmotic pressure works out to:

$$P = M R T = (1.1 \text{ moles / L}) [0.082 \text{ L atm / (K moles)}] (300 \text{ K}) = 27 \text{ atm}.$$

Thus, the osmotic pressure of seawater is about 27 atmospheres! This means that if you have a semi-permeable membrane (impermeable to the sodium and chlorine ions) separating fresh water from seawater, there is a very large pressure that drives the pure water through the membrane to the seawater side.

In the reverse-osmosis desalination process, however, we want the flow to go in the other direction, driving pure water from the seawater to the freshwater side. This can be done if the seawater is placed under hydrostatic pressure larger than the 27 atmospheres of osmotic pressure - thus, typical pressures for the seawater in a reverse-osmosis desalination facility are in the range of 40 - 80 atmospheres. A very recent development, in 2013, was the announcement of a new type of membrane, just one atom thick - this is made from graphene (a single sheet of carbon), with holes in it just the right size to pass water molecules but not the salt molecules. Such a very thin sheet offers a lot of promise, as it should be much easier for the water molecules to diffuse through than through the membranes that are currently used, which are many atomic layers thick.

Reverse osmosis is also used in the maple syrup industry, to remove most of the water from the sap before boiling down the rest to make maple syrup and maple sugar.

A related process, **active transport**, is at work in the cells of living things. For instance, the concentration of potassium ions (K+) inside a cell may be 20 times larger than the concentration outside the cell. Just the opposite happens for sodium ions (Na+), for which the outside concentration may be 15 times higher than the concentration inside the cell. Normally, we would expect these ions to diffuse from the high-concentration region to the low-concentration region, but active transport, through the action of a **sodium-potassium pump**, works to maintain the significant imbalance in concentrations. This takes energy, which comes from hydrolyzing ATP. The net result of the sodium-potassium pump is that for every three sodium ions that are pumped out of the cell, two potassium ions are pumped in. This is a key part of why there is generally a potential difference across the cell membrane (positive outside, negative inside).

Essential Question 14.7: As fresh water is being removed from seawater in a desalination plant, what happens to the molarity of the seawater? Does this make it harder or easier to remove the pure water? How do you think this issue is addressed in a desalination facility?

Answer to Essential Question 14.6: As the pure water is removed, the salt concentration increases (the same number of sodium and chlorine molecules are now in a smaller amount of water. This increases the molarity, which will cause a corresponding increase in the osmotic pressure of the seawater, making it less likely for water molecules to diffuse through the semi-permeable membrane.

This issue is generally addressed by removing the concentrated seawater from the system and replacing it with seawater of the usual molarity.

Chapter Summary

Essential Idea regarding looking at thermodynamic systems on a microscopic level
We can apply basic principles of physics to a system of gas molecules, at the microscopic level, and get important insights into macroscopic properties such as temperature. Temperature is a measure of the average kinetic energy of the atoms or molecules of the gas.

The Ideal Gas Law
The ideal gas law can be written in two equivalent forms.

In terms of n, the number of moles of gas , $PV = nRT$, (Equation 14.1)

where $R = 8.31$ J/(mol K) is the universal gas constant.

In terms of N, the number of molecules, $PV = NkT$, (Equation 14.2) where

$k = 1.38 \times 10^{-23}$ J/K is Boltzmann's constant.

What Temperature Means

$K_{av} = \dfrac{3}{2}kT$. (Equation 14.14: **Average kinetic energy is directly related to temperature**)

As Equation 14.14 shows, temperature is a direct measure of the average kinetic energy of the atoms or molecules in the ideal gas.

The Maxwell-Boltzmann Distribution

The Maxwell-Boltzmann distribution is the distribution of molecular speeds in a container of ideal gas, which depends on the molar mass M of the molecules and on the absolute temperature, T. The distribution is characterized by three speeds. In decreasing order, these are the root-mean-square speed; the average speed; and the most-probable speed. These are given by equations 14.16 – 14.18:

$$v_{rms} = \sqrt{\frac{3RT}{M}} \; ; \qquad v_{av} = \sqrt{\frac{8RT}{\pi M}} \; ; \qquad v_{prob} = \sqrt{\frac{2RT}{M}} \; .$$

A Cylinder Sealed by a Piston that can Move Without Friction

A common example of an ideal gas system is ideal gas sealed inside a cylinder by means of a piston that is free to move without friction. When the piston is at its equilibrium position the pressure of the gas is generally determined by balancing the forces on the piston's free-body diagram, rather than from the volume or temperature of the gas. The diagram at right illustrates this idea for a cylinder sealed at the top by a piston of area A. The combined forces directed down, the force and gravity and the force associated with atmospheric pressure acting on the top of the piston, must be balanced by the upward force associated with the gas pressure acting on the bottom of the piston.

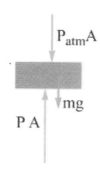

The Equipartition Theorem

The equipartition theorem is the idea that each contribution to the internal energy (energy associated with the motion of the molecules) of an ideal gas contributes equally. Each contribution is known as a degree of freedom.

$$\text{The energy from each degree of freedom} = \frac{1}{2} Nkt = \frac{1}{2} nRt .\qquad \text{(Equation 14.20)}$$

A monatomic ideal gas can experience translational motion in three dimensions. With three degrees of freedom the internal energy is given by:

$$E_{int} = NK_{av} = \frac{3}{2} NkT = \frac{3}{2} nRT .\qquad \text{(Eq. 14.19: \textbf{Internal energy of a monatomic ideal gas})}$$

At intermediate temperatures molecules in a diatomic ideal gas have two additional degrees of freedom, associated with rotation about two axes.

$$E_{int} = \frac{5}{2} NkT = \frac{5}{2} nRT .\quad \text{(Eq. 14.21: \textbf{Internal energy of a diatomic ideal gas})}$$

Molecules in a polyatomic ideal gas can rotate about three axes.

$$E_{int} = \frac{6}{2} NkT = 3NkT = 3nRT .\qquad \text{(Eq. 14.22: \textbf{Internal energy of a polyatomic ideal gas})}$$

The P-V Diagram

A graph of pressure versus volume (a P-V diagram) can be very helpful in understanding an ideal gas system. We will exploit these even more in the next chapter. The ideal gas law tells us that the product of pressure and volume (which has units of energy) is proportional to the temperature of a system. Lines of constant temperature are known as isotherms. The diagram at right shows two isotherms. The isotherm that is farther from the origin has twice the absolute temperature as the isotherm closer to the origin.

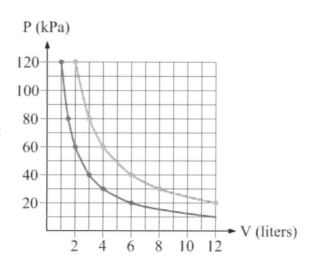

End of Chapter Exercises

Exercises 1 – 12 are conceptual questions that are designed to see if you have understood the main concepts of the chapter.

1. While on an airplane, you take a drink from your water bottle, and then screw the cap tightly back on the bottle. After landing, you notice that the bottle is a funny shape, as if someone is deforming it by squeezing it. Explain what has happened.

2. A common lecture demonstration involves placing a gob of shaving cream in a bell jar, which is a device that can be sealed off from the atmosphere. Much of the air is then pumped out of the bell jar. What do you expect to happen when this is done? Why?

3. As shown in Figure 14.15, a sealed cylinder of ideal gas is divided into two parts by a piston that can move left and right without friction. There is ideal gas in both parts, but the parts are isolated from one another by the piston. The piston is in its equilibrium position. The volume occupied by the gas on the left side is larger than that occupied by the gas on the right side. On which side is the gas pressure higher? Explain your answer.

Figure 14.15: A sealed cylinder that is divided into two parts by a piston that is free to move left or right without friction. For Exercises 3 and 4.

4. Return to the situation described in Exercise 3 and shown in Figure 14.15. Initially, the temperature of the gas on the right side is larger that that of the gas on the left side. As time goes by, the two sides approach the same equilibrium temperature, as the temperature gradually decreases on the right side and gradually increases on the left side. Describe what happens to the piston as the system progresses toward equilibrium.

5. As shown in Figure 14.16, a sealed cylinder of ideal gas is divided into two parts by a piston that can move up and down without friction. There is ideal gas in both parts, but the gas from one part is isolated from that in the other part by the piston. The piston, which has a weight of 50.0 N, is shown in its equilibrium position. The volume occupied by the gas in the lower part is larger than that occupied by the gas in the upper part. On which side is the gas pressure higher? Explain your answer.

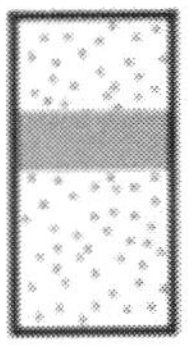

Figure 14.16: A sealed cylinder divided into two parts by a piston that is free to slide up and down without friction, for Exercise 5.

6. Three identical cylinders are sealed with identical pistons that are free to slide up and down the cylinder without friction. Each cylinder contains ideal gas, and the temperature is the same in each case, but the volumes occupied by the gases differ. In each cylinder the piston is above the gas, and the top of each piston is exposed to the atmosphere. As shown in Figure 14.17, the volume occupied by the gas is largest in case 1 and smallest in case 3. Rank the cylinders in terms of (a) the pressure of the gas, and (b) the number of moles of gas inside the cylinder.

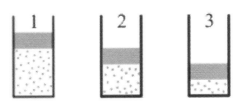

Figure 14.17: Three cylinders containing different volumes of gas at the same temperature, for Exercise 6.

7. Three identical cylinders are sealed with identical pistons that are free to slide up and down the cylinder without friction, as shown in Figure 14.18. Each cylinder contains ideal gas, and the gas occupies the same volume in each case, but the temperatures differ. In each cylinder the piston is above the gas, and the top of each piston is exposed to the atmosphere. In cylinders 1, 2, and 3 the temperatures are 0°C, 50°C, and 100°C, respectively. Rank the cylinders in terms of (a) the pressure of the gas, and (b) the number of moles of gas inside the cylinder.

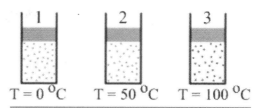

Figure 14.18: Three cylinders containing equal volumes of gas at different temperatures, for Exercise 7.

8. Is it possible for the total kinetic energy of the atoms in one container of ideal gas to be the same as the total kinetic energy of the atoms in a second container of ideal gas, but for their temperatures to be different? If so, describe how you could achieve this.

9. Consider a sealed box of ideal gas that is separated into two parts of equal volume by a partition. All the gas molecules are in one half of the box and there is nothing at all in the other half of the box. The partition consists of two sliding doors, which can be opened quickly and automatically like the doors of an elevator. When the sliding doors open, allowing the molecules to expand into the other half of the box, do you expect either the pressure or the temperature to remain constant? Basing your argument on kinetic theory, state which of these parameters (pressure or temperature) you expect to stay constant, explain why, and explain what happens to the other parameter.

10. Two containers of ideal gas have the same volume, the same pressure, and have the same number of moles of gas, but the type of molecule in each container is different. To be specific, one container contains argon atoms while the other contains xenon atoms, which are both monatomic ideal gases. Which of the following are the same for the two containers and which are different? In cases where there is a difference, state how they differ. (a) Temperature, (b) average kinetic energy of the atoms, (c) total kinetic energy of the atoms, (d) rms speed of the atoms, (e) most probable speed of the atoms.

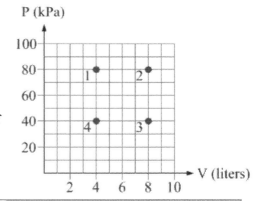

11. Four states, labeled 1 through 4, of a particular thermodynamic system, are shown on the P-V diagram in Figure 14.19. The number of moles of gas in the system is constant. Rank the states based on their temperature, from largest to smallest.

Figure 14.19: Four states are shown on the P-V diagram for a particular thermodynamic system. For Exercise 11.

12. Consider the P-V diagram in Figure 14.20. Find three other points on the diagram in which the system would have the same temperature as it has in state 1.

Figure 14.20: A P-V diagram for a particular thermodynamic system. For Exercise 12.

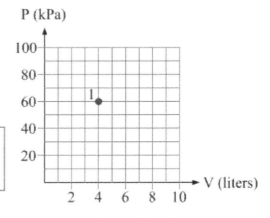

Exercises 13 – 16 deal with the ideal gas law.

13. A sample of monatomic ideal gas is kept in a container that keeps a constant volume while the temperature of the gas is raised from 10°C to 20°C. If the pressure of the gas is P at 10°C, what is the pressure at 20°C?

14. A particular cylindrical bucket has a height of 50 cm, while the radius of its circular cross-section is 15 cm. The bucket is empty, aside from containing air. The bucket is then inverted and, being careful not to lose any of the air trapped inside, lowered 20 m below the surface of a fresh-water lake. (a) If the temperature is the same at both points, what fraction of the bucket's volume is occupied by the air when the bucket is 20 m down? (b) Would the fraction be larger, smaller, or the same if the temperature drops from 25°C at the surface to 5.0°C 20 m below the surface? Why? (c) Calculate the fraction of the bucket's volume that is occupied by air when the bucket is 20 m down and the temperature changes as described in (b).

15. An empty metal can is initially open to the atmosphere at 20°C. The can is then sealed tightly, and heated to 100°C. While at 100°C, the lid of the can is loosened, opening the can to the atmosphere, and then the can is sealed tightly again. When the can cools to 20°C again, what is the pressure inside?

16. In 1992, a Danish study concluded that a standard toy balloon, made from latex and filled with helium, could rise to 10000 m (where the pressure is about 1/3 of that at sea level) in the atmosphere before bursting. In the study, a number of balloons were filled with helium, and then placed in a chamber maintained at –20°C. The pressure was gradually reduced until the balloons exploded, and then the researchers determined the height above sea level corresponding to that pressure. Assuming the balloons were filled with helium at +20°C and about atmospheric pressure, determine the ratio of the balloon's volume when it exploded to its volume when it was filled.

Questions 17 – 19 deal with calculating the rms average.

17. Consider the set of numbers –3, –2, –1, 1, 3, 5. (a) What is the average of this set of numbers? (b) What is the rms average of this set of numbers?

18. Is it possible for a set of four numbers to have an average of zero but an rms average that is non-zero? If so, come up with a set of four numbers for which this is true.

19. (a) Is it possible for the average of a set of four numbers to be equal to the rms average of those numbers? If so, find a set of four numbers for which this is true. (b) Is it possible for the average of a set of four numbers to be larger than the rms average of those numbers? If so, find a set of four numbers for which this is true.

Questions 20 – 25 are a sequence of ranking tasks that relate to cylinders that are sealed by pistons that are free to slide without friction. In all cases, the piston is at its equilibrium position.

20. Three identical cylinders are sealed with pistons that are free to slide up and down the cylinder without friction, but the masses of the pistons differ. As shown in Figure 14.21, Piston 1 has the largest mass, while piston 2 has the smallest mass. Each cylinder contains ideal gas at the same temperature and occupying the same volume. In each cylinder, the piston is above the gas, and the top of each piston is exposed to the atmosphere. Rank the cylinders in terms of (a) the pressure of the gas, and (b) the number of moles of gas inside the cylinder.

Figure 14.21: Three cylinders containing the same volume of gas at the same temperature, but the pistons have different mass. For Exercise 20.

21. As in Exercise 20, three identical cylinders are sealed with pistons that are free to slide up and down the cylinder without friction, but the masses of the pistons differ. As shown in Figure 14.22, piston 1 has the largest mass, while piston 2 has the smallest mass. The gas in each cylinder occupies the same volume, but in cylinders 1, 2, and 3 the temperatures are 0°C, 50°C, and 100°C, respectively. In each cylinder the piston is above the gas, and the top of each piston is exposed to the atmosphere. Is it possible to rank the cylinders, from largest to smallest, in terms of the pressure of the gas based on the information given here? If so, state the ranking, and compare the ranking to that in the previous exercise, explaining either why you expect the rankings to be the same or why you expect the rankings to differ. If it is not possible to rank the cylinders based on their pressures, clearly explain why not.

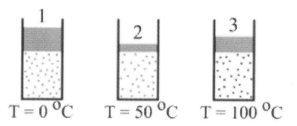

Figure 14.22: Three cylinders containing the same volume of gas at different temperatures, and with pistons of different mass. For Exercise 21.

22. Three identical cylinders are sealed with identical pistons that are free to slide up and down the cylinder without friction. Each cylinder contains ideal gas, and the gas occupies the same volume in each case, but the temperatures differ. As shown in Figure 14.23, each cylinder is inverted so the piston is below the gas, and the bottom of each piston is exposed to the atmosphere. In cylinders 1, 2, and 3 the temperatures are 0°C, 50°C, and 100°C, respectively. Rank the cylinders, from largest to smallest, in terms of (a) the pressure of the gas (b) the number of moles of gas inside the cylinder.

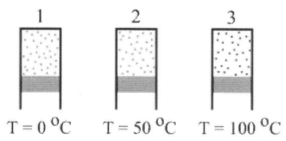

Figure 14.23: Three inverted cylinders containing the same volume of gas at different temperatures. For Exercise 22.

23. Three identical cylinders are sealed with pistons that are free to slide up and down the cylinder without friction, but the masses of the pistons differ. Piston 1 has the largest mass, while piston 2 has the smallest mass. Each cylinder contains ideal gas at the same temperature and occupying the same volume. As shown in Figure 14.24, each cylinder is inverted so the piston is below the gas, and the bottom of each piston is exposed to the atmosphere. Rank the cylinders, from largest to smallest, in terms of (a) the pressure of the gas, and (b) the number of moles of gas inside the cylinder.

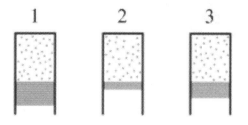

Figure 14.24: Three inverted cylinders containing the same volume of gas at equal temperatures, but having pistons of different mass. For Exercise 23.

24. Three identical cylinders are sealed with pistons that are free to slide up and down the cylinder without friction. Each cylinder contains ideal gas, and the gas occupies the same volume in each case and is at the same temperature. As shown in Figure 14.25, the pistons in cylinders 1 and 3 are identical but the piston in cylinder 1 is above the gas while cylinder 3 is inverted so the piston is below the gas. The piston in cylinder 2, which is above the gas, has more mass than the other two pistons. The top surfaces of the pistons in cylinders 1 and 2, and the bottom surface of the piston in cylinder 3, are exposed to the atmosphere. Rank the cylinders, from largest to smallest, in terms of (a) the pressure of the gas, and (b) the number of moles of gas inside the cylinder.

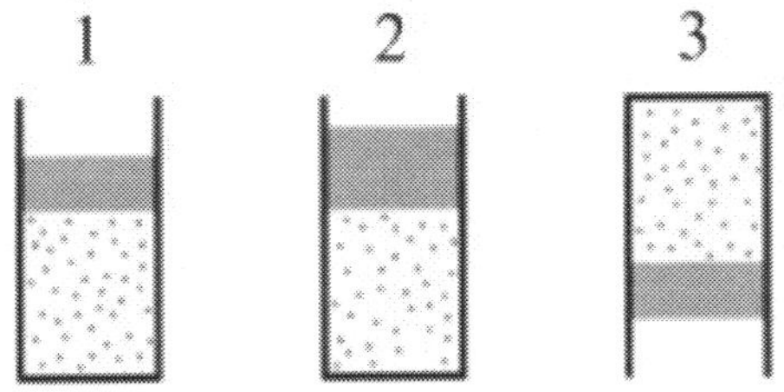

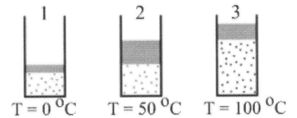

Figure 14.25: Three cylinders containing the same volume of gas at equal temperatures, but cylinder 3 is inverted while the piston in cylinder 2 has a larger mass than the other two pistons. For Exercise 24.

25. Three identical cylinders are sealed with pistons that are free to slide up and down the cylinder without friction. Each cylinder contains ideal gas. As shown in Figure 14.26, the volume occupied by the gas is different in each case, and the temperatures are also different. The pistons also have different masses. The piston in cylinder 1 has the smallest mass while the piston in cylinder 2 has the largest mass. In each cylinder the piston is above the gas, and the top of each piston is exposed to the atmosphere. Is it possible to rank the cylinders, from largest to smallest, in terms of the pressure of the gas? If so, provide the ranking. If not, explain why not.

Figure 14.26: Three cylinders containing different volumes of gas at different temperatures, and with pistons of different mass. For Exercise 25.

Exercises 26 – 35 relate to cylinders that are sealed by pistons that are free to slide without friction.

26. In Exploration 14.4, we analyzed a cylinder filled with ideal gas that is sealed by a piston that is above the gas. The piston is a cylindrical object, with a weight of 20.0 N, that can slide up or down in the cylinder without friction. The inner radius of the cylinder, and the radius of the piston, is 10.0 cm. With the top of the piston exposed to the atmosphere, at a pressure of 101.3 kPa, we determined the pressure inside the cylinder. Now, very slowly, so that the gas inside the cylinder stays constant at the temperature of its surroundings, sand is poured onto the top of the piston. When 30.0 N of sand has been added to the piston, determine: (a) the pressure inside the cylinder, and (b) the ratio of the final volume occupied by the gas after the sand is added to the volume occupied by the gas before the sand is added.

27. A cylinder filled with ideal gas is sealed by a piston that is above the gas. The piston is a cylindrical object, with a weight of 10.0 N, that can slide up or down in the cylinder without friction. The inner radius of the cylinder, and the radius of the piston, is 10.0 cm. (a) If the top of the piston is exposed to the atmosphere, and the atmospheric pressure is 101.3 kPa, what is the pressure inside the cylinder? (b) What happens to the gas pressure, and the volume occupied by the gas, if the cylinder's temperature is gradually increased from 20°C to 60°C? Be as quantitative as possible.

28. Consider the cylinder in Exercise 27. If the cylinder is inverted, so the piston lies below the gas, what will happen to the piston? Will it remain in the cylinder, or will it fall out? Explain qualitatively what will happen, and explain why. Assume the piston is initially located about halfway down the cylinder.

29. Return to Exercise 28, but now analyze it quantitatively. Assuming the piston is in equilibrium below the gas, determine (a) the gas pressure, and (b) the ratio of the volume occupied by the gas when the piston is below the gas to the volume occupied by the gas when the piston is above the gas.

30. A cylinder filled with ideal gas is sealed by a piston that is above the gas. The piston is a cylindrical object, with a weight of 50.0 N, that can slide up or down in the cylinder without friction. The inner radius of the cylinder, and the radius of the piston, is 5.00 cm. The top of the piston is exposed to the atmosphere, and the atmospheric pressure is 101.3 kPa. The cylinder has a height of 30.0 cm, and, when the temperature of the gas is 20°C, the bottom of the piston is 15.0 cm above the bottom of the cylinder. Find the number of moles of ideal gas in the cylinder.

31. Return to the cylinder described in Exercise 30. When the temperature of the gas is raised from 20°C to 200°C, find the distance between the bottom of the cylinder and the bottom of the piston when the piston comes to its new equilibrium position.

32. Consider again the cylinder described in Exercise 30. What is the maximum temperature the ideal gas can have for the cylinder to remain sealed?

33. Consider again the cylinder described in Exercise 30. Now, the entire cylinder is sealed in a vacuum chamber and air is gradually pumped out of the chamber. Assume the temperature of the gas remains the same. (a) Describe qualitatively what, if anything, happens to the cylinder as the air is removed from the chamber. (b) What is the lowest pressure the chamber can have for the cylinder to remain sealed?

34. Consider again the cylinder described in Exercise 30. As the temperature of the gas is gradually raised from 20°C to 220°C you, wearing insulating gloves to prevent a nasty burn, push down on the top of the piston so that the piston remains in the same position at all times. How much downward force do you have to exert on the piston when the gas temperature is (a) 120°C? (b) 220°C?

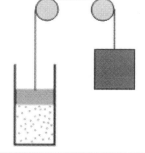

35. Consider again the cylinder described in Exercise 30, except this time the piston is tied to a string that passes over a pulley system and is tied to a block that has a weight of 90.0 N, as shown in Figure 14.27. Both the block and the piston are in equilibrium. What is the pressure in the cylinder?

Figure 14.27: The 50.0 N piston is tied to a 90.0 N block by a string passing over a pulley system. The system is in static equilibrium. For Exercise 35.

General exercises and conceptual questions

36. Kids love to bounce around inside a moonwalk, which consists of a floor inflated with air and four walls made of elastic mesh. In a particular moonwalk, there are 8 children, each with a mass of 15 kg. Each wall of the moonwalk measures 3.0 m by 3.0 m and, on average, each child bounces off a wall once every 5.0 s. Assume that each child has a speed of 2.0 m/s when he or she hits a wall; that the child's velocity is directed perpendicular to the plane of the wall; and that the collision with the wall simply reverses the direction of the child's velocity. What is the average pressure experienced by a wall because of the children in the moonwalk?

37. A box of ideal gas contains two kinds of atoms, which have different masses. The atoms are in thermal equilibrium. You observe that the average speed of one of the kinds of atoms is 50% larger than the average speed of the other. What is the ratio of the masses of the atoms?

38. Equation 14.16 gives an expression for the root-mean-square speed of the Maxwell-Boltzmann distribution. Derive this equation by starting from Equation 14.14,

$$K_{av} = \frac{3}{2}kT .$$

39. You have two identical cylinders that are sealed with identical pistons that are free to slide up and down the cylinder without friction. Each cylinder contains ideal gas, and the gas occupies the same volume and is at the same temperature in each case. In each cylinder the piston is above the gas. Cylinder A contains argon gas (atomic mass = 40 g), while cylinder B contains xenon (atomic mass = 131 g). (a) In which cylinder is the pressure larger? Explain your answer. (b) In which cylinder is the number of moles of gas larger? Explain your answer.

40. You have two identical cylinders that are sealed with identical pistons that are free to slide up and down the cylinder without friction. Each cylinder contains the same number of moles of a diatomic ideal gas, and the gas is at the same temperature in each case. In each cylinder the piston is above the gas. Cylinder A contains oxygen gas (molecular mass = 32 g), while cylinder B contains nitrogen (molecular mass = 28 g). (a) In which cylinder is the pressure larger? Explain your answer. (b) In which cylinder is the volume occupied by the gas larger? Explain your answer.

41. As shown in Figure 14.28, a sealed cylinder of ideal gas is divided into two parts by a piston that can move up and down without friction. There is ideal gas in both parts, but the gas from one part is isolated from that in the other part by the piston. The volume occupied by the gas in the lower part is twice that occupied by the gas in the upper part, while the temperature is the same in both parts. The pressure in the lower part is 2000 Pa, and the weight of the piston is 50.0 N. The piston is in its equilibrium position. The cross-sectional area of the cylinder is 100 cm². Determine (a) the pressure in the upper part, and (b) the ratio of the number of moles of gas in the lower part to the number of moles in the upper part.

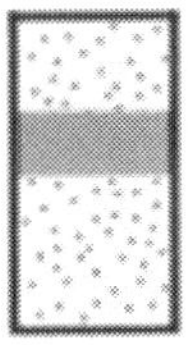

Figure 14.28: A sealed cylinder divided into two parts by a piston that is free to slide up and down without friction, for Exercise 41.

42. A cylinder sealed by a movable piston contains a certain number of molecules of air, which we can treat as an ideal gas. The pressure is initially 1×10^5 Pa in the cylinder, and the temperature is 20°C. Very slowly, so the temperature of the gas remains constant, you push on the piston so that the volume occupied by the gas changes from V, its original value, to $V/4$. Plot a graph of the pressure in the cylinder as a function of the volume occupied by the gas.

43. Return to the situation described in Exercise 42. If you carry out the compression very quickly, instead of slowly, the temperature of the gas can change significantly. (a) Would you expect the temperature in the cylinder to increase or decrease? (b) Thinking about the interaction between the piston and the individual gas molecules, come up with an explanation regarding why and how the average kinetic energy of the gas molecules changes.

44. On a hot summer day, you are sitting at a café drinking a carbonated beverage from a tall glass. As you watch bubbles rise from the bottom to the top of the glass you start thinking that they should be changing in volume as they rise. (a) Why, and in what way, would you expect the bubbles to change in volume as they rise? (b) Assuming the beverage has the same density as water, 1000 kg/m³, estimate the ratio of the volume of a bubble at the surface to its volume at the bottom of the glass, 30 cm below the surface.

45. A spherical copper container with a radius of 8.0 cm is sealed when the air inside is at atmospheric pressure, 101.3 kPa, and the temperature is 20°C. (a) How many moles of gas does the sphere contain? (b) Neglecting any change in volume in the copper sphere itself, plot a graph of the pressure in the container as a function of temperature over the range of –150°C to +150°C. (c) What is the slope of the graph equal to? State your answer in terms of variables as well as giving a numerical value.

46. Consider the set of pressures, volumes, and temperatures for a sealed container of ideal gas shown in Table 14.3. Complete the table, and then rank the four states, from largest to smallest, based on their (a) pressure, (b) volume, and (c) temperature.

State	Pressure	Volume	Temperature
A	P_i	V_i	T_i
B	$3P_i$	$2V_i$	
C	$\frac{1}{2}P_i$		$2T_i$
D		$3V_i$	$\frac{1}{2}T_i$

Table 14.3: A table showing the pressure, volume, and temperature of a sealed ideal gas in four different states, for Exercise 46.

47. You have a cubical box measuring 30 cm on each side. Sealed in this box is monatomic argon gas at a temperature of 20°C and at a pressure of 100 kPa. (a) Determine the number of moles of argon in the box. (b) Determine the number of atoms of argon in the box. For these argon atoms, determine the (c) most probable speed (d) average speed (e) rms speed.

48. Return to the box of argon gas described in Exercise 47. Let's make some simple calculations to work out approximately how many collisions each side of the box experiences every second, and to determine the average force exerted on one side of the box by a single colliding atom. (a) Using the given pressure and the area of one side of the box, determine the average force applied by the atoms to one side of the box. (b) Use the value of v_{rms} and the relationship $v_{rms}^2 = 3v_x^2$ to find the value of v_x. (c) Use Equation 14.5 to determine the time interval between successive collisions of one atom with one side of the box. (d) Determine how many collisions one atom makes with one side of the box every second. (e) Multiply by the number of atoms to determine the total number of collisions that one side of the box experiences every second. (f) Divide your answer from part (a) by your answer from part (e) to determine the average force associated with one collision.

49. A particular ideal gas system contains a fixed number of moles of ideal gas. The number of moles of gas is such that the product $nR = 0.300$ J/K. Values of the volume and temperature for particular states of the system are shown in Table 14.4. (a) Find the pressure for each of the states shown in the table. (b) Plot these points on a P-V diagram. (c) Describe a system that these states could correspond to, and explain how you could move the system from state 1 through the other states listed to state 5.

State	P (kPa)	V (liters)	T (K)
1		1.0	150
2		2.0	300
3		3.0	450
4		4.0	600
5		5.0	750

Table 14.4: Volume and temperature readings in five states of an ideal gas system, for Exercise 49.

50. A particular ideal gas system contains a fixed number of moles of ideal gas. The number of moles of gas is such that the product $nR = 0.100$ J/K. Values of the pressure and temperature for particular states of the system are shown in Table 14.5. (a) Find the volume for each of the states shown in the table. (b) Plot these points on a P-V diagram. (c) Describe a system that these states could correspond to, and explain how you could move the system from state 1 through the other states listed to state 5.

State	P (kPa)	V (liters)	T (K)
1	40		80
2	80		160
3	120		240
4	160		320
5	200		400

Table 14.5: Pressure and temperature readings in five states of an ideal gas system, for Exercise 50.

51. Four states, labeled 1 through 4, of a particular thermodynamic system, are shown on the P-V diagram in Figure 14.29. The number of moles of gas in the system is constant. (a) Rank the states based on their temperature, from largest to smallest. (b) If the number of moles of gas is chosen such that the product $nR = 10.0$ J/K, find the absolute temperature of the system in the various states.

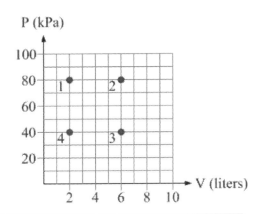

Figure 14.29: Four states are shown on the P-V diagram for a particular thermodynamic system. For Exercise 51.

52. Consider the situations represented by the two P-V diagrams shown in Figures 14.29 and 14.30. Let's say Figure 14.29 shows four different states of thermodynamic system A, and Figure 14.30 shows four different states of thermodynamic system B. The number of moles of ideal gas is constant in both systems, and the two systems happen to have the same temperature when each system is in its own state 2. Find the ratio of the absolute temperatures of the two systems when both systems are in (a) state 1, and (b) state 3. (c) Find the ratio of the number of moles of gas in system A to that in system B.

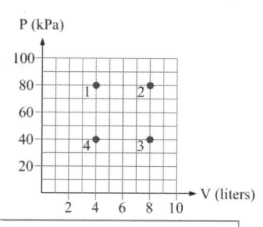

Figure 14.30: Four states are shown on the P-V diagram for a particular thermodynamic system. For Exercise 52.

53. You have two containers of the same volume. Both containers are full of ideal gas. The containers contain the same number of moles of gas, and their temperatures are identical. The internal energy of the gas in one container is 600 J, while in the other container it is only 360 J. How do the pressures compare?

54. Comment on each statement in the following conversation between two students, pertaining to the P-V diagram in Figure 14.31. In addition, state the correct answer to the question the students are trying to answer.

Valentina: This question wants to know by what factor the temperature of an ideal gas system increases when the system goes from state 1 to state 2, as shown in Figure 14.31. Let's see, the volume doubles, so does that mean the temperature doubles?

Brandon: I'm not sure. Don't we have to know how the system moves from state 1 to state 2? The diagram doesn't show us that at all.

Valentina: You might be right about that. Hmmm...also, I don't think my first answer can be right, because the pressure changes by a factor of 1.5. That, by itself, means the temperature goes up by 1.5, too.

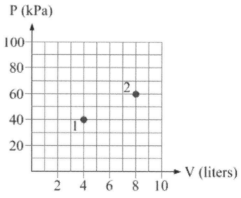

Figure 14.31: Two states are shown on the P-V diagram for a particular thermodynamic system. For Exercise 54.

55. This problem is adapted from a problem created by Todd Cooke and Joe Redish at the University of Maryland. A common earthworm requires about 1.0 micromoles of oxygen per gram of body mass per hour. This oxygen diffuses through the skin of the worm at a rate of about 0.25 micromoles per square centimeter of surface area. (a) We will first assume the worm is a cylinder 10 cm long and has a mass of 3.0 g, and we will neglect the area of the end-caps of the cylinder. Assuming just enough oxygen diffuses through the skin to satisfy the metabolic needs of the worm, what is the worm's radius? (b) If the radius doubles, how do you expect the mass and surface area of the worm to change? Would this thicker worm be able to take in enough oxygen to meet its needs?

Steam engines, similar to the one shown here powering a tractor, were a key component of the Industrial Revolution. A steam engine is a heat engine. In a heat engine, energy is extracted at a relatively high temperature, and some of that energy is used to do useful work. As we will learn in this chapter, by the laws of thermodynamics, some fraction of the energy also has to be wasted.

The same basic principles of heat engines are exploited today in modern car engines and in power plants that generate electricity. In this chapter, we will examine the principles of physics on which such devices are based.

Photo credit: Pamela Hodson /iStockPhoto

Chapter 15 – The Laws of Thermodynamics

CHAPTER CONTENTS

Thermodynamics is the branch of physics that deals with systems in which energy is in the form of heat, work, and internal energy. Such systems have a wide variety of practical applications, ranging from gasoline-fueled car engines to refrigerators and air conditioners. Some of these devices are rather complex; however, underlying their operation are some basic rules known as the laws of thermodynamics. Understanding these rules will be the theme of the chapter.

15-1 The First Law of Thermodynamics

Consider a cylinder of ideal gas sealed with a piston, a system we examined in Chapter 14. Let's explore what happens to the system's energy when the temperature increases.

EXPLORATION 15.1 – Adding heat to a cylinder

A cylinder of ideal gas, as in Figure 15.1, is sealed with a piston that can move up or down without friction. The top of the piston is exposed to the atmosphere. The ideal gas in the cylinder is initially at room temperature, 20°C. The cylinder is then placed in a water bath that is maintained at a constant temperature of 90°C.

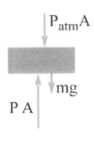

Figure 15.1: A cylinder sealed with a piston that can move up or down without friction. The cylinder is initially at room temperature, 20°C.

Step 1 – *Describe qualitatively what happens to this gas system when it is placed in the water bath.*
Because the temperature of the gas in the cylinder is lower than that of the water bath, heat will be transferred from the water into the gas. This is shown in Figure 15.2. This transfer of energy will continue until the gas in the cylinder reaches 90°C, the same temperature as the water bath. At that point, thermal equilibrium will be reached and the energy transfer will stop.

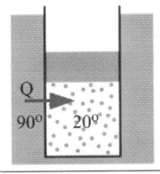

Step 2 – *Describe what happens to the energy transferred as heat from the water bath to the ideal gas.* Increasing the temperature of the gas means increasing the internal energy of the gas. Thus, some of the heat Q transferred to the gas causes a change in internal energy, ΔE_{int}. However, something else

Figure 15.2: When the cylinder is placed in a water bath that is maintained at a temperature higher than the gas, heat is naturally transferred from the water bath into the cylinder until thermal equilibrium is reached.

must change for the gas to satisfy the condition set by the ideal gas law, $PV = nRT$. As we discussed in Chapter 14, in a cylinder in which the piston is free to move, the pressure is determined by the forces on the piston's free-body diagram (see Figure 15.3).

When the gas temperature reaches 90°C, the piston is at a new equilibrium position, but the pressure in the cylinder is the same as it was before – nothing has changed on the free-body diagram. Thus, to satisfy the ideal gas law, the volume occupied by the gas increases when the pressure increases. This is shown in Figure 15.4. The expanding gas does a positive amount of work W, exerting an upward force and causing an upward displacement of the piston. Like the change in internal energy, the work W comes from the heat Q transferred from the water bath into the gas. Thus, some of the transferred heat goes into raising the internal energy of the system, and some goes into doing work.

Figure 15.3: Applying Newton's second law to the free-body diagram tells us that, with the piston at equilibrium, the pressure of the gas is given by $P = P_{atm} + mg/A$, where A is the area of the top or bottom of the piston.

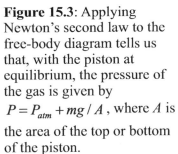

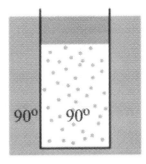

Figure 15.4: When thermal equilibrium is reached, the gas occupies a larger volume because the temperature is higher, maintaining a constant pressure.

Key Idea: Heat transferred between a system and its surroundings can be transformed into two other forms of energy. Some (or even all, in certain cases) of the heat goes into changing the internal energy of the gas (corresponding to a temperature change) while the remaining heat goes into doing work. This statement of energy conservation is the first law of thermodynamics, which is defined more formally below. **Related End-of-Chapter Exercises: 1 and 13.**

The first law of thermodynamics is a statement of energy conservation as it relates to a thermodynamic system. Heat, which is energy transferred into or out of a system, can be transformed into (or come from) some combination of a change in internal energy of the system and the work done by (or on) the system.

$$Q = \Delta E_{int} + W \ .$$ (Equation 15.1: **The First Law of Thermodynamics**)

Q is positive when heat is added to a system, and negative when heat is removed.

ΔE_{int} is positive when the temperature of a system increases, and negative when it decreases.

W is positive when a system expands and does work, and negative when the system is compressed.

EXAMPLE 15.1 – Some numerical calculations

Let's do two numerical calculations related to Exploration 15.1. Let's say that 3500 J is transferred to the cylinder as heat. The work done by the gas while the heat is being transferred is 1400 J. (a) Calculate the change in internal energy experienced by the gas in the cylinder. (b) If the change in internal energy can be calculated, in this case, using the equation $\Delta E_{int} = (3/2)nR\Delta T$, calculate the number of moles of gas in the cylinder.

SOLUTION

(a) We can calculate the change in internal energy by re-arranging the equation for the first law.

$$\Delta E_{int} = Q - W = 3500 \text{ J} - 1400 \text{ J} = 2100 \text{ J} \ .$$

(b) Now, we can solve for the number of moles by re-arranging the equation that was given for the change in internal energy.

$$n = \frac{(2/3)\Delta E_{int}}{R\Delta T} = \frac{(2/3)\,2100 \text{ J}}{(8.31 \text{ J/mol K})(+70\text{K})} = \frac{1400 \text{ J}}{(8.31 \text{ J/mol K})(+70\text{K})}$$

$$n = \frac{20 \text{ J/K}}{8.31 \text{ J/mol K}} = 2.4 \text{ moles}$$

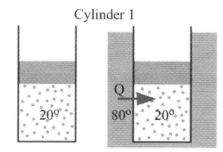

Cylinder 1

Essential Question 15.1: Figure 15.5 shows two cylinders containing equal amounts of the same ideal gas in the same initial state – the gas in the two cylinders has the same pressure, volume, and temperature. The difference between them is that, in cylinder 1, the piston sealing the cylinder is free to move up or down without friction, while the piston in cylinder 2 is fixed so the volume occupied by the gas is constant. Both cylinders are initially at 20°C, but are then placed in a water bath that is maintained at a constant temperature of 80°C and allowed to come to equilibrium. In this process, which cylinder has more heat transferred to it? Which cylinder experiences a larger change in internal energy?

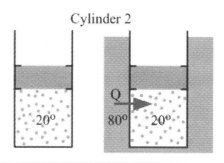

Cylinder 2

Figure 15.5: The piston in cylinder 1 is free to move without friction, while the piston in cylinder 2 is fixed in place so the gas occupies a constant volume.

Answer to Essential Question 15.1: Coming to equilibrium means reaching the same temperature as the water bath. Thus, both cylinders will experience the same change in temperature, +60°C. Internal energy is directly related to temperature. Because the same number of moles of the same type of gas experiences the same temperature change, the change in internal energy is the same in both cases. The difference between the cylinders is that the gas in cylinder 1 will expand when the temperature increases, so the gas does work moving the piston. No work is done by the gas in cylinder 2. Thus $Q_1 = \Delta E_{int} + W$ is larger than $Q_2 = \Delta E_{int}$. More heat needs to be added to cylinder 1 than cylinder 2, the difference corresponding to the work done by the gas in cylinder 1.

15-2 Work, and Internal Energy

The first law involves three parameters. Heat (Q) involves a transfer of energy into or out of a system. Let's now explore the ideas of work (W) and change in internal energy (ΔE_{int}).

For the case of the cylinder in section 15-1, the work done by the gas on the piston is the magnitude of the force the gas exerts on the piston multiplied by the magnitude of the piston's displacement, Δh. Using the fact that $F = PA$, and that the volume and height are related by area:

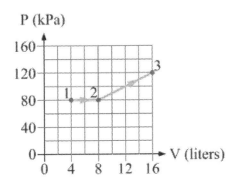

$$W = F\Delta h = (PA)\Delta h = P(A\Delta h) = P\Delta V .$$

Thus, for a constant pressure process we have:
$W = P\Delta V$.   (Eq. 15.2: **Work done at constant pressure**)

If the pressure changes, Equation 15.2 does not apply. Is there a general method of finding work that is valid in all cases? Consider now the two processes shown in the P-V diagram in Figure 15.6.

Figure 15.6: The P-V diagram shows an expansion from state 1 to state 2 at constant pressure, followed by another expansion that takes the system to state 3 along the path indicated.

For the constant pressure process, in which the system expands from state 1 to state 2, Equation 15.2 tells us that the work done by the gas in that process is $P\Delta V$. This is the area under the curve defining the process. For the expansion from state 2 to state 3, which is not at constant pressure, the work is still equal to the area under the curve defining the process. This gives us our general method of finding work. The two shaded areas shown in Figure 15.7 represent the work done by the gas in the two processes.

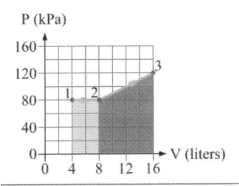

Work: A practical way to calculate the work done by a gas in a particular thermodynamic process is to find the area under the curve for that process on the P-V diagram.

W = area under the curve on the P-V diagram.

(Equation 15.3: **Work done by a gas**)

Figure 15.7: The area of the rectangular region under the 1 to 2 process is the work done by the gas in the expansion from state 1 to state 2. The area of the region under the 2 to 3 process is the work done by the gas in moving from state 2 to state 3.

EXAMPLE 15.2A – Calculating the work
Find the work done by the gas in the two processes shown in Figure 15.6.

SOLUTION
For the constant-pressure 1 to 2 process, the work is the area of a rectangle, as in Figure 15.7.
$$W_{1\rightarrow2} = P\Delta V = (80 \text{ kPa})(8.0 \text{ L} - 4.0 \text{ L}) = (80\times10^3 \text{ Pa})(4.0\times10^{-3} \text{ m}^3) = +320 \text{ J.}$$

For the expansion from state 2 to 3, the work is the area under the 2 to 3 line in Figure 15.7. This is equal to the area of a rectangle, with the top of the rectangle at the average pressure.
$$W_{2\rightarrow3} = P_{av}\Delta V = (100 \text{ kPa})(16 \text{ L} - 8.0 \text{ L}) = (100\times10^3 \text{ Pa})(8.0\times10^{-3} \text{ m}^3) = +800 \text{ J.}$$

Let's turn now to the change in internal energy. The change in internal energy is independent of the process that moves a system from one state to another. Thus, if we know what the change in internal energy is for one process, we can apply that to all processes.

Change in internal energy: If the temperature of an ideal gas changes, the change in internal energy of the gas is proportional to the change in temperature. If there is no change in temperature, there is no change in internal energy (as long as the number of moles of gas remains constant).

$$\Delta E_{int} = n C_V \Delta T \text{ .} \qquad \text{(Equation 15.4: \textbf{Change in internal energy})}$$

Monatomic: $C_V = \dfrac{3}{2}R$ Diatomic: $C_V = \dfrac{5}{2}R$ Polyatomic: $C_V = 3R$

C_V is the heat capacity at constant volume, which we will examine in Section 15-3.

If we just want the internal energy, we remove the deltas.
$$E_{int} = n C_V T \text{ .} \qquad \text{(Equation 15.5: \textbf{Internal energy of an ideal gas})}$$

EXAMPLE 15.2B – Calculating the change in internal energy
Consider again the P-V diagram shown in Figure 15.6. If the gas is diatomic, find the change in internal energy associated with the two processes shown on the diagram.

SOLUTION
To do this, we will combine the ideal gas law with the information on the graph.
$$\Delta E_{int} = \frac{5}{2}nR\Delta T = \frac{5}{2}nR(T_2 - T_1) = \frac{5}{2}(nRT_2 - nRT_1) = \frac{5}{2}(P_2V_2 - P_1V_1).$$

Plugging in the values for the pressures and volumes shown on the P-V diagram gives:
$$\Delta E_{int,1\rightarrow2} = \frac{5}{2}\left[(80 \text{ kPa})(8.0 \text{ L}) - (80 \text{ kPa})(4.0 \text{ L})\right] = 1600 \text{ J} - 800 \text{ J} = +800 \text{ J.}$$

Using a similar process for the expansion from state 2 to state 3 gives:
$$\Delta E_{int,2\rightarrow3} = \frac{5}{2}\left[(120 \text{ kPa})(16.0 \text{ L}) - (80 \text{ kPa})(8.0 \text{ L})\right] = 3800 \text{ J} - 1600 \text{ J} = +2200 \text{ J.}$$

Related End-of-Chapter Exercises: 14, 16, 46, 47.

Essential Question 15.2: If we did not know the path taken on the P-V diagram, in Figure 15.6, from state 2 to state 3, could we still find the work or the change in internal energy? Explain.

Answer to Essential Question 15.2: An important distinction between work and the change in internal energy is that the work depends on the process involved in taking a system from one state to another, while the change in internal energy depends only on the initial and final states. Thus, if we do not know the path taken on the P-V diagram, we can not find the work – different processes have different amounts of work associated with them. On the other hand, we can find the change in internal energy, because we know the pressure and volume of the initial and final states. No matter what the process, for a diatomic ideal gas the change in internal energy in moving from state 2 to state 3 will be the +2200 J we calculated in Example 15.2B.

15-3 Constant Volume and Constant Pressure Processes

Let's consider once again two different thermodynamic processes, one in which heat is added to a system at constant volume, and the other when heat is added at constant pressure.

EXPLORATION 15.3A – A constant-volume process
A sample of monatomic ideal gas is initially at a temperature of 200 K. The gas occupies a constant volume. Heat is then added to the gas until the temperature reaches 400 K. This process is shown on the P-V diagram in Figure 15.8, where the system moves from state 1 to state 2 by the process indicated. The diagram also shows the cylinder in state 1 and again in state 2. The figure also shows the 200 K isotherm (lower) and the 400 K isotherm (higher).

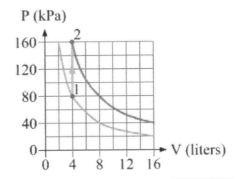

Step 1 – Find the number of moles of gas in the cylinder.
Applying the ideal gas law to state 1 gives:

Figure 15.8: A P-V diagram showing a constant-volume process that moves a system of monatomic ideal gas from state 1 to state 2.

$$n = \frac{PV}{RT} = \frac{(80 \text{ kPa})(4.0 \text{ L})}{(8.31 \text{ J/mol K})(200 \text{ K})} = 0.19 \text{ moles} .$$

Step 2 – Find the work done in this process.
The work done is the area under the curve for the process. Because there is no area under the curve in a constant-volume process the work done by the gas is zero: $W = 0$.

Step 3 – Find the change in internal energy for this process.
In a constant-volume process all the heat added goes into changing the internal energy of the gas. Because the gas is monatomic we have $C_V = 3R/2$. This gives:

$$\Delta E_{int} = \frac{3}{2} n R \Delta T = \frac{3}{2}(0.19 \text{ moles})(8.31 \text{ J/mol K})(400 \text{ K} - 200 \text{ K}) = +480 \text{ J} .$$

Step 4 – Find the heat added to the gas in this process. The First Law of Thermodynamics tells us that $Q = \Delta E_{int} + W$, but if the work done by the gas is zero we have $Q = \Delta E_{int}$. In this case we have $Q = +480$ J.

Key ideas for a constant-volume process: There is no work done by the gas: $W = 0$.
The heat added to the gas is equal to the change in internal energy: $Q = n C_V \Delta T$.
Related End-of-Chapter Exercises: 17, 18.

EXPLORATION 15.3B – A constant-pressure process
     The system from the previous Exploration is now taken from state 1 to state 3 (in which, like state 2, the system is at 400 K) by the constant-pressure process indicated in Figure 15.9.

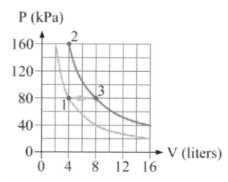

P (kPa)

Step 1 – *Find the work done in this process.* The work done by the gas in this process is the area under the curve on the P-V diagram. Because the pressure is constant we can use Equation 15.2:

$$W = P\Delta V = (80 \text{ kPa})(8.0 \text{ L} - 4.0 \text{ L}) = (80 \text{ kPa})(+4.0 \text{ L}) = +320 \text{ J}.$$

Figure 15.9: A P-V diagram showing a constant-pressure process that moves a system of monatomic ideal gas from state 1 to state 3.

Step 2 – *Find the change in internal energy for this process.* Because the temperature change is the same, the change in internal energy is the same as it is in the constant-volume process:

$$\Delta E_{\text{int}} = \frac{3}{2}nR\Delta T = \frac{3}{2}(0.19 \text{ moles})(8.31 \text{ J/mol K})(400 \text{ K} - 200 \text{ K}) = +480 \text{ J}.$$

Step 3 – *Find the heat added to the gas.* Applying the First Law of Thermodynamics gives:
$$Q = \Delta E_{\text{int}} + W = +480 \text{ J} + 320 \text{ J} = +800 \text{ J}$$

Key ideas for a constant-pressure process: The work done by the gas is given by Equation 15.2: $W = P\Delta V$.   **Related End-of-Chapter Exercises: 3, 19 – 21.**

Heat Capacity
     In Chapter 13, we used $Q = mc\Delta T$ to find the heat needed to change the temperature of a substance of mass m and specific heat c. For gases, a more convenient equation is $Q = nC\Delta T$,   where C is known as the heat capacity. The value of the heat capacity depends on the process the gas follows when the heat is added. At constant volume, when the work done is zero:

$$Q = \Delta E_{\text{int}} = nC_V\Delta T, \quad \text{(Eq. 15.6: \textbf{Heat needed to change temperature at constant volume})}$$

where C_V is the heat capacity at constant volume. As mentioned in Section 15-2, we apply $\Delta E_{\text{int}} = nC_V\Delta T$ to all processes, because the change in internal energy is process independent.

Monatomic: $C_V = \frac{3}{2}R$ Diatomic: $C_V = \frac{5}{2}R$ Polyatomic: $C_V = 3R$

     In contrast, the **heat needed to change temperature at constant pressure** is given by:
$$Q = W + \Delta E_{\text{int}} = P\Delta V + nC_V\Delta T = nR\Delta T + nC_V\Delta T = n(R + C_V)\Delta T = nC_P\Delta T, \quad \text{(Eq. 15.7)}$$

where $C_P = R + C_V$ is the heat capacity at constant pressure.

Monatomic: $C_P = \frac{5}{2}R$ Diatomic: $C_P = \frac{7}{2}R$ Polyatomic: $C_P = 4R$

Essential Question 15.3: In Exploration 15.3B, 800 J of heat added to the system of monatomic ideal gas at 200 K increases the system's temperature to 400 K via a constant-pressure process. If the same 800 J of heat is added to the system, starting in state 1, but the volume is kept constant what will the final temperature of the system be? What about the final pressure?

Answer to Essential Question 15.3: In a constant-volume process, for a monatomic ideal gas:

$$Q = \Delta E_{\text{int}} = \frac{3}{2} n R \Delta T = \frac{3}{2}(0.19 \text{ moles})(8.31 \text{ J/mol K})(400 \text{ K} - 200 \text{ K}) = +480 \text{ J}.$$

Solving for the temperature change when 800 J of heat is added, we get:

$$\Delta T = \frac{2Q}{3nR} = \frac{2 \times 800 \text{ J}}{3(0.19 \text{ moles})(8.31 \text{ J/mol K})} = +333 \text{ K}.$$

Because the initial temperature is 200 K, the final temperature is 200 K + 333 K = 533 K. Applying the ideal gas law, gives the corresponding pressure:

$$P = \frac{nRT}{V} = \frac{(0.19 \text{ moles})(8.31 \text{ J/mol K})(533 \text{ K})}{4.0 \text{ L}} = 213 \text{ kPa}.$$

15-4 Constant Temperature and Adiabatic Processes

Let's now consider two more thermodynamic processes, the constant temperature (also known as isothermal) process and the adiabatic process.

A constant-temperature (isothermal process: Because the temperature is constant there is no change in internal energy. The First Law of Thermodynamics tells us that, in this case, $Q = W$, and it can be shown (using calculus is the most straightforward way to prove this) that:

$$Q = W = nRT \ln\left(\frac{V_f}{V_i}\right). \qquad \text{(Eq. 15.8: \textbf{Heat and work for an isothermal process})}$$

EXAMPLE 15.4A – Add heat at constant temperature

700 J of heat is added to a system of ideal gas, while the temperature is kept constant at 400 K. The system initially has a pressure of 160 kPa and occupies a volume of 4.0 liters.

(a) Is this possible? Can temperature remain constant while heat is added? Explain.
(b) Sketch this process on a P-V diagram, keeping in mind the following question: When heat is added at constant temperature does the gas pressure increase, as in a constant-volume process, or does the volume increase, as in a constant-pressure process?
(c) What are the final values of the gas pressure and the volume of the gas?

SOLUTION

(a) This is possible. The first law of thermodynamics tells us that heat Q is converted into some combination of internal energy and/or work. When the temperature is constant we have the special case of no change in internal energy, so all the heat is converted into work.

(b) The temperature is constant so the process proceeds along an isotherm, shown in Figure 15.10. Because Q is positive, and $Q = W$, the work is also positive. Positive work means that the volume must increase, so the pressure drops to keep the temperature constant.

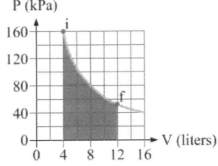

Figure 15.10: A P-V diagram corresponding to heat being added to a system at constant temperature. The final pressure and volume are calculated in part (c). The area corresponding to the work done by the gas is shown shaded.

(c) First, let's re-write Eq. 15.8 using the ideal gas law: $Q = nRT \ln\left(\dfrac{V_f}{V_i}\right) = P_i V_i \ln\left(\dfrac{V_f}{V_i}\right)$.

Isolating the logarithm on the right side gives: $\dfrac{Q}{P_i V_i} = \ln\left(\dfrac{V_f}{V_i}\right)$.

Taking the exponential of both sides: $e^{Q/(P_i V_i)} = e^{\ln(V_f/V_i)} = \dfrac{V_f}{V_i}$.

The final volume is thus: $V_f = V_i\, e^{Q/(P_i V_i)} = (4.0\,\text{L})\, e^{700\,\text{J}/(160\,\text{kPa} \times 4.0\,\text{L})} = 11.9\,\text{L}$.

Solving for the final pressure gives: $P_f = \dfrac{nRT}{V_f} = \dfrac{P_i V_i}{V_f} = \dfrac{(160\,\text{kPa})(4.0\,\text{L})}{11.94\,\text{L}} = 54\,\text{kPa}$.

An adiabatic process: In an adiabatic process no heat is added to or removed from the gas (i.e., $Q = 0$). Examples include systems insulated so no heat is exchanged with the surroundings, and systems in which processes happen so fast that there is no time to add or remove heat. Because $Q = 0$ for an adiabatic process the First Law of Thermodynamics tells us that $\Delta E_{int} = -W$. The energy for any work done comes from the change in the system's internal energy.

$PV^{\gamma} = \text{constant}$,   (Equation 15.9: **Equation for an adiabatic process on the P-V diagram**)

where γ is the ratio of the heat capacity at constant pressure to the heat capacity at constant volume:

$\gamma = \dfrac{C_P}{C_V}$ (Equation 15.10: The constant γ for an adiabatic process.

EXAMPLE 15.4B – Analyzing an adiabatic process

A system of monatomic ideal gas experiences an adiabatic expansion that moves it from an initial state, at 400 K, to a final state at a temperature of 200 K. The process is shown on the P-V diagram in Figure 15.11. Calculate the values of the final pressure and volume.

SOLUTION

Because the gas is monatomic, we have: $\gamma = \dfrac{C_P}{C_V} = \dfrac{5R/2}{3R/2} = \dfrac{5}{3}$.

From Equation 15.9, we know that: $P_f V_f^{\gamma} = P_i V_i^{\gamma}$

From the ideal gas law, we have $P_f = nRT_f/V_f$, as well as $n = P_i V_i/(RT_i)$.

Combining these results leads to $T_f V_f^{\gamma-1} = T_i V_i^{\gamma-1}$.   (Equation 15.11)

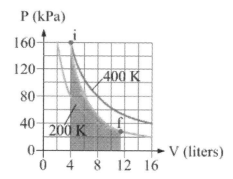

P (kPa)

Figure 15.11: The P-V diagram for the adiabatic expansion.

Solving for the final volume gives: $V_f = V_i\left(T_i/T_f\right)^{1/(\gamma-1)} = (4.0\,\text{L})(2)^{3/2} = 11.3\,\text{L}$.

The final pressure is thus: $P_f = P_i(V_i/V_f)^{\gamma} = (160\,\text{kPa})\left(\dfrac{4.0\,\text{L}}{11.31\,\text{L}}\right)^{5/3} = 28\,\text{kPa}$.

Related End-of-Chapter Exercises for this section: 4, 22 – 27.

Essential Question 15.4: How much work is done by the gas in Example 15.4B?

Answer to Essential Question 15.4: The work done by the gas in the expansion is shown by the area shaded in red in Figure 15.11. We can find a numerical value for the work done by applying the First Law of Thermodynamics, $Q = \Delta E_{int} + W$. Because Q = 0, we have:

$$W = -\Delta E_{int} = -nC_V \Delta T = -\frac{3}{2}nR\left(T_f - T_i\right) = -\frac{3}{2}nRT_f + \frac{3}{2}nRT_i.$$

We could solve for the number of moles of gas, but let's instead apply the ideal gas law:

$$W = -\frac{3}{2}P_f V_f + \frac{3}{2}P_i V_i = \frac{3}{2}\left(P_i V_i - P_f V_f\right) = \frac{3}{2}(160 \text{ kPa} \times 4.0 \text{ L} - 28.3 \text{ kPa} \times 11.31 \text{ L}) = 480 \text{ J}$$

15-5 A Summary of Thermodynamic Processes

There is no single step-by-step strategy that can be applied to solve every problem involving a thermodynamic process. Instead let's summarize the tools we have to work with. These tools can be applied in whatever order is appropriate to solve a particular problem.

Tools for Solving Thermodynamics Problems
- The P-V diagram can help us to visualize what is going on. In addition, the work done by a gas in a process is the area under the curve defining that process on the P-V diagram.
- The ideal gas law, $PV = nRT$.
- The first law of thermodynamics, $Q = \Delta E_{int} + W$.
- The general expression for the change in internal energy, $\Delta E_{int} = nC_V \Delta T$.
- In specific special cases (see the summary in Figure 15.14), there are additional relationships that can be used to relate the different parameters.

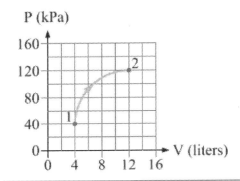

EXAMPLE 15.5 – Applying the tools

A system of monatomic ideal gas is taken through the process shown in Figure 15.12. For this process find (a) the work done by the gas, (b) the change in internal energy, and (c) the heat added to the gas.

Figure 15.12: The process that moves the system from state 1 to state 2 follows a circular arc on the P-V diagram that covers ¼ of a circle.

SOLUTION

(a) The area under the curve has been split into two parts in Figure 15.13, a ¼-circle and a rectangle. Each box on the P-V diagram measures 20 kPa × 2.0 L, representing an area of 40 J. The rectangular area covers 8 boxes, for a total of 320 J of work. The radius of the quarter-circle is four boxes, so the area of that quarter circle is given by:

$$W_{1/4} = \frac{1}{4}\pi r^2 = \frac{1}{4}\pi\left(4 \text{ units}\right)^2 = 4\pi \text{ boxes} = 4\pi \text{ boxes} \times 40 \text{ J/box} = 500 \text{ J}.$$

Thus, the total work done by the gas is 320 J + 500 J = 820 J.

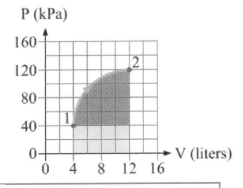

Figure 15.13: To find the work done by the gas here, it is simplest to split the area under the curve into two pieces, a ¼-circle and a rectangle.

(b) The gas is monatomic, so $C_V = 3R/2$.

Combining Eq. 15.4 with the ideal gas law:

$$\Delta E_{int} = \frac{3}{2} nR\Delta T = \frac{3}{2}\left(nRT_f - nRT_i\right) = \frac{3}{2}\left(P_f V_f - P_i V_i\right) = \frac{3}{2}\left(120 \text{ kPa} \times 12 \text{ L} - 4.0 \text{ kPa} \times 4.0 \text{ L}\right) = 1920 \text{ J}$$

(c) Using the first law of thermodynamics: $Q = \Delta E_{int} + W = 1920 \text{ J} + 820 \text{ J} = 2740 \text{ J}$.

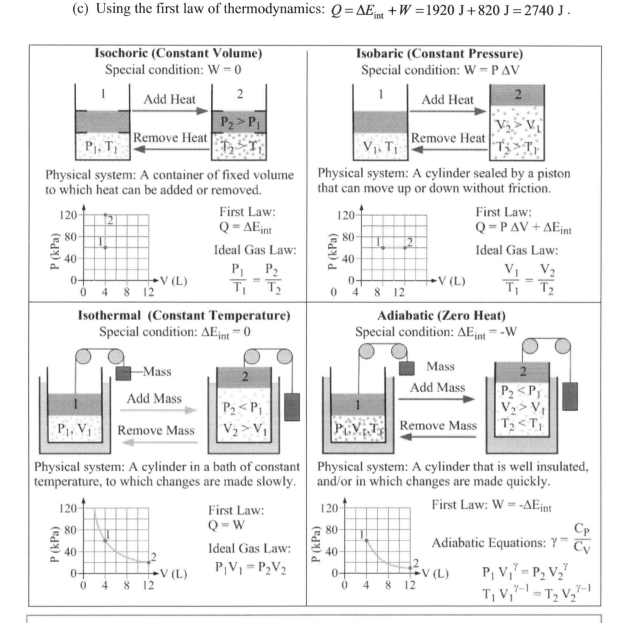

Figure 15.14: A summary of four special-case thermodynamic processes. For each, we see the special condition associated with that process; a pictorial representation and description in words of a corresponding physical system; a P-V diagram for the process; and equations we can apply to solve problems associated with the process.

Related End-of-Chapter Exercises for this section: 6, 7, 28 – 31.

Essential Question 15.5: Compare the P-V diagrams for the isochoric and isobaric processes in Figure 15.14. Assuming these pertain to the same system, and state 1 is the same in both cases, in which case is the change in internal energy larger? In which case is more heat involved?

Answer to Essential Question 15.5: The ideal gas law tells us that temperature is proportional to PV. $P_2V_2 = 480$ J for state 2 in both processes we are considering, so the temperature in state 2 is the same in both cases. $\Delta E_{int} = nC_V\Delta T$, and all three factors on the right-hand side are the same for the two processes, so the change in internal energy is the same (+360 J, in fact). Because the gas does no work in the isochoric process, and a positive amount of work in the isobaric process, the First Law tells us that more heat is required for the isobaric process (+600 J versus +360 J).

15-6 Thermodynamic Cycles

Many devices, such as car engines and refrigerators, involve taking a thermodynamic system through a series of processes before returning the system to its initial state. Such a cycle allows the system to do work (e.g., to move a car) or to have work done on it so the system can do something useful (e.g., removing heat from a fridge). Let's investigate this idea.

EXPLORATION 15.6 – Investigate a thermodynamic cycle

One cycle of a monatomic ideal gas system is represented by the series of four processes in Figure 15.15. The process taking the system from state 4 to state 1 is an isothermal compression at a temperature of 400 K. Complete Table 15.1 to find Q, W, and ΔE_{int} for each process, and for the entire cycle.

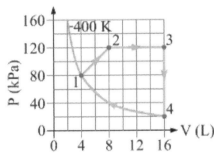

Process	Special process?	Q (J)	W (J)	ΔE_{int} (J)
1 → 2	No	+1360		
2 → 3	Isobaric			
3 → 4	Isochoric		0	
4 → 1	Isothermal			0
Entire Cycle	No			0

Table 15.1: Table to be filled in to analyze the cycle. See Step 1 for a justification of the 0's in the table.

Figure 15.15: The series of four processes making up the cycle.

Step 1 – *Fill in any zeroes in the table.* There is no work done in the 3 → 4 process, and the change in internal energy is zero in the 4 → 1 process. However, ΔE_{int} for the entire cycle is also zero (this is always true), because the system returns to its original state, and therefore its original temperature.

Step 2 – *Analyze the 1 → 2 process.* The work done by the gas is
$$W_{1\to2} = P_{av}\Delta V = 100 \text{ kPa}\times(8.0 \text{ L} - 4.0 \text{ L}) = 400 \text{ J}.$$

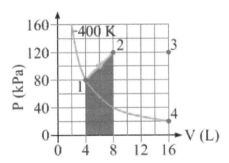

Figure 15.16: The work done by the gas in the 1 → 2 process equals the area under the curve, shown shaded.

From the ideal gas law, PV is proportional to temperature. Because $P_1V_1 = 320$ J corresponds to $T_1 = 400$ K (so $nR = 0.80$ J/K in this case), then $P_2V_2 = 960$ J $= 3(P_1V_1)$ corresponds to a tripling of the temperature to $T_2 = 1200$ K. We know that $C_V = 3R/2$, so:

$$\Delta E_{int,1\to2} = \frac{3}{2}nR\Delta T = \frac{3}{2}nR(T_2 - T_1) = \frac{3}{2}(0.80 \text{ J/K})(1200 \text{ K} - 400 \text{ K}) = +960 \text{ J}.$$

Using the first law: $Q_{1\to2} = \Delta E_{int,1\to2} + W_{1\to2} = +960 \text{ J} + 400 \text{ J} = +1360 \text{ J}.$

Step 3 – Analyze the 2 → 3 process. The work done by the gas in the 2 → 3 process is the area of the shaded region in Figure 15.17:
$$W_{2\to3} = P\Delta V = 120 \text{ kPa} \times (16.0 \text{ L} - 8.0 \text{ L}) = 960 \text{ J}.$$

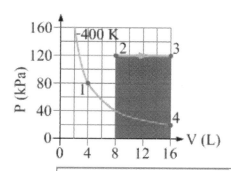

Doubling the volume while keeping the pressure constant doubles the absolute temperature. Thus $T_3 = 2T_2 = 2400$ K, and:

$$\Delta E_{int,2\to3} = nC_V \Delta T = \frac{3}{2}nR(T_3 - T_2) = \frac{3}{2}(0.80 \text{ J/K})(2400 \text{ K} - 1200 \text{ K}) = +1440 \text{ J}.$$

Applying the first law tells us that

$$Q_{2\to3} = \Delta E_{int,2\to3} + W_{2\to3} = +1440 \text{ J} + 960 \text{ J} = +2400 \text{ J}.$$

Figure 15.17: The work done by the gas in the 2 → 3 process equals the shaded area under the curve.

Step 4 – Analyze the 3 → 4 process. With no work done in this process:
$$Q_{3\to4} = \Delta E_{int,3\to4} = \frac{3}{2}nR(T_4 - T_3) = \frac{3}{2}(0.80 \text{ J/K})(400 \text{ K} - 2400 \text{ K}) = -2400 \text{ J}.$$

Step 5 – Analyze the 4 → 1 process. $\Delta E_{int} = 0$ for an isothermal process. The work done by the gas is the area of the shaded region in Figure 15.18, in blue to remind us that it is negative. Using Equation 15.8:

$$Q_{4\to1} = W_{4\to1} = nRT \ln(V_f / V_i) = (0.80 \text{ J/K})(400 \text{ K}) \ln(4.0 \text{ L} / 16 \text{ L}) = -444 \text{ J}.$$

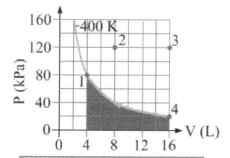

Figure 15.18: The work done by the gas in the 4 → 1 is negative, and is the shaded area under the curve.

Step 6 – Complete the table. Each process, and the entire cycle, satisfies the first law, and values in each column sum to the value for the cycle.

Process	Special process?	Q (J)	W (J)	ΔE_{int} (J)
1 → 2	No	+1360	+400	+960
2 → 3	Isobaric	+2400	+960	+1440
3 → 4	Isochoric	−2400	0	−2400
4 → 1	Isothermal	−444	−444	0
Entire Cycle	No	+916	+916	0

Table 15.2: The completed table. Check that each row satisfies the first law and each column adds up.

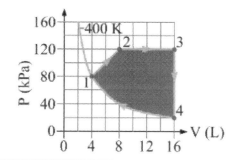

Figure 15.19: The enclosed area is the net work done in the cycle.

Key Ideas for a Thermodynamic Cycle: The cycle satisfies the first law of thermodynamics, as does each of its processes. The change in internal energy for any cycle is always zero because the system returns to its original state, and the area of the enclosed region on the P-V diagram is the net work done by the gas in the cycle.   **Related End-of-Chapter Exercises: 32 – 38.**

Essential Question 15.6: Complete Table 15.3, for a cycle consisting of three processes. The process taking the system from state 1 to state 2 is adiabatic.

Process	Q (J)	W (J)	ΔE_{int} (J)
1 → 2			−300
2 → 3	+800	0	
3 → 1			
Entire Cycle		−200	

Table 15.3: Fill in the missing values.

Answer to Essential Question15.6: We can immediately fill in the following values: $Q_{1 \to 2} = 0$ because there is no heat in an adiabatic process; $\Delta E_{int, 2 \to 3} = +800$ J so the 2 → 3 process satisfies the first law; and $\Delta E_{int, cycle} = 0$ because that is always true (the system returns to its initial state,

with no change in temperature). With those three values in the table it is straightforward to fill in the remaining values by (a) making sure that each row satisfies the first law ($Q = W + \Delta E_{int}$) and, (b) that in every column the sum of the values for the individual processes is equal to the value for the entire cycle. The completed table is shown in Table 15.4.

Process	Q (J)	W (J)	ΔE_{int} (J)
1 → 2	0	+300	−300
2 → 3	+800	0	+800
3 → 1	−1000	−500	−500
Entire Cycle	−200	−200	0

Table 15.4: The completed table.

15-7 Entropy and the Second Law of Thermodynamics

A system of ideal gas in a particular state has an entropy, just as it has a pressure, a volume, and a temperature. Unlike pressure, volume, and temperature, which are easy to determine, the entropy of a system can be difficult to find. On the other hand, changes in entropy can be quite straight-forward to calculate.

> **Entropy:** Entropy is in some sense a measure of disorder. The symbol for entropy is S, and the units are J/K.
>
> **Change in entropy:** In certain cases the change in entropy, ΔS, is easy to determine. An example is in an isothermal process in which an amount of heat Q is transferred to a system:
>
> $$\Delta S = \frac{Q}{T} .$$ (Equation 15.12: **Change in entropy for an isothermal process**)
>
> In cases where heat is transferred while the temperature changes, the change in entropy can be approximated if the temperature change is small compared to the absolute temperature. In this case:
>
> $$\Delta S \approx \frac{Q}{T_{av}} ,$$ (Equation 15.13: **Approximate change in entropy**)
>
> where T_{av} is the average temperature of the system while the heat is being transferred.

EXAMPLE 15.7 – Mixing water
You have two containers of water. In one container there is 1.0 kg of water at 17°C, while in the second container there is 1.0 kg of water at 37°C. You then pour the water in one container into the other container and allow the system to come to equilibrium. Assuming no heat is transferred from the water to the container or the surroundings, determine the change in entropy for (a) the water that is initially cooler, (b) the water that is initially warmer, and (c) the system.

SOLUTION
Because the two samples of water have equal mass the equilibrium temperature will be 27°C, halfway between the initial temperatures of the two samples. Thus, while heat is being transferred from the warmer water to the cooler water the average temperature of the cooler water will be 22°C, or 295K, and the average temperature of the warmer water will be 32°C (305K).

(a) Because a temperature change of 10° is small compared to the absolute temperature of about 300K, we can use Equation 15.13 to find the change in entropy. To find the heat, use:

$$Q = mc\Delta T = (1.0 \text{ kg})(4186 \text{ J/kg °C})(10°\text{C}) = 41860 \text{ J} \,.$$

This heat is added to the cooler water, so the heat is a positive quantity. Thus:
$$\Delta S_1 \approx \frac{Q}{T_{av}} = \frac{+41860 \text{ J}}{295\text{K}} = +141.90 \text{ J/K} \,.$$

(b) The heat added to the cooler water comes from the warmer water, so the heat involved for the warmer water is also 41860 J, but negative because it is removed. Thus:
$$\Delta S_2 \approx \frac{Q}{T_{av}} = \frac{-41860 \text{ J}}{305\text{K}} = -137.25 \text{ J/K} \,.$$

(c) Although the entropy of the warmer water decreases, on the whole, the system's entropy increases by an amount:
$$\Delta S = \Delta S_1 + \Delta S_2 = +141.90 \text{ J/K} - 137.25 \text{ J/K} = +4.7 \text{ J/K} \,.$$

The situation above is an example of an irreversible process. Even though energy in the 2.0 kg of water at 27°C after mixing is equal to the total energy in the two separate containers before mixing, the entropy of the mixture at one temperature is larger than that of the system of two separate containers at different temperatures. The mixture will not spontaneously separate back into two halves with a 20°C difference, making the process irreversible. This leads us to the second law of thermodynamics.

The Second Law of Thermodynamics: The entropy of a closed system never decreases as time goes by. Reversible processes do not change the entropy of a system, while irreversible processes increase a system's entropy.

$\Delta S \geq 0$.   (Equation 15.14: **The Second Law of Thermodynamics**)

Entropy is sometimes referred to as time's arrow – time proceeds in the direction of increasing entropy. Imagine watching a science fiction movie in which a spacecraft in deep space explodes into a million pieces. Then you play the film backwards, and see the million pieces magically come together to form the spacecraft. You know without a doubt that the film is running backwards – what is it that gives it away? Both momentum and energy are conserved in the explosion, whether you view it forwards or backwards. What gives it away that you are viewing the film backwards is that the process of the million pieces coming together to form the spacecraft decreases the entropy of the system. Our experience is that systems obey the second law of thermodynamics, and proceed in a direction that tends to increase entropy.

What if you view a film forwards and backwards and you can not tell which direction corresponds to time moving forwards? An example would be an elastic collision between two objects. In such a case, the process on film is most likely reversible, with no change (or negligible change) in entropy, giving us nothing to go by to determine the direction of increasing time.

Related End-of-Chapter Exercises: 39, 40, and 57.

Essential Question 15.7: When you were younger, you were probably asked to clean up your room. Let's say that you cleaned your room up on Saturday, and over the course of the next week it gradually got messy again – the second law of thermodynamics at work! The next Saturday you had to clean your room again. What happened to the entropy of your room when you cleaned it up? Does this violate the second law of thermodynamics? Explain.

Answer to Essential Question 15.7: Decreasing the disorder of the room decreases its entropy. If the cleaning up process happened spontaneously that would violate the Second Law. However, in the system of you and the room, the decrease in entropy of the room is more than offset by the increase in entropy in your body, as measured by things like waste products that accumulated in your muscles as you did work. The room does not represent a closed system; examining a closed system (you and the room) we see that the Second Law of Thermodynamics is satisfied.

15-8 Heat Engines

Heat engines use heat to do work. Examples are car engines and heat engines that run in reverse, such as refrigerators and air conditioners. All heat engines require two temperatures. Adding heat at a higher temperature expands the system, while removing it at a lower temperature contracts the system, re-setting the engine so that a new cycle can occur.

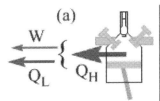

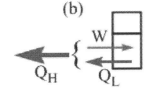

Diagrams of the energy flow in a heat engine are shown in Figure 15.20. Q_H is the magnitude of the heat added or removed at a higher temperature, while Q_L is the magnitude of the heat added or removed at a lower temperature. W is the magnitude of the work involved, which is negative for a cooling device.

Figure 15.20: Energy flow diagrams. In (a), some heat from the cylinder of a car engine does useful work and the rest is discarded into the atmosphere. In (b), the combination of the heat removed from inside a refrigerator, and the work needed to extract it, is discarded into the room by the refrigerator's cooling coils.

The energy equation to accompany the diagram above (which is simply the First Law of Thermodynamics applied to a cycle, or a number of cycles) is:

$$Q_H - Q_L = W .$$ (Eq. 15.15: **Energy equation for a heat engine or cooling device**)

The efficiency, *e*, of a heat engine is the work done by the engine divided by the heat added to cause that work to be done:

$$e = \frac{W}{Q_H} = \frac{Q_H - Q_L}{Q_H} = 1 - \frac{Q_L}{Q_H} .$$ (Equation 15.16: **Efficiency for a heat engine**)

Can an ideal engine have an efficiency of 1, by eliminating losses from things like friction? No, in fact, as was proved by the French mathematician Sadi Carnot (1796 – 1832).

EXPLORATION 15.8 – An ideal (Carnot) engine

Carnot showed that an engine runs at maximum efficiency when it operates on the four-process cycle described in Table 15.5, and shown in the P-V diagram in Figure 15.21.

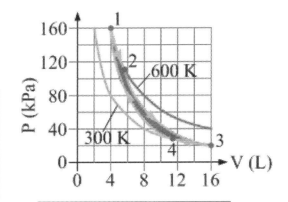

Process	Description	Heat	ΔS
1 → 2	Isothermal expansion at T_H	$+Q_H$	$+Q_H / T_H$
2 → 3	Adiabatic expansion to T_L	0	0
3 → 4	Isothermal compression at T_L	$-Q_L$	$-Q_L / T_L$
4 → 1	Adiabatic compression to T_H	0	0
Cycle	The Carnot Cycle	$Q_H - Q_L$	0

Table 15.5: The sequence of processes in a Carnot cycle.

Figure 15.21: The P-V diagram for the Carnot cycle.

Step 1 – *Equate the sum of the individual changes in entropy to the change in entropy for the cycle.* Entropy is a state function. Because the cycle returns the system to its initial state the system returns to its original entropy. The change in entropy for a complete cycle is always zero:

$$\sum \Delta S = 0 \qquad \text{which gives, in this case:} \qquad \frac{Q_H}{T_H} = \frac{Q_L}{T_L}.$$

This gives: $\qquad \dfrac{T_L}{T_H} = \dfrac{Q_L}{Q_H}.$ $\qquad$ (Equation 15.17: **The Carnot relationship for an ideal engine**)

Step 2 – *Write the efficiency equation in terms of temperatures.* Substituting Equation 15.17 into the efficiency equation, Equation 15.16 gives the efficiency of an ideal engine:

$$e_{ideal} = 1 - \frac{T_L}{T_H}. \text{ (Equation 15.18: } \textbf{Efficiency for an ideal engine)}$$

> **Key ideas for an ideal engine:** The maximum efficiency of an engine is determined by the two temperatures the engine operates between. The third law of thermodynamics states that it is impossible to reach absolute zero, so even an ideal engine can never achieve 100% efficiency.
> **Related End-of-Chapter Exercises: 41 – 43 and 58.**

EXAMPLE 15.8 – A heat pump

If you heat your home using electric heat, 1000 J of electrical energy can be transformed into 1000 J of heat. An alternate heating system is a heat pump, which extracts heat from a lower-temperature region (outside the house) and transfers it to a higher-temperature region (inside the house). The work done by the heat pump is 1000 J, and the temperatures are $T_H = 17°C = 290K$ and $T_L = -23°C = 250K$. (a) Predict whether the maximum amount of heat delivered to the house in this situation is more than, less than, or equal to 1000 J. (b) Calculate this maximum heat.

SOLUTION

(a) Many people predict that the heat pump delivers less than 1000 J of heat to the house, perhaps because of the condition that the efficiency is less than 1. However, the heat pump can be viewed as a cooling device because it is cooling the outside – it acts like an air conditioner in reverse. The heat delivered to the house is Q_H, which from the energy-flow diagram in Figure 15.20 (b) is larger than W. Thus, the pump delivers more than 1000 J of heat to the house.

(b) To find the maximum possible amount of heat we will treat the heat pump as an ideal device, and apply the Carnot relationship (Equation 15.17). Re-arranging this equation gives:

$$Q_L = \frac{T_L}{T_H} Q_H.$$

Substituting this into Equation 15.15, $Q_H - Q_L = W$, gives:

$$Q_H - \frac{T_L}{T_H} Q_H = W$$

Solving for Q_H, the heat delivered to the house, gives:

$$Q_H = \frac{T_H}{T_H - T_L} W = \frac{290K}{290K - 250K}(1000 \text{ J}) = 7250 \text{ J}.$$

Thus, 1000 J of work extracts 6250 J of heat from the outside air, delivering a total of 7250 J of heat to the indoors. This is why heat pumps are far superior to electric heaters!

Related End-of-Chapter Exercises: 44, 45, 59, 60.

Chapter Summary

Essential Idea: Heat can be used to do work.
In this chapter we looked at the connection between heat, work, and the change in internal energy, and we saw how the laws of thermodynamics can be applied to understand the basic operation of practical devices such as engines, refrigerators, and air conditioners.

The First Law of Thermodynamics
The first law of thermodynamics is a statement of energy conservation as it relates to a thermodynamic system. Heat, Q, can be transformed into (or come from) some combination of a change in internal energy, ΔE_{int}, of the system and the work, W, done by the system.

$$Q = \Delta E_{int} + W \ .$$ (Equation 15.1: **The First Law of Thermodynamics**)

Q is positive when heat is added to a system, and negative when heat is removed.

ΔE_{int} is positive when the temperature of a system increases, and negative when it decreases.

W is positive when a system expands and does work, and negative when the system is compressed.

Work, and Change in Internal Energy
The work done by a system during a thermodynamic process is equal to the area under the curve corresponding to that process on a P-V diagram.

Change in internal energy: If the temperature of an ideal gas changes the change in internal energy of the gas is proportional to the change in temperature.
$$\Delta E_{int} = nC_V \Delta T \ .$$ (Equation 15.4: **Change in internal energy**)

Heat Capacity
C_V is known as the heat capacity at constant volume.

Monatomic: $C_V = \dfrac{3}{2}R$ Diatomic: $C_V = \dfrac{5}{2}R$ Polyatomic: $C_V = 3R$

$C_P = R + C_V$ is the heat capacity at constant pressure.

Monatomic: $C_P = \dfrac{5}{2}R$ Diatomic: $C_P = \dfrac{7}{2}R$ Polyatomic: $C_P = 4R$

Special Thermodynamic Processes (see also Figure 15.14 in Section 15-5)
1. **An isochoric (constant volume) process.** There is no work done by the gas: $W = 0$. The heat added to the gas is equal to the change in internal energy: $Q = \Delta E_{int} = nC_V \Delta T$.

2. **An isobaric (constant pressure) process.** The work done by the gas is $W = P \Delta V$.

3. **An isothermal (constant temperature) process.** There is no change in internal energy, $\Delta E_{int} = 0$, so $Q = W$. The work done by the gas is given by:

$$Q = W = nRT \ln\left(\frac{V_f}{V_i}\right)$$ (Eq. 15.8: **Heat and work for an isothermal process**)

4. **An adiabatic process.** Such a process is characterized by no heat transfer: $Q = 0$. On the P-V diagram an adiabatic process moves along a line given by:

$$PV^\gamma = \text{constant} \qquad \text{(Eq. 15.9: \textbf{Equation for an adiabatic process on the P-V diagram})}$$

$$\gamma = \frac{C_P}{C_V} \qquad \text{(Equation 15.10: \textbf{The constant } \gamma \text{ \textbf{for an adiabatic process}})}$$

Another useful equation in an adiabatic situation is: $T_f V_f^{\gamma-1} = T_i V_i^{\gamma-1}$. (Eq. 15.11)

Tools for Solving Thermodynamics Problems

- The P-V diagram can help us to visualize what is going on. In addition, the work done by a gas in a process is the area under the curve defining that process on the P-V diagram.
- The ideal gas law, $PV = nRT$, and the first law of thermodynamics, $Q = \Delta E_{int} + W$.
- The general expression for the change in internal energy, $\Delta E_{int} = nC_V \Delta T$.
- In specific special cases (see Figure 15.14, and the preceding section in this summary), there are additional relationships that can be used to relate the different parameters.

A Thermodynamic Cycle

A cycle is a sequence of processes that returns a system to its original state. The cycle as a whole satisfies the first law of thermodynamics, as does each of its processes. The change in internal energy for any cycle is always zero, because the system returns to its initial state, and the area of the enclosed region on the P-V diagram is the net work done in the cycle.

Entropy and the Second Law of Thermodynamics

Entropy is in some sense a measure of disorder. The symbol for entropy is S, and the units are J/K. Entropy is sometimes called time's arrow, because entropy tends to increase with time.

$$\Delta S \approx \frac{Q}{T_{av}}, \qquad \text{(Equation 15.13: \textbf{Approximate change in entropy})}$$

where T_{av} is the average temperature of the system while the heat is being transferred.

$$\Delta S \geq 0. \qquad \text{(Equation 15.14: \textbf{The Second Law of Thermodynamics})}$$

The entropy of a closed system tends to increase as time goes by.

Heat Engines

A heat engine, such as a car engine, uses a thermodynamic cycle to do work. Cooling devices such as refrigerators and air conditioners are heat engines run in reverse, having work done on them to pump heat from a cooler place to a warmer place.

$$Q_H - Q_L = W. \qquad \text{(Eq. 15.15: \textbf{Energy equation for a heat engine or cooling device})}$$

$$e = \frac{W}{Q_H} = \frac{Q_H - Q_L}{Q_H} = 1 - \frac{Q_L}{Q_H}. \qquad \text{(Equation 15.16: \textbf{Efficiency for a heat engine})}$$

$$\frac{T_L}{T_H} = \frac{Q_L}{Q_H}. \qquad \text{(Equation 15.17: \textbf{The Carnot relationship for an ideal engine})}$$

$$e_{ideal} = 1 - \frac{T_L}{T_H}. \qquad \text{(Equation 15.18: \textbf{Efficiency for an ideal engine})}$$

End-of-Chapter Exercises

Exercises 1 – 12 are conceptual questions that are designed to see if you have understood the main concepts of the chapter.

1. Two cylinders of ideal gas are initially identical (same pressure, volume, number of moles, and temperature) except that, in cylinder 1, the piston is free to move up or down without friction while, in cylinder 2, the piston is fixed in place so the volume occupied by the gas is constant. The same amount of heat is then added to both cylinders. Rank the cylinders based on (a) the final temperature of the gas, (b) the change in internal energy for the gas, and (c) the final pressure of the gas.

2. A system of ideal gas can be taken from state 3 to state 4 by three different paths labeled A, B, and C on the P-V diagram in Figure 15.22. Rank these paths based on (a) the work done by the gas; (b) the change in internal energy of the gas; (c) the heat transferred to the gas.

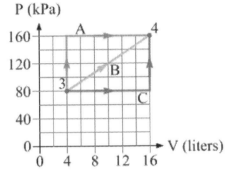

Figure 15.22: A system is taken from state 3 to state 4 by one of three possible paths. For Exercise 2.

3. Consider two systems, each containing the same number of moles of ideal gas at the same initial pressure, volume, and temperature. In system A, the temperature is then doubled by adding heat at constant volume. In system B, the temperature is doubled by adding heat at constant pressure. (a) Under these conditions, which system would you expect to need more heat to be transferred to it, to achieve the doubling of the temperature? Why? (b) It turns out that exactly the same amount of heat is added to each system to double the temperature. Upon closer inspection, you find that the gases are different in the two systems. In what way are they different?

4. A system of monatomic ideal gas is in a particular initial state. You now have a choice between three different processes, each of which double the volume occupied by the gas. Process A is an isothermal (constant temperature) expansion, process B is an adiabatic expansion, and process C is an isobaric (constant pressure) expansion. Rank these three processes, from most positive to most negative, based on (a) the work done by the gas, (b) the change in internal energy of the gas, and (c) the heat transferred to the gas during the process.

5. Is it possible to accomplish any of the following with a sample of ideal gas? If so, explain how you would do it. If not, explain why not. (a) Add heat to a system without changing the system's temperature. (b) Increase the temperature of a system without adding heat. (c) Add heat to a system while decreasing the temperature. (d) Change the temperature of a system that is so well-insulated that no heat is transferred between the system and its surroundings.

6. A system of ideal gas is in a particular initial state, and you will add a particular amount of heat Q to the system.
   Process A: Add the heat at constant temperature.
   Process B: Add the heat at constant volume.
   Process C: Add the heat at constant pressure.
   Which process should you use if you want the system to end up (a) doing the most work? (b) with the maximum possible final temperature? (c) with the maximum possible final volume? (d) with the maximum possible final pressure?

7. Return to the situation described in Exercise 6. Rank the three processes based on (a) the work done by the gas in the process. (b) the final temperature of the system. (c) the final volume of the system. (d) the final pressure of the system.

8. A system of ideal gas is in a particular initial state, and you want the system to do a particular amount of work, W. You have three different processes to choose from. Process A is a constant pressure process; process B is done at constant temperature; while process C is an adiabatic process. (a) Sketch a P-V diagram showing the three different processes. Which process should you use if you want the system to end up (b) using the smallest amount of heat? (c) with the maximum possible final temperature? (d) with the maximum possible final volume? (e) with the minimum possible final pressure?

9. Return to the situation described in Exercise 8. Rank the three processes based on (a) the heat required by the process. (b) the final temperature of the system. (c) the final volume of the system. (d) the final pressure of the system.

10. A system of ideal gas is in a particular initial state, and you want to double the absolute temperature of the system. You have three different processes to choose from. Process A is a constant pressure process; process B is done at constant volume; while process C is an adiabatic process. (a) Sketch a P-V diagram showing the three different processes. Which process should you use if you want the system to end up (b) with the maximum possible final volume? (c) with the minimum possible final pressure?

11. Return to the situation described in Exercise 10. Rank the three processes based on (a) the heat required by the process. (b) the final temperature of the system. (c) the final volume of the system. (d) the final pressure of the system.

12. On a hot day, you decide to open your fridge door to be cooled by the cold air inside the fridge. In the short term, this is effective for cooling you down. If you left the fridge door open for a long time, however, with the refrigerator running, would the room end up warmer, cooler, or the same temperature as it was when you first opened the door? Explain.

Exercises 13 – 16 deal with heat, change in internal energy, and work, the three parameters that are connected by the first law of thermodynamics.

13. Two cylinders of ideal gas are initially identical (same pressure, volume, number of moles, and temperature) except that in cylinder 1 the piston is free to move up or down without friction, while in cylinder 2 the piston is fixed in place so the volume occupied by the gas is constant. The temperature in each cylinder is initially 50°C. Each cylinder is then placed in a container of ice water that is maintained at 0°C, and allowed to come to equilibrium. (a) Which cylinder has more heat removed from it in this process? (b) Which cylinder experiences the larger change in internal energy in this process? Justify your answers.

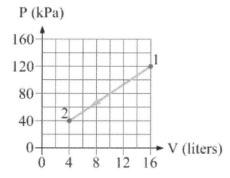

14. (a) How much work is done by the gas in the process shown in Figure 15.23? (b) If the gas is monatomic, what is the change in internal energy associated with the process? (c) How much heat is involved in this process? (d) Is heat transferred into the gas or transferred out of the gas?

Figure 15.23: A system is taken from state 1 to state 2 by the process shown. For Exercise 14.

15. A system of ideal gas can be taken from state 3 to state 4 by three different paths labeled A, B, and C on the P-V diagram in Figure 15.24. Assume the gas is diatomic. For each of the three paths, calculate (a) the work done by the gas; (b) the change in internal energy of the gas; (c) the heat transferred to the gas.

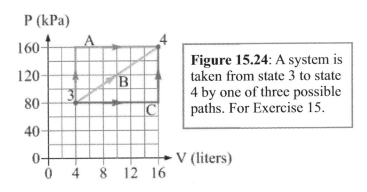

Figure 15.24: A system is taken from state 3 to state 4 by one of three possible paths. For Exercise 15.

16. A system of monatomic ideal gas contains a certain number of moles of gas so that $nR = 0.200$ J/K . The system is taken from state 5 to state 6 (see Figure 15.25) by an unknown sequence of processes. (a) What is the temperature of the system in state 6? (b) What is the work done by the gas as the system moves from state 5 to state 6? (c) What is the change in internal energy of the gas as it moves from state 5 to state 6?

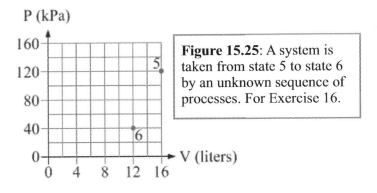

Figure 15.25: A system is taken from state 5 to state 6 by an unknown sequence of processes. For Exercise 16.

Exercises 17 – 21 deal with constant volume and constant pressure processes.

17. A cylinder of diatomic ideal gas is sealed in a cylinder by a piston that is fixed in place so the gas occupies a constant volume of 2.0 L. The pressure is initially 100 kPa, and the temperature is 400 K. The temperature is then decreased to 200 K by means of a constant volume process. (a) Sketch this process on a P-V diagram, and mark the final state of the system accurately on the diagram. (b) Find the work done by the gas in this process. (c) Find the change in internal energy of the gas in this process. (d) Find the amount of heat removed from the gas in this process. (e) Find the final internal energy of the gas.

18. Repeat Exercise 17, but this time the gas is monatomic instead of diatomic.

19. Repeat Exercise 17, but this time the piston is free to move without friction, and the decrease in temperature from 400 K to 200 K is done via a process in which the pressure remains constant at 100 kPa.

20. Four different processes are shown on the P-V diagram in Figure 15.26. The processes are carried out on a system of diatomic ideal gas. Rank these processes, from most positive to most negative, based on (a) the work done by the gas during the process, and (b) the change in internal energy of the gas during the process. Express your rankings in a form like 2>1=3>4.

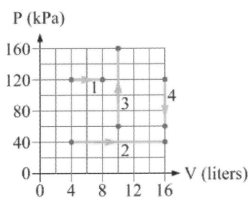

Figure 15.26: A P-V diagram showing four different processes, for Exercises 20 and 21.

21. Return to the situation described in Exercise 20 and shown in Figure 15.26. (a) For the process in which the work done by the gas is most positive, calculate the work. (b) For the process in which the internal energy of the gas increases by the largest amount, calculate that change in internal energy.

Exercises 22 – 27 deal with constant temperature and adiabatic processes.

22. A cylinder of monatomic ideal gas is sealed in a cylinder by a piston. The gas occupies a volume of 2.0 L. The pressure is initially 100 kPa, and cylinder is placed in an oven that maintains the temperature at 400 K. 100 J of work is then done on the piston, compressing the gas (in other words, the gas does –100 J of work). The work is done very slowly so that the gas maintains a constant temperature. (a) Sketch this process on a P-V diagram, shading in the region corresponding to the work done by the gas. (b) Find the change in internal energy of the gas in this process. (c) Find the final volume occupied by the gas. (d) Find the final pressure of the gas.

23. Repeat Exercise 22, but now the compression of the system is done very quickly so there is no time for heat to be transferred between the system and its surroundings. In other words, the process is adiabatic.

24. A system of diatomic ideal gas is in an initial state such that the pressure is 80 kPa and the volume occupied by the gas is 6.0 L. The system then experiences a compression at constant temperature that raises the pressure to 150 kPa. (a) Find the final volume occupied by the gas. (b) Find the work done by the gas in this process. (c) Find the ratio of the final temperature to the initial temperature.

25. Repeat Exercise 24, but now the compression that raises the pressure of the system to 150 kPa is an adiabatic process.

26. A system of monatomic ideal gas is in an initial state such that the pressure is 120 kPa and the volume occupied by the gas is 3.0 L. The system then experiences an expansion at constant temperature that increases the volume occupied by the gas to 8.0 L. (a) Find the final pressure of the gas. (b) Find the work done by the gas in this process. (c) Find the ratio of the final temperature to the initial temperature.

27. Repeat Exercise 26, but now the expansion that increases the volume to 8.0 L is an adiabatic process.

Exercises 28 – 31 deal with thermodynamic processes in general. Make use of the various tools for solving thermodynamic problems, outlined in section 15-5.

28. A system of diatomic ideal gas with an initial temperature of 500K is taken from an initial state *i* to a final state *f* by the process shown on the P-V diagram in Figure 15.27. Determine (a) the work done by the gas in the process. (b) the final temperature of the system. (c) the change in internal energy of the gas. (d) the heat associated with the process.

29. A system of monatomic ideal gas has an initial pressure of 100 kPa, a temperature of 400K, and a number of moles such that the product $nR = 2.0$ J/K. 1000 J of heat is then added to the system, while the pressure remains constant. Determine (a) the work done by the gas in the process, and (b) the final temperature of the gas. (c) Sketch an accurate P-V diagram for this process.

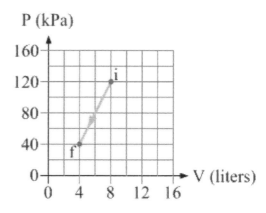

Figure 15.27: A process carried out on a system of diatomic ideal gas, for Exercise 28.

30. A system of monatomic ideal gas has an initial volume of 12 liters, a temperature of 600K, and a number of moles such that the product $nR = 0.50$ J/K. 200 J of heat is then removed from the system at constant temperature. Determine (a) the work done by the gas in the process, and (b) the final volume occupied by the gas. (c) Sketch an accurate P-V diagram for this process.

31. A system of diatomic ideal gas is taken through the process shown in Figure 15.28. The process has the shape of a ¼-circle on the P-V diagram. For this process, find (a) the work done by the gas, (b) the change in internal energy, and (c) the heat added to the gas.

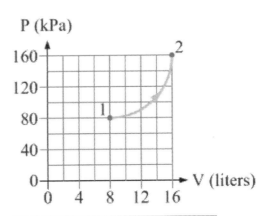

Figure 15.28: P-V diagram for a process taking a diatomic ideal gas from state 1 to state 2, for Exercise 31.

Exercises 32 – 38 deal with thermodynamic cycles.

32. Complete Table 15.6, where the process taking the system from state 2 to state 3 is isothermal.

Process	Q (J)	W (J)	ΔE_{int} (J)
1 → 2			-300
2 → 3	+200		
3 → 1		+100	
Entire Cycle	+500		

Table 15.6: A table for a three-process cycle, for Exercise 32.

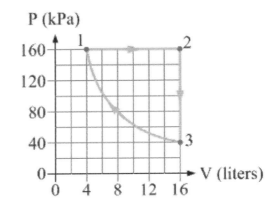

Figure 15.29: A three-process cycle, for Exercises 33 and 34.

33. Consider the P-V diagram for a three-process cycle in Figure 15.29. The cycle consists of a constant pressure process, a constant volume process, and an isothermal process at a temperature of 500 K. The system consists of a particular amount of a diatomic ideal gas. For this cycle, construct a table like that shown in Table 15.6.

34. Consider again the situation described in Exercise 33, and shown in Figure 15.29. What happens to all the values in the table when the cycle is carried out in the direction opposite to that shown in the P-V diagram, going from state 1 to 3 to 2 to 1?

35. Consider the following four-process cycle that is carried out on a system of monatomic ideal gas, starting from state 1 in which the pressure is 120 kPa and the volume is 4.0 liters. Process A is a constant-pressure process that doubles the volume; process B is an isothermal process, doubling the volume from the previous state; process C is a constant volume process that returns the system to its initial temperature; and process D is an isothermal process that returns the system to state 1. (a) Sketch a P-V diagram for this cycle. (b) Calculate the net work done by the gas in this cycle.

36. Consider the following four-process cycle that is carried out on a system of monatomic ideal gas, starting from state 1 in which the pressure is 80 kPa and the volume is 4.0 liters. Process A is an isothermal process that triples the volume; process B is a constant volume process that returns the system to a pressure of 80 kPa; process C is an isothermal process that returns the system to a volume of 4.0 liters; and process D is a constant volume process that returns the system to state 1. (a) Sketch a P-V diagram for this cycle. (b) For this cycle, construct a table like that shown in Table 15.6.

37. Consider the four-process cycle shown in the P-V diagram in Figure 15.30. (a) What does the area of the enclosed region, shown in red, represent? (b) What is the area of one of the boxes on the P-V diagram? (c) Estimate how many such boxes it would take to equal the area of the enclosed region. (d) Estimate the work done by the gas in one cycle.

38. Consider the four-process cycle shown in the P-V diagram in Figure 15.30. Find the work done by the gas in the process taking the system from (a) state 1 to state 2, (b) state 2 to state 3, (c) state 3 to state 4, (d) state 4 to state 1. For the entire cycle determine (e) the net work done by the gas, (f) the net change in internal energy of the gas, and (g) the net heat added to the gas.

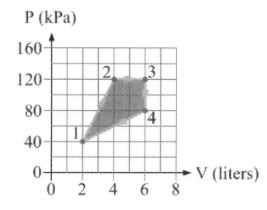

Figure 15.30: The P-V diagram for a four-process cycle, for Exercises 37 and 38.

Exercises 39 – 45 deal with entropy, the second law of thermodynamics, and heat engines.

39. You have two containers of water. In one container there is 1.0 kg of water at 7.0°C, while in the second container there is 5.0 kg of water at 27°C. You then pour the water in the first container into the second container and allow the system to come to equilibrium. Assuming no heat is transferred from the water to the container or the surroundings, determine the change in entropy for (a) the water that is initially cooler, (b) the water that is initially warmer, and (c) the system.

40. You have two containers of water. In one container there is 5.0 kg of water at 7.0°C, while in the second container there is 1.0 kg of water at 27°C. You then pour the water in the first container into the second container and allow the system to come to equilibrium. Assuming no heat is transferred from the water to the container or the surroundings, determine the change in entropy for (a) the water that is initially cooler, (b) the water that is initially warmer, and (c) the system.

41. In Section 15-8, we discussed the Carnot cycle, the cycle for a Carnot engine. Let's now examine the Stirling cycle, the cycle for a Stirling engine. The four processes are:
    Process A: Expansion at constant temperature T_H.
    Process B: Removing heat at a constant volume V_2.
    Process C: Contraction at constant temperature T_L (say $T_H = 2T_L$).
    Process D: Adding heat at constant volume V_1 (say $V_2 = 2V_1$).

    Process D returns the system to its initial state. (a) Sketch a P-V diagram for this cycle. (b) Create a table to show, for each process and the entire cycle, whether the heat, work, change in internal energy, and change in entropy, is positive, negative, or zero.

42. Return to the situation described in Exercise 41. Let's say the system uses monatomic ideal gas, that $V_1 = 8.0\,\text{L}$, $T_L = 300\,\text{K}$, and the pressure in the initial state is $P_1 = 150\,\text{kPa}$. (a) Determine the net work done by the engine in one cycle. (b) Calculate the efficiency of the engine by dividing the net work done by the total amount of heat added to the system during processes A and D. (c) Compare your answer in part (b) to the ideal efficiency of an engine operating between temperatures of T_L and T_H.

43. A particular car engine operates between temperatures of $T_H = 400°C$ (inside the cylinders of the engine) and $T_L = 20°C$ (the temperature of the surroundings). (a) Given these two temperatures, what is the maximum possible efficiency the car can have? (b) With this maximum efficiency, determine how much heat must be obtained by burning fuel in the car's engine to accelerate a 1500 kg car from rest to a speed of 100 km/h (neglect resistive forces).

44. A refrigerator is characterized by a number called the coefficient of performance (COP), which is defined as the heat removed from the cool place (inside the fridge) divided by the work required to remove that heat: $COP = Q_L / W$. (a) If you are buying a new fridge is it better to buy one with a higher COP or a lower COP? Let's say that a fridge operates between temperatures of $T_L = 4°C$ and $T_H = 30°C$. A particular fridge has a COP of 6.0. (b) If 500 J of work is done by the compressor in the fridge to extract heat from inside the fridge, how much heat is removed from inside the fridge? (c) How much heat is dumped by the fridge into the surroundings, in this case? (d) What is the maximum possible COP that would be achieved by an ideal fridge operating between the two temperatures specified above?

45. An air conditioner and a refrigerator operate on the same principles, so the definition of coefficient of performance given in the previous exercise applies to air conditioners as well as refrigerators. Let's say a particular air conditioner has a COP of 7.5 when operating between temperatures of $T_L = 20°C$ (inside your house) and $T_H = 30°C$ (outside your house). (a) Is this an ideal device? Explain. (b) If the outside temperature increased to 35°C, would you expect this to change the coefficient of performance? Explain.

General problems and conceptual questions

46. In Section 15-2, we calculated the work done by the gas, and the change in internal energy of the gas, for a system of diatomic ideal gas that went from state 1 to state 2 and then to state 3 (see Figure 15.6). Determine the heat Q associated with the process that takes the system from (a) state 1 to state 2 (b) state 2 to state 3.

47. In section 15-2, we examined a system of diatomic ideal gas that was moved from state 1 to state 3 via state 2. Let's now bring the same system from state 1 to state 3 via state 4, as shown in Figure 15.31. Draw your own version of the P-V diagram, and then shade in the region that corresponds to the work done in moving the system from (a) state 1 to state 4 (b) state 4 to state 3. (c) Calculate the work done by the gas in these two processes.

48. Consider the system of diatomic ideal gas described in Exercise 47, and the processes shown on the P-V diagram in Figure 15.31. For the two processes shown, calculate (a) the change in internal energy of the gas, and (b) the heat added to the gas.

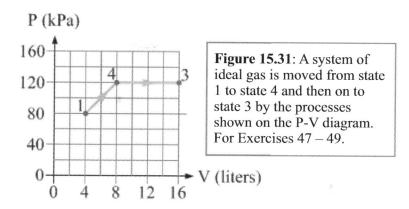

Figure 15.31: A system of ideal gas is moved from state 1 to state 4 and then on to state 3 by the processes shown on the P-V diagram. For Exercises 47 – 49.

49. Re-do Exercise 48. This time, the gas is monatomic instead of diatomic.

50. A system of monatomic ideal gas is moved from state 1 to state 2 by the process shown in Figure 15.32. (a) Determine the work done by the gas in this process. (b) Determine the change in internal energy for this process. (c) Determine the amount of heat associated with this process, and state whether heat was transferred into the gas or out of the gas.

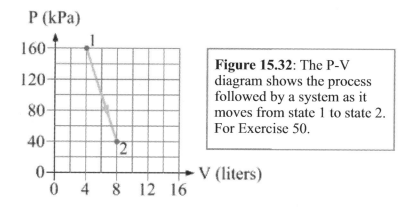

Figure 15.32: The P-V diagram shows the process followed by a system as it moves from state 1 to state 2. For Exercise 50.

51. A system of monatomic ideal gas is initially in state *i*, as shown in Figure 15.33. Compare the following situations, in which the system starts in the initial state *i* in each case. In case 1, the system's volume is increased by 4.0 L in a constant-pressure process. In case 2, the volume is decreased by 4.0 L in a constant-pressure process. (a) Sketch these two cases on a P-V diagram. (b) In which case is the magnitude of the work done by the gas larger? Explain your answer. (c) Calculate the work done by the gas in case 2.

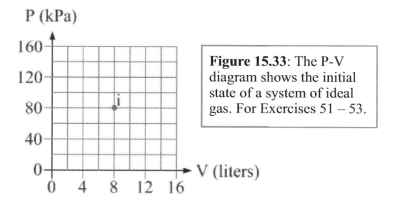

Figure 15.33: The P-V diagram shows the initial state of a system of ideal gas. For Exercises 51 – 53.

52. Repeat Exercise 51, but this time the changes in volume occur at constant temperature instead of at constant pressure.

53. Repeat Exercise 51 but, this time, the changes in volume have magnitudes of 2.0 L and occur during adiabatic processes.

54. (a) Why does a bicycle pump get hot as you pump it? (b) The pump is first filled with air, which we can approximate as a diatomic ideal gas, at room temperature and atmospheric pressure. You then quickly compress the gas, increasing the pressure by a factor of 5. Estimate the final temperature of the gas in this process.

55. A system of ideal gas has an initial pressure of 100 kPa and occupies a volume of 8.0 liters. Doubling the system's absolute temperature by means of a constant pressure process would require an amount of work W. Instead, you decide to double the absolute temperature by carrying out two processes in sequence, a constant volume process followed by a constant pressure process. Find the final pressure and volume of the system if the total work done in the sequence of processes is (a) $W/2$, or (b) $1.5\ W$.

56. Repeat Exercise 55 but, this time, the sequence of processes used to double the temperature is done in reverse order. First, you do a constant pressure process, and then you do a constant volume process.

57. Consider what happens to entropy when a house is built. The construction involves taking a large number of bricks, pieces of wood, nails, cans of paint, wires, pipes, etc., and putting them together into a well-ordered system. Does this process violate the second law of thermodynamics? Explain.

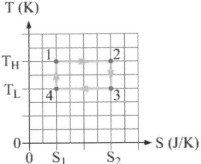

58. Return to the description of the Carnot cycle in Section 15-8. The P-V diagram for the Carnot cycle can be hard to follow, because none of the processes is a straight line. Let's consider the T-S (temperature-entropy) diagram for the Carnot cycle instead, shown in Figure 15.34. For the Carnot cycle, the T-S diagram is a rectangle. (a) The area under the curve on a P-V diagram represents the work done by the gas. What does the area under the curve represent on the T-S diagram? (b) The area of the region enclosed by a cycle on a P-V diagram represents the net work done by the gas in one cycle. What does the area of the region enclosed by a cycle on the T-S diagram represent?

Figure 15.34: T-S (temperature-entropy) diagram for the Carnot cycle.

59. Ocean Thermal Energy Conversion (OTEC) generates electricity using the difference between the temperatures at the surface of the ocean, where the water is warm, and far below the surface, where the water is cool. A test system off Hawaii operated between temperatures of 80°F (the water temperature at the surface) and 40°F (the temperature far below the surface), and operated with an efficiency of about 2.3%. (a) What is the maximum efficiency we could expect from this system? (b) At the maximum efficiency, let's say the plant generates electricity at the rate of 100 MW. At what rate is energy being removed from the warm surface water? (c) At what rate is energy being deposited into the cooler sub-surface water?

60. Do some research on the Ocean Thermal Energy Conversion process described in Exercise 59, and write a paragraph or two describing how it works. Comment, in particular, on whether this idea is feasible for generating electricity, as well as on the environmental impact of such a system.

61. Comment on each statement in the following conversation between two students.

Jesse: This question talks about a thermodynamic system that is so well insulated that no heat is transferred between the system and the surrounding environment. That must mean that the temperature of the system stays constant, right?

Scott: I think so. That's the way it works in your house - the more insulation you have, the more constant the temperature of your house is.

Jesse: So, because the temperature is constant, there is no change in internal energy, and with no heat transferred there must also be no work done, right?

Answers to selected problems from Essential Physics, Chapter 1

1. (a) 1 mile (b) 20 miles (c) 80 km

3. (a) 256 cm (b) 2000 cm^2

5. You agree on (a) and (b), but you disagree on (c), (d), and (e).

7. 4 m in the positive y-direction.

11. (a) 3.16×10^7 s (b) 9.46×10^{15} m (c) 1600 m (d) 180 kg (e) 42195 m

13. 10 grams

15. (a) 1 hectare (b) 100 m (c) 500 m

17. (a) 6.3 L/s. US gallons are smaller than imperial gallons, and we assume we use US gallons in this situation.

19. First, we use the inverse tangent to find that vector A makes an angle of 21.8° with the x-axis. Second, we know that vector B makes an angle of 63.8° with the x-axis, so we can determine that the angle between the vectors A and B is 63.8° − 21.8° = 42.0°. Third, we can determine that the magnitude of vector A is $\sqrt{29}$ m. At this point, we can use the cosine law to find the magnitude of vector C:

$$C^2 = A^2 + B^2 - 2AB\cos\theta_C$$
$$= (29 \text{ m}^2) + (16 \text{ m}^2) - 2 \times (\sqrt{29} \text{ m}) \times (4 \text{ m}) \times \cos(42.0°)$$

This gives the magnitude of C as 3.60 m.

Now, we can use the sine law to determine θ_B, the angle between the vectors A and C.

$$\frac{\sin\theta_B}{B} = \frac{\sin\theta_C}{C} \quad \Rightarrow \quad \theta_B = \sin^{-1}\left(\frac{B\sin\theta_C}{C}\right) = \sin^{-1}\left(\frac{(4.00 \text{ m})\sin(42.0°)}{3.60 \text{ m}}\right) = 48.0°$$

Finally, we can find the angle between C and the x-axis, which is 48.0° − 21.8° = 26.2°. So, the vector C has a magnitude of 3.60 m and is at an angle of 26.2° below the x-axis.

21. (a) 30 m at an angle of 36.9° above the positive x-axis.
(b) 52 m at an angle of 87.8° above the negative x-axis.
(c) 52 m at an angle of 87.8° below the positive x-axis.

23. (a) 4.2 m at an angle of 45° above the negative x-axis.
(b) 3.2 m at an angle of 18.4° above the positive x-axis.
(c) 2.0 m in the negative x direction.

25. (a)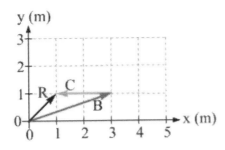

(b) $+(1.0 \text{ m})\hat{x}+(1.0 \text{ m})\hat{y}$

27. (a) + (b)

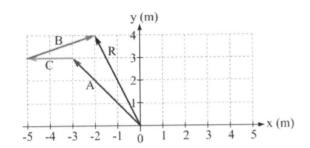

(c) + (d)

Vector	*x*-component	*y*-component
A	−3 m	+3 m
B	+3 m	+1 m
C	−2 m	0
R=A+B+C	−2 m	+4 m

(e) 4.5 m at an angle of 63.4° above the negative *x*-axis.

29. (a) + (b)

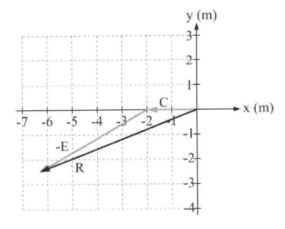

(c) + (d)

Vector	*x*-component	*y*-component
C	−2 m	0
−E	−4.33 m	−2.5 m
R=C−E	−6.33 m	−2.5 m

(e) 6.81 m at an angle of 21.6° below the negative *x*-axis.

31. (a) + (b)

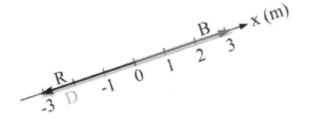

(c) + (d)

Vector	*x*-component	*y*-component
B	$+\sqrt{10}$ m	0
D	$-\sqrt{40}$ m	0
R=B+D	−3.16 m	0

(e) 3.16 m in the negative *x*-direction.

35. 425 m at an angle of 41.7° north of west

37. 62.5 miles is awfully precise – it has three significant figures, which seems like too many. Most likely, Nike said that the shoe was good up to 100 km (note that this has 1 significant figure), and somewhere along the line the distance was converted to miles using the approximation that 1 km is 5/8 of a mile, with three significant figures (too many) being kept in the number printed in the paper.

41. (a) Anya could first travel $+(3 \text{ blocks})\hat{x} + 0\hat{y}$, followed by $0\hat{x} + (4 \text{ blocks})\hat{y}$ (or the reverse order). (b) As the crow flies, her net displacement has a magnitude of 5 blocks, but she traveled a distance of 7 blocks. (c) $(+2.4 \text{ blocks})\hat{x} + (1.8 \text{ blocks})\hat{y}$, followed by $(-2.4 \text{ blocks})\hat{x} + (3.2 \text{ blocks})\hat{y}$ (or the reverse order). (d) $0\hat{x} + (5 \text{ blocks})\hat{y}$.

43. (a) 4500 m   (b) 2800 m   (c) She did not travel in a straight-line path between her starting point and her current location. Her route could have done a lot of zig-zagging, for instance.

45. (a) $x = 4$ (b) $x = 2$

47. (a) Equation 1 is incorrect. It has units of length on the left, and inverse length on the right. A valid equation has the same units on both sides. Equation 2 is dimensionally correct, with units of length on the right and left. Equation 3 is incorrect – in the numerator on the right side, units of length are being subtracted from units of length squared. This is not allowed – quantities that are added or subtracted must have the same units. (b) No, dimensional analysis alone does not guarantee that Equation 2 is correct (although, as we will see in the optics section, it is correct). By simply doing dimensional analysis, we cannot be sure of the + and – signs, and we also don't know if there are any numerical factors missing from the equation. For example, the following equation is also dimensionally correct, but it is not the correct equation:

$$d_i = \frac{3 d_o f}{d_o + f}.$$

Answers to selected problems from Essential Physics, Chapter 2

1. (a) 4 > 1=2=3   (b) 1=2=3=4

3. (a) This motion could be for you as you walk along a sidewalk. Let's say you're walking away from your house. The graph starts with you 20 m away from your house. For the first 5 seconds, you are tying your shoe, so you remain at rest. Then, you walk slowly, away from your house, for the next 10 seconds to a street, and then walk quickly for 5 seconds across the street. Just as you reach the far side, you realize that you forgot your iPod, so you decide to go back home to get it. You have to wait for 10 seconds for some cars to go by, and then you walk quickly 60 meters back home again.
    (b) You're on the subway in Boston, which is notoriously slow, For the first 5 seconds, the train moves slowly along the track. It then accelerates, first gradually, then more quickly, until the train's speed is 6 m/s. It keeps moving at that speed for 10 seconds, until the driver sees a red signal light on the track up ahead. The driver lets the train coast to a stop, coming to rest right by the signal.

5. We don't even know in which direction the object was moving. At the very least, it would be helpful to have each of the positions of the object labeled with the time the object passed through these positions. It would also be helpful to know how far apart the black lines are in the diagram.

7. (a) 2.0 m (b) 2.0 m (c) 12 m (d) +(2.0 m)$\hat{x}$ (e) −(2.0 m)$\hat{x}$ (f) +(2.0 m)$\hat{x}$

9. (a) Any constant-velocity motion, such as traveling at constant velocity along a straight highway in a car. (b) Not possible.

13. 16.8 m/s

15. (a) Kirsty Coventry. This is generally true in a race like this – the winner covers the distance in the least time, so the winner has the largest average speed.
    (b) Aya Terakawa. At the instant Kirsty Coventry touches the wall, her net displacement is very close to zero, while Aya Terakawa's has a magnitude of a few meters at that time. The average velocity is the net displacement divided by the time, so Aya Terakawa's average velocity has the larger magnitude at that time.

17. (a) 4.8 m/s.

19. (a) −(20 m)$\hat{x}$ (b) 100 m

21. (a)

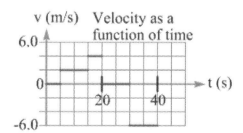

(b) 0   (c) +(1.6 m/s)$\hat{x}$

(d) −(6 m/s)$\hat{x}$

(e) +(0.29 m/s)$\hat{x}$

23. (a) $x = +42.5$ m, $v = +3$ m/s, $a = +0.2$ m/s^2
    (b) $x = +115$ m, $v = +6$ m/s, $a = 0$
    (c) $x = +167.5$ m, $v = +3$ m/s, $a = -0.6$ m/s^2

25. (a) 155 m   (b) $+(155$ m$)\hat{x}$ (c) 3.88 m/s   (d) $+(3.88$ m/s$)\hat{x}$

27. (a) We can't say.   (b) The acceleration is directed to the right.
    (c) If the motion is left-to-right, it could be a car accelerating from rest with constant acceleration. If the motion is right-to-left, it could be a car slowing down to rest at a red light.

29. (a)

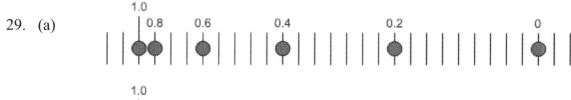

    (b)

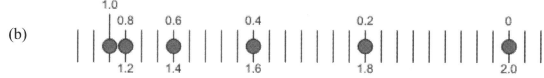

        The object is at the same positions on the way back.

    (c) $a = 10$ m/s^2, to the right   (d) 10 m/s, directed left   (e) 5 m/s, directed left

31. (a)

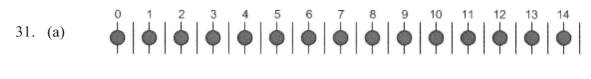

    (b)

33. (a) 0 (b) 1.33 m/s^2, to the right   (c) 0 (d) 0.4 m/s^2, to the right

35.

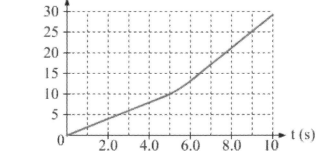

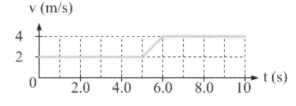

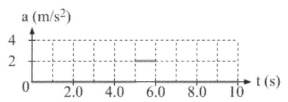

37.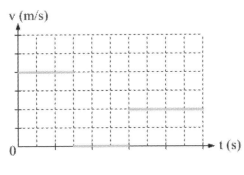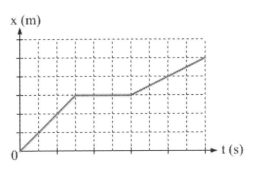

39. (a) 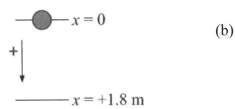 $x = 0$

$+$

$x = +1.8$ m

(b)

Parameter	Value
Initial position	$x_i = 0$
Final position	$x_f = +1.8$ m
Initial velocity	$v_i = 0$
Final velocity	$v_f = ?$
Acceleration	$a = +10$ m/s^2
Time	$t = ?$

(c) $v_f^2 = v_i^2 + 2a(x_f - x_i)$

(d) 6.0 m/s down

(e) One possibility is $v_f = v_i + at$

(f) 0.60 s

41. (a)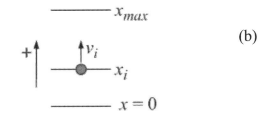

x_{max}

$+$ $\uparrow v_i$

x_i

$x = 0$

(b)

Parameter	Value
Initial position	$x_i = ?$
Maximum height	$x_{max} = ?$
Final position	$x_f = 0$
Initial velocity	$v_i = +12$ m/s
Velocity at maximum height	$v_{max} = 0$
Final velocity	$v_f = ?$
Acceleration	$a = -10$ m/s^2
Time	$t_f = 2.6$ s

(c) $x_f = x_i + v_i t_f + \frac{1}{2} a t_f^2$

(d) 2.6 m above the ground

(e) $v_{max}^2 = v_i^2 + 2a(x_{max} - x_i)$

(f) 9.8 m above the ground

43. (a)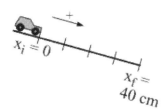

(b)

Parameter	Value
Initial position	$x_i = 0$
Final position	$x_f = +40$ cm
Initial velocity	$v_i = +20$ cm/s
Final velocity	$v_f = ?$
Acceleration	$a = ?$
Time	$t_f = 1.0$ s

(c) $x_f = x_i + v_i t_f + \frac{1}{2} a t_f^2$

(d) 40 cm/s² down the ramp

45. For this problem, we'll analyze half the motion, from the initial point to the point where the car is farthest from you. This takes exactly half of the time.

(a)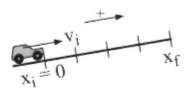

(b)

Parameter	Value
Initial position	$x_i = 0$
Final position	$x_f = ?$
Initial velocity	$v_i = +1.00$ m/s
Final velocity	$v_f = 0$
Acceleration	$a = ?$
Time	$t_f = 2.00$ s

(c) $x_f - x_i = v_{average} t_f$

(d) 1.0 m

47.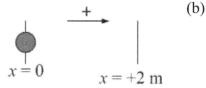

(b)

Parameter	Value
Initial position	$x_i = 0$
Final position	$x_f = 2.0$ m
Initial velocity	$v_i = 0$
Final velocity	$v_f = 150$ km/h $= 41.67$ m/s
Acceleration	$a = ?$
Time	$t_f = 0$

(c) 430 m/s²
(d) 0.096 s

49. Because the velocity of each car is constant, the acceleration is zero for all three cars.

a (m/s²)

[graph with vertical axis a (m/s²) marked 1, 2, 3 and horizontal axis t (s) marked 2.0, 4.0]

51. about 40 cm

55. (a) 0 (b) 10 m/s (c) 10 m/s^2 down

57. (a) equal for both (b) equal for both (c) 3H/4 above the ground
    (d)

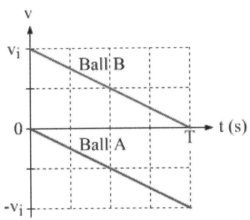

59.

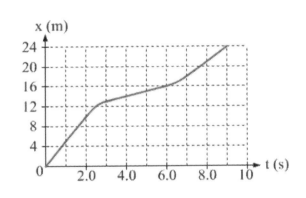

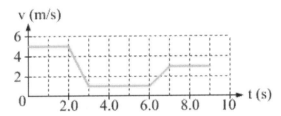

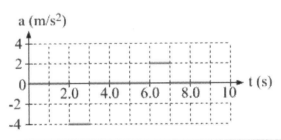

Answers to selected problems from Essential Physics, Chapter 3

1. FBD 1 is the correct free-body diagram in all five cases. As far as forces are concerned, "at rest" and "constant velocity" situations are equivalent.

3. (a) FBD 1 (b) FBD 3 (c) FBD 2

5. The penny and the feather fall at the same rate. In the absence of air resistance, the only force that matters is the force of gravity. Thus, the acceleration of both objects is the acceleration due to gravity, which is the same for any object.

7. (a) The net force on each of the blocks is zero. (b) Both block 2 and block 3 experience three forces. 1 – they both experience a force of gravity, applied by the Earth, which is the same for both blocks. 2 – They both experience a downward normal force, applied by the block above it. The downward normal force applied by block 2 on block 3 is larger than that applied by block 1 on block 2, however, because this normal force is equal to the weight of the blocks sitting on top of the block feeling the normal force. 3 – They both experience an upward normal force, applied by the object immediately below them. The normal force applied by the floor on block 3, however, has a magnitude equal to the weight of three blocks, while the upward normal force applied by block 3 on block 2 has a magnitude equal to the weight of two blocks.

9. (a) Any constant-velocity motion, such as traveling at constant velocity along a straight highway in a car, a hockey puck sliding across a frictionless ice surface, or a space probe drifting through space, trillions of kilometers from anything else. (b) Not possible. (c) This is possible as long as there are 2 or more forces acting on the object, and all the forces cancel. An example could be you sitting on a chair.

11. Yuri could throw the wrench directly away from the space station. To do this, he would exert a force on the wrench in the direction he throws the wrench. By Newton's third law, the wrench would exert an equal-and-opposite force back on Yuri. Once Yuri let go of the wrench, he would drift at constant velocity back toward the space station.

13.   15.

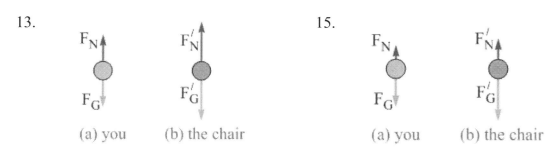

(a) you   (b) the chair   (a) you   (b) the chair

17.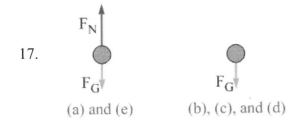

(a) and (e)   (b), (c), and (d)

(f) We just need two different free-body diagrams. In both (a) and (e), the ball's acceleration is directed up, so the upward normal force applied by your hand is larger than the downward force of gravity. In (b) – (d), the only thing acting on the ball is the force of gravity, and whether the ball is moving up, down, at instantaneously at rest is irrelevant.

19.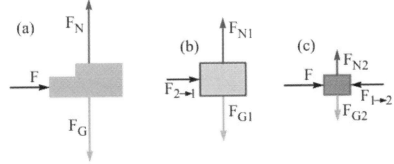

21.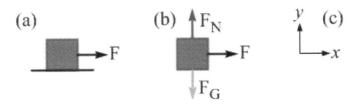

(d)
$$\Sigma \vec{F}_x = m\vec{a}$$
$$+6.0 \text{ N} = m \times (3.0 \text{ m/s}^2)$$

(e) 2.0 kg

23.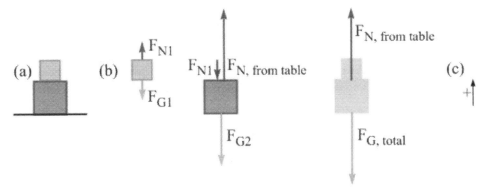

From the third FBD

(d) $\Sigma \vec{F} = (M + m)\vec{a} = 0$

$+F_{N,\text{from table}} - F_{G,\text{total}} = 0$

From the second FBD

$\Sigma \vec{F} = (M)\vec{a} = 0$

$+F_{N,\text{from table}} - F_{G2} - F_{N1} = 0$

(e) 10 N, directed down   (f) 50 N, directed up

25.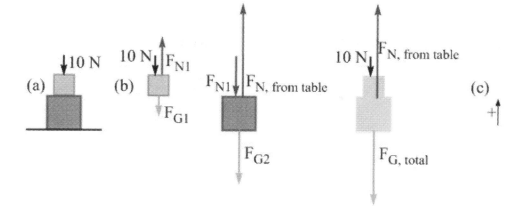

From the third FBD   From the second FBD

(d) $\Sigma \vec{F} = (M+m)\vec{a} = 0$ $\Sigma \vec{F} = (M)\vec{a} = 0$

$+F_{N,\text{from table}} - F_{G,\text{total}} - 10\text{ N} = 0$ $+F_{N,\text{from table}} - F_{G2} - F_{N1} = 0$

(e) 20 N, directed down   (f) 60 N, directed up

27.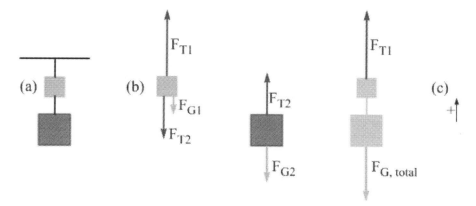

From the third FBD   From the second FBD

$\Sigma \vec{F} = (M+m)\vec{a}$ $\Sigma \vec{F} = (M)\vec{a}$

(d) $+F_{T1} - F_{G,\text{total}} = (M+m)\dfrac{g}{4}$ $+F_{T2} - F_{G2} = M\dfrac{g}{4}$

$+F_{T1} - (M+m)g = (M+m)\dfrac{g}{4}$ $+F_{T2} - Mg = M\dfrac{g}{4}$

$F_{T1} = \dfrac{5}{4}(M+m)g$ $F_{T2} = \dfrac{5}{4}Mg$

(e) 15 N   (f) 10 N

29.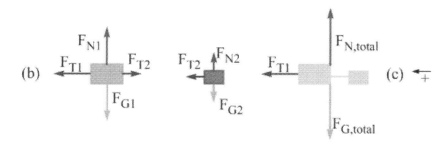

From the third FBD

$$\Sigma \vec{F} = (2m + m)\vec{a}$$

(d) $+30\ \text{N} = 3ma$

$$a = \frac{10\ \text{N}}{m}$$

From the second FBD

$$\Sigma \vec{F} = (m)\vec{a}$$

$$+F_{T2} = m\left(\frac{10\ \text{N}}{m}\right)$$

$$F_{T2} = 10\ \text{N}$$

(e) 10 N

31. (a) 20 m/s east   (b) 30 m/s east   (c) 0

33. (a) $3.3 \times 10^4\ \text{N}$ (b) 3800 N

35. 28.5 N

37. Up until $t = 5$ s, there is no net force applied to the object. For 1 s, between $t = 5$ s and $t = 6$ s, a constant force directed to the right is applied. After $t = 6$ s, no net force is applied.

39. 100 N

41. (a) The fifth floor. (b) 20% of g (a little under 2.0 m/s^2)

43. (a) 0.004 m/s^2 (b) 350 m/s (c) 2400 m/s

45. The maximum acceleration, of 3.6 m/s^2, is when all three forces are in the same direction. The minimum acceleration is zero, because the forces can be arranged so that they cancel out.

47. 27 N

49. (a) $8T/5$ (b) $2T$

51. (a) equal in the two cases   (b) $F/5$ in both cases (c) in case 2
    (d) (i) $F/2$ (ii) $4F/5$

53. (a) 1.0 N (b) 2.0 N (c) 5.0 N

55. (a) No, it's the same free-body diagram.   (b) No, it's also the same. In each case, the ball is being acted upon only by gravity.

57. (a) $(100\ \text{m/s}^2)\, m$ (b) 2 m/s, in the direction the puck was traveling initially

59. (a) $F_{Mm} = \dfrac{m}{M+m} F$ (b) $F_{Mm} = \dfrac{M}{M+m} F$ (c) $\dfrac{M}{m} = 4$

61. (a) The elevator is at rest for the first 2 seconds, and then accelerates down for 2 seconds. This is followed by a 4-second period of constant velocity, directed down, followed by a 2-second period of upward acceleration, while the elevator is moving down but in the process of coming to rest. The elevator remains at rest after that.

(b) 2.0 m/s^2 (c) The elevator moved down a total distance of 24 m.

Answers to selected problems from Essential Physics, Chapter 4

1. (a) Point your kayak perpendicular to the riverbank. To cross the river in the shortest time, you need to maximize the component of your velocity that is directed across the river. The current in the river affects only the component of your velocity that is directed parallel to the river, so the current can neither add to nor subtract from your velocity directed across the river. To maximize your velocity across the river, therefore, you must direct as much of your velocity relative to the water across the river, which means aiming your kayak perpendicular to the river.

(b) To land directly across from your starting point, you must aim the kayak upstream, so that the component of your velocity relative to the water that is directed parallel to the river exactly cancels the velocity of the water relative to the riverbank (the current, in other words).

3. Any origin can be used to answer the question. Assuming that you both use a traditional coordinate system with coordinate axes directed horizontally and vertically, you would agree on the value of the *x*-coordinate of the initial position, and the *x*-coordinate of the final position. You would disagree on the value of the *y*-coordinate of the initial position, and the *y*-coordinate of the final position, but the key thing is that you would agree on the magnitude of the displacement in the *y*-direction. You would also agree on the answer to the time it takes for the object to reach the water, assuming that you both did the calculations correctly.

5. (a) A=B=C   (b) A=B=C   (c) C>B>A   (d) C>B>A

7. (a) G>F>E   (b) G>F>E   (c) E>F>G   (d) G>F>E

9.

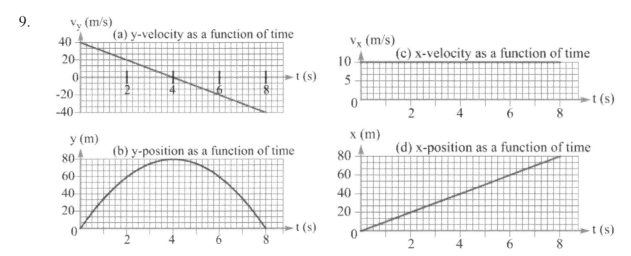

11. (a) $4H$ (b) $2T$ (c) $4R$

13. (a) 20 km/h north (b) 20 km/h south (c) 210 km/h north (d) 190 km/h south
    (e) The answers stay the same even after the vehicles change positions. Finding relative velocities simply involves adding or subtracting two velocity vectors, and the velocity vectors are the same no matter what the relative positions of the car, truck, and motorcycle are.

15. (a) 270 s (b) 279 s

17. (a) 3 m/s east (b) 1 m/s east (c) We can't say. All we have is information about the velocities of these people relative to one another. We have no information about the velocity of any of them with respect to the lamppost.

19. (a) 1.5 hours (b) 150 km/h west (c) You have to fly due east for 4.5 more hours.

21. (a) 2.0 s (b) 2.0 s (c) 2.0 m (east of the base of the mast)

23. (a) 1.8 m (b) 6.0 m/s, directed straight up (c) zero

25.

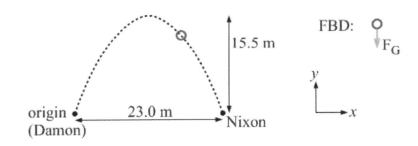

(e)

	x-direction	y-direction
initial position	$x_i = 0$	$y_i = 0$
at max. height		$y_{max} = +15.5$ m
final position	$x_f = +23.0$ m	$y_f = 0$
initial velocity	$v_{ix} = ?$	$v_{iy} = ?$
acceleration	$a_x = 0$	$a_y = -9.81$ m/s^2

(f) $v_{iy} = +17.44$ m/s; time of flight is 3.555 s; $v_{ix} = +6.469$ m/s.
This gives an initial velocity of 18.6 m/s at an angle of 69.6° above the horizontal.

27.

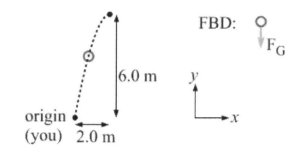

(e)

	x-direction	y-direction
initial position	$x_i = 0$	$y_i = 0$
final position	$x_f = +2.0 \text{ m}$	$y_f = +6.0 \text{ m}$
initial velocity	$v_{ix} = ?$	$v_{iy} = ?$
final velocity		$v_{fy} = 0$
acceleration	$a_x = 0$	$a_y = -9.8 \text{ m/s}^2$

(f) $v_{iy} = +10.84 \text{ m/s}$; time of flight is 1.11 s; $v_{ix} = +1.81 \text{ m/s}$.

This gives an initial velocity of 11.0 m/s at an angle of 80.5° above the horizontal.

29.

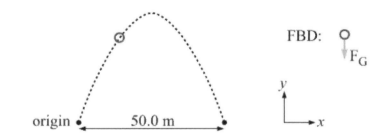

(e)

	x-direction	y-direction
initial position	$x_i = 0$	$y_i = 0$
final position	$x_f = +50.0 \text{ m}$	$y_f = 0$
initial velocity	$v_{ix} = ?$	$v_{iy} = ?$
acceleration	$a_x = 0$	$a_y = -9.81 \text{ m/s}^2$
time	4.56 s	4.56 s

(f) $v_{iy} = +22.37 \text{ m/s}$; $v_{ix} = +10.96 \text{ m/s}$.

This gives an initial velocity of 24.9 m/s at an angle of 63.9° above the horizontal.

31.

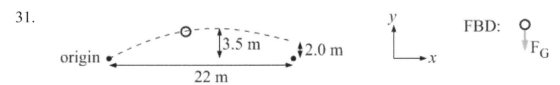

(e)

	x-direction	y-direction
initial position	$x_i = 0$	$y_i = 0$
max. height		$y_{max} = +3.5$ m
final position	$x_f = +22$ m	$y_f = +2.0$ m
initial velocity	$v_{ix} = ?$	$v_{iy} = ?$
acceleration	$a_x = 0$	$a_y = -9.8$ m/s^2

(f) $v_{iy} = +8.28$ m/s; $t = 1.40$ s; $v_{ix} = +15.73$ m/s.

(g)

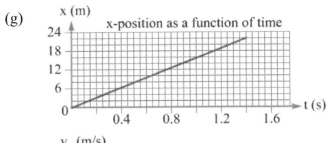

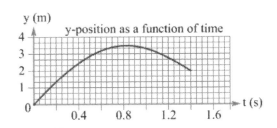

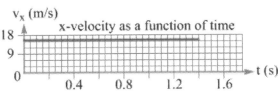

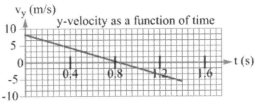

(h)

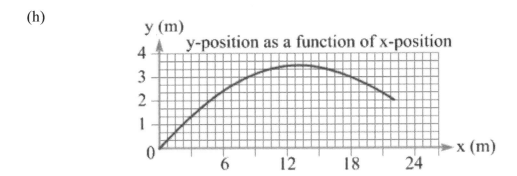

33.

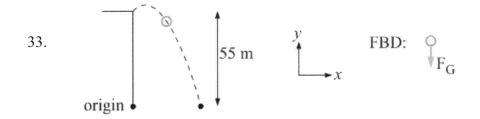

(e)

	x-direction	y-direction
initial position	$x_i = 0$	$y_i = +55.0$ m
final position	$x_f = ?$	$y_f = 0$
initial velocity	$v_{ix} = +(12.0 \text{ m/s})\cos(34.0°)$ $v_{ix} = +9.948$ m/s	$v_{iy} = +(12.0 \text{ m/s})\sin(34.0°)$ $v_{iy} = +6.71$ m/s
acceleration	$a_x = 0$	$a_y = -9.81 \text{ m/s}^2$

(f) $t = 4.10$ s   (g) 40.8 m

35.

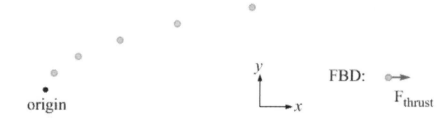

origin

FBD: F_{thrust}

(e)

	x-direction	y-direction
initial position	$x_i = 0$	$y_i = 0$ m
final position	$x_f = ?$	$y_f = ?$
initial velocity	$v_{ix} = 0$	$v_{iy} = +5.0$ m/s
final velocity (acceleration phase)	$v_{fx} = a_x t = +20$ m/s	$v_{fy} = +5.0$ m/s
acceleration	$a_x = +4.0 \text{ m/s}^2$	$a_y = 0$

(f) 56 m   (g) 160 m

37. (a) 19.6 m/s, up   (b) 12.5 m/s, directed downfield
    (c) 23.2 m/s, at an angle of 57.5° above the horizontal

39. 10.2°

41. 12 m

43. (a) 125 km/h at an angle of 28.6° west of north (b) 138 km/h at an angle of 24.8° west of north

45. Powell must have been running at about 9.4 m/s when he took off. World-class sprinters run at about 10 m/s, so Powell's speed is just a little less than that of a world-class sprinter. One part of being a long-jumper, in fact, is excellent speed.

47. 71.6°

49. (a) The ball was caught at a higher level than where it was launched. You can tell this from the asymmetry in the graph of the y-component of the velocity. If the ball came down to the same level from which it was launched, the final speed would have matched the initial speed. Because the final speed is lower than the initial speed, the ball must be higher than the launch point. (b) 30 m (c) At $t = 2.0$ s. The ball's y-component of velocity is zero at the highest point, and $t = 2.0$ s is when the y-velocity passes through zero in this case.

51.

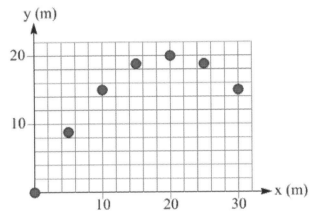

53. With these numbers, the ball is 2.8 m below the launch point.

55. (a) The shortest time is zero. The longest time is 4.0 seconds. (b) 20 m (c) 40 m

57. (a) 13.9 m/s (b) 16.0 m/s (c) You can't do it! A launch angle of 45° gets the closest to your friend, but the range in that case is 19.6 m, a little short of reaching your friend. You can't reach your friend at any angle unless you increase the speed of the ball from what it is in (a).

59. 24.1° and 65.9°

61. (a) 1.24 m (b) 1.35 m (c) 1.22 m

Answers to selected problems from Essential Physics, Chapter 5

1. (a) FBD 3 (b) FBD 3 (c) FBD 4 (d) FBD 1 (e) FBD 2 (f) FBD 5

3. (a) Mg is the force of gravity, applied on the box by the Earth. F_N is the normal force, which is the perpendicular component (perpendicular to the surfaces in contact) of the contact force applied on the box by the ramp. F_S is the force of static friction, the parallel component of the contact force that acts to keep the box at rest on the ramp. F_K is the force of kinetic friction, the parallel component of the contact force that opposes the motion if the box is sliding up or down the ramp. (b) (i) FBD 4 (ii) FBD 5 (iii) FBD 1.

5. (a) FBD 4 (b) FBD 4 (c) FBD 4 (d) FBD 4, although the force of static friction would have to be larger in this case than it is in cases (a) – (c), to give a net force acting up the hill. In the other three cases, the net force is zero. (e) FBD 5

7.

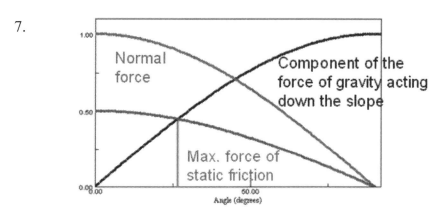

(a) See the graph in red. Note that the *y*-axis is in units of mg. (b) The graph could also represent the component of the force of gravity that acts perpendicular to the slope, and the maximum possible force of static friction if the coefficient of static friction equals 1.0. (c) See the graph in black. (d) That graph can also represent the force of static friction needed to act on the box to prevent it from slipping. (e) See the blue graph. (f) The box starts to slide when the angle of the incline is just to the right of where the blue line meets the black line (see the vertical orange line on the graph above). Beyond that angle, the force of friction needed to keep the box at rest is larger than the maximum possible force of friction.

9. (a) equal in both cases (b) equal in both cases (c) both forces have a magnitude of 8.0 N (d) We can conclude that the coefficient of static friction between the box and the surface is at least 0.4.

11. If the turntable rotates at constant angular velocity (that is, the rate at which the turntable spins is constant), then the force of static friction gives rises to the centripetal acceleration. If the turntable is speeding up or slowing down, then the component of the force of static friction that is directed toward the center of the turntable gives rise to the centripetal acceleration.

13.

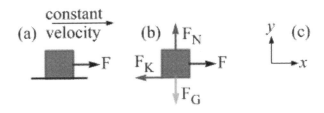

(a) constant velocity →F

(b) F_N ↑ ; F_K ← block →F ; F_G ↓

(c) y ↑ → x

$$\sum \vec{F}_y = m\vec{a}_y = 0 \qquad\qquad \sum \vec{F}_x = m\vec{a}_x = 0$$

(e) $$+F_N - mg = 0 \qquad\qquad +F - F_K = 0$$

$$F_N = mg \qquad\qquad F = F_K = \mu_K F_N = \mu_K mg$$

(f) $\mu_K = \dfrac{F}{mg} = \dfrac{10\text{ N}}{20\text{ N}} = 0.5$

15.

(a) block with F at 45°, acceleration →

(b) F_N ↑, F at angle, F_G ↓

(c) y ↑ → x

(d) F_N ↑, $F\sin(45°)$ ↑, $F\cos(45°)$ →, F_G ↓

$$\sum \vec{F}_y = m\vec{a}_y = 0 \qquad\qquad \sum \vec{F}_x = m\vec{a}_x$$

(e) $$+F_N + F\sin(45°) - mg = 0 \qquad\qquad +F\cos(45°) = ma$$

$$F_N = mg - F\sin(45°) \qquad\qquad F = ma / \cos(45°)$$

(f) $a = 2.0$ m/s^2 (g) 7.1 N   (h) 19.5 N

17.

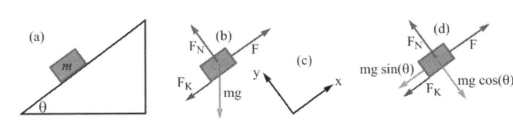

(a) inclined plane with block m at angle θ

(b) F_N ↑, F →, F_K ←, mg ↓

(c) y, x axes

(d) F_N, F, $mg\sin(\theta)$, F_K, $mg\cos(\theta)$

(e)

$$\sum \vec{F}_y = m\vec{a}_y = 0 \qquad\qquad \sum \vec{F}_x = m\vec{a}_x = 0$$

$$+F_N - mg\cos\theta = 0 \qquad\qquad +F - F_K - mg\sin\theta = 0$$

$$F_N = mg\cos\theta \qquad\qquad F = F_K + mg\sin\theta$$

(f) $F = \mu_K mg\cos\theta + mg\sin\theta = (0.2)(20\text{ kg})(9.8\text{ m/s}^2)\cos 30° + (20\text{ kg})(9.8\text{ m/s}^2)\sin 30°$

$F = 132$ N

19.

(a)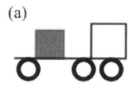

(b)

(c)

$$\Sigma \vec{F}_y = m\vec{a}_y = 0$$

(e) $+F_N - mg = 0$

$F_N = mg$

$$\Sigma \vec{F}_x = m\vec{a}_x$$

$F_S = ma$

(f) The force is a static force of friction, acting in the direction of the truck's acceleration, with a magnitude of 6.0 N.

(g) 3.9 m/s^2 (h) We don't need to know the coefficient of kinetic friction

21. (a)

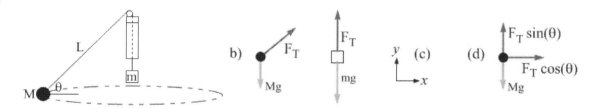

$$\Sigma \vec{F}_y = M\vec{a}_y = 0$$

(e) $+F_T \sin\theta - Mg = 0$

$F_T \sin\theta = Mg$

$$\Sigma \vec{F}_x = M\vec{a}_x$$

$F_T \cos\theta = M\dfrac{v^2}{r}$

$F_T \cos\theta = M\dfrac{v^2}{L\cos\theta}$

$$\Sigma \vec{F} = m\vec{a} = 0$$

$F_T = mg$

(f) $F_T = mg$ (g) $v = \sqrt{\dfrac{gL\cos^2\theta}{\sin\theta}}$ or $v = \sqrt{\dfrac{mgL\cos^2\theta}{M}}$

23. 40 N

25. (a) 47 N, at an angle of 58° with respect to the horizontal. (b) 47 N (c) 1.6 N

27.

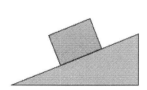

(a) Using $g = 9.8$ m/s^2, the contact force exactly balances the force of gravity, and is thus 49 N directed straight up. (b) 47 N (c) 13 N.

29. (a) 6.3 N, down the slope (b) 16 N, in the direction of the normal force
    (c) 5.8 N, so a component of this force is directed down the slope

31. 4.1 m

33. This is true when the force has a downward vertical component.
    (b) $\tan\theta_c = \mu_S$ (c) 26.6°

35. (a) 8.5 N   (b) We can say that the coefficient of static friction between the box and the surface is at least 0.35.

37. (a) 0.8 m (b) 0.4 s (c) 0.89 s

39. (a) 2.1 m/s² (b) 3.0 m/s²

41. 2.9 m/s²

43. 2.1 m/s²

45. There is only one solution, which is $\mu_K = 0.33$

47. (a) 2.0 m/s² , directed down the ramp (b) 25 N
    (c) The minimum value of F is 6.67 N, while the maximum value is 55.6 N.

49. (a) 60 N   (b) 45 N   (c) 105 N

51. (a) 295 N (b) 590 N

53. (a) 2.7 m/s² , directed left (b) 5.3 m/s² , directed left
    (c) 5.3 m/s² , directed right (d) 2.7 m/s² , directed right

55. (a) 2.0 m/s² , directed right (b) 4.0 N, directed right (c) 32 N, directed left

57. $v = \sqrt{gr\tan\theta} = 26.7$ m/s

59. (a) The horizontal component of the normal force.
    (b) $F_N = \dfrac{mg}{\sin\theta}$ (c) $v = \sqrt{\dfrac{gr}{\tan\theta}}$

Answers to selected problems from Essential Physics, Chapter 6

1. B > C > A

3. It does not make any difference. As far as you are concerned, the situations are equivalent. It would make a difference if your evil twin was driving a car that had either more mass, or less mass, than your car. Because of the symmetry of this collision with your evil twin, however, the very front of your car stays at rest during both collisions, so the end result is the same, for you.

5. No, it does not. A good example is something circular, like a car tire, a donut, or a bagel. For those objects, the center of mass is at the center of the hole, where there is no actual material.

7. (a) $1 = 2 = 3 > 4$ (b) $2 > 1 > 4 > 3$ (c) $1 > 2 = 4 > 3$ (d) $1 > 2 > 4 > 3$

9. (a) $K_A = K_B$ (b) $v_A > v_B$ (c) $p_B > p_A$

11. (a) Yes. A good example is any uniform circular motion situation, in which the force changes the direction of the velocity but does not change its magnitude – the speed is constant. The momentum changes because momentum is a vector, and the direction of the vector changes. The kinetic energy stays the same, however. (b) No, this is not possible. We'll assume the mass of the object is constant, so changing the kinetic energy produces a change in speed. That produces a corresponding change in momentum, because the magnitude of the momentum is proportional to the speed.

13. (a) 480 kg m/s south (b) 960 kg m/s south (c) $480\sqrt{2} = 680$ kg m/s southeast

15. (a) 0.96 kg m/s to the right   (b) 0.64 kg m/s to the right

17. (a) 20 N east (b) 10 N west (c) 40 N west

19.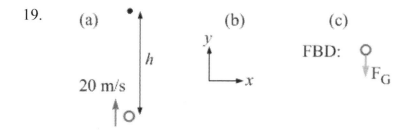

(d) 4.0 kg m/s directed up   (e) zero (f) 4.0 kg m/s directed down
(g) $mg = 2.0$ N directed down   (h) 2.0 s

21.

(a) 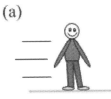 side view

(b) y x

(c) FBD: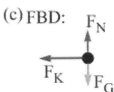

$F_{y, \text{racket}}$

$F_{x, \text{racket}}$

F_G

(d) 200 N, in the direction of the ball's velocity when it leaves the racket. (e) Yes. The force of gravity also acts on the ball over this time interval, so if the ball does not acquire a component of momentum directed down, the net vertical force on the ball during this time interval must be zero. The racket must exert an upward vertical force that exactly balances the force of gravity.

23. (a)

(b) y x

(c) FBD:

F_N

F_K F_G

(d) $\vec{F}_{net} = -\mu_K mg$ (e) $\Delta \vec{p} = m\Delta \vec{v} = m(v_f - v_i)$

(f) $\vec{F}_{net}\Delta t = \Delta \vec{p} \Rightarrow -\mu_K mg\Delta t = m(v_f - v_i)$

(g) $\mu_K = \dfrac{v_f - v_i}{-g\Delta t} = \dfrac{-2.0 \text{ m/s}}{-(10 \text{ m/s}^2)(2.0 \text{ s})} = 0.10$

25. (a) –16 J   (b) In the first 2 seconds, the displacement is +11 m. In the next 5 seconds, the displacement is +5 m. In the final second, the displacement is –3.5 m. This gives a net displacement of +12.5 m, so the cart is at $x = +12.5$ m at $t = 8$ s.

27. (a) at $t = 4$ s   (b) 10 m/s   (c) 5 m/s in the +x direction

29. (a) 7.5 m/s in the +x direction   (b) 6 m/s in the +x direction   (c) 0

31. (a) $X_{cm} = +0.3$ m ; $Y_{cm} = +0.7$ m   (b) The center of mass shifts toward the location of ball 3.

33. 1.0 m

35.   (a)   (b)   (c)

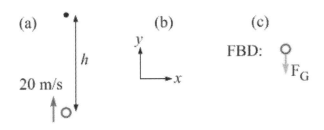

(a) h   20 m/s

(b) y x

(c) FBD: F_G

(d) 40 J   (e) zero   (f) –40 J   (g) $mg = 2.0$ N directed down   (h) 20 m

37. (a)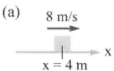

8 m/s

x = 4 m

(b) $\longrightarrow x$

(c)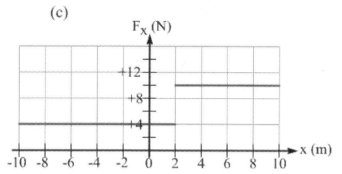

(d) 64 J

(e) 100 J

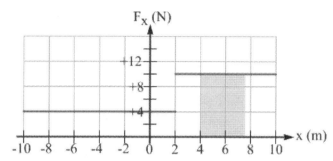

(f) +36 J

(g) at $x = +7.6$ m

(h) A speed of +2.0 m/s corresponds to a kinetic energy of 4.0 J. Thus, +60 J of work was done from the position where the object had 4.0 J to $x = +4$ m, where it has 64 J of kinetic energy. An area of +60 J on the graph takes the object back to a position of $x = -8$ m.

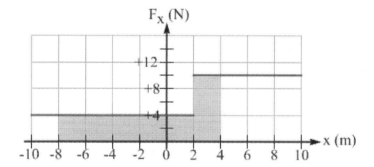

39.

(a)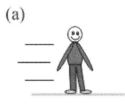

(b)

y

x

(c) FBD: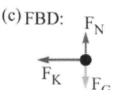

F_N

F_K F_G

(d) $\vec{F}_{net} = -\mu_K mg$ (e) $\Delta K = \frac{1}{2}mv_f^2 - \frac{1}{2}mv_i^2 = \frac{1}{2}m(v_f^2 - v_i^2)$

(f) $\vec{F}_{net}\Delta x = \Delta K \Rightarrow -\mu_K mg\Delta x = \frac{1}{2}m(v_f^2 - v_i^2)$

(g) $\mu_K = \frac{v_f^2 - v_i^2}{-2g\Delta x} = \frac{(4.0 \text{ m/s})^2 - (6.0 \text{ m/s})^2}{-2(10 \text{ m/s}^2)(5.0 \text{ m})} = 0.20$

41. (a) 7 m/s in the +x direction   (b) 1 m/s in the –x direction   (c) 5 m/s, with components of 3 m/s in the +x direction and 4 m/s in the +y direction. This is equivalent to 5 m/s at an angle of 37° above the positive x axis.

43. The x-component of the initial momentum is 2.0 kg m/s. The y-component of the initial momentum is 2.95 kg m/s. Thus, the net momentum has a magnitude of 3.56 kg m/s, at an angle of 55.9° above the horizontal.

45. (a) 18 m/s   (b) 50 m/s, to the right

47. (a) Equal in both cases   (b) +225 J   (c) More work is done in case 2   (d) No work is done in case 1, while +23.0 J of work is done in case 2.

49. (a) 1.0 s   (b) 4.0 m/s   (c) 2.5 s

(d)

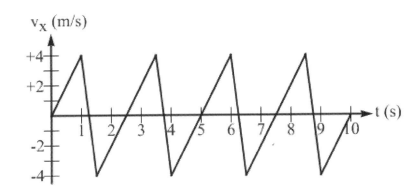

51. (a) 2.8 m   (b) 4.0 m

53. (a) The stopping time is unchanged   (b) The stopping time is doubled
    (c) The stopping time is halved

55. (a) 160 m, 4 times longer than the original distance   (b) The stopping time is twice that of the original case.

57. (a) $2T$ (b) $4T$

59. (a) The kilowatt-hour is a unit of energy.   (b) $1 \text{ kW h} = 3.6 \times 10^6 \text{ J}$ (c) About 55 cents.

61. (a) 3.6×10^{29} kg m/s   (b) The Sun does no work on the Earth – the Earth's kinetic energy is constant.

Answers to selected problems from Essential Physics, Chapter 7

1. It is more tiring to walk uphill for an hour than it is to walk over level ground, because when you walk uphill you are doing work against gravity.

3. The kinetic energy comes from the gravitational potential energy of the snow or mud, stored in the position of the snow or mud high up at the top of the slope.

5. The speeds are equal in cases B and C, and smaller in case A. The final kinetic energy is equal to the initial kinetic energy (which is zero in case A and equal to the same non-zero value in cases B and C) plus the difference in gravitational potential energy between the initial and final positions.

7. (a) Mechanical energy is not conserved. The rock does work on the can, crushing it. The rock's mechanical energy gets transformed into other forms of energy, such as thermal energy and sound energy. (b) Energy is conserved in this process. This is consistent with the law of conservation of energy. If you account for all the different forms of energy before and after the crushing of the can, there is the same total amount of energy beforehand and afterwards.

9. (a) 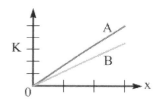 (b) 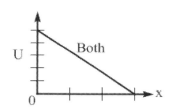 (c)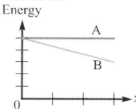

11. The energy is stored temporarily in the spring. The more the spring is compressed, the more energy is stored in it. So, as the block compresses the spring, the spring slows down, and the block's kinetic energy is transformed into potential energy in the spring. After the block comes momentarily to rest, the spring pushes on the block, transforming the potential energy back into the block's kinetic energy.

13. (a) 4.95 m/s (b) The ball reaches its maximum speed when it passes through its lowest point, immediately below the hook. (c) Because of energy conservation, the ball goes to 1.25 m higher than the equilibrium position. (d) The only numerical value we need is the vertical distance that the ball starts above equilibrium. The mass cancels out of the energy equation, and the string length is not needed.

15. (a) This is a reasonable assumption. A typical observation is that, in one swing, the ball reaches almost the same vertical position on one side that it started from on the other, meaning that only a small fraction of the ball's mechanical energy is removed in one swing. (b) After many swings, the ball comes to rest. All of the initial gravitational potential energy is converted to thermal energy because of the negative work done on the ball by the resistive forces. This work is negative because the resistive forces always oppose the ball's motion. (c) –2.45 J

17.

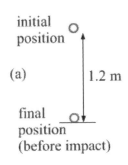

initial position

(a)   1.2 m

final position (before impact)

(b) $U_i + K_i + W_{nc} = U_f + K_f$, and define the zero level for gravitational energy to be at the level of the floor.

(c) $K_i = 0$, because the keys start at rest; $W_{nc} = 0$, because we assume that air resistance is negligible; $U_f = 0$, because the keys end up on the floor, which is where we defined the zero for gravitational potential energy to be.

Our equation thus becomes $U_i = K_f \implies mgh = \dfrac{1}{2}mv_f^2$

(d) $v_f = \sqrt{2gh} = 4.8$ m/s

19.

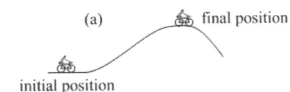

(a)   final position

initial position

(b) $U_i + K_i + W_{nc} = U_f + K_f$, and define the zero level for gravitational energy to be at the starting point.

(c) $K_i = 0$, because you start from rest; $U_i = 0$, because you start from the zero level for gravitational potential energy. Let's also assume that the number given in the problem for the work represents the net work done by non-conservative forces.

Our equation thus becomes $W_{nc} = U_f + K_f \implies 1.5\times10^4 \text{ J} = mgh + \dfrac{1}{2}mv_f^2$

(d) $h = \dfrac{1.5\times10^4 \text{ J} - \dfrac{1}{2}mv_f^2}{mg} = 22$ m

21.

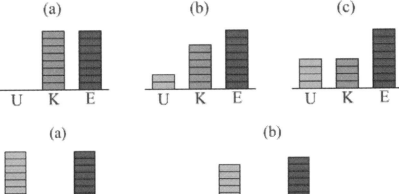

23.

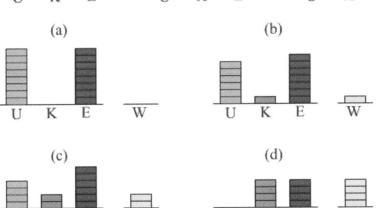

25.
(a)

Before   After

(b) $m(50 \text{ km/h}) + M(0) = (m+M)(10 \text{ km/h})$, with the positive direction being the direction of the car's velocity before the collision (which is the same direction as the velocity of the car and the truck after the collision).

(c) $M = \dfrac{m(50 \text{ km/h}) - m(10 \text{ km/h})}{10 \text{ km/h}} = 4m = 8000 \text{ kg}$

27.
(a)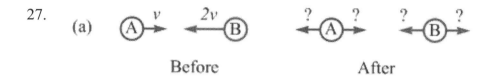

Before   After

(b) Let's define left as the positive direction, because the net momentum of the system is in that direction:
$-mv + 2mv = mv_{Af} + mv_{Bf}$

$v = v_{Af} + v_{Bf}$

(c) $v_{Bf} - v_{Af} = v_{Ai} - v_{Bi} = -3v$

(d) $2v_{Bf} = -2v \implies v_{Bf} = -v$ $\qquad\qquad v_{Af} = +2v$

Puck B is moving to the right after the collision, and puck A is moving left after the collision.

29. (a)

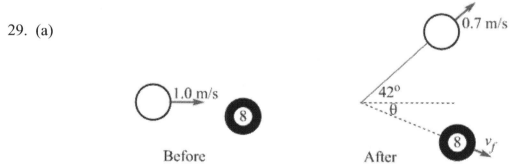

Before   After

(b)

	Cue ball (initial)	8-ball (initial)	Cue ball (final)	8-ball (final)
x-direction	$m \times 1.0 \text{ m/s}$	0	$m \times (0.70 \text{ m/s}) \cos(42°)$ $= m \times 0.520 \text{ m/s}$	$mv_f \cos\theta$
y-direction	0	0	$m \times (0.70 \text{ m/s}) \sin(42°)$ $= m \times 0.468$	$-mv_f \sin\theta$

(c) From the y-direction, $mv_f \sin \theta = m \times 0.468$ m/s $\Rightarrow v_f \sin \theta = 0.468$ m/s

From the x-direction, $mv_f \cos \theta = m \times 0.480$ m/s $\Rightarrow v_f \cos \theta = 0.480$ m/s

Dividing one equation by the other gives $\tan \theta = \dfrac{0.468}{0.480} \Rightarrow \theta = 44.3°$

Solving for v_f gives $v_f = 0.67$ m/s.

The velocity of the 8-ball after the collision is 0.67 m/s at an angle of 44° from the direction of the cue ball's velocity before the collision.

31. (a) $v_{2f} = \sqrt{2\mu_K g d} = 1.0$ m/s   (b) $h = 0.293$ m; $v_{1i} = \sqrt{2gh} = 2.42$ m/s;

$v_{1f} = v_{1i} - 2v_{2f} = 0.42$ m/s, in the same direction as the ball's velocity just before the collision.

33. (a) 4×10^{10} J   (b) For such a large object going that fast, neglecting air resistance is probably not a good idea. If air resistance could be neglected, the plane's engines could be switched off after reaching cruising altitude and the plane would continue at constant velocity after that. So, the energy is a good deal larger than the value we calculated in (a).

35. (a) Because mechanical energy (the sum of the potential and kinetic energies) is conserved in this situation, the mechanical energy at the beginning, at the lowest point, and in the final situation are all the same, so we can relate the energy at the beginning to the energy at the end.   (b) $v_f = \sqrt{2gh} = \sqrt{2 \times (9.8 \text{ m/s}^2) \times (1.0 \text{ m})} = 4.4$ m/s

37. (a) There must be a kinetic force of friction acting on the box. If there was no friction force, the box's speed would increase as it went down the incline. The kinetic force of friction must exactly balance the component of the force of gravity that acts down the slope.   (b) $\mu_K mg \cos \theta = mg \sin \theta \Rightarrow \mu_K = \tan \theta = 0.75$ (c) We don't need to know either the mass or the speed. The mass cancels out of the equation, and all we need to know about the velocity is that it is constant – the exact value does not matter.

39. (a) $v = \sqrt{2gh} = \sqrt{16 \text{ m}^2/\text{s}^2} = 4.0$ m/s

(b) $W_{nc} = \dfrac{1}{2}mv_f^2 - mgh = m\left(\dfrac{v_f^2}{2} - gh\right) = (0.050 \text{ kg})\left(\dfrac{(3.0 \text{ m/s})^2}{2} - (10 \text{ m/s}^2)(0.8 \text{ m})\right)$

$W_{nc} = -0.175$ J

(c) The work done by air resistance is negative, which makes sense because the force of air resistance is opposite in direction to the ball's displacement.

41. The motion would be similar, it would just happen more slowly than it would on the Earth. On the Moon, the block would reach the same positions on either side that it did on the Earth, and it would stop in the same place that it did on the Earth, but the speed during the motion would be smaller on the Moon than it is on the Earth.

43. For the first block, $mgh = K_1$. For the second block, $mgh + W_{nc} = 0.8K_1$. Combining these gives $W_{nc} = -0.2mgh$. However, the work done by non-conservative forces is the work done by the kinetic friction force, which is $W_{nc} = -\mu_K F_N d$, where $F_N = mg\cos\theta$, and d, the distance moved along the slope, is related to h by $d = h/\sin\theta$. Bringing everything together gives:

$-\dfrac{\mu_K mgh\cos\theta}{\sin\theta} = -0.2mgh$ which gives $\tan\theta = 1.25$, which gives an angle of 51°.

45. $m_B = \dfrac{0.5m_A v_f^2 + m_A g(2.0\text{ m})\sin\theta + \mu_K m_A g\cos\theta(2.0\text{ m})}{g(2.0\text{ m}) - 0.5v_f^2} = \dfrac{36 + 24 + 20}{20 - 18} = 40$ kg

47. (a) Energy conservation: $mgh = \dfrac{1}{2}mv_b^2 + 2mgR$

Forces and circular motion: $2mg = \dfrac{mv_b^2}{R}$

Combining these gives $h = 3R$.

(b) Energy conservation: $mgh = 3mgR = \dfrac{1}{2}mv_a^2$

Forces and circular motion: $F_N - mg = \dfrac{mv_a^2}{R}$

Combining these gives $F_N = 7mg$

49. Energy conservation: $mgh = \dfrac{1}{2}mv_b^2 + 2mgR$

Forces and circular motion: $F_N = 0 \implies mg = \dfrac{mv_b^2}{R}$

Combining these gives $h = 2.5R$.

51. (a) 6 m/s to the right
(b) The collision is inelastic. One way to justify this is to work out the elasticity, which is 0.5 in this case. That is less than 1, implying an inelastic collision. A second way to justify this is to work out the kinetic energy before and after the collision. There is a lot less kinetic energy in the system after the collision ($m\times 80$ m^2/s^2) than before ($m\times 200$ m^2/s^2), so the collision is inelastic. It is not completely inelastic, because the two pucks do not stick together after the collision.

53. 0.5v, directed left

55. (a) After the collision, the object of mass $3m$ is at rest, while the object of mass m is $-2v$ (b) $4h$

57. (a) Before the collision, the Jetta was traveling at 4/3 the speed of the Acura. The Jetta was traveling faster, in other words.

(b) Carrying out an energy analysis after the collision allows us to get a range of speeds for the cars immediately after the collision: $v = \sqrt{2\mu_K gd}$, so this speed is somewhere between 6.96 m/s and 7.41 m/s. Doing a momentum analysis, we can then find a range of speeds for the two cars after the collision. The range for the Acura is 8.61 m/s to 9.17 m/s, while the range for the Jetta is 11.5 m/s to 12.2 m/s. Converting these values to mph gives a range for the Acura of 19.3 mph to 20.5 mph and a range for the Jetta of 25.7 mph to 27.4 mph. Thus, the Jetta was clearly speeding, so that driver should get a speeding ticket. The Acura may or may not have been slightly speeding, so we can't really say that driver was entirely blameless. From these results, we can't say definitively that the driver of the Jetta was entirely to blame.

59. Yes, the result does match the result we got using the component method.

61. (a) (i) $v_i/5$ (ii) $4v_i/5$ (b) $4T$ and $16D$

Answers to selected problems from Essential Physics, Chapter 8

1. The unknown mass is larger than M. Both large objects are attracting the small object, but the net force is pointing more toward the ball with the unknown mass, so the unknown mass is exerting a large force on the small object than is the ball of mass M. The distances involved are the same, so the larger force requires a larger mass.

3. (a) More than M (b) $2M$

5. This is not possible. The two large masses would have to be repelling the smaller mass, but gravitational forces can only be attractive.

7. $\dfrac{Gm^2}{r^2}$, straight down

9. (a) 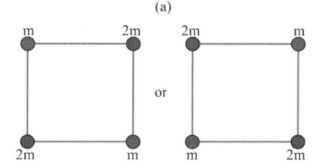 (b)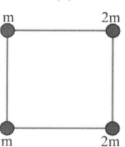

11. (a) Kinetic energy is conserved – the speed is constant. (b) Gravitational potential energy is conserved – the distance between the centers-of-mass of the objects remains the same. (c) Total mechanical energy is conserved – this is based on the fact that kinetic energy and potential energy are each conserved, and also because no non-conservative forces act. (d) Linear momentum of the smaller object is not conserved. The magnitude of the linear momentum stays the same, but its direction steadily changes because of the gravitational force exerted on it by the massive object.

13. (a) No, we can not arrange the balls so that all three simultaneously experience no force. With the balls placed in a line, the left-most ball will experience two forces directed right, and the right-most ball will experience two forces directed left. Thus, those two balls each have a net force acting on them. (b) If we place the third ball carefully, then the ball that is between the other two balls can experience no net force, with the forces it experiences from the other two balls canceling out.

15. $F_{2m} > F_m > F_{3m}$

17. $F_{Venus} > F_{Earth} > F_{Mercury} > F_{Mars}$

19. (a) zero   (b) $\dfrac{GM}{R^2}$, directed straight down

21. The two solutions are $\dfrac{m}{2}$ or $2.5m$.

23. The two solutions are $x = -\dfrac{a}{2}$ or $x = +\dfrac{a}{2\sqrt{2}}$

25. (a) $2 > 1 > 3$ (b) $-\dfrac{8Gm^2}{r}$

27. (a) $\dfrac{2Gm^2}{r^2}$, directly toward the ball of mass $4m$

(b) $\dfrac{\sqrt{2}Gm^2}{r^2}$, in the positive x direction

(c) $\dfrac{(1.40)Gm^2}{r^2}$, at an angle of $30.4°$ above the positive x-axis

29. (a) $\dfrac{2\sqrt{2}Gm^2}{r}$ (b) $\dfrac{2\sqrt{2}Gm^2}{r}$ (c) $\dfrac{\left(1+\sqrt{2}\right)Gm^2}{r}$

31. (a) At only one location.   (b) $x = 5a$

33. Most years have 365 days. However, it take the Earth about 365.25 days to travel once around the Sun, so every 4th year we add an extra day to the calendar so that we are, every 4th year, at the same point in the orbit on a particular time and day. The Earth's orbital period is actually slightly less than 365.25 days (currently 365.242374 days), so adding one day every 4th year is slightly more than we need to correct the problem. Thus, almost every 100 years, we don't add a 366th day to the calendar. Leap years are every year that is a multiple of 4, except for those that are multiples of 100 that are not multiples of 400. The years 1700, 1800, and 1900 were not leap years, for example, but the year 2000 was a regular leap year.

35. (a) 3×10^4 m/s   (b) 0.006 m/s^2 (c) The acceleration of the Sun because of the force the Earth exerts on it is about 1.8×10^{-8} m/s^2

37. (a) $F_{3m} > F_{2m} > F_m$ (b) $\sqrt{52}\ Gm^2$ in units of m^{-2}

39. (a) The ball of mass m at the top of the picture experiences a larger gravitational force than does the ball of mass m at the center. The forces on the ball at the center mostly cancel – the net force acting on the ball at the center is the same as the force the

ball at the center applies to the ball at the top. However, the ball at the top experiences three additional forces, all of which have a downward component, so the ball at the top experiences a larger net force. (b) The ball of mass $2m$ on the negative y-axis experiences a net force of magnitude $5.33\dfrac{Gm^2}{r^2}$. The other two balls of mass $2m$ experience forces of magnitude $5.17\dfrac{Gm^2}{r^2}$.

41. (a) $v = \sqrt{\dfrac{2GM_{moon}}{R_{moon}}} = \sqrt{\dfrac{2\left(6.67\times10^{-11}\ \text{Nm}^2/\text{kg}^2\right)\left(7.35\times10^{22}\ \text{kg}\right)}{1.737\times10^6\ \text{m}}} = 2380$ m/s

(b) There is some logic to this idea. For one thing, it would take a tremendous amount of fuel, and impressive engines, to launch a manned mission to Mars from the surface of the Earth, with the mass of the fuel adding significantly to the mass of the spacecraft. Ferrying everything to the Moon in a few trips, constructing the spacecraft there, and then blasting off from the Moon would take more total energy, but there are definite advantages in terms of the lower mass of the spacecraft and the smaller energy needed for any one trip. There are also obvious technological challenges associated with setting everything up on the Moon, however!

43. $v = \sqrt{\dfrac{2Gm}{L}}$

45. (a) They will not come back together. Because G is such a small number, the initial kinetic energy is a much larger number than the initial gravitational potential energy, which is negative, so the total mechanical energy is positive. That means that there is enough energy in the system for the objects to escape from one another.
(b) Only very slightly less than 0.10 m/s.

47. (a) $v = \dfrac{2\pi r}{T}$ (b) $\Sigma\vec{F} = ma \implies \dfrac{GmM}{r^2} = \dfrac{mv^2}{r}$

(c) First, in our equation in (b) we can cancel m, the mass of the orbiting object, as well as one factor of r, which gives us: $\dfrac{GM}{r} = v^2 = \dfrac{4\pi^2 r^2}{T^2}$.

Re-arranging this equation gives us the required result: $T^2 = \dfrac{4\pi^2 r^3}{GM}$

49. (a) 3.88×10^8 m
(b) 6.2×10^{24} kg (this is only slightly higher than the accepted value)

51. (a) 4.22×10^7 m

53.   There are two possible solutions. The mass of the second ball is $3m$, and its location is either at $x = -4a$ or at $x = +2a$.

55.   The mass of the second ball is $4m$, and its location is at $x = +2a$.

57.   (a) $\left(4 + 6\sqrt{2}\right)\dfrac{Gm^2}{L^2}$, toward the center of the square.

(b) $17.7\dfrac{Gm^2}{L^2}$, at an angle of $31.2°$ above the negative x-direction, defining the negative x-direction as pointing to the left.

59.   (a) 3.0×10^4 m/s   (b) The speed is independent of the mass of the orbiting object, so the speed would not change at all.

Answers to selected problems from Essential Physics, Chapter 9

1.   The two buoyant forces are equal. In both cases, the buoyant force acting on the block must balance the force of gravity acting on the block, so the buoyant force in both cases is equal to the force of gravity that acts on the block.

3.   (a) The aluminum ball has less mass than the steel ball, even though the balls are the same size, because aluminum is less dense than steel. (b) The balls displace equal volumes of water – because the balls are completely submerged, each ball displaces a volume of water that is equal to its own volume.

5.   (a) When you suck on the straw, you remove air from the straw, reducing the pressure inside the straw. This produces a pressure imbalance – it is really the effect of atmospheric pressure acting on the surface of the liquid that forces the fluid to flow up through the straw and into your mouth. You do not suck the liquid up – you just reduce the pressure in the straw, and atmospheric pressure pushes the fluid up the straw.
(b) Once again, atmospheric pressure is responsible for this. When you seal the top of the straw, the pressure at the top of the straw is equal to atmospheric pressure. When you remove the straw from the liquid, the fluid at the bottom of the straw is also exposed to atmospheric pressure. With the pressures equal at the top and bottom, there is nothing to balance the weight of the column of liquid in the straw, so some of the liquid leaks out. The air at the top of the straw, above the column of liquid, then occupies a larger volume, and when the volume increases, the pressure decreases. The pressure at the bottom is still atmospheric pressure, however. The liquid is in equilibrium when the force of gravity acting down on the liquid is balanced by the force coming from the pressure difference between the top and bottom, multiplied by the cross-sectional area of the straw.

7.   (a) C > B > A   (b) A = B = C

9. The scale reading increases. When you put your finger in the water, the water exerts an upward buoyant force on you. By Newton's third law, you exert an equal-and-opposite force down on the water, which is passed on by the water to the beaker, which passes it on to the scale.

11. This all comes down to pressure. When you lie on a bed of nails, your weight is distributed over a large number of nails, so the upward normal force exerted on you by any one nail is small, and the pressure associated with that force is also relatively small. If you tried to support yourself on a single nail, however, the nail would have to exert a force on you equal to your weight. That force would be exerted over a very small area, leading to a very high pressure, most likely high enough to break the skin and cause plenty of pain!

13. (a) 300 kg/m^3 (b) 1500 kg/m^3

15. The ice will float higher in the glass. When the oil is poured in, the ice displaces some of the oil, so the oil exerts an upward buoyant force on the ice. This means that the water exerts a smaller buoyant force on the ice than before, which requires the ice to displace less water – the ice must be floating higher in the glass than before to displace less water.

17.

(a)   (b)   (c) $\sum \vec{F} = m\vec{a} = 0$
$+F_B - mg = 0$
$F_B = mg$

(d) $F_B = 8.0 \text{ N}$

19.

(a)   (b)   (c) $\sum \vec{F} = m\vec{a} = 0$
$+F_B - mg = 0$
$F_B = mg$

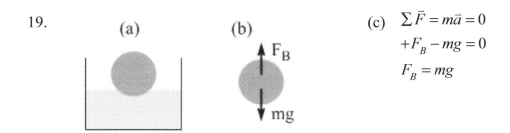

(d) $5.0 \times 10^{-4} \text{ m}^3$ (e) $5.5 \times 10^{-3} \text{ m}^3$ (f) $V = \frac{4}{3}\pi r^3$ (g) 11 cm

(h) There is no need to use a particular value of g – the value of g cancels out in the equation.

21.

(a)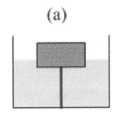

(b)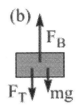

(c)
$$\sum \vec{F} = m\vec{a} = 0$$
$$+F_B - F_T - mg = 0$$
$$F_B = mg + F_T$$

(d) 75%   (e) 10%

23.

(a)

(b)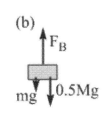

(c) Each styrofoam "shoe" has a weight of mg, and supports half of Mg, the weight of the person.
$$\sum \vec{F} = m\vec{a} = 0$$
$$+F_B - 0.5Mg - mg = 0$$
$$F_B = mg + 0.5Mg$$

(d) 8.4×10^{-2} m^3 (e) Estimating a shoebox to be 30 cm × 15 cm × 15 cm, with a volume of 6.75×10^{-3} m^3, each Styrofoam "shoe" is as big as 12 regular shoeboxes.

25.

(a)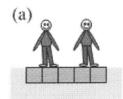

(b)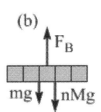

(c) Assume the raft (mass m) is fully submerged, and that there are n kids standing on the raft, each with a mass M.

$$\sum \vec{F} = m\vec{a} = 0$$
$$+F_B - nMg - mg = 0$$
$$F_B = mg + nMg$$
$$\rho_{water}Vg = \rho_{wood}Vg + nMg$$

(d) 22 kids

27. 5.1 cm

29. (a) The liquid will fall in the tube.   (b) 1.20 cm (c) 16.3 cm

31. (a) As the water drops, its speed increases. By the continuity equation, the product of the area and the speed must remain the same, so as the speed increases the cross-sectional area of the stream gets reduced. (b) 2.0 m/s (c) 0.10 cm^2

33. (a) $v = \sqrt{2gh}$ (b) $t = \sqrt{\dfrac{2(H-h)}{g}}$ (c) $\Delta x = vt = 2\sqrt{(H-h)h}$

35. (a) 116000 N/m^2 (b) 118000 N/m^2

37. (a) The wooden cube experiences a larger buoyant force – it is displacing more fluid. (b) 300 kg/m^3 (c) 39.2 N (d) 3700 kg/m^3

39. (a) The wooden block is less dense than the fluid, and it starts lower in the fluid than its equilibrium position, so the wooden block will accelerate upward. The metal cube is more dense than the fluid, so it will sink toward the bottom of the container. (b) The wooden block has an acceleration of 0.79 m/s^2 up, while the metal cube has an acceleration of 3.2 m/s^2 down. (c) After a long time, the wooden block floats at the surface, with 74% of its volume below the surface, while the metal cube is at rest on the bottom of the container.

41. 4.7 g

43. Using a constant density for air gives a height for the atmosphere of about 9 km. This is a very rough estimate, because the density of air is not constant with height – in general, the higher you go, the less dense the air is. Thus, because this calculation uses a value for density that is too large, the estimate for the atmosphere's height is too small – the number represents a lower bound on the height.

45. 1 atm = 14.7 psi (pounds per square inch) = 760 torr = 760 mm of mercury
    = 1.01325 bar

47. (a) The normal force is the same no matter what orientation the brick has. The normal force is 150 N. (b) The maximum pressure is 12300 N/m^2, while the minimum pressure is 3270 N/m^2.

49. The boiling point of water depends on pressure. At high altitude, the pressure is lower than it is at sea level, so water boils at a lower temperature than it does at standard atmospheric pressure. As any British person will tell you, when making a cup of tea it is important that the water be very hot – at high altitude, the boiling water simply does not get hot enough to make a good cup of tea.

51. (a) 1960 N/m^2. (b) 0 (c) –2940 N/m^2.

53. (a) 980 Pa (b) 41 m/s

55. 8.5 m

57. (a) 17 m/s. (b) 346000 N/m^2

59. The problem does not have a valid solution, with the numbers given. If we set the absolute pressure at the pump to be 1.5 atmospheres, however, we can get a value for the area of the tube, so let's do that. This gives 2.0 × 10^{-5} m^2

61. Jaime has the right ideas. The other two have some misconceptions about how it all works.

Answers to selected problems from Essential Physics, Chapter 10

1. (a) The red ones have the same speed as one another. The blue ones also have the same speed as one another, with a value twice the speed of the red ones. (b) None of them have the same velocity. Velocity is a vector, and the velocity of each object is in a different direction. (c) All four objects have the same angular velocity. (d) Once again, the magnitude of the acceleration of the two red objects is the same, and the blue objects also have the same magnitude acceleration as one another. However, none of the objects have the same acceleration because the directions of the accelerations are all different. (e) They all have the same angular acceleration, which is zero for all four objects.

3. (a) – (c) clockwise   (d) counterclockwise

5. The answer is Yes in all three cases. The pictures below show just one of many examples for each situation.

7. As the angle θ decreases, the torque associated with the force F must **stay the same** while the magnitude of the force F must **increase** so that the rod remains in equilibrium.

9. The cube has a mass of 300 g; the ellipsoid has a mass of 100 g; and the sphere has a mass of 1200 g.

11. $M = 2m$

13. (a) – (c)

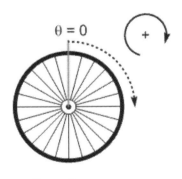

$\theta = 0$

(d)

Parameter	Value
Initial angular position	$\theta_i = 0$
Final angular position	$\theta_f = \dfrac{\pi}{2}$ rad/s
Initial angular velocity	$\omega_i = 0$
Final angular velocity	$\omega_f = ?$
Angular acceleration	$\alpha = +5.0$ rad/s^2
Time	$t = ?$

(e) $\omega_f^2 = \omega_i^2 + 2\alpha(\theta_f - \theta_i)$

(f) 4.0 rad/s, clockwise

(g) One possibility is $\Delta\theta = \omega_i t + \dfrac{1}{2}\alpha t^2$

(h) 0.79 s

15. (a) – (c)

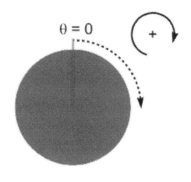

$\theta = 0$

(d)

Parameter	Value
Initial angular position	$\theta_i = 0$
Final angular position	$\theta_f = ?$
Initial angular velocity	$\omega_i = +2.4$ rad/s
Final angular velocity	$\omega_f = 0$
Angular acceleration	$\alpha = -1.2$ rad/s^2
Time	$t = ?$

(e) $\omega_f^2 = \omega_i^2 + 2\alpha(\theta_f - \theta_i)$

(f) 2.4 rad, clockwise

17. (a) – (c)

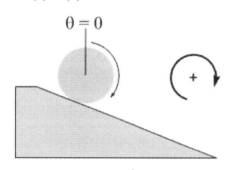

$\theta = 0$

(d)

Parameter	Value
Initial angular position	$\theta_i = 0$
Final angular position	$\theta_f = +80$ rad
Initial angular velocity	$\omega_i = +30$ rad/s
Final angular velocity	$\omega_f = ?$
Angular acceleration	$\alpha = ?$
Time	$t = 2$ s

(e) $\Delta\theta = \omega_i t + \dfrac{1}{2}\alpha t^2$

(f) +10 rad/s^2

19. (a) – (c)

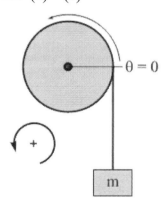

$\theta = 0$

m

(d)

Parameter	Pulley	Block
Initial position	$\theta_i = 0$	$y_i = 0$
Final position	$\theta_f = ?$	$y_f = ?$
Initial velocity	$\omega_i = +0.5$ rad/s	$v_i = +0.1$ m/s
Final velocity	$\omega_f = 0$	$v_f = 0$
acceleration	$\alpha = ?$	$a = ?$
Time	$t = 3$ s	$t = 3$ s

(e) You have a lot of freedom about how to solve this problem. You could immediately convert the angular velocity of the pulley into the linear velocity of the block, and then do everything with the block's straight-line motion, or, you can solve everything for the pulley first, and then convert for the block at the end. Note that we're using half the time, 3 seconds, because the block will move up for the first half of the time, and then fall back down during the second half of the time.

$$\omega_f = \omega_i + \alpha t, \quad \omega_f^2 = \omega_i^2 + 2\alpha(\theta_f - \theta_i) \quad \text{and} \quad \Delta y = r(\Delta \theta)$$

(f) $\alpha = -(0.5/3)$ rad/s^2; $\theta_f = +0.75$ rad; $\Delta y = +0.15$ m

21. (a) – (c)

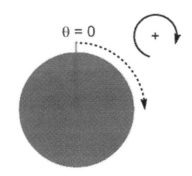

(d)

Parameter	Value
Initial angular position	$\theta_i = 0$
Final angular position	$\theta_f = 75° = \dfrac{5\pi}{12}$ rad
Initial angular velocity	$\omega_i = +0$
Final angular velocity	$\omega_f = 8$ rev/s $= 16\pi$ rad/s
Angular acceleration	$\alpha = ?$
Time	$t = ?$

(e) $\omega_f^2 = \omega_i^2 + 2\alpha(\theta_f - \theta_i)$ gives $\alpha = +965$ rad/s^2

(f) $\omega_f = \omega_i + \alpha t$ gives 0.052 s

23. (a) counterclockwise (b) $\tau = (4.0 \text{ m}) F \sin 30°$ (c) $\tau = (4.0 \text{ m})(F \sin 30°) \sin 90°$
    (d) $\tau = (4.0 \text{ m} \times \sin 30°) F \sin 90°$

25. (a) 8.0 N m (clockwise) (b) zero (c) 8.0 N m (counterclockwise)
    (d) 16 N m (counterclockwise) (e) 16 N m, counterclockwise

27. (a) 18 kg m^2 (b) 45 kg m^2

29. (a) The rotational inertia is minimized for an axis that passes through the center-of-mass of the system. Thus, the axis should pass through the center-of-mass, which is located between the balls at a distance of 0.40 m away from the ball of mass $2M$.
    (b) The rotational inertia in this case is $0.32 + 0.64 = 0.96$ kg m^2.

31. (a) $2Md^2$ (b) $4Md^2$ (c) $6Md^2$

33. (a) + (b)

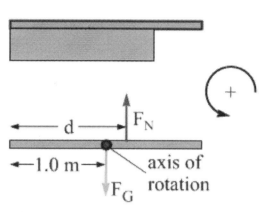

(c) The normal force has to balance the weight, so it must be 40 N up. (d) Let's take torques about the center of gravity of the board, because that eliminates mg from the torque equation. (e) $r \times F_N = 0$, so $r = 0$, where r is the distance of the normal force from the center of the board. The only way to make the torques balance is for the normal force to act at the center of gravity of the board.

35. (a) + (b)

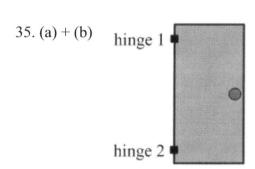

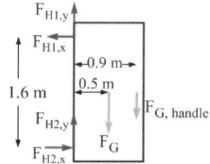

(c) The horizontal component of one hinge force is equal and opposite to the horizontal component of the other hinge force – they are the only two horizontal forces acting on the door. (d) Taking torques around the upper hinge eliminates both components of the force acting on the upper hinge, as well as the vertical component of the force applied to the lower hinge. $+(0.5 \text{ m})(20 \text{ N}) + (0.9 \text{ m})(10 \text{ N}) - (1.6 \text{ m})(F_{H2,x}) = 0$

(e) Rounding to 2 significant figures, we get 12 N, toward the door.
(f) 12 N, away from the door.

37.

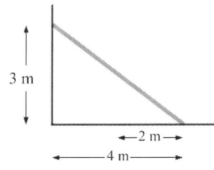

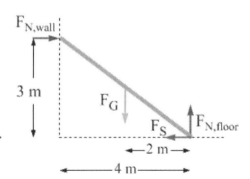

(c) There are four forces here, two vertical and two horizontal. Thus, the two vertical forces must balance one another, and the two horizontal forces must balance one another. Vertically, the upward normal force applied by the ground on the ladder must balance the weight of the ladder, so that normal force must be 600 N. Horizontally, the normal force from the wall on the ladder must balance the force of static friction applied by the ground on the ladder.

(d) Let's sum torques about the axis that passes through the point where the ladder touches the ground. That eliminates, from the torque equation, both the normal force applied by the ground and the force of friction applied by the ground on the ladder (that's especially critical, because we don't know the size of that force).

(e) Taking counterclockwise to be positive, $+(2.0 \text{ m})(600 \text{ N}) - (3.0 \text{ m})(F_{N,wall}) = 0,$

which gives a normal force, applied by the wall on the ladder, of 400 N.
(f) The force of static friction is also 400 N, and the normal force applied by the ground on the ladder is 600 N. $F_{S,max} = \mu_S F_N$, and, for the ladder to remain in equilibrium, the maximum possible force of static friction must be greater than or equal to the force of static friction we need in this case. Thus:

$$\mu_S F_N \geq 400 \text{ N} \quad \Rightarrow \quad \mu_S \geq \frac{400 \text{ N}}{F_N} = \frac{400 \text{ N}}{600 \text{ N}} = \frac{2}{3}.$$

Thus, the minimum coefficient of static friction we need is 2/3.

39. In many ways, the motions are very similar. At any instant in time, for instance, the velocity of the ball, in m/s, has the same numerical value as the angular velocity of the disk, in rad/s. At any instant in time, the displacement of the ball, in m, has the same numerical value as the angular displacement of the disk, in radians. A big difference is that the center-of-mass of the disk does not move in this situation, but if we compare the straight-line motion of the ball to the rotational motion of the disk, we see many parallels.

41. 1.0 m

43. Let's estimate that the distance from Boston to Seattle is 3000 miles, or about 5000 km, which is 5×10^6 m. A tire has a diameter of about 60 cm, which is a radius of 30 cm. In each revolution of the tire, the car covers a distance of $2\pi r \approx 2$ m. Dividing the total distance of the trip by the distance per revolution gives something like 2 – 3 million revolutions.

45. (a) 0.2 rad/s (b) 4 m/s^2 (c) 4 m/s^2 (d) 0.04 rad/s^2

47.   With a lever, there are basically three forces to consider. A see-saw is an example of a lever in action. On a see-saw, the forces are the force of gravity acting on the person on one side, the force of gravity acting on the person on the other side, and the support force provided by the fulcrum, in the middle. Taking torques about the center, there is no torque from the support force, so we conclude that the torque from the force of gravity on one side is balanced by the torque from the force of gravity acting on the other side. This is how a light person can balance a heavy person – the light person sits far from the center, and/or the heavy person sits close to the center, so their torques can balance. Archimedes simply took this logic to an extreme. If we think of the heavy person as the Earth, and the light person as Archimedes, then with the Earth very close to the fulcrum, and Archimedes very far away (hence, the need for a very long lever/seesaw), Archimedes could balance the Earth and, by moving a little farther out, could move the Earth.

49. Note that, in calculating the torque, the mass of the rod is irrelevant.
 (a) 5.0 N m clockwise (b) 2.5 N m clockwise (c) 5.0 N m clockwise
 (d) 2.5 N m clockwise

51. (a) 200 N   (b) 160 N to the left   (c) 60 N down

53. (a) 48 N   (b) 27 N to the right   (c) 20 N up

55. (a) The biceps force increases by 200 N (b) Yes, the humerus force changes. The humerus force is still directed down, and its magnitude increases by 180 N.

57. The force is 2 N, directed horizontally to the left, and it is applied at the lower end.

59. (a) The mobile is clearly not in equilibrium. At the very least, the section at the bottom right is not in equilibrium, because 2 balls 10 cm from the supporting string cannot balance 1 ball 30 cm away. (b) No, adding one ball is not sufficient. We can add one ball to the system at the bottom right, making a chain of 3 balls 10 cm from the supporting string to balance the 1 ball 30 cm away. However, that gives 8 balls on the right hanging from the top rod, and that can't be balanced by the three balls, twice as far from the support string, on the left of the top rod. (c) We can add two balls to bring the system into equilibrium. One ball is added to the system at the bottom right, as explained in part (b). Then, the other ball needs to be added to the chain of three at the top left, giving a chain of four that can balance the other eight balls.

61. No, the answer does not change. In this situation, the torque due to the force of gravity acting on Julie balances the torque due to the force of gravity acting on the board. If the system is moved to the Moon, both these torques are reduced by the same factor, so they are still equal. In other words, g cancels out from the torque equation.

Answers to selected problems from Essential Physics, Chapter 11

1. Calculating the angular acceleration in the four cases gives:

(a) $\alpha = \dfrac{3F}{ML}$ (b) $\alpha = \dfrac{3F}{2ML}$ (c) $\alpha = \dfrac{12F}{ML}$ (d) $\alpha = \dfrac{6F}{ML}$

Thus, ranking by angular acceleration gives $c > d > a > b$

3. (a) Yes, the spool will move (and accelerate) to the right   (b) Yes, the spool will rotate (and have an angular acceleration) counterclockwise

5. No, because the net velocity of any point is the vector sum of the translational velocity and the velocity associated with pure rotation. The rotational velocity can be at most equal in magnitude to the translational velocity. If the vectors are in opposite directions, the net velocity can be zero, but not in the opposite direction as the translational velocity.

7. (a) Note that the force of static friction acts up the slope. The object rotates counterclockwise as it rolls up the slope, but the angular speed decreases. This requires a clockwise torque, about the center of the object. A torque about the center can only come from the force of friction, and we can only get a clockwise torque if the force of friction is directed up the slope.

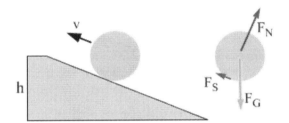

(b) The ring will travel farther up the ramp. Both objects have the same translational kinetic energy at the bottom but the ring, because of its larger rotational inertia, will have more rotational kinetic energy (the angular speeds at the bottom are equal, because the angular speed is the speed divided by the radius). Thus, the ring has a larger total kinetic energy than the sphere. All of the kinetic energy gets turned into gravitational potential energy at the highest point, so the ring ends up with a larger gravitational potential energy, which means it goes higher up the slope.

9. (a) The figure skater's angular speed increases. This is all because of angular momentum conservation – pulling her arms in reduces the skater's angular momentum, so, to keep the angular momentum (the product of the rotational inertia and the angular velocity) constant, the angular velocity increases in magnitude. (b) In this process, the skater's rotational kinetic energy increases. The extra energy comes from the work done by the skater on her arms and hands to pull them in closer to her body.

11.

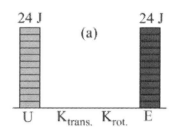

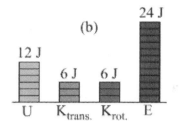

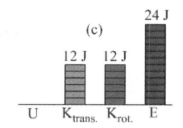

13.

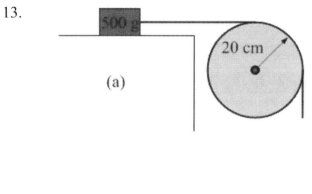

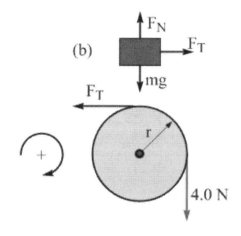

(c) $\sum \tau = I\alpha$

$(4\text{ N})r - F_T r = \dfrac{1}{2}Mr^2\dfrac{a}{r}$

$(4\text{ N}) - F_T = \dfrac{1}{2}Ma$

(d) $\sum \vec{F} = m\vec{a}$

$F_T = ma$

(e) $a = \dfrac{4\text{ N}}{\dfrac{1}{2}M + m} = 2.7\text{ m/s}^2$

15.

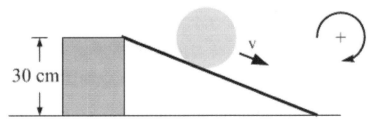

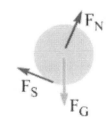

(c) $\sum \vec{\tau} = I\vec{\alpha}$

$F_S r = \dfrac{2}{5}mr^2\dfrac{a}{r}$

$F_S = \dfrac{2}{5}ma$

(d) $\sum \vec{F} = m\vec{a}$

$mg\sin\theta - F_S = ma$

(e)
$$mg \sin \theta = ma + F_S = \frac{7}{5}ma$$

$$a = \frac{5g \sin \theta}{7} = \frac{5(9.8 \text{ m/s}^2)(0.3 \text{ m})}{7(2.0 \text{ m})} = 1.05 \text{ m/s}^2$$

(f) $\quad 2.0 \text{ m} = \frac{1}{2}at^2$

$$t = \sqrt{\frac{4.0 \text{ m}}{a}} = 1.95 \text{ s}$$

17. (a) $\frac{\pi}{2}$ m (b) $\frac{\pi}{2}$ m (c) No. First, we have to add the displacements as vectors, and the displacements are in different directions. Second, the displacement associated with the rotational motion has a smaller magnitude than the distance associated with the rotational motion, because the direction of motion changes as the wheel rotates. (d) The displacement has components of 1.0 m vertically and $\frac{\pi}{2}$ m horizontally. Using the Pythagorean theorem with these components gives 1.9 m for the magnitude of the displacement.

19. 22.5 m/s

21. The speeds are the same in the two cases. Because the sphere rolls without slipping, mechanical energy is conserved, so the gravitational potential energy is transformed into rotational kinetic energy and translational kinetic energy. The gravitational potential energy is the same in the two cases, and the fraction of the total kinetic energy that is translational kinetic energy is the same in both cases, so the speed must be the same.

23. (a) A's kinetic energy is larger by a factor of two. (b) A's angular speed is larger by a factor of two. (c) They have the same angular momentum.

25. (a) the stopping time would double (b) the stopping time would double
(c) the stopping time would be half as long

27. We actually don't have enough information to solve this problem. Let's take the wheel's mass to be 800 grams and its radius to be 40 cm, with all the wheel's mass concentrated around the rim of the wheel.

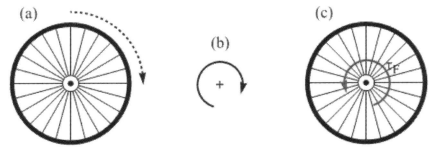

(d) The net torque is simply the frictional torque, τ_F (e) $\Delta L = I\Delta\omega = mr^2\left(\omega_f - \omega_i\right)$

(f) $\tau_F \Delta t = mr^2\left(\omega_f - \omega_i\right)$ (g) $\tau_F = 0.026$ N m, in the direction opposite to the wheel's direction of rotation.

29. (a) At any instant in time, the speed of block A is equal to the speed of block B, because they are connected by the same fixed-length string. (b) 34.5 kg

31. The assumption here is that the ball rolls without slipping. In that case, $h = 2.7\,R$. This is a little larger than the result of $2.5\,R$ we got previously for a frictionless block. The extra height in this case gives us a little more gravitational potential energy to start with, which goes into the ball's rotational kinetic energy at the top of the loop.

33. 39 m/s

35. (a) $U_i + K_i + W_{nc} = U_f + K_f$, and define the zero level for gravitational potential energy to be the lowest point (the final point).
    (b) $K_i = 0$, because the disk starts from rest; $U_f = 0$, because the disk ends up at the zero level for gravitational potential energy; $W_{nc} = 0$, because no non-conservative forces act on the disk (there is a friction force acting, but because it is static friction the energy associated with friction is the rotational kinetic energy of the disk).

(c) $U_i = K_f$

$$mgh = \frac{1}{2}mv^2 + \frac{1}{2}I\omega^2$$
$$= \frac{1}{2}mv^2 + \frac{1}{2}\left(\frac{1}{2}mr^2\right)\frac{v^2}{r^2}$$
$$= \frac{3}{4}mv^2$$

(d) $v = \sqrt{\frac{4}{3}gh} = 2.6$ m/s

37. Let's assume that the sphere rotates only, and that there is no translational motion.
    (a) $U_i + K_i + W_{nc} = U_f + K_f$, and define the zero level for gravitational potential energy to be the level of the sphere's center of mass.
    (b) $K_i = 0$, because the sphere starts from rest; $U_i = 0$ and $U_f = 0$, because the sphere's center of mass does not move, and the center of mass starts and ends at the zero level for gravitational potential energy.

(c) $W_{nc} = K_f$

$$Fd = \frac{1}{2}I\omega^2 = \frac{1}{5}mr^2\omega^2$$

(d) $\omega = \sqrt{\frac{5Fd}{mr^2}} = 18$ rad/s

39. (a) 5.0 rad/s^2 , clockwise (b) 2.5 rad/s^2 , clockwise
    (c) 20 rad/s^2 , clockwise (d) 10 rad/s^2 , clockwise

41. The tension only needs to be different when the system accelerates. In that case, the pulley has an angular acceleration, which requires a net torque. The only way to have a net torque acting on the pulley is for the two tensions to be different. (a) With everything at rest, the two tensions are the same. (b) In this case, the system accelerates. In particular, the pulley's angular acceleration is directed counterclockwise, so the tension above the block of mass M is larger than the tension in the part of the string that is above the block of mass m. (c) In this case, there is no acceleration, so the tension is the same at all points in the string.

43. (a) $a = g/5$. If we use $g = 9.8$ m/s^2, we get $a = 1.96$ m/s^2. (b) The speed of either block at this time is 3.92 m/s. To get the angular velocity, we need to know the radius of the pulley. If $r = 10$ cm, then the angular velocity is 39 rad/s, directed counterclockwise.

45. (a) 4.1 m/s^2 , down (b) 1.6 m/s^2 , up

47. 1.1 m/s^2, down

49. (a) 2.0 m/s^2 (b) The tension is larger between block B and the pulley. In this situation, the pulley has an angular acceleration directed clockwise, so there must be a net torque acting clockwise on the pulley. The net torque is proportional to the difference between the two tensions, with the clockwise torque from the tension in the vertical part of the string being larger than the counterclockwise torque from the tension in the horizontal part of the string. This is true only if the tension in the vertical part of the string is larger than the tension in the horizontal part of the string.
 (c) 24 N in the vertical part of the string, and 21 N in the horizontal part

51. Yes, the spool can remain motionless in this case, if there is an appropriate force of static friction acting down the slope. Then, the force you exert on the spool up the slope would balance the sum of the force of static friction and the component of the force of gravity that acts down the slope. The torque you exert about the center of the spool would balance the torque the force of static friction exerts about the center of the spool.

53. (a) $\dfrac{9}{8} MR^2$ (b) $\dfrac{33}{19} Mg$

55. The bicycle wheel will spin for twice as long as the solid disk, because the rotational inertia of the wheel is two times larger than the rotational inertia of the disk.

57. (a) The pennies farther from the axis are more likely to lose contact with the meter stick than are the pennies close to the axis. All points on the meter stick will have the same angular acceleration, but the linear acceleration of any point is the angular acceleration multiplied by the distance from the axis to the point. Thus, points farther from the axis will have the greatest linear acceleration and, if the linear acceleration exceeds g, the pennies won't keep up with the meter stick. (b) The acceleration at the far end of the stick is $a = \dfrac{3}{2} g$, while at the center it is only half of this, $\dfrac{3}{4} g$.

59. (a) The system's angular momentum stays the same – there is no net external torque being applied to the system. When you get farther from the center, the system's rotational inertia increases - the angular speed decreases so the angular momentum stays the same.
(b) The system's rotational kinetic energy decreases. The merry-go-round applies an inward directed force to you, so when you move out from the center negative work is done by the turntable on you. This accounts for the loss of mechanical energy.
(c) When you get to the outer edge and you are rotating with the turntable, the system's angular velocity is 1.8 rad/s directed clockwise. To conserve this angular velocity with the merry-go-round at rest, you would have to run with an angular velocity of 3.6 rad/s, directed clockwise.

61. (a) 3.7 m/s (b) 3.8 s (c) 3.8 revolutions

63. (a) Angular momentum is conserved. (b) Translational kinetic energy is not conserved. (c) Gravitational potential energy is not conserved. (d) Total mechanical energy is conserved. The total mechanical energy (the sum of the kinetic energy and the potential energy) is conserved, but the gravitational potential energy increases with the comet's distance from the Sun. Thus, the translational kinetic energy must decrease to compensate for the increase in potential energy as the comet moves farther from the Sun. On its return journey toward the Sun, the kinetic energy increases as the potential energy decreases.

Answers to selected problems from Essential Physics, Chapter 12

1. The force is still 5 N, but in the opposite direction.

3. The block has been moved either to $x = -20$ cm, or to $x = +20$ cm. So, the block has either been moved 10 cm to the left, or 30 cm to the right.

5.

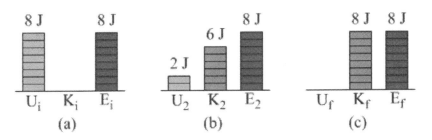

7. (a) $2 > 1 = 4 > 3$ (b) $2 = 4 > 1 = 3$ (c) $2 > 1 = 4 > 3$

9. (a) No. The initial energy is all stored in the spring, so changing the mass has no effect on the energy, in this situation.   (b) Yes, the energy is proportional to the spring constant, so doubling the spring constant will double the energy.   (c) Yes, the energy is proportional to the square of the amplitude, so increasing the amplitude by a factor of $\sqrt{2}$ will double the energy.

11. (a) $1 = 3 > 2 = 4$ (b) $4 > 2 = 3 > 1$ (c) $2 = 4 > 1 = 3$

13. 30.0 N

15. The time-of-flight is 0.535 s. The launch speed of the ball is 4.68 m/s. The spring constant is 76 N/m.

17. (a) 3.43 m   (b) 1.2 rad/s

19. (a) 3.09 m   (b) 5.73 m/s

21. The maximum speed will be reached at a point higher up the ramp then when there was no friction. Maximum speed will be reached at the point where the forces balance. With friction directed up the slope, the block does not need to compress the spring as much as it did before for there to be no net force acting on the block.

23. 1.48 m

25. $\omega = \pi$ rad/s; $k = \dfrac{\pi^2}{2}$ N/m; total mechanical energy = 0.582 J;

   amplitude = 0.4857 m; $x = 0.3434$ m

(a) 0.291 J   (b) 0.291 J

27. (a) The bar graphs apply at two locations, one at a particular distance from the equilibrium point on one side of the equilibrium position, and the other at the same distance from equilibrium on the other side of the equilibrium position.
    (b) Both locations are 0.462 m away from the equilibrium position.

29. (a) 0.319 m   (b) 42.8°

31. 1.71 s

33. (a) 0.67 s   (b) 52.8 N/m   (c) 4.47 m/s in the negative direction
    (d) 13.2 m/s^2 in the positive direction

35. There are four possible answers for this question, one answer for each of the four times the block is a distance of 30 cm from equilibrium in the first complete oscillation. The four possible values of k are 1.67 N/m, 18.7 N/m, 47.8 N/m, and 98.9 N/m.

37. (a) The collision cannot be elastic, because the two carts stick together – this is a completely inelastic collision. Immediately after the collision, the two carts are moving at 1.0 m/s, with a total kinetic energy of 0.5 J. All that energy goes into the spring at maximum compression, so the spring must compress 50 cm.

39. 3.00 m

41. The red (solid) lines are for the first case, and the blue (dashed) lines are for the second case. The period and angular frequency referenced on the graphs are for the first case.

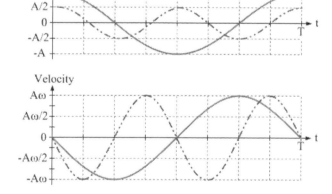

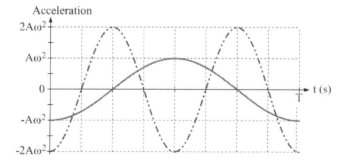

43. (a) 2.51 rad/s   (b) We need to find the new angular frequency of the turntable, which should now be 3.55 rad/s.

45. (a) 0.31 m/s   (b) 0.25 m/s^2

47. $x = A\sin(\omega t)$; $v = A\omega\cos(\omega t)$; $a = -A\omega^2\sin(\omega t)$, where $A = 1.00$ m and $\omega = 2.00$ rad/s

49. 34 N/m

51. 0.684 kg

55.

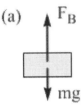

(a)

(b) 3.3 m/s^2, up   (c) As the block moves, the force of gravity remains the same, but the buoyant force changes, depending on how much of the block is submerged. The buoyant force starts off larger than the force of gravity, goes through a place where it is equal to the force of gravity, and then becomes less than the force of gravity. After the block comes to rest for an instant, it falls back down, eventually returning to the initial state (if we neglect all resistive forces) and the cycle continues.

57. (a) 2.0 m.   (b) 4.43 m/s   (c) 4.64 m/s
    (d) Not surprisingly, the small-angle approximation works best for small angles. 60° is not a small angle, so the approximation leads to a result for the speed which is larger than it should be. The speed calculated using energy conservation in (b) is accurate in this case.

59. 4.9 J

1. (a) In England, temperatures are measured in Celsius. Bill, coming from the United States, is used to hearing temperatures stated in Fahrenheit. (b) 77°F

3. The disk expands when heated, increasing the radius of the disk, and increasing the disk's rotational inertia. Because the disk's angular momentum should be conserved, the angular velocity of the disk decreases when the rotational inertia increases.

5. (a) Brass has a larger thermal expansion coefficient than iron. When the temperature increases from 20°C, the brass will expand more than the iron, so the brass strip will be longer than the iron strip. This causes the bimetallic strip to curve, with the longer strip on the outside (the longer side) of the curve. Thus, in the picture, the brass is the upper strip and the iron is the lower strip.
    (b) One way to do this is to have the strip bend when it cools, so the strip makes electrical contact with a switch, turning on the heating system. When the room warms up, so does the strip, straightening the strip out so that it no longer makes contact with the switch, turning the heating system off again.

7. $m_B > m_C > m_A$

9. $m_B > m_C > m_A$

11. (a) The temperature of the black one increases more quickly, because the black object absorbs energy more efficiently from the Sun. (b) The temperature of the black one also decreases more quickly. Black objects also emit energy more quickly than do shiny objects.

13. (a) $T_K = \left(\dfrac{5°K}{9°F}\right)\left(T_F + 459.4°F\right)$ (b) $T_F = \left(\dfrac{9°F}{5°K}\right)T_K - 459.4°F$

15. (a) Canada uses the metric system, so the temperature given on the box would be in Celsius, instead of in Fahrenheit as James is used to. James set the temperature on his stove to 250°F, but the cake should have been baked at a temperature of 250°C. (b) The cake was underdone, because 250°F is a lower temperature than 250°C. (c) 482°F.

17. 1.0007 cm

19. (a) Water will spill out of the pot as the temperature increases. This is because the thermal expansion coefficient for water is larger than the thermal expansion coefficient for aluminum, so the water will expand to a larger volume than the volume of the pot. (b) 0.84% of the water has spilled out by the time the temperature reaches 80°C.

21. 19.98 rad/s

23. (a) There are three heat terms here. Two of these terms are positive, representing the heat gained by the water (Q_w) and the container (Q_c) as their temperatures increase. The third term is negative, as it represents the heat lost by the block (Q_b) as it cools down.

(b) $Q_w + Q_c + Q_b = 0$

(c) $m_w c_w (T_F - 10°C) + m_c c_c (T_F - 10°C) + m_b c_b (T_F - 80°C) = 0$

(d) $(m_w c_w + m_c c_c + m_b c_b) T_F = m_w c_w \times 10°C + m_c c_c \times 10°C + m_b c_b \times 80°C$

$$T_F = \frac{m_w c_w \times 10°C + m_c c_c \times 10°C + m_b c_b \times 80°C}{m_w c_w + m_c c_c + m_b c_b}$$

$$T_F = \frac{(500 \text{ g}) \times [4186 \text{ J/(kg °C)}] \times 10°C + (200 \text{ g}) \times [900 \text{ J/(kg °C)}] \times 10°C + (300 \text{ g}) \times [900 \text{ J/(kg °C)}] \times 80°C}{(500 \text{ g}) \times [4186 \text{ J/(kg °C)}] + (200 \text{ g}) \times [900 \text{ J/(kg °C)}] + (300 \text{ g}) \times [900 \text{ J/(kg °C)}]}$$

$$T_F = 17.4°C$$

25. (a) There are four heat terms here. One of these terms is positive, representing the heat gained by the block (Q_b) as its temperature increases. The other three terms are negative, representing the heat lost by the water (Q_{w1}) as it cools down as a liquid from +50°C to 0°C; the heat lost by the water (Q_{w2}) as it changes phase from liquid to solid at 0°C; and the heat lost by the water (Q_{w3}) as it cools down as a solid from 0°C to –20°C.

(b) $Q_b + Q_{w1} + Q_{w2} + Q_{w3} = 0$

(c)
$$m_b c_b [-20°C - (-196°C)] + m_w c_{w,liquid} (0°C - 50°C) - m_w L_f + m_w c_{w,solid} (-20°C - 0°C) = 0$$

(d)
$$m_w = \frac{-m_b c_b [-20°C - (-196°C)]}{c_{w,liquid} (0°C - 50°C) - L_f + c_{w,solid} (-20°C - 0°C)}$$

$$m_w = \frac{-(500 \text{ g}) \times [385 \text{ J/(kg °C)}] \times (+176°C)}{[4186 \text{ J/(kg °C)}] \times (-50°C) - (335000 \text{ J/kg}) + [2060 \text{ J/(kg °C)}](-20°C)}$$

$$m_w = 57.9 \text{ g}$$

27. (a) There are three heat terms here. One of these terms is positive, representing the heat gained by the block (Q_b) as its temperature increases. The other two terms are negative, representing the heat lost by the water (Q_w) as it cools down as a liquid from +50°C to +20°C, and the heat lost by the cup (Q_c) as it cools down from +50°C to +20°C

(b) $Q_b + Q_w + Q_c = 0$

(c) $m_b c_b [+20°C - (-196°C)] + m_w c_w (+20°C - 50°C) + m_c c_c (+20°C - 50°C) = 0$

(d)
$$m_w = \frac{-m_b c_b [+20°C - (-196°C)] - m_c c_c (+20°C - 50°C)}{c_w (+20°C - 50°C)}$$

$$m_w = \frac{-(500 \text{ g}) \times [385 \text{ J/(kg °C)}] \times (+216°C) - (300 \text{ g}) \times [900 \text{ J/(kg °C)}] \times (-30°C)}{[4186 \text{ J/(kg °C)}] \times (-30°C)}$$

$m_w = 267$ g

29. 17.1 minutes

31. (a) 19.8 s (b) 11.5 s   (c)

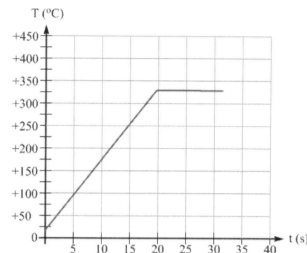

33. 250 W

35.

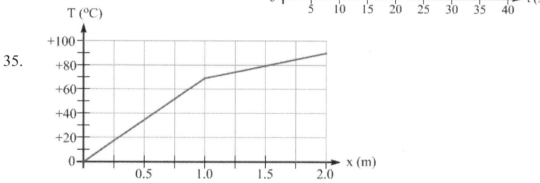

37. The property of the liquid that is being exploited here is the fact that the density of the liquid depends on temperature. To be specific, the density of the liquid decreases as the temperature of the liquid increases. The balls in the thermometer all have slightly different densities from one another, and their densities are close to that of the liquid. Take, for example, a Galileo thermometer with 5 balls in it, marked 18°C, 19°C, 20°C, 21°C, and 22°C, and imagine that the thermometer starts off at a temperature of 17°C. At this temperature, the density of the liquid would be higher than that of the balls, and the balls would all float. If the thermometer's temperature was raised above 18°C, however, the density of the liquid would decrease below that of the 18°C ball, and that ball would sink to the bottom. Raising the temperature further, if the liquid's temperature rises above 19°C, the 19°C ball sinks, etc. This thermometer gives the temperature within 1°C – if the 19°C ball sinks while the 20°C ball floats, all we know is that the temperature is somewhere between 19°C and 20°C.

39. 250000 J

41. (a) The final temperature should be quite a bit lower than 60°C. The masses of the lead and water are the same, so the final temperature would be halfway between 100°C and 20°C only if the specific heats of water and lead were the same. In a case like this, $mc\Delta T$ for the lead has the same magnitude as $mc\Delta T$ for the water. However, the specific heat of water is significantly larger than the specific heat of lead, so the temperature change for the lead must be significantly larger than the temperature change for the water. The final temperature will be a lot closer to the initial temperature of the water than it is to the initial temperature of the lead. (b) 22.4°C.

43. 325 g

45. (a) 0.447 kg represents the smallest amount of ice that can be added to the punch to produce a final temperature of 0°C. In this case, all the ice melts, so the punch transfers the maximum amount of heat possible to the ice. (b) The other extreme is to use a large amount of ice, so large that the punch cools down to 0°C and then all the punch freezes, with the situation at equilibrium being that everything is solid at 0°C. The mass of ice required to do this is 27.1 kg of ice.

47. 24.6°C

49. 26.1°C

51. There is not quite enough information given here to state the ranking by final temperature. In general, we can say that the temperature of the cup with the gold ball will be greater than or equal to the temperature of the cup with the copper ball, which will be greater than or equal to the temperature of the cup with the aluminum ball. However, we would need to know how the mass of the balls compares to the mass of the water in the cup to determine the exact ranking by final temperature. The difficulty arises because the water could experience a phase change, so there are wide ranges of ball masses that can result in a final temperature of 0°C for the three cases, and these ranges overlap.

55. Inverting the 1.2 m tube 100 times is like dropping the balls from a height of 120 m. If we assume that all the gravitational potential energy is converted into heat, and all the heat is transferred to the lead balls, we have $mgh = mc\Delta T$. The mass cancels, so we don't even need the mass that is given. This gives a temperature difference of:

$$\Delta T = \frac{gh}{c} = \frac{(9.8 \text{ m/s}^2) \times (120 \text{ m})}{129 \text{ J/(kg °C)}} \approx 9°C$$

57. (a) 3.65×10^{26} W (b) 1360 W

59. (a) It takes longer than T to heat the water with the aluminum pot. Aluminum has a smaller thermal conductivity than copper, so the rate at which energy is transferred through the base of the aluminum pot to the water is less than the rate at which energy is transferred through the base of the copper pot. (b) 1.7 T

61. (a) 6.84×10^{10} J (b) 6.22×10^{10} J (c) 1 kW h = 3.6×10^6 J, which means you would save about $350. (Note that the house in this example has an especially low R-value – with a higher R value, the amount of energy lost through the walls is reduced, as is the cost and the savings from reducing the temperature.)

63. 359 m of aluminum

Answers to selected problems from Essential Physics, Chapter 14

1. When you are several km above the surface of the Earth, the pressure in the plane is substantially lower than it is at ground level. Thus, when you opened the bottle when you were in the plane, and the plane was high above the Earth, and then sealed the bottle again, the pressure inside the bottle was quite a bit lower than standard atmospheric pressure. When you are back at ground level, the atmospheric pressure applies inward directed forces to the bottle that are not balanced by the forces from the lower pressure in the bottle, pushing out, so the bottle tends to collapse. Note that this decrease in volume of the bottle increases the pressure inside the bottle.

3. The pressure is the same on both sides. If the pressure was higher on one side than the other, the piston would not be in equilibrium – it would experience a net force acting to move it toward the side that had lower pressure.

5. The pressure is higher in the lower section of the container. If you draw the free-body diagram of the piston, you can see that the upward force applied on the piston from below, associated with the pressure in the lower section, has to balance the weight of the piston plus the downward force associated with the pressure in the upper section. For everything to balance, the pressure in the lower section must be larger than the pressure in the upper section.

7. (a) 1 = 2 = 3 (b) 1 > 2 > 3

9. The average kinetic energy of the gas molecules should stay the same – there is no work being done. This means that we expect the temperature to remain the same. With the temperature remaining the same while the volume doubles, the pressure should be reduced by a factor of 2.

11. 2 > 1 = 3 > 4

13. $P \times 293/283 = 1.04\, P$

15. $P_{atm} \times 293/373 = 0.79\, P_{atm}$

17. (a) 0.5 (b) 2.9

19. (a) Yes, if the numbers are all the same, such as (2,2,2,2), in which the average is 2 and the rms average is also 2. (b) No, the rms average of a set of numbers is always greater than or equal to the average of that set of numbers.

21. The ranking by pressure does not change with temperature – it just depends on the free-body diagram of the piston. In both Exercise 20 and 21, the ranking by pressure is $1 > 3 > 2$.

23. (a) $2 > 3 > 1$ (b) $2 > 3 > 1$

25. $2 > 3 > 1$

27. (a) The pressure is 101.6 kPa (b) The piston's free-body diagram does not change, so the pressure remains the same. The absolute temperature increases by 14%, so the volume increases by 14% (coming from upward movement of the piston).

29. (a) The pressure is 101.0 kPa (b) 1.003 : 1

31. 24.2 cm

33. (a) As the pressure is reduced outside the cylinder, the piston in the cylinder will move up in the cylinder. (b) 47.5 kPa

35. 96.2 kPa

37. 1 : 2.25

39. (a) The pressure is the same in the two cylinders. The pressure is determined by the free-body diagram, which is the same in both cases. (b) The number of moles of gas is the same in each cylinder. The number of moles of gas can be determined by applying the ideal gas law. With the same pressure, volume, and temperature, the number of moles is the same.

41. (a) 7000 Pa (b) 7 : 1

43. (a) The volume decreases, so the work done by the gas is a negative quantity. If the change happens so quickly that there is no time for heat to be transferred, then, by the first law of thermodynamics, the change in internal energy must be a positive value. The change in internal energy is proportional to the temperature change, so the temperature change is positive – the temperature increases. (b) As the piston is moving, the average relative speed between the piston and the gas molecules that collide with the piston increases. The collisions with the piston are elastic, and a property of elastic collisions is that the magnitude of the relative speed between the colliding objects is the same before and after the collision. This means that the gas molecules have a higher speed, after the collision, when they collide with a piston that is moving toward them before the collision,

than they do when they collide with a stationary piston. Giving the molecules more kinetic energy is consistent with raising the temperature.

45. (a) 0.089 moles (b)

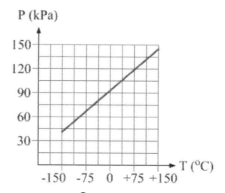

(c) The slope of the graph is equal to nR/V, which has a value of 346 Pa/K (or 0.346 kPa/K).

47. (a) 1.1 moles (b) 6.7×10^{23} atoms (c) 349 m/s (d) 394 m/s (e) 427 m/s

49. (a) The pressure in all five states is 45 kPa.
(b)

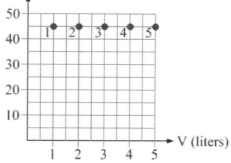

(c) One example of a system in which the pressure remains constant is a cylinder that is sealed by a piston which is free to slide up or down without friction. Let's say that the system starts in state 1, with a temperature of 150 K. Heat can then be added to the system – this is done relatively slowly, so that the pressure is almost constant. The more heat is added, the higher the temperature gets.

51. (a) $2 > 3 > 1 > 4$

(b) $T_1 = 16$ K, $T_2 = 48$ K, $T_3 = 24$ K, $T_4 = 8$ K

53. The pressures are equal.

Answers to selected problems from Essential Physics, Chapter 15

1. (a) $2 > 1$ (b) $2 > 1$ (c) $2 > 1$

3. (a) All other factors being equal, we would expect system B to require more heat. In system B, some of the heat goes to doing work, and the rest goes to increasing the internal energy (increasing the temperature). In system A, all the heat added goes to increasing the internal energy. Thus, system B needs all the heat that A needs, plus an additional amount that goes into the work done by the gas. (b) A possible explanation for this result (that both systems actually require the same amount of heat) is that system B is a system of monatomic ideal gas, while system A is a system of diatomic ideal gas. The extra energy needed to change the internal energy of the diatomic ideal gas, compared to that of the monatomic ideal gas in system B, is exactly equal to the work done by the gas in system B.

5. (a) Yes, if the process is isothermal. All the heat added goes into doing work. One way to do this is to have a piston free to move in a cylinder. If mass is slowly and gradually removed from the piston, the piston will slowly expand at constant temperature. (b) Yes, this can be done with an adiabatic compression. An example is a bicycle pump, in which the gas is compressed quickly, with no time for heat to be transferred, resulting in a significant gain in the temperature of the gas.   (c) Yes, if the work done by the system is larger than the heat transferred. This is essentially the process in a fridge, in which a gas expands and cools, with heat coming out of the inside of the fridge into the cool gas. The work done by the gas is generally larger than the heat transferred to it.   (d) Yes. If such a system expands, it will cool, and if it contracts, its temperature increases.

7. (a) $A > C > B$ (b) $B > C > A$ (c) $A > C > B$ (d) $B > C > A$

9. (a) $A > B > C$ (b) $A > B > C$ (c) $C > B > A$ (d) $A > B > C$

11. (a) $A > B > C$ (b) $A = B = C$ (c) $A > B > C$ (d) $C > B > A$

13. (a) Cylinder 1 has more heat removed from it. Cylinder 1 experiences a decrease of internal energy and negative work, while cylinder 2 only has a decrease in internal energy. (b) The changes in internal energy are the same, because the temperature difference is the same in both cases.

15. (a) $W_A = 1920$ J; $W_B = 1440$ J; $W_C = 960$ J   (b) 5600 J in each case
     (c) $Q_A = 7520$ J; $Q_B = 7040$ J; $Q_C = 6560$ J

17.

(b) 0
(c) –250 J
(d) 250 J of heat was removed
(e) 250 J

(a)

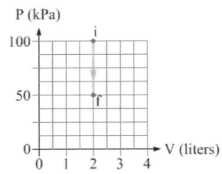

19.   (a)

(b) –100 J
(c) –250 J
(d) 350 J of heat was removed
(e) 250 J

21. (a) +480 J   (b) +1200 J

23. (a) 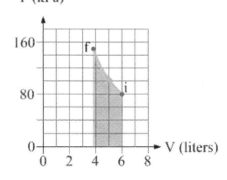   (b) 533 K   (c) 1.3 L   (d) 205 kPa

25. (a) 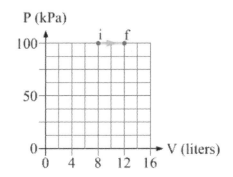   (b) 3.8 L   (c) –236 J   (d) 1.2 : 1

27. (a) 23.4 kPa   (b) +259 J   (c) 0.52

29. (a) +400 J   (b) 600 K   (c)

31. (a) +777 J
    (b) +4800 J
    (c) +5577 J

33.

Process	Q (J)	W (J)	ΔE_{int} (J)
1 → 2	+6720	+1920	+4800
2 → 3	−4800	0	−4800
3 → 1	−887	−887	0
Entire Cycle	+1033	+1033	0

35. (a)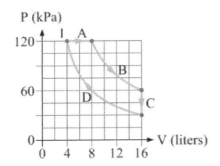

(b) +480 J

37. (a) The area represents both the net work done by the system in one cycle, as well as the heat added to the system in one cycle (these are always equal to one another, because the net change in internal energy for one cycle is zero). (b) 20 J (c) 8 boxes (d) +160 J

39. (a) +146.9 J/K (b) −141.9 J/K (c) +5.0 J/K

41. (a)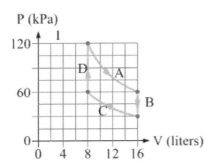

(b)

Process	Q	W	ΔE_{int}	ΔS
A	+	+	0	+
B	−	0	−	−
C	−	−	0	−
D	+	0	+	+
Cycle	+	+	0	0

43. (a) 0.56 (b) 1.0×10^6 J

45. (a) This air conditioner is not ideal. The coefficient of performance can be written as

$$COP = \frac{Q_L}{W} = \frac{Q_L}{Q_H - Q_L}$$. If the air conditioner was ideal, then we can replace the heats in

the equation by temperatures, like so: $$COP = \frac{T_L}{T_H - T_L}$$. If we work out the coefficient of

performance using the absolute temperatures, we get 29.3, significantly higher than the value of 7.5 that is given. Thus, the air conditioner must be less than ideal (this is the case for all real-life air conditioners, of course).
 (b) Yes, we would expect that an increase in the outside temperature would change the coefficient of performance. If the outside temperature increases, the air conditioner has to do more work to extract the same amount of heat from inside your house and release that heat, plus the work, outside. Thus, when the outside temperature increases, the coefficient of performance decreases.

47.   (a) (in red) + (b) (in blue)

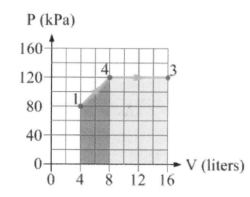

(c) $W_{1\rightarrow4} = +400$ J ;
    $W_{4\rightarrow3} = +960$ J

49. (a) $\Delta E_{int,1\rightarrow4} = +960$ J ;
        $\Delta E_{int, 4\rightarrow3} = +1440$ J

    (b) $Q_{1\rightarrow4} = +1360$ J ;
        $Q_{4\rightarrow3} = +2400$ J

51.   (a)
(b) The magnitude of the work done by the gas is the
magnitude of the area under the curve for the process
on the P-V diagram. The two areas have the same
magnitude, so the magnitude of the work done in case 1
is equal to the magnitude of the work done in case 2.
(c) –320 J

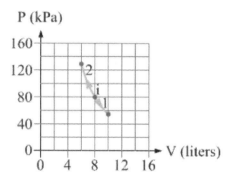

53.   (a)
    (b) The work done clearly has a larger magnitude
for process 2 – the area under the curve for that process
is larger than it is for process 1.
    (c) –203 J

55. (a) The pressure is 150 kPa, and the volume is
(32/3) L
    (b) The pressure is 50 kPa, and the volume is 32 L

57. The entropy of the set of building materials decreases during construction, because a
whole collection of items is assembled into a highly ordered structure by the time the
house is done. This does not violate the second law, however, because the set of building
materials is not a closed system. If you calculated the entropy of the system of building
materials, all the light and heavy machinery involved in building the house, and all the
people who worked on the house, the entropy of that system would increase.

59. (a) 11.3%   (b) 880 MW (that's 880 million joules every second)   (c) 780 MW

61. These students have some incorrect ideas. What is described here is an adiabatic
process – no heat is transferred, because the system is so well insulated. However, this
does not mean that the temperature remains the same – the temperature can change, with
any change in internal energy being offset by the work.

Made in the USA
Charleston, SC
13 July 2013